BIOCHEMISTRY OF MAMMALIAN REPRODUCTION

BIOCHEMISTRY OF MAMMALIAN REPRODUCTION

I   GAMETES AND GENITAL TRACT FLUIDS

II REPRODUCTIVE ENDOCRINOLOGY

EDITED BY

LOURENS J. D. ZANEVELD

University of Illinois, College of Medicine

ROBERT T. CHATTERTON

Northwestern University Medical School

1807   1982

A WILEY-INTERSCIENCE PUBLICATION

JOHN WILEY & SONS

New York • Chichester • Brisbane • Toronto • Singapore

Library of Congress Cataloging in Publication Data:

Zaneveld, Lourens J. D.
  Biochemistry of mammalian reproduction

  "A Wiley-Interscience publication."
  Includes index.
1. Reproduction.   2. Endocrinology.   3. Mammals—
Reproduction.   I. Zaneveld, Lourens J. D.
II. Chatterton, Robert T.

QP251.B56   599.01'6   82-1914
ISBN 0-471-05731-2   AACR2

Printed in the United States of America

10 9 8 7 6 5 4 3 2 1

CONTRIBUTORS

Stan A. Beyler

Research Associate
Department of Physiology and Biophysics
College of Medicine
University of Illinois at the Medical Center
Chicago, Illinois

Craig W. Beattie

Associate Professor
Department of Surgery
College of Medicine
University of Illinois at the Medical Center
Chicago, Illinois

Barbara J. Bentwood

Assistant Professor
Department of Obstetrics and Gynecology
John A. Burns School of Medicine and Pacific Biomedical Research Center
University of Hawaii
Honolulu, Hawaii

Asok K. Bhattacharyya

Assistant Professor
Department of Physiology and Biophysics
College of Medicine
University of Illinois at the Medical Center
Chicago, Illinois

Robert T. Chatterton, Jr.

Professor
Department of Obstetrics and Gynecology
Northwestern University Medical School
Chicago, Illinois

Kerry L. Cheesman

Research Associate
Department of Obstetrics and Gynecology
Northwestern University Medical School
Chicago, Illinois

Geula Gibori

Associate Professor
Department of Physiology and Biophysics
College of Medicine
University of Illinois at the Medical Center
Chicago, Illinois

Paula L. Hoffman

Associate Professor
Department of Physiology and Biophysics
College of Medicine
University of Illinois at the Medical Center
Chicago, Illinois

Gary L. Jackson

Professor
Department of Veterinary Biosciences
College of Veterinary Medicine
University of Illinois
Urbana, Illinois

Randal C. Jaffe

Assistant Professor
Department of Physiology and Biophysics
College of Medicine
University of Illinois at the Medical Center
Chicago, Illinois

John B. Josimovich

Professor
Department of Obstetrics and Gynecology
College of Medicine and Dentistry of New Jersey
Newark, New Jersey

Sheila M. Judge

Research Associate
Ben May Laboratory for Cancer Research
University of Chicago
Chicago, Illinois

William J. King

Research Associate
Ben May Laboratory for Cancer Research
University of Chicago
Chicago, Illinois

Mazie Kopta

Research Associate
Department of Obstetrics and Gynecology
Washington University
St. Louis, Missouri

Tsuei-Chu Liu

Research Endocrinologist
Department of Veterinary Biosciences
College of Veterinary Medicine
University of Illinois
Urbana, Illinois

Josephine Miller

Instructor
Departments of Obstetrics and Gynecology, and Anatomy
College of Medicine
University of Illinois at the Medical Center
Chicago, Illinois

R. Nicholas Peterson

Professor
Department of Medical Physiology and Pharmacology
School of Medicine
Southern Illinois University
Carbondale, Illinois

Kenneth L. Polakoski

Associate Professor
Department of Obstetrics and Gynecology
Washington University
St. Louis, Missouri

H. G. Madhwa Raj

Assistant Professor
Departments of Obstetrics/Gynecology and Pharmacology
University of North Carolina
School of Medicine
Chapel Hill, North Carolina

Luis J. Rodriguez-Rigau

Assistant Professor
Department of Reproductive Medicine and Biology
The University of Texas Medical School at Houston
Houston, Texas

B. Jane Rogers

Associate Professor
Department of Obstetrics and Gynecology
John A. Burns School of Medicine and Pacific Biomedical Research Center
University of Hawaii
Honolulu, Hawaii

Gebhard F. B. Schumacher

Professor
Department of Obstetrics and Gynecology
University of Chicago
Chicago, Illinois

Emil Steinberger

Professor and Chairman
Department of Reproductive Medicine and Biology
The University of Texas Medical School at Houston
Houston, Texas

Peter F. Tauber

Professor
Department of Obstetrics and Gynecology
University of Essen
Essen, West Germany

Kantilal H. Thanki

Adjunct Assistant Professor
Departments of Obstetrics and Gynecology, and Physiology
College of Medicine and Dentistry of New Jersey
Newark, New Jersey

Judith L. Vaitukaitis

Professor
Section of Endocrinology and Metabolism
Thorndike Memorial Laboratory
Boston City Hospital
Departments of Medicine and Physiology
Boston University School of Medicine
Boston, Massachusetts

Don P. Wolf

Professor
Department of Obstetrics and Gynecology
The University of Texas Medical School at Houston
Houston, Texas

Lourens J. D. Zaneveld

Professor
Departments of Physiology and Biophysics, and Obstetrics and Gynecology
College of Medicine
University of Illinois at the Medical Center
Chicago, Illinois

Panayiotis M. Zavos

Assistant Professor
Department of Animal Science
University of Kentucky
Lexington, Kentucky

Stephen J. Zimniski

Research Associate
Department of Biochemistry
Vanderbilt University
Nashville, Tennessee

PREFACE

This volume represents a summary of our knowledge regarding the biochemical aspects of the reproductive processes. Both the reproductive tract and the endocrinological aspects are covered. Although review articles, book chapters, or books are available on a number of isolated topics covered here, the authors are not familiar with any other monograph that, in a single volume, covers the entire field of reproduction from a biochemical standpoint. To understand the significance of the biochemistry of a particular process, each chapter also incorporates some discussion of the physiological aspects of the topic. Reviews are particularly scarce in the area of genital tract biochemistry. The wide scope of this volume should make it of interest to many people, ranging from students new to the field to scientists and clinicians who would like to advance their knowledge in areas other than that of their own expertise. It is hoped that the book will find use as a review; as a textbook for graduate students in biological and clinical sciences and for students of human and veterinary medicine; and in the laboratory or clinic as a summary of research in the many areas of reproduction. We believe an overview such as we have attempted to assemble will not only be useful from an educational standpoint, but will also help to define areas where further study is necessary.

An attempt was made to keep the chapters relatively brief and to the point, but still detailed enough so that a good understanding of the topic can be obtained. It is accepted that each one of the chapters could easily be expanded to form a book by itself. The authors were requested to cover the most important facets, not all the arguments surrounding a particular subject. For this reason, only the most important references were quoted or, preferably, review articles or other book chapters dealing with the topic. These articles and reviews are representative of the material covered in the chapter, and the reader may find these useful starting points for a more detailed review of the literature. Needless to say, the references represent only an extremely small number of the articles available on the topic. We apologize to those who feel slighted because their names were omitted, but this could not be helped if the objectives of this book were to be met.

We truly appreciate the efforts put forth by the authors. It was not easy to summarize the large amount of information available on each topic and to condense the information into a logical discourse. This discussion is by no means complete, however, and the reader is referred to other books dealing with the physiology and morphology of the reproductive tract if more detail is desired.

We would like to dedicate the volume to all those who have contributed to our knowledge of reproductive biochemistry.

LOURENS J. D. ZANEVELD
ROBERT T. CHATTERTON

Chicago, Illinois
June 1982

CONTENTS

BIOCHEMISTRY OF MAMMALIAN REPRODUCTION

PART ONE

GAMETES AND GENITAL TRACT FLUIDS

Edited by

LOURENS J. D. ZANEVELD

CHAPTER ONE

THE TESTIS AND SPERMATOGENESIS

LUIS J. RODRIGUEZ-RIGAU

EMIL STEINBERGER

Department of Reproductive Medicine and Biology
The University of Texas Medical School at Houston,
Houston, Texas

1   GENERAL INTRODUCTION

The testis is a unique organ. It is the source of two totally different products: a highly specialized motile cell (the spermatozoon), and several powerful hormones. The origin of these products is compartmentalized within specific anatomical sites in the testis. This localization of specific functions has to be kept in mind and addressed in appropriate fashion when biochemical analyses are conducted in whole testicular tissue. Exciting new experimental approaches, useful in the study of testicular biochemistry, have been developed in the past two decades. These include advances in organ culture, application of cell separation techniques, collection of testicular fluids, and development of new, or improvement of existing, experimental techniques (e.g., radioimmunoassay, isotopic methods for evaluation of the synthesis of steroids, proteins, nucleic acids and lipids, and electron microscopy). The scope of this chapter does not permit the presentation of a comprehensive review of testicular biochemistry, since any one topic would form a book by itself. It is meant to provide the reader with an overview of the salient aspects of this exciting and relatively new topic.

2   FUNCTIONAL ANATOMY OF THE TESTIS

The parenchyma of the testis is enclosed by a capsule composed of three layers: the visceral layer of the tunica vaginalis, the tunica albuginea and the tunica vasculosa. The cavity of the tunica vaginalis separates the testis from the scrotum. The parenchyma of the testis is composed of seminiferous tubules and interstitial tissue. The seminiferous tubules form loops that terminate at both ends in the tubuli recti, which in turn empty into the rete testis. The rete testis connects with the epididymis by means of the ductuli efferentes. Spermatogenesis takes place in the seminiferous tubules; the seminiferous epithelium, composed of somatic cells (the Sertoli cells) and germ cells (Fig. 1), sheds the newly formed spermatozoa into the lumen of the tubules. The seminiferous tubules are surrounded by a limiting membrane composed of fibroblasts, contractile myoid cells, and collagen fibers embedded in a mucopolysaccharide matrix (1).

2.1 Interstitial Tissue

Interstitial tissue is located between the seminiferous tubules and contains Leydig cells (interstitial cells), blood vessels, lymphatics, and numerous macrophages. The amount of interstitial tissue varies from approximately 10% of the total testicular volume in the rat to 60% in the boar. In most species, Leydig cells occur as clusters or strands in the spaces between the seminiferous tubules. Leydig cells are involved in the synthesis of androgens. Thus their cytoplasm is rich in smooth endoplasmic reticulum (see Chapters 16 and 19). The Leydig cells apparently do not store testosterone, but secrete it into the interstitial area, from where it diffuses into the blood and lymph, and into the seminiferous tubules.

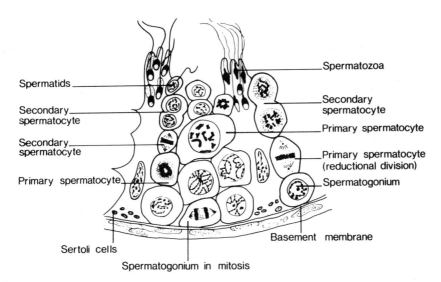

Figure 1 Schematic representation of a cross section of the seminiferous tubule, illustrating the process of spermatogenesis. (Prepared by the Biocommunications Art Department of the University of Illinois at the Medical Center, Chicago, Illinois.)

2.2   Seminiferous Tubules

2.2.1   Germ Cells

The spermatogonia are the least differentiated germ cells. They reside near the basement membrane of the seminiferous tubules and divide mitotically by way of a number of germ cell types to form spermatocytes. By the process of meiosis, the spermatocytes first acquire a 4n complement of DNA and then form four haploid cells: the spermatids. The spermatids do not divide further but enter a lengthy process of metamorphosis (spermiogenesis), resulting in the formation of flagellated, motile cells: the spermatozoa. The entire process from spermatogonia to spermatozoa is known as spermatogenesis (Fig. 1).

2.2.2   Stem (Germ) Cell Renewal

Since germ cells proliferate continuously during adult life, it is necessary to assure constant supply of spermatogonia. The process responsible for this has been termed stem (germ) cell renewal, and although its details are far from being understood, the best evidence points toward the concept of a "self-renewing" stem cell (2). In other words, some spermatogonia divide to form more spermatogonia, whereas others proliferate to give rise to spermatocytes.

2.2.3   Kinetics of Spermatogenesis (3,4)

The duration of spermatogenesis in various mammalian species is as follows: mouse, 34.5 days; hamster, 35 days; Long-Evans and Sherman strains of rat, 48 days; Sprague-Dawley rat, 51.6 days; Wistar rat, 53.2 days; rabbit, 43 days; bull, 49 days; and man, 64 days.

The different steps of development of each germ cell type are timed precisely. The various types of germ cells are distributed in a specific fashion throughout the seminiferous epithelium. The pattern of distribution is predicated by the timing of the development of each germ cell type. Consequently, not all the germ cells can be observed at the same time in any given area of the seminiferous tubule. Instead, precisely defined cellular associations or stages are formed. The cellular composition of each stage is based on the duration of each developmental step (Fig. 2). The stages reappear at regular intervals in segments of the seminiferous tubule. One *cycle of the seminiferous epithelium* is defined as the entire sequence of stages between two appearances of the same cellular association. In the rat, the spermatogenic cycle consists of 14 stages (Fig. 2), whereas in the mouse and monkey, 12 stages have been defined.

The duration of the cycle of the seminiferous epithelium varies from species to species: boar, 8.3 days; mouse, 8.6 days; hamster, 8.7 days; ram, 10.4 days; monkey, 10.5 days; rabbit, 10.7 days; Long-Evans and Sherman rats, 12 days; Sprague-Dawley rats, 12.9 days; Wistar rats, 13.3 days; bull, 13.5 days; man, 16 days. If we could trace a single spermatogonium entering the spermatogenic process through mitotic divisions, formation of meiotic cells, formation of spermatids, and metamorphosis of the spermatids to spermatozoa, we would

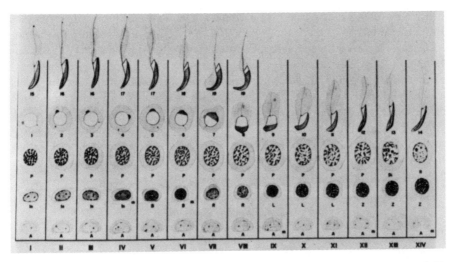

Figure 2 Composition of the 14 cellular associations or stages observed in the seminiferous epithelium of the rat. Each column (identified by roman numbers I to XIV at the
base) consists of the various cell types constituting a cellular association. The stages
succeed one another in time according to the sequence indicated from left to right in the
figure. The succession of the 14 stages constitutes the spermatogenic cycle, whereas the
entire process in the figure from the right side of the bottom row (type A spermatogonia)
to the left side of the top row (step 19 spermatid) illustrates the entire process of
spermatogenesis. *A*, type A spermatogonia, *In*, intermediate-type spermatogonia, *B*,
type B spermatogonia, *R*, resting primary spermatocytes, *L*, leptotene primary spermatocytes, *Z*, zygotene primary spermatocytes, *P*, pachytene primary spermatocytes, *Di*,
diakinesis of primary spermatocytes, *II*, secondary spermatocytes, *1-19*, steps 1 to 19 of
spermiogenesis. Subscript *m* next to a spermatogonium indicates occurrence of mitosis.
From Perey *et al.*, *Am. J. Anat.* **108**, 47 (1961).

find that, at least in the rat, the cells go through almost four complete spermatogenic cycles (four sequences of the 14 stages or cellular associations) before
spermatogenesis is completed. This is because of the synchronization of the
entry of spermatogonia into spermatogenesis, and because of the precise duration of each developmental step. The first cycle of spermatogenesis results in
formation of pachytene primary spermatocytes (see Fig. 2). A complete sequence of the 14 stages is again required to form the most immature spermatid.
A third complete spermatogenic cycle results in formation of step-15 spermatids. Maturation of these spermatids to spermatozoa requires eight stages of
spermatogenesis. As spermatogenesis proceeds, the cells move from close to the
basement membrane to the lumen, and become associated with other cells in
other steps of development (i.e., a resting primary spermatocyte is always found
in association with a pachytene primary spermatocyte, a step-7 or -8 spermatid
and a step-19 spermatid).

The various stages of the spermatogenic cycle are arranged in consecutive
order along the length of the seminiferous tubule, so that contiguous areas are

either less or more advanced by a single stage. The development progresses along the length of the tubule, forming a *wave of spermatogenesis*. In the rat, each seminiferous tubule contains 12 complete spermatogenic waves. The maintenance of the wave is due to the precise synchronization of the entry of spermatogonia into the spermatogenic process in different areas of the seminiferous tubules. In man it is difficult to evaluate the kinetics of the spermatogenic process; the cellular associations are highly irregular in both cellular composition and topographic distribution. It appears that human spermatogenesis proceeds in "clones" rather than "waves"; in other words, spermatogonia that enter the spermatogenic process apparently form clones of germ cells that develop independently from the surrounding cells (5).

2.2.4   *Sertoli Cells (6)*

The Sertoli cells line the basement membrane of the seminiferous tubules. Their cytoplasm extends toward the lumen of the tubules, enveloping the adjacent germ cells. They proliferate during the early stages of testicular development, but do not divide in the adult. The cytoplasm of each Sertoli cell comes in close contact with that of its neighboring Sertoli cells through specialized junctions. This contact creates a blood–testis barrier and divides the germinal epithelium into two compartments: basal, containing the spermatogonia; and adluminal, containing the spermatocytes and spermatids. The cytoplasm of the Sertoli cell contains both smooth and rough endoplasmic reticulum, mitochondria, a Golgi apparatus, and lysosomes. The nucleus is irregular and complex, with a homogeneous nucleoplasm and a single tripartite nucleolus.

The Sertoli cells probably have many functions, including phagocytosis of degenerating germ cells and residual bodies, maintenance of the blood–testis barrier, secretion of seminiferous tubule fluid, production of androgen-binding protein (ABP), secretion of a factor called Sertoli cell factor (SCF or *inhibin*) that selectively suppresses follicle-stimulating hormone (FSH), steroidogenesis, and so forth. In addition, the Sertoli cells are the primary site for FSH action in the testis, and are androgen-target cells. Thus it seems clear that the Sertoli cells are more than just "supporting" or "nursing" cells for the germ cells. Their many functions, possibly some of which are still unknown, appear to be of paramount importance for the process of spermatogenesis. Details of the involvement of the Sertoli cell in the secretion of seminiferous tubule fluid as well as on the maintenance of a biochemical milieu favorable to spermatogenesis are covered in subsequent sections of this chapter. Protein markers of Sertoli cell function are also discussed.

2.3   Hormonal Control of Spermatogenesis (7, 8)

Gonadotropins secreted by the pituitary gland are essential for testicular function. Early studies suggested that follicle-stimulating hormone (FSH) was primarily responsible for stimulation of the functional activity of the seminiferous tubules, whereas luteinizing hormone (LH) regulated Leydig cell function. Later studies demonstrated that LH is indeed the major regulatory hormone for interstitial cell steroidogenesis (for review see Chapter 18) but showed that the primary hormone responsible for maintenance of spermatogenesis is testosterone. FSH appears to be required only for the completion of spermiogenesis during the first wave of spermatogenesis, i.e., at puberty; after this, spermatogenesis proceeds as long as an adequate supply of testosterone is available. If spermatogenesis is allowed to regress following hypophysectomy, FSH is again required for reinitiation of spermatogenesis.

The Sertoli cells are the only cells in the testis that bind FSH, and thus are considered to be a primary and possibly the exclusive target for FSH action. Sertoli cell production of cAMP and ABP is stimulated *in vitro* by FSH. The Sertoli cells have further been demonstrated to possess an androgen receptor, but such receptors are also present in some germinal elements. Consequently, it is probable that both Sertoli cells and some germ cells are androgen-target cells. It has been demonstrated that Sertoli cell production of ABP increases in response to androgens.

On the basis of the above information, the following hypothesis for the molecular aspects of hormonal regulation of spermatogenesis in the mammalian testis was proposed (Fig. 3). FSH delivered by the arterial blood supply diffuses into the seminiferous tubule and binds to a Sertoli cell membrane receptor. This results in activation of adenylyl cyclase with formation of cAMP, promoting DNA-dependent RNA synthesis, which in turn causes synthesis of proteins, including ABP. Androgens produced by the Leydig cells also diffuse into the seminiferous tubule and are concentrated in the vicinity of the target cells (germ cells and Sertoli cells) by ABP. In the target cells the androgen molecules bind to the specific cytoplasmic androgen receptor. The androgen-receptor complex is translocated into the nucleus and binds to an acceptor site in the chromatin. This probably results in specific changes still largely unknown, which are essential for Sertoli cell function and progression of spermatogenesis. Since ABP is transported to the epididymis, it seems probable that ABP also causes androgen concentration near the androgen-target cells in this organ.

3 NUCLEIC ACIDS AND NUCLEOPROTEINS (9–11)

The early steps of spermatogenesis involve several mitotic divisions of type A spermatogonia ultimately to form type B spermatogonia. The mitotic cycle can be divided into four periods, designated as G_1, S, G_2, and M. S is the period of DNA synthesis. G_1 and G_2 (gaps) are the periods from the previous cell division to the start of DNA synthesis and from the end of DNA synthesis to the beginning of mitosis (M), respectively. In type A (the most primitive) spermatogonia the S period is short and of constant duration. In type B spermatogonia, (derived through mitotic divisions from the type A spermatogonia), it is long

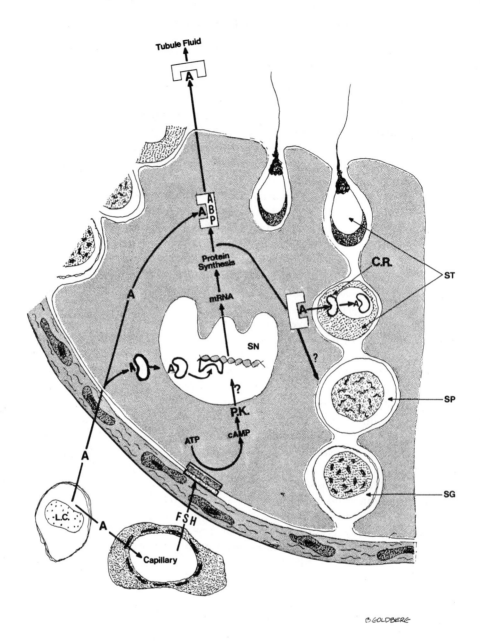

Figure 3 Molecular mechanisms of hormonal control of spermatogenesis. *Sn*, Sertoli cell nucleus, *SG*, spermatogonium, *SP*, spermatocyte, *ST*, spermatid, *LC*, Leydig cell, *PK*, cAMP-dependent phosphokinase, *A*, androgen, *FSH*, follicle-stimulating hormone, *CR*, cytoplasmic receptor, *ABP*, androgen-binding protein, $\int$, receptor, $\mathfrak{M}$, chromatin acceptor. From Steinberger, *Clin. Obstet. Gynecol.* **22**, 187 (1979).

and variable. The G_2 period, however, changes from being variable and long in type A spermatogonia to short and constant in type B spermatogonia. The durations of the cell cycle and the G_1 phase do not change. In addition, there is a decline in the rate of RNA synthesis in type B spermatogonia. These changes are apparently related to increased nuclear condensation of the spermatogonia during maturation.

B-type spermatogonia divide to form primary spermatocytes, which undergo meiosis. The prophase of the first meiotic division is very long and can be divided into preleptotene, leptotene, zygotene, and pachytene stages. Prelepto-tene primary spermatocytes undergo active DNA synthesis. Leptotene primary spermatocytes are tetraploid. In the animal kingdom, DNA synthesis ceases at the leptotene stage. In the *Lilium* (plant), further DNA synthesis appears to occur during zygotene and pachytene stages (12, 13). This additional DNA synthesis in the *Lilium* has been interpreted as representing delayed replication of definite chromosomal regions involved in chromosome pairing, formation of the synaptinemal complex, and repair of chromosome breaks correlated with crossing over. In addition to DNA synthesis, there is also synthesis of certain nucleoproteins in the zygotene-pachytene stage in the *Lilium*. These nuclear proteins are physically associated with the newly synthesized DNA, and are probably involved in synapsis and crossing over (14). There is no clear evidence in animals for similar DNA synthetic activity during the meiotic prophase.

After the long meiotic prophase, the first meiotic division gives rise to di-ploid secondary spermatocytes, which undergo a reduction division to form haploid spermatids. The spermatids do not divide further, but differentiate to form spermatozoa.

The sex chromosome bivalent condenses during the meiotic prophase and forms the sex vesicle. These chromosomes become inactivated and do not label with [³H]uridine. The autosomes show very active RNA synthesis during mei-osis. It has been suggested that the activity of the genes of the autosomes necesary for spermatogenesis requires the inactivation of the X chromosome. Indeed, there is evidence to suggest that a functional X chromosome is not compatible with normal spermatogenesis (e.g., Klinefelter syndrome in man).

The rate of RNA synthesis is low in preleptotene primary spermatocytes, declines further during the leptotene and zygotene stages, increases markedly in the pachytene stage, and declines again during the diplotene stage and diakinesis (Fig. 4). Still, RNA synthesis during the diplotene stage is higher than during the leptotene or zygotene stages. The synthesis of nucleoprotein follows a sim-ilar pattern (Fig. 4). Ribosomal RNA synthesis is almost nonexistent during meiosis. Since cytoplasmic continuity between Sertoli cells and contiguous sper-matocytes has been demonstrated, it has been suggested that transfer of the ribo-somal RNA from the Sertoli cell to the spermatocytes may take place, and may play an important role in the regulation of spermatogenesis.

The RNA synthesized during meiosis remains in the nucleus for a consider-able period of time and is possibly involved in the regulation of gene transcrip-tion or chromosome pairing. During diakinesis, this RNA is transferred to the

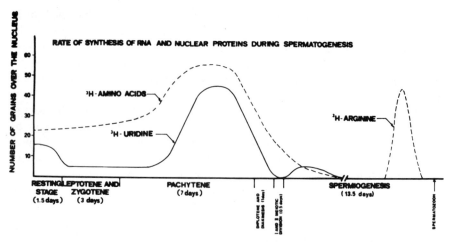

Figure 4   Pattern of [³H]uridine and [³H]amino acid incorporation into RNA and nuclear proteins during spermatogenesis in the mouse, one hour after administration of the labeled precursors. From Monesi, *The Regulation of Mammalian Spermatogenesis*, Thomas, Springfield, Illinois (1973).

cytoplasm. In the spermatid it probably plays an important role in the regulation of protein synthesis. RNA and nucleoprotein synthesis ceases after the second meiotic division, approximately at the time when the nucleus of the spermatid begins to elongate. At this point in spermiogenesis, all components of the nucleus of the spermatid except for the DNA and histones are eliminated, and thus no RNA remains in the nucleus, whereas the cytoplasm contains RNA originally synthesized in the spermatocyte. Late in spermiogenesis, the cytoplasmic RNA in spermatids is also eliminated and becomes a part of the residual body (cytoplasmic droplet), leaving spermatozoa free of RNA.

As mentioned above, synthesis of nucleoprotein ceases at the same time as RNA synthesis stops. However, during late spermiogenesis there is a sudden resurgence of nucleoprotein synthesis (Fig. 4). This synthetic activity consists entirely of production of basic, arginine-rich proteins (considered to be protamines) that replace the previously existing lysine-rich histone. The lysine-rich histone is eliminated from the spermatids in the residual bodies. The histone-to-protamine transition produces an increase in the electrostatic binding of the nucleoprotein to the DNA, resulting in condensation of the chromatin in the nucleus of the spermatid. The condensation commences at the acrosomal pole of the nucleus and progresses distally. It is probably responsible for the changes in the cytochemical reactivity and chemical properties of the chromatin (e.g., decrease in Feulgen reactivity, increased staining with the Sakaguchi reaction, increased resistance of the DNA to mild acid hydrolysis and to heat denaturation). The chromatin condensation may protect the genome during passage of the spermatozoa through the male and female reproductive tracts and may be necessary for fertilization. During the fertilization process, in various mammalian species (e.g., rat, rabbit, hamster), dispersion of the nuclear material occurs after entry of the spermatozoon into the egg, prior to the formation of the pronucleus, apparently reversing the changes that occurred during spermiogenesis.

In summary, during the process of spermatogenesis, DNA synthesis is completed prior to the meiotic prophase. The sex chromosomes are genetically inactive during meiosis; however, active RNA synthesis occurs in the autosomes. The nuclear RNA synthesis in the spermatocyte continues throughout the prophase, and the RNA is transferred to the cytoplasm at diakinesis. Thus the cytoplasmic RNA of the spermatid originates primarily from the nucleus. This cytoplasmic RNA probably serves an important role in the regulation of protein synthesis during spermiogenesis. During spermiogenesis, RNA and nucleoprotein synthesis ceases, and all components of the nucleus except for the DNA-nucleoprotein complex are eliminated. The somatic-type, lysine-rich histone is replaced during late spermiogenesis by a basic arginine-rich nucleoprotein, a change that apparently is responsible for chromatin condensation in the mature spermatid.

4 PROTEINS AND ENZYMES (15, 16)

4.1 Total Proteins

In rats, the total protein concentration of the testis remains constant between 21 and 35 days of age and decreases between 35 and 45 days of age. Between 42 and 56 days, coinciding with the appearance of spermatozoa, there is a sharp increase in protein concentration; this is followed by a slow and small decline as spermatozoa are transported to the epididymis.

Rates of protein synthesis of the various cell types in the testis have been investigated utilizing *in vitro* techniques, involving incorporation of radio-labeled amino acids. These studies suggest that in the rat testis, the highest rates of protein synthesis occur in pachytene primary spermatocytes, and the lowest rates in secondary spermatocytes and spermatids (Fig. 4). Protein synthesis rates appear to be higher in type A than in type B spermatogonia. Except for nuclear incorporation of labeled arginine, the rate of incorporation of labeled amino acids into the proteins of the spermatid is low. By contrast, both the nucleus and the cytoplasm of the Sertoli cells readily incorporate labeled amino acids into proteins.

Rates of protein incorporation of labeled amino acids are affected by glucose (this is also discussed in section 6.1 of this chapter), which stimulates incorporation of labeled lysine into proteins in all types of germ cells (Fig. 5). In the absence of spermatids (e.g., after hypophysectomy or cryptorchidism), this stimulatory effect of glucose on protein synthesis is decreased. In the rat, the effect of glucose first appears at 20 days of age and increases between 30 and 50 days (spermatids are first present in the testis between 20 and 30 days of age).

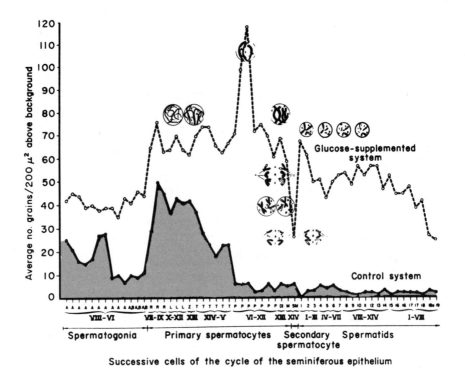

Figure 5   Incorporation of [³H]ʟ-lysine into protein of cells of the seminiferous epithelium of the rat testis in the presence and absence of glucose. *I–XIV* refer to the stages of the spermatogenic cycle. From Davis and Firlit, *Am. J. Physiol.* **209**, 425 (1965).

Thus the rate of protein synthesis in spermatids appears to have the greatest sensitivity to stimulation by glucose, even though protein synthesis is relatively low in spermatids when compared to other germ cell types (e.g., spermatocytes).

Testosterone added *in vitro* decreases the incorporation of labeled lysine into proteins by testicular tissue, whereas LH, prolactin or thyroid-stimulating hormone (TSH) have no effect. FSH stimulates protein synthesis in the testes of immature animals, but not in testicular tissue from adult animals.

Testicular protein synthesis appears to be temperature dependent. In rat, mouse, and hamster, maximal rates of lysine incorporation into protein by tissue slices *in vitro* were noted at 32°C, whereas in rabbit, guinea pig, and dog, maximal protein synthesis rates were observed at 37.5°C. These differences may be related to higher scrotal temperatures in the latter species. Increased temperatures result in inhibition of testicular protein synthesis *in vitro*, but this effect can be prevented by the addition of glucose. Elevated temperature may cause depletion of glucose by the spermatid, which in turn leads to decreased activity of the enzymes required for protein synthesis.

4.2 Protein Markers of Testicular Function

4.2.1 Introduction

A number of proteins, most of which are enzymes, have been identified as markers of the presence or function of specific cell types in the testis. Although a few testicular enzymes are specific to the testis, or even to a specific cell type in the testis, most are involved in general metabolic processes and are not unique to the testis. Most enzymes of the steroid biosynthetic pathways are markers for the function of the Leydig cells (see Chapter 9).

4.2.2 Sertoli-Cell-Specific Proteins

There is no conclusive evidence for a highly specific protein marker of Sertoli cell function. However, several proteins have been demonstrated to be produced specifically by the Sertoli cells and apparently not by other testicular cells. The activity of some enzyme systems persists or increases after depletion of germ cells from the testis, a finding suggesting that these enzymes may be specific for Sertoli cells. Thus these proteins and enzymes have been used as markers of Sertoli cell function. γ-Glutamyltranspeptidase (GTP) activity in developing rat testis parallels the pattern of Sertoli cell replication and maturation. Furthermore, the activity increases after germ cell depletion caused by vitamin A deficiency in the rat. β-Glucuronidase activity is also highest in testes depleted of spermatocytes and spermatids. Catalase activity has been correlated positively with Sertoli cell numbers. It persists after hypophysectomy and x-irradiation and declines in the immature testis at the onset of spermatogenesis, paralleling the decline in relative Sertoli cell density of the tissue.

Androgen-binding protein (ABP) is produced by isolated rat Sertoli cells in culture, and its synthesis is stimulated by FSH (6, 17). X-irradiation and administration of drugs like nitrofurazone and ethionine, resulting in complete depletion of germ cells from the seminiferous tubules, have little or no effect on testicular levels of ABP in the rat (18). Thus in the rat, the evidence is strong for Sertoli cell origin of ABP. Although ABP has been demonstrated in the testes and testicular fluids of several mammalian species (e.g., rat, rabbit, ram, bull), no convincing evidence for production of ABP by testis from other mammalian species, including man, has been provided. Rat ABP appears to be a glycoprotein, which has been characterized using gel filtration, sucrose-gradient centrifugation, polyacrylamide gel electrophoresis, isoelectric focusing, Sephadex gel equilibration techniques, and so forth (17–19). It appears to have a sedimentation coefficient 4.6 S, a Stokes radius of about 47 Å, and a molecular weight of about 90,000 (19). These values are different from those of the testicular androgen receptor.

ABP binds testosterone and dihydrotestosterone with high affinity, but it has a rapid dissociation rate (17). Thus the androgen-protein complex stores testosterone in such a way that it can be rapidly released when required by the target cell. In the rabbit, ABP is found in testicular efferent duct fluid, but a similar androgen-binding protein (testosterone-binding globulin, or TeBG) is also found in serum. These proteins are very similar in steroid specificity, sedimentation coefficient, and Stokes radius, and cannot be separated by electrophoresis in polyacrylamide gels, but they can be separated by isoelectric focusing and ion-exchange chromatography (19). In addition, the ABP concentration in efferent duct fluid is much higher than that of TeBG in serum, and no TeBG is found in efferent duct fluid by isoelectric focusing. The concentration of ABP in rete testis fluid is much lower in the bull than in the rat. As does the rabbit, the bull has a serum TeBG. The main function of ABP has been suggested to be the concentration of androgen molecules in the vicinity of androgen-target cells in the testis and epididymis (see section 2.3 of this chapter). It seems possible that in animals with extremely high intratesticular testosterone concentrations (e.g., man, boar), there is no need for ABP, whereas in species with lower androgen concentration in the testis (e.g., rat, rabbit, bull), there is a need for this binding protein to maximize the supply of androgen molecules to the target cells.

Sertoli cell factor (SCF), or inhibin, is produced by cultured Sertoli cells; it selectively suppresses FSH release and synthesis by pituitary cells in culture (6, 20). Although the existence of a FSH-inhibiting substance, or "inhibin," was postulated by McCullagh in 1932 (21), and inhibin activity was demonstrated by several investigators in extracts of spermatozoa, seminal plasma, and rete testis fluid of several mammals as early as 1972 (for review see Ref. 22), the Sertoli cell origin of inhibin activity was not conclusively demonstrated until 1976 (20). There is good evidence for the protein nature of inhibin (22, 23). Its activity is lost after heating or treatment with proteolytic agents. With the use of several purification procedures (e.g., ammonium sulphate fractionation, precipitation with ethanol or acetone, gel filtration and chromatography, ion-exchange chromatography), "inhibins" from several sources (e.g., bovine and ovine testis and rete testis fluids, ovine testicular lymph, human and bovine seminal plasma) have been studied. However, no consensus regarding the molecular weight or amino acid composition of inhibin has been reached, probably because pure inhibin has not yet been isolated and because it is difficult to compare "inhibins" from different sources. Apparently, inhibin activity is present in both low- and high-molecular-weight fractions following gel chromatography. Inhibin may exist in polymeric form or as a combination of native inhibin and a carrier protein (24).

4.2.3 Protein Markers of Spermatogenesis (16)

A number of enzymes have been reported to be markers of germ cell maturation. These include lactate dehydrogenase-X (LDH-X), sorbitol dehydrogenase (SDH), carnitine acetyltransferase (CAT), acid phosphatase, certain phosphorylases, certain aminopeptidases, amylase, alkaline phosphatase, a trypsin-like proteinase, and adenosine triphosphatase (ATPase). Of these, LDH-X has been studied in greatest detail. LDH-X is a unique enzyme, present in the testes of most species at times when spermotozoa are actively being formed. It is almost absent from the testes of immature animals, during the nonbreeding season in the testes of seasonal animals, or after experimentally induced cessation of spermiogenesis in the rat (e.g., after vitamin A deficiency). This enzyme is not present in Leydig cells; it appears to be restricted to the seminiferous tubule, and specifically to the midpiece of spermatozoa and the mitochondrial fraction of testicular homogenates. The precise function of LDH-X is not known, although it has been suggested that it is important for the oxidation of lactate by spermatozoa (25).

SDH appears to be related to the primary spermatocyte, at least in the rat and guinea pig. In immature animals and following germ cell depletion in the rat (e.g., after vitamin A deficiency), testicular SDH activity is low. CAT activity has been reported to be associated with primary spermatocytes in the rat, but this does not appear to be true for the ram. Acid phosphatase is present in both interstitial cells and germ cells; two of the four acid phosphatases of the rat testis appear to be specifically associated with the germinal epithelium.

5 LIPIDS (26, 27)

5.1 Introduction

The testis is rich in lipids, which are involved in generation of energy, formation of membranes, steroidogenesis, prostaglandin synthesis, and other functions. There is a paucity of information regarding the role of specific lipids in the process of spermatogenesis.

5.2 Lipid Composition of the Testis

The concentrations of total lipids, triglycerides, phospholipids, and total cholesterol in the testes of various mammalian species are summarized in Table 1. Total lipids are low in man and ram, and high in the guinea pig. Triglycerides are highest in the ram and lowest in the rat. The concentrations of phospholipids are similar in all species studied, with the exception of the ram testis, in which they are low. Although the concentration of total cholesterol varies little between species, the ratio of free to esterified cholesterol is variable. Indeed, although rat and ram testes contain similar amounts of total cholesterol (Table 1), the rat testis contains higher levels of esterified cholesterol than the ram testis. These species differences appear to be due to different rates of synthesis and catabolism of esterified cholesterol. The concentration of total cholesterol (per gram) in the rat is higher in the testis than in the liver.

The lipid composition of human testes obtained at autopsy is shown in Table 2. The phospholipids and neutral lipids of the human testis contain relatively large amounts of unsaturated fatty acids. Saturated fatty acids are present in lower concentrations. In the rat testis, the concentration of 22:5 (22-carbon fatty acid with five double bonds) is high, whereas it is almost absent from the

Table 1   Testicular Lipids in Various Mammalian Species[a]

Species[b]	Total Lipids (mg/gm)	Triglycerides (μmol/gm)	Phospholipids (mg/gm)	Total Cholesterol (mg/gm)
Rat (1)	21.7–22.9	—	11.8–14.8	2.4
Rat (2)	22.9	8.2	11.2	2.4
Mouse (1)	20.7	—	14.1	2.4
Hamster (1)	18.7	—	16.1	2.7
Guinea pig (1)	34.3	—	14.1	2.2
Rabbit (1)	21.2	—	15.9	2.3
Rabbit (3)	28.1	17.8	9.7	2.3
Ram (4)	17.6	21.0	2.5	2.3
Man (1)	15.9	—	8.9	3.4

[a] Adapted from Johnson, in *The Testis*, Vol. 2, Chapter 4, Academic Press (1970).
[b] References: (1) Bieri and Prival, *Comp. Biochem. Physiol.*, **15**, 275 (1965). (2) Butler *et al.*, *J. Reprod. Fert.*, **15**, 157 (1968). (3) Johnson *et al.*, *J. Reprod. Fert.*, **16**, 409 (1968). (4) Johnson *et al.*, *J. Anim. Sci.*, **26**, 945 (1967).

human testis. Although there are considerable species differences in total concentration of fatty acids in the testis, the same fatty acids predominate in each lipid fraction of all species.

The concentration of lipids in germ cells and interstitial cells of all mammalian testes is high. Lipids of spermatozoa are discussed in detail in section 7.7 of this chapter. Interstitial cells contain the major amount of testicular cholesterol. Sertoli cells of rat, bull, ram, and human testis contain lipid droplets, which are apparently absent in Sertoli cells of guinea pig and boar. The lipid composition of the subcellular fractions of the rat testis is summarized in Table 3. The Golgi apparatus appears to have the highest total lipid concentration. Neutral lipids are highest in lipid vesicles, followed by residual bodies, Golgi apparatus and whole testis. The microsomes contain the lowest levels of neutral lipids. Approximately 60% of mitochondrial and microsomal lipids is phospholipid. However, these data should be interpreted with caution, since cell fractions from whole testicular tissue were analyzed.

Table 2   Lipid Composition of Human Testes Removed at Autopsy[a]

Lipid (mg/gm)	Age			
	< 3 da	4 mo	19–51 yr	56–89 yr
Total lipid	45.5	22.9	28.5	26.7
Phospholipid	20.5	15.2	15.9	15.6
Acylglycerol	3.87	3.83	3.48	3.42
Free cholesterol	2.66	1.43	1.46	1.49
Esterified cholesterol	0.96	1.05	0.59	0.83

[a] From Coniglio *et al.*, *Biol. Reprod.* **12**, 255 (1975).

Table 3 Lipid Composition of Subcellular Fractions of the Rat Testis[a]

Fraction	Total Lipid (mg/mg protein)	Lipid Phosphorus (μg/mg protein)	Phospholipids (% of total lipids)	Neutral Lipids (% of total lipids)	Cerebroside (% of total lipids)
Whole testis	0.40	62.88	62.9	34.6	2.5
Golgi	1.26	18.49	58.3	39.4	2.3
Microsomes	0.63	53.21	83.8	14.5	1.7
Residual bodies	0.60	19.67	29.5	68.0	2.5
Lipid vesicles	> 19	12	1.5	98.5	—

[a] From Keenan et al., Biochim. Biophys. Acta **270**, 433 (1972).

5.3 Lipid Changes With Age

In man, interstitial cell lipids increase gradually between 1 and 18 years of age, and then decline slowly. Sertoli cell lipids increase with age. In rat testis, total lipids, phospholipids, and neutral lipids increase from 21 to 28 days of age and decline between 35 and 63 days of age. In the bull, total lipids and phospholipids increase with age. In most species, esterified cholesterol increases with age, whereas free cholesterol declines. Sexual maturation in many species is associated with accumulation of a 22-carbon fatty acid, particularly 22:5, while levels of shorter fatty acids decrease. Long-chain fatty acids associated with cholesterol or phospholipids increase during development, coinciding with the appearance in the testis of advanced germinal cells. The in vitro capacity of 14- and 24-day-old rat testes for oxidation of labeled palmitate to CO_2 is much higher than that of adult testes. These data suggest that spermatids have low rates of fatty acid oxidation. Immature bovine testicular tissue exhibits higher linoleic acid desaturase activity and greater capacity to incorporate labeled fatty acids into lipids in vitro than do adult testes.

5.4 Effects of Hormones on Testicular Lipids

Administration of LH results in an increase of total and free cholesterol levels of the testis. In hypophysectomized animals, testicular concentrations of total lipids, cholesterol, cholesteryl esters, and 22:5 also increase. Administration of FSH to hypophysectomized animals results in even higher lipid concentrations. Neither the incorporation of labeled phosphorus into lipids nor the biosynthesis of fatty acids from acetate appears to be stimulated by gonadotropins. However, the incorporation of choline into phosphatidylcholine in testicular tissue of hypophysectomized rats is stimulated specifically by LH. Hypophysectomy increases the rate of catabolism of phospholipids and triglycerides, and impairs the conversion of linoleate to arachidonate. Gonadotropin treatment results in partial reversal of these effects. In the rat, testosterone treatment for two weeks results in decreased incorporation of labeled glucose into lipids in vitro, whereas *in vivo* it causes increases in total testicular cholesterol and accumulation of lipids in the Leydig cells. Estradiol treatment results in increased Sertoli cell lipids. Although thyroid hormones do not appear to be directly involved in the regulation of testicular lipids, hypothyroidism has been reported to be associated with increased testicular lipid concentrations, and hyperthyroidism with decreased levels of total lipids, phospholipids, and free cholesterol. Alloxan diabetes results in impairment of the conversion of linoleate to arachidonate and 22:5 fatty acid.

5.5 Effects of Nutrition on Testicular Lipids

Starvation results in increased testicular lipid concentrations. Specifically, it causes accumulation of lipids in the Sertoli cells. In the rat, the amounts of fatty acids in the diet do not appear to influence the concentrations of testicular fatty acids markedly. However, *in vitro* and *in vivo* studies have suggested an increase in the rate of synthesis of fatty acids by testicular tissue of fat-deficient rats. In the boar, an increase in dietary linoleate does not result in higher levels of this fatty acid in the testis. In the rabbit, a fat-free diet for several months causes a decrease in testicular levels of total lipids, phospholipids, and free cholesterol. Since the testis relies predominantly on endogenous sources of cholesterol rather than on exogenous supply, restriction or increase of dietary cholesterol alone does not result in marked changes in testicular levels of cholesterol.

5.6 Effects of Testicular Injury on Lipid Concentrations

Increased temperature, cryptorchidism, radiation, vitamin A deficiency, and other causes of damage to the spermatogenic process are associated with an increase in testicular lipid concentrations, predominantly in the Sertoli cells, and decreased levels of 22:5 fatty acid. The decline in testicular levels of 22:5 fatty acid may be related to the loss of mature spermatids, whereas the increase in Sertoli cell lipids could be the result of a metabolic disturbance in the Sertoli cells, or to phagocytosis of degenerating spermatids that are rich in lipids. Impaired steroidogenesis results in increased lipid content of the Leydig cells.

   In the rat, vitamin A deficiency causes testicular degeneration associated with an increase in total testicular lipids and cholesterol. *In vitro* studies have suggested that cholesterol side-chain cleavage activity is decreased in these animals. Increased temperature and cryptorchidism cause an increase in concentrations of total lipids, phospholipids, triglycerides, and cholesterol in the testis, although there are considerable species differences. The conversion of acetate to esterified cholesterol appears to be increased following thermal injury, whereas the utilization of free cholesterol is inhibited. The *in vitro* incorporation of fatty acids into lipids is enhanced. The effect of cryptorchidism on testicular lipids may not be entirely due to a temperature increase, since in unilateral cryptorchidism, lipid changes have also been noted in the scrotal testis. Thus it seems possible that in response to thermal injury, the abdominal testis may produce a substance or substances that are responsible for the effect on the scrotal testis.

Little is known about the effect of noxious agents on lipids of the testis. Administration of cadmium chloride, an agent that causes injury to the blood–testis barrier (see section 7.4 of this chapter), results in increased testicular concentrations of cholesterol. Zinc prevents the effect of cadmium chloride on the blood–testis barrier, and also the effect of cadmium chloride on testicular cholesterol levels.

6 CARBOHYDRATE METABOLISM (28)

6.1 Introduction

Spermatogenesis requires an adequate supply of exogenous metabolic nutrients. A brief discussion of testicular carbohydrate metabolism, its cellular localization and pathways is presented in this section. Developmental changes and effects of hormones or injurious agents are also briefly reviewed. The concentrations of carbohydrates in seminiferous tubule and rete testis fluids and the pathways of carbohydrate metabolism by testicular spermatozoa are discussed in sections 7.4 and 7.8 of this chapter.

6.2 Cellular Localization and Pathways of Carbohydrate Metabolism

The germ cells may receive metabolic nutrients by diffusion from the interstitial blood and fluid, from the Sertoli cells, and/or from the seminiferous tubule fluid. In most mammalian species, glycogen is present in higher concentration in testicular interstitial fluid than in blood. In some species (e.g., man, bull), the Sertoli cells and spermatogonia contain glycogen, but in other species (e.g., rat, mouse), they do not. Spermatocytes and spermatids are the major sites of testicular carbohydrate metabolism. Glucose appears to be the major substrate for these cells, although in species in which Sertoli cells contain glycogen, spermatocytes and spermatids may utilize it as an alternative substrate. Indeed, in the latter species, Sertoli cell glycogen appears to be depleted as spermatogenesis advances. In the rat and the mouse, whose Sertoli cells contain little or no glycogen, it has been suggested that developing spermatids utilize glucose that diffuse through the Sertoli cells from the peritubular space. The concentration of glucose in seminiferous tubule fluid is low (see section 7.4.2 of this chapter). Although this fluid contains relatively large amounts of inositol and amino acids, spermatids and spermatozoa do not appear to utilize them as energy sources. Thus although not conclusively proved, it seems likely that lipids are important energy sources for spermatids and spermatozoa, as they are for the nongerminal elements (e.g., Leydig cells, Sertoli cells).

The ability of testicular tissue *in vitro* to utilize added substrates such as glucose, pyruvate, and oxaloacetate appears to be well correlated with the presence of an intact germinal epithelium (29, 30). In the presence of glucose, oxygen uptake by testicular tissue from the rat and rabbit increases. The rates of oxygen and glucose uptake by the testes from several mammalian species are summarized in Table 4. It has been suggested that only substrates of at least three carbons are able to maintain the respiratory processes of the testis; acetate fails to stimulate testicular oxygen uptake.

In the absence of glucose, the rat testis utilizes lipids almost exclusively. In the absence of oxygen, testicular tissue forms lactate from glucose, as is the case with other tissues. However, testicular tissue is unique in that it is also capable of aerobic lactate formation. The activity of citric-acid-cycle enzymes is high in the testis. Carboxylation of pyruvate appears to be important in the synthesis of

Table 4 *In Vitro* **Uptakes of glucose and Oxygen by Testicular Tissue from Various Mammalian Species**[a,b]

Uptakes	Rat	Rabbit	Dog	Ram	Man
Glucose (μg/mg/hr)	10.5–19.2 (1)	12.9–54.7 (2)	—	2.0–4.5 (3)	—
Oxygen no substrate (μl/mg/hr)	2.5–4.4 (4)	3.1–6.0 (2)	4.6 (5)	—	4.1 (5)
Oxygen with glucose (μl/mg/hr)	4.6–7.5 (6)	3.3–7.1 (2)	5.5 (5)	2.5–5.0 (3)	3.8 (5)

[a] Adapted from Free, *in The Testis*, Vol. 2, Chapter 3, Academic, New York (1970).
[b] References: (1) Tepperman and Tepperman, *Endocrinology* **47**, 459 (1950); N. C. Vera Cruz, Ph.D. dissertation, Ohio State University (1968); E. D. Massie, M. S. thesis, Ohio State University (1968). (2) Ewing and Van Demark, *J. Reprod. Fert.* **6**, 9 (1963); Ewing *et al.*, *Can J. Physiol. Pharmacol.* **42**, 527 (1964); Ewing *et al.*, *J. Reprod. Fert.* **12**, 295 (1966); Ewing, *Am. J. Physiol.* **212**, 1261 (1967); Van Demark *et al.*, *Am. J. Physiol.* **215**, 977 (1968); Zogg *et al.*, *Am. J. Physiol.* **215**, 985 (1968). (3) Annison *et al.*, *Biochem. J.* **88**, 482 (1963); Setchell and Waites, *J. Physiol. London* **171**, 411 (1964); Waites and Setchell, *J. Reprod. Fert.* **8**, 339 (1964); Setchell and Hinks, *Biochem. J.* **102**, 623 (1967). (4) Dickens and Simer, *Biochem. J.* **24**, 1301 (1930); Dickens and Greville, *Biochem. J.* **26**, 1546 (1932); Dickens and Greville, *Biochem. J.* **27**, 1123 (1933); Elliot *et al.*, *Biochem. J.* **31**, 1003 (1937); Von Schuler, *Helv. Chim. Acta* **24**, 119 (1941); Von Schuler, *Helv. Chim. Acta* **27**, 1796 (1944); Tepperman *et al.*, *Endocrinology* **45**, 491 (1949); Paul *et al.*, *Proc. Soc. Exp Med. Biol.* **79**, 555 (1952); Paul *et al.*, *Endocrinology* **53**, 585 (1953); Steinberger and Wagner, *Endocrinology* **69**, 305 (1961). (5) Kato, *Acta Urol. Japan.* **49**, 659 (1958). (6) Dickens and Simer, *Biochem. J.* **23**, 936 (1929); Dickens and Simer, *Biochem. J.* **24**, 1301 (1930); Dickens and Simer, *Biochem. J.* **25**, 985 (1931); Dickens and Greville, *Biochem. J.* **26**, 1546 (1932); Dickens and Greville, *Biochem. J.* **27**, 832 (1933); Dickens *Biochem. J.* **28**, 537 (1934); Elliot *et al.*, *Biochem. J.* **31**, 1003 (1937); Von Schuler, *Helv. Chim. Acta* **24**, 119 (1941); Von Schuler, *Helv. Chim. Acta* **27**, 1796 (1944); Paul *et al.*, *Endocrinology* **53**, 585 (1953); Fitko, *Endokrynol. Polska* **16**, 177 (1965); E. D. Massie, M. S. thesis, Ohio State University (1968); N. C. Verz Cruz, Ph.D. dissertation, Ohio State University (1968).

citric-acid-cycle intermediates when these become depleted by lipolysis, protein synthesis, and nucleic acid synthesis. Glucose-6-phosphate cyclase activity is present in the testis; this is probably responsible for the high concentrations of myoinositol in the seminiferous tubule fluid. Hexokinase and phosphogluco-mutase activities have also been demonstrated in the testis. The pentose-cycle enzymes, associated primarily with steroidogenesis and generation of NADPH, are probably present only in some testicular cells, primarily the Leydig cells. Leydig cells are also rich in lactate and malate dehydrogenase activities. Lactate, isocitrate, glucose-6-phosphate, glutamate, and malate dehydrogenase activities are present in spermatogonia. Spermatocytes and spermatids contain lactate, isocitrate, succinate, β-hydroxybutyrate and sorbitol dehydrogenases. Hyaluronidase activity is associated primarily with spermatids and spermatozoa.

In vitro incubation of testicular tissue with labeled glucose has demonstrated that glucose carbons are incorporated into lipids, fatty acids, cholesterol, and steroids. A relatively large amount of glucose is metabolized by the testis to amino acids and nucleic acid bases. As discussed in section 4.1 of this chapter, glucose stimulates the incorporation of labeled amino acids into testicular proteins (Fig. 4). This effect appears to be primarily on spermatid protein synthesis, since in the absence of spermatids (immature testis, cryptorchidism, hypophysectomy) the stimulatory effect of glucose on protein synthesis is decreased.

6.3 Developmental Changes in Carbohydrate Metabolism

Glycolytic and respiratory processes are more active in immature than in adult testes. Oxygen uptake is high in immature animals, drops gradually with age, and remains constant after the animal reaches maturity (29, 31). Oxygen uptake is not stimulated by glucose in immature rat testis. The stimulatory effect of glucose on respiration that is characteristic of the adult testis is initiated at the time when spermatids first appear in the testis, coinciding with the decrease in the respiration rate in the absence of added substrate. This is the time when the testis also becomes capable of aerobic and anaerobic fructolysis.

6.4 Effects of Testicular Injury, Hypophysectomy, and Hormones on Carbohydrate Metabolism

Following an injury to the testis (e.g., heat, cryptorchidism, x-irradiation, nitrofuran treatment), the uptake of oxygen by rat testicular tissue rises (29, 30, 32, 33). This is associated with a decrease in the stimulatory effect of glucose on oxygen uptake. Thus a respiratory pattern similar to that of the immature testis occurs. Since these injuries cause a depletion of spermatids and spermatocytes from the testis, the results have been interpreted as suggesting that these cells have lower rates of respiration than the nongerminal elements; a decrease in germ cells, causing a relative increase in nongerminal elements with higher rates of oxygen uptake, would explain the increase in testicular respiration rates after injury (29, 30). However, Steinberger and Wagner (32) demonstrated that the increase in testicular oxygen uptake following x-irradiation can be prevented by hypophysectomy, whereas treatment of hypophysectomized and x-irradiated animals with gonadotropins increases oxygen uptake. These results suggest that the changes in respiration that result from testicular injury may be secondary to increased secretion of gonadotropins by the pituitary. Subsequent studies demonstrated that treatment of x-irradiated animals with estradiol benzoate or testosterone propionate to block the rise in gonadotropins also prevents the rise in endogenous respiration induced by x-irradiation (33). In intact animals, estradiol treatment causes a rise in respiration, and this effect is prevented by the concomitant administration of testosterone.

In addition to the effect of testicular injury on oxygen uptake, metabolism of labeled glucose to amino acids decreases and lactate formation increases in testes depleted of most germinal elements (e.g., after cryptorchidism). In the rabbit, cryptorchidism for 11 days does not affect glucose uptake by testicular tissue.

Hypophysectomy causes a decrease in hexokinase activity of rat testicular tissue. This effect can be reversed by LH treatment. Glucose-6-phosphate dehydrogenase activity does not change after hypophysectomy, although it is stimulated by human chorionic gonadotropin (hCG) and inhibited by steroids in some species. FSH or pregnant mare serum gonadotropin (PMSG) stimulate and testosterone inhibits glucose uptake by rat testicular tissue *in vitro*.

7   TESTICULAR FLUIDS AND SPERMATOZOA (34–36)

7.1   Introduction

The anatomical structure of the testis provides several compartments dealing with fluid production and transport (Fig. 6). The blood delivered to the testis by the arterial system is drained from the testis by complex venous and lymphatic systems. The arteriolar and capillary trees are localized exclusively in the interstitial area and do not enter the seminiferous tubules. Thus transudates from the capillary bed escape into the interstitial area, forming the *interstitial fluid compartment*. Some of this fluid may pass into the seminiferous tubules, but most is drained by the lymphatic and venous systems. The *seminiferous tubule fluid compartment* is the result of the activity of the seminiferous epithelium on one hand, and the contribution from the interstitial fluid on the other. This fluid drains through the lumen of the seminiferous tubules into the rete testis, and then ultimately into the epididymis.

7.2   Collection of Testicular Fluids

Several techniques can be utilized to obtain testicular fluids. It must be emphasized that the fluids obtained by each technique are different, and thus compari-

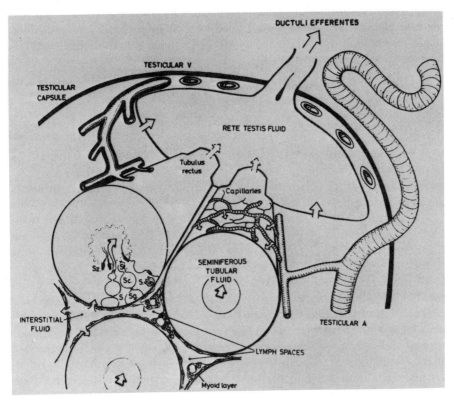

Figure 6 The fluid compartments of the testis. From Waites, *The Testis*, Academic Press, New York (1977).

sons are difficult and often inappropriate. By decapsulating the testis postmortem, a mixture of fluids and cells can be obtained. This mixture is inadequate for evaluation of the composition of seminiferous tubule fluid. Rete testis fluid can be collected *in vivo* from several mammals (e.g., rabbit, ram, bull, boar) by insertion of catheters through the ductuli efferentes. Ligation of the ductuli efferentes provides for an easier collection of rete testis fluid, either through a catheter or by micropuncture. However, this procedure results in disturbances of testicular function with time.

It has been established that the biochemical composition of rete testis fluid differs from that of seminiferous tubule fluid. Thus data derived from analyses of rete testis fluid should not automatically be assumed to be true for seminiferous tubule fluid. Seminiferous tubule fluid can be collected specifically by micropuncture and aspiration. However, since backflow from the rete testis has been demonstrated, some contamination of this fluid with rete testis fluid can be expected. A technique involving micropuncture, injection of oil into the lumen of a seminiferous tubule, and collection of newly secreted "primary" fluid has been described. In subsequent sections of this chapter dealing with biochemical composition, origin, and function of testicular fluids, an attempt is made to identify the type of fluid studied by each author (e.g., rete testis fluid, seminiferous tubule fluid). When mixtures of fluids were obtained, they are referred to as "testicular fluids."

7.3   Fluid Flow

Absorption of water into the seminiferous tubules increases the fluid pressure in the tubules. Resorption of water in the epididymis creates a pressure gradient, with fluid flowing into the rete testis, efferent ducts, and epididymis. In addition to this passive flow, contractions of the seminiferous tubules or the testicular capsule, as well as the ciliary activity of the cells in the vasa efferentia, probably aid in fluid transport.

After catheterization of the rete testis, fluid flow has been reported to be 15–33 μl/gm testis/hr in the rat, 14–50 μl/gm/hr in the rabbit, 10 μl/gm/hr in the monkey, and 4–10 μl/gm/hr in the ram and bull. The rate of spermatogenesis has not been found to have any influence on fluid flow. Except for potassium levels in the rat and inositol and glutamate levels in the ram (see section 7.4 of the chapter), spermatogenesis does not appear to influence the biochemical composition of testicular fluid significantly. Secretion of tubule fluid becomes detectable at the time when the specialized junctions between Sertoli cells are formed and spermatids first appear in the tubules (20–30 days of age in the rat). In the rat and ram, the fluid flow is not affected by hypophysectomy or hormonal treatments (gonadotropins, growth hormone, estradiol, testosterone, insulin). Oxytocin administration results in a small increase in fluid flow in the rat. Heat or cryptorchidism significantly inhibit fluid flow, whereas changes in blood pressure have no effect. Unilateral castration results in a small increase in fluid flow in the bull, whereas it has no effect in the rat. In the ram, electrical stimulation increases fluid flow but, denervation of the testis, sympathetic stimulation, or injection of atropine or pilocarpine have no effect on fluid flow. In rams and rats, administration of acetazolamide, sodium acetate, or sodium bicarbonate results in decreased testicular fluid secretion. In isolated and perfused ram or goat testes, increase of the perfusion pressure causes no change in the rate of fluid production.

7.4   Biochemical Composition of Testicular Fluids (34–36)

7.4.1   Amino Acids, Proteins, and Enzymes

Except for glutamic acid, aspartic acid, glycine, alanine, and proline, the concentration of most amino acids in rete testis fluid is lower than in plasma. Glutamic acid is the amino acid present in highest concentration in rete testis fluid of ram and bull. Total protein concentration in ram and bull rete testis fluid is 3–5% that of plasma (10% in the rat). By electrophoresis it can be demonstrated that several proteins found in rete testis fluid are absent from plasma, and some plasma proteins are absent from rete testis fluid or seminiferous tubule fluid. Bull rete testis fluid shares at least 13 protein components with plasma. In the rat, at least 10 electrophoretic bands found in rete testis fluid are also found in plasma. Free-flow seminiferous tubule fluid of the rat contains less albumin (3% of total protein) than rete testis fluid (14%) or plasma (41%). Rete testis fluid contains globulins, which are absent from seminiferous tubule fluid, suggesting that the blood–testis barrier is weakest at the level of the rete testis.

In man and rhesus monkey, composition of testicular fluids was analyzed after collecting fluid into a container after removing the tunica albuginea (Table 5). Although these fluids were probably contaminated with other fluids (e.g., lymph, blood), biochemical analyses of fluids collected by this same technique in goats and rats gave similar results to those of fluids collected by cannulation of the rete testis (37, 38).

Some enzymes are present in higher concentration in rete testis fluid than in plasma, especially those associated with the acrosome (e.g., N-acetyl-β-glucosaminidase, acid phosphatase, hyaluronidase). Lysosomal enzymes (e.g., β-glucuronidase, alkaline phosphatase, aminopeptidase, α-amylase) and most

Table 5 Composition of Testicular Fluids From Man and Rhesus Monkey[a]

Parameter	Man (1)	Rhesus Monkey (2)
Total protein (gm/100 ml)	3.60–3.77	3.70–5.58
Glucose-6-phosphatase (Racker unit/100 ml)	3.0–9.0	2.0–6.0
Glucose-6-phosphate dehydrogenase (units/ml)	4950–5550	6400–6600
Acid phosphatase (mg phosphorus/100 ml/hr)	103.6–116.8	78.6–109.8
Alkaline phosphatase (mg phosphorus/100 ml/hr)	16.7–20.5	24.6–38.2
Glucose (mg/100 ml)	20.4–22.5	27.9–33.0
Glycogen (mg/100 ml)	19.6–21.0	23.5–28.0
Lactic acid (mg/100 ml)	544–639	588–600
Ascorbic acid (mg/100 ml)	24.0–28.0	7.6–10.2
Total lipids (mg/100 ml)	281.3	136.3–144.8
Free sterols (mg/100 ml)	42.1	17.9–22.6
Sterol esters (mg/100 ml)	79.2	39.9–45.2
Phospholipids (mg/100 ml)	19.9	15.1–19.8
Sodium (mEq/l)	183.1	187.6–296.2
Potassium (mEq/l)	7.8	7.4–10.3
Chloride (mEq/l)	139.6	123.7–131.4
Bicarbonate (mmol/l)	25.4–40.4	—
Calcium (mEq/l)	3.8	1.9–4.9
Zinc (μg/100 ml)	55–80	—
pH	7.0	7.2–7.4

[a] References: (1) Pande *et al.*, *J. Clin. Endocrinol. Metab.* **27**, 892 (1967). (2) Pande *et al.*, *Indian J. Exptl. Biol.* **5**, 65 (1967) and **6**, 135 (1968).

metabolic enzymes are found in lower concentrations in rete testis fluid than in blood. In the rat, two of the four esterases found in circulation are also present in rete testis fluid, but this fluid also contains four esterases not found in blood. Seminiferous tubule fluid contains five esterases, two of which are specific. Trypsin inhibitor activity is present in the rete testis fluid of ram and boar, in lower concentrations than in epididymal fluid.

ABP has been detected in testicular fluids from several mammals (see section 4.2.2 of this chapter). LH was found in ram rete testis fluid, but neither FSH nor LH were found in rat rete testis fluid. Inhibin activity has been reported in ram, boar, and bull rete testis fluids.

7.4.2   Carbohydrates

Very low concentrations of glucose and fructose are found in rete testis fluid. The concentration of lactic acid in ram rete testis fluid is approximately 50% of the plasma concentration. Except for D-myoinositol, present in high concentrations (50–100 times the concentration in blood), no significant amounts of other carbohydrates are present in the rete testis fluid of most mammals.

7.4.3   Ions, Osmolality, and pH (Table 6)

Testicular fluids contain higher concentrations of sodium and chloride than potassium, calcium, and bicarbonate. The sodium–potassium ratio is approximately 9:1. The concentration of potassium in testicular fluids is approximately three times that of plasma, whereas the sodium and chloride concentrations are close to those of plasma. The concentrations of calcium, magnesium, and inor-

Table 6   Ions, Osmolality, and pH of Testicular Fluids From Various Mammalian Species[a]

Parameter	Man (1)	Rhesus (2)	Bull (3)	Ram (3)	Rat (4) Tubule	Rat (4) Rete Testis
pH	7.0	7.3	7.3	7.1	7.3	—
Osmolality (mOsm/l)	—	—	381	298	338	328
Ions (mEq/l)						
Sodium	183	191	118	118	109	134
Potassium	7.8	8.2	13.6	12.5	45.2	17.0
Calcium	3.8	3.1	2.2	2.1	—	—
Chloride	140	128	133	128	118	137
Bicarbonate	28.7	—	4.0	8.1	19.6	21.0

[a] References: (1) Pande et al., J. Clin. Endocrinol. Metab. **27**, 892 (1967). (2) Pande et al., Indian J. Exptl. Biol., **5**, 65 (1967) and **6**, 135 (1968). (3) Crabo, Acta Vet. Scand. **6**, Suppl. 5, 1 (1965); Voglmayr et al., Nature **210**, 861 (1966); Voglmayr et al., J. Reprod. Fert. **21**, 449 (1970); Setchell et al., J. Reprod. Fert. **24**, 81 (1971). (4) Tuck et al., Australian J. Exptl. Biol. Med. Sci. **47**, 32 (1969); Tuck et al., Eur. J. Physiol. **318**, 225 (1970); Setchell, J. Reprod. Fert. **23**, 79 (1970); Levine and Marsh, J. Physiol. London **213**, 557 (1971); Henning and Young, Experientia **27**, 1037 (1971).

ganic phosphorous are considerably lower in testicular fluids than in plasma. The ionic compositions of seminiferous tubule and rete testis fluids have been noted to be different in the rat: seminiferous tubule fluid contains much higher potassium concentrations and lower sodium and chloride levels than rete testis fluid. Both fluids are approximately isotonic with plasma and are somewhat more acidic than plasma. In the rat, destruction of the seminiferous epithelium by busulphan treatment results in a decrease of testicular fluid potassium concentration without change in sodium levels. This suggests that the germinal epithelium contributes to the levels of potassium, but not significantly to the levels of sodium in testicular fluids.

7.4.4 Steroids

High testosterone concentrations are present in rete testis fluid, but they are lower than those in testicular venous blood (10–50% in the ram, bull, and monkey, 60–90% in the rabbit and rat). Testosterone levels in seminiferous tubule fluid are higher than those in rete testis fluid, but lower than those in interstitial fluid. Through the rete testis, approximately 300 ng of testosterone reaches the epididymis daily in the ram, 900 ng in the bull, and 20 ng in the rat. Besides testosterone, dehydroepiandrosterone, 5-androstenediol, 3 α- and 3β-androstanediols, androstenedione, 17-hydroxyprogesterone, and progesterone are present in relatively high concentrations in rete testis fluid. In monkeys and bulls, estrogens appear to be present in rete testis fluid at similar concentrations as in peripheral blood.

7.5 The Blood–Testis Barrier (34, 35, 39)

Differences in concentrations of ions, proteins, steroids, and so forth between testicular fluids and plasma suggest that most substances do not diffuse freely into and out of the seminiferous tubules. A strong blood–testis barrier, formed by the tight junctions between Sertoli cells, effectively divides the seminiferous tubule into a basal compartment (containing spermatogonia and preleptotene spermatocytes) and an adluminal compartment (containing the remaining germ cell types and tubular fluid). This barrier limits the entry of certain blood constituents into the seminiferous tubule fluid, and apparently maintains within the adluminal compartment an environment favorable to the normal progression of the spermatogenic process.

The blood–testis barrier has been investigated primarily by a method involving injection or infusion of radioactively labeled substances into the general circulation and investigation of their passage into the adluminal compartment of the seminiferous tubules. Water, ethanol, urea, glycerol, and bicarbonate diffuse freely into the seminiferous tubules. Labeled sucrose and inulin also pass into the tubules both *in vivo* and *in vitro*, although more slowly than water, whereas p-aminohippurate and EDTA do not. 3-o-Methylglucose enters the rete testis fluid rapidly in the ram and rat, apparently not by active transport but by facilitated diffusion. Inositol exchanges slowly between the testis and the blood. Sodium, potassium, chloride, and rubidium pass into the tubules with relatively slow entry rates. Most amino acids and bases enter the seminiferous tubule readily. Fibrinogen and γ-globulin are excluded by the blood–testis barrier. Labeled albumin enters testicular fluids of rams and rats very slowly, and the passage occurs predominantly at the level of the rete testis. This and the higher concentration of proteins in rete testis fluid compared with seminiferous tubule fluid suggest that the blood-testis barrier is weakest at the rete testis. This is confirmed by studies demonstrating penetration of tryptan blue, 5-hydroxytryptamine, 5-hydroxytryptophan, γ-globulin, acriflavine, and so forth into the rete testis, with no penetration into the seminiferous tubules. Also, in allergic orchitis experimentally induced, immune inflammatory lesions are seen in the rete testis earlier than in the seminiferous tubules.

FSH, LH, and growth hormone pass into the seminiferous tubule, whereas insulin and prolactin do not. Labeled steroids are readily transferred, expecially testosterone and its metabolic precursors. 5α-Reduced androgens, estrogens, and corticosteroids enter the tubules at lower rates than testosterone. Labeled cholesterol is excluded from the seminiferous tubules.

The blood–testis barrier is somewhat less effective in hypophysectomized rats and is affected by temperature. It can be destroyed by treatment with cadmium chloride and irradiation, or after an inflammatory process such as allergic orchitis. Zinc can prevent the effect of cadmium chloride on the blood-testis barrier.

7.6   Origin of Testicular Fluids

Passive filtration from the blood is not the only origin of testicular fluid. Active secretion and reflux from the rete testis are other possibilities. Available evidence suggests that the Sertoli cells are major regulators of the flow and composition of seminiferous tubule fluid. Supporting this is the fact that after destruction of the germinal epithelium, fluid flow is not affected, and only minor changes in biochemical composition of testicular fluid occur.

Potassium and bicarbonate are actively transported into the tubule against their electrochemical gradients, whereas the entry of sodium and chloride is favored by the gradient. It is possible that the membrane of the Sertoli cell has a conventional sodium–potassium exchange pump. Carbonic anhydrase inhibitors decrease testicular fluid production, suggesting that this enzyme plays an important role. Most substances that pass the blood–testis barrier appear to do so through the Sertoli cells rather than through an extracellular route.

Of the total volume of testicular fluid, it has been demonstrated that over 50% is secreted by the rete testis epithelium. Since the blood–testis barrier is weakest at the rete testis, it is possible that compounds excluded from the tubules by the barrier enter the rete testis fluid and then pass into the seminiferous tubule fluid by retrograde flow.

7.7   Functions of Testicular Fluids

Functions of the fluids include sperm transport to the epididymis and maintenance of the spermatozoa during this transport. It has been suggested that testicular fluids have an inhibitory effect on sperm metabolism, to protect the spermatozoal energy stores for as long as possible during their transport. The influence of the Sertoli cell on spermatogenesis is possibly exerted through the testicular fluids. ABP and steroids are transferred to the epididymis by the testicular fluids. Inhibin activity is present in testicular fluids; it may be transported to the rete testis, vasa efferentia, and/or caput epididymis, where it is resorbed into the bloodstream. To date, functional explanations for the high concentrations of glutamic acid, inositol, and potassium in testicular fluids have not been presented, although glutamic acid was shown in *in vitro* experiments to be necessary for normal spermatogenesis.

7.8   Testicular Spermatozoa

The concentration of spermatozoa in the rete testis fluid obtained by catheterization has been reported to range from 14 to 200 million/ml in the ram. In the bull, values of 50–100 million/ml have been reported, and in the rat they averaged 33 million/ml. After ligation of the ductuli efferentes, sperm concentrations close to 100 million/ml were measured in rat and hamster rete testis fluids. Although testicular spermatozoa are not capable of forward movement and lack fertilizing ability, they are relatively similar in biochemical composition and structure to spermatozoa in other regions of the reproductive tract and ejaculated spermatozoa. A summary of some of the biochemical differences between testicular and ejaculated spermatozoa is presented in Table 7 (for review see Ref. 40). Testicular spermatozoa appear to contain a higher amount of protein than ejaculated spermatozoa. As discussed in Section 3 of this chapter, the nucleoproteins of the developing spermatid undergo a transition, which results in increased strength of the electrostratic binding of the DNA to the nucleoprotein, and chromatin condensation. During passage through the epididymis (see Chapter 2) further changes in the sperm head occur, possibly explaining the differences in protein concentration in the heads of testicular and ejaculated spermatozoa.

Testicular spermatozoa contain less ATPase than ejaculated spermatozoa. However, the lack of forward motility of testicular spermatozoa is apparently not caused by ATPase deficiency. In testicular spermatozoa, ATPase is activated to the same degree by calcium and magnesium, and by sodium and potassium at alkaline pH. In ejaculated spermatozoa, ATPase is more effectively stimulated by calcium than by magnesium, and sodium and potassium result in higher stimulation of ATPase in acid medium (Table 7). The sodium and calcium levels of testicular spermatozoa are higher than those in ejaculated spermatozoa, whereas potassium and magnesium levels are similar. The uptake of

Table 7 Biochemical Difference Between Ram Testicular and Ejaculated Spermatozoa

Parameter[a]	Testicular Spermatozoa	Ejaculated Spermatozoa
Proteins (1)	High	Low
ATPase (1)	Activated to same degree by calcium and magnesium	More effectively stimulated by calcium than magnesium
	More stimulated by sodium and potassium in alkaline pH	More stimulated by sodium and potassium in acid pH
Sodium and calcium levels (2)	High	Low
Oxygen uptake (3)	Linear in presence or absence of phosphate	Linear only in absence of phosphate
	Linear stimulation by glucose only in presence of phosphate	Linear stimulation by glucose only in absence of phosphate
Glucose oxidation (3)	High	Low
Lactic acid production (3)	High	Low
Incorporation of glucose into (4)		
Fatty acids	Low	High
Neutral lipids	High	Low
Phospholipids	High	Low

[a] References: (1) Voglmayr et al., Biol. Reprod., 1, 121 (1969). (2) Setchell et al., Biol. Reprod. Suppl. 1, 40 (1969). (3) Volgmayr et al., Nature 210, 861 (1966) J. Reprod. Fert. 14, (1967), and J. Reprod. Fert. 21, 449 (1970). (4) Scott et al., Biochem. J. 102, 456 (1967); Scott and Setchell, Biochem. J. 107, 273 (1968).

oxygen by testicular spermatozoa is linear in the presence or absence of phosphate. Glucose stimulates this uptake in a linear fashion in the presence, but not in the absence, of phosphate. In ejaculated spermatozoa, oxygen uptake and its stimulation by glucose are only linear in the absence of phosphate (Table 7). The oxidation of glucose to CO_2 by testicular spermatozoa is higher than that of ejaculated spermatozoa, and lactic acid production is also higher. This suggests that the oxidative pathway of utilization of glucose predominates in testicular spermatozoa of the species studied (ram and bull).

Pentose-cycle enzyme activity is almost absent in spermatozoa. Myoinositol and glutamic acid, present in testicular fluids in relatively high concentrations, are apparently not used as substrates by testicular spermatozoa. Thus it appears that spermatozoa predominantly metabolize glucose, and in its absence, fructose.

Increased temperature inhibits the utilization of oxygen and glucose by testicular spermatozoa, but appears to have no effect on the generation of lactate. Ram testicular spermatozoa remain viable when stored at cool temperatures (1°C) for several days. In addition, during this storage the cytoplasmic droplet migrates, and the respiratory pattern becomes similar to that of ejaculated

Table 8 Phospholipids of Ram Testicular Spermatozoa[a]

Phospholipid	Amount (mg/10^9 spermatozoa)
Total phospholipids	1908.6
Phosphatidylcholine	215.3
Phosphatidylethanolamine	88.6
Phosphatidylinositol	41.6
Phosphatidylserine	52.6
Phosphatidylglycerol	15.0
Phosphatidic acid	24.6
Cardiolipin	116.0
Serine plasmalogen	31.6
Ethanolamine plasmalogen	170.0
Choline plasmalogen	564.3
Sphingomyelin	244.6
Alkyl ether phospholipid	230.3

[a] From Scott et al., Biochem. J. **102**, 456 (1967).

spermatozoa. Thus some of the maturation processes that occur in the epididymis can be reproduced by storage of testicular spermatozoa at cool temperatures.

Lipogenesis by testicular spermatozoa is not affected by heat. Under aerobic conditions, the concentrations of free fatty acids, acyl esters, and phosphatides in testicular spermatozoa decrease. Labeled glucose is incorporated readily into neutral lipids and phospholipids by testicular spermatozoa. Although phosphatidylinositol and plasmalogens are present in relatively high concentrations in ram testicular spermatozoa (Table 8), they are not synthesized from labeled glucose by these cells. Compared with ejaculated spermatozoa, testicular spermatozoa incorporate smaller amounts of labeled glucose into fatty acids, but higher amounts into neutral lipids and phospholipids (Table 7).

REFERENCES

1 Steinberger, E., Structural Considerations of the Male Reproductive System, in DeGroot, L. (ed.), *Endocrinology*, Vol. 3, Chapter 118, Grune & Stratton, New York, (1979).

2 Clermont, Y., Huckins, C, Renewal of spermatogonia in the rat. *Am. J. Anat.* **93**, 475 (1953).

3 Steinberger, E., and Steinberger, A., Spermatogenic Function of the Testis, in Hamilton, D. W., and Greep, R. O. (eds.), *Handbook of Physiology*, sect. 7, Vol. 5, Chapter 1, American Physiological Society, Washington, DC (1975).

4 Clermont, Y., Spermatogenesis, in Greep, R. O., and Koblinsky, M. A. (eds.), *Frontiers in Reproduction and Fertility Control*, Part 2, Chapter 27, MIT Press, Cambridge (1977).

5 Chowdhury, A. K., Thymidine-³H labeling of spermatogonia in rat and human seminiferous tubules mounted in toto. *Anat. Rec.* **169**, 296 (1971).

6 Steinberger, A., and Steinberger, E., The Sertoli Cell, in Johnson, A. D., and Gomes, W. R. (eds.), *The Testis*, Vol. 4, Chapter 11, Academic, New York (1975).

7 Steinberger, E., Hormonal Regulation of the Seminiferous Tubule Function, *in* French, F. S., Hansson. V., Ritzen, E. M., and Nayfeh, S. N. (eds.), *Hormonal Regulation of Spermatogenesis*, p. 337, Plenum, New York (1975).

8 Steinberger, E., Hormonal Control of Spermatogenesis, *in* DeGroot, L. (ed.), *Endocrinology*, Vol. 3, Chapter 122, Grune & Stratton, New York (1979).

9 Gledhill, B. L., Nucleic Acids of the Testis, *in* Johnson, A. D., Gomes, W. R., and Vandemark. N. L. (eds.), *The Testis*, Vol. 2, Chapter 6, Academic, New York (1970).

10 Monesi, V., Synthetic Activities During Spermatogenesis, *in* Segal. S. J., Crozier, R., Corfman, P. A. and Condliffe, P. G. (eds.), *The Regulation of Mammalian Reproduction*, Chapter 9, Charles C Thomas, Springfield, Illinois (1973).

11 Monesi, V., Nucleoprotein Synthesis in Spermatogenesis, *in* Mancini, R. E., and Martini. L. (eds.), *Male Fertility and Sterility*, p. 59. Academic, New York (1974).

12 Ito, M., Hotta, Y., and Stern, H., Studies of meiosis *in vitro* II. Effect of inhibiting DNA synthesis during meiotic prophase on chromosome structure and behavior. *Develop. Biol.* **16**, 54 (1967).

13 Stern, H., and Hotta, Y., DNA Synthesis in relation to chromosome pairing and chiasma formation. *Genetics*, Suppl. **61**, 27 (1969)

14 Hotta, Y., Parchman, L.G., Stern, H., Protein synthesis during meiosis. *Proc. Nat. Acad. Sci. USA* **60**, 575 (1968).

15 Davis, J. R., Langford, G. A., Testicular Proteins, *in* Johnson, A. D., Gomes, W. R., and Vandenmark, N. L. (eds.), *The Testis*, Vol. 2, Chapter 5. Academic, New York, (1970).

16 Hodgen, G.D., Enzyme Markers of Testicular Function, *in* Johnson, A. D., and Gomes, W. R. (eds.), *The Testis*, Vol. 4, Chapter 12. Academic, New York, (1977).

17 Sanborn, B. M., Elkington, J. S. H., Steinberger, A., *et al.*, Androphilic Proteins in the Testis, *in* Spilman, G. H., Lobl, T. L., Kirton, K. T. (eds.), *Regulatory Mechanisms of Male Reproductive Physiology*, p. 45. *Excerpta Medica*, Amsterdam (1976).

18 Hansson, V., Weddington, S. C., Naess, O., *et al.*, Testicular Androgen Binding Protein (ABP)—a Parameter of Sertoli Cell Secretory Function, *in* French, F. S., Hansson, V., Ritzen, E. M., and Nayfeh, S. H. (eds.), *Hormonal Regulation of Spermatogenesis*, Vol. 2, p. 323. Plenum, New York (1975)

19 Hansson, V., Ritzen, E. M., French, F. S., and Nayfeh, S. N., Androgen Transport and Receptor Mechanisms in Testis and Epididymis, *in* Hamilton, D. W., and Greep, R. O. (eds.), *Handbook of Physiology*, Sec. 7, Vol. 5, Chapter 7. American Physiological Society, Washington DC (1975).

20 Steinberger, A., and Steinberger, E., Secretion of an FSH-inhibiting factor by cultured Sertoli cells. *Endocrinology* **99**, 918 (1976).

21 McCullagh, D. R., Dual endocrine activity of the testes. *Science* **76**, 19 (1932).

22 Franchimont, P., Chari, S., Hazee-Hagelstein, M. T., *et al.*, Evidence for the Existence of Inhibin, *in* Troen, P., and Nankin, H. R. (eds.), *The Testis in Normal and Infertile Men*, p. 253. Raven, New York (1977).

23 Steinberger, A., and Steinberger, E., Inhibition of FSH by a Sertoli Cell Factor *in vitro*, *in* Troen, P., and Nankin, H. R. (eds.), *The Testis in Normal and Infertile Men*. p. 271. Raven, New York (1977).

24 Franchimont, P., Demoulin. A., Verstraelen-Proyard, J., *et al.*, Nature and Mechanisms of Action of Inhibin: Perspective in Regulation of Male Fertility, *Endocrine Approaches to Male Contraception, Int. J. Androl.* (1978), Suppl. 2.

25 Machado de Domenech, E., Domenech, C. E., Aoki, A., *et al.*, Association of the testicular lactate dehydrogenase isozyme with a special type of mitochondria. *Biol. Reprod.* **6**, 136 (1972).

26 Johnson, A. D., Testicular Lipids, *in* Johnson, A. D., Gomes, W. R., and Vandenmark, N. L. (eds.), *The Testis*, Vol. 2, Chapter 4. Academic, New York (1970).

27 Coniglio, J. G., Testicular Lipids, *in* Johnson, A. D., and Gomes, W. R. (eds.), *The Testis*, Vol. 4, Chapter 13. Academic, New York (1977).

28 Free, M. J., Carbohydrate Metabolism in the Testis, *in* Johnson, A. D., Gomes, W. R., and Vandenmark, N. L. (eds.), *The Testis*, Vol. 2, Chapter 3. Academic, New York (1970).

29 Tepperman, J., Tepperman, H. M., and Dick, H. J., A study of the metabolism of rat testis *in vitro*. *Endocrinology* **45**, 491 (1949).

30 Featherstone, R. M., Nelson, W. O., Welden, F., *et al.*, Pyruvate oxidation in testicular tissues during furadroxyl-induced spermatogenic arrest. *Endocrinology* **56**, 727 (1955).

31 Von Schuler, W., Beziehungen Zwischen Organstoffwechsel und Levensalte von Ratten. *Helv. Physiol. Pharmacol. Acta* **1**, 105 (1943).

32 Steinberger, E., and Wagner, C., Observations on the endogenous respiration of rat testicular tissue. *Endocrinology* **69**, 305 (1961).

33 Steinberger, E., The Effect of Sex Steroids in Endogenous Respiration of X-Irradiated Testis, in *Proceedings of an International Symposium on the Effects of Ionizing Radiation in the Reproductive System*, p. 213. Pergamon, New York (1963).

34 Setchell, B. P., The Entry of Substances into the Seminiferous Tubules, *in* Mancini, R. E., and Martini, L. (eds.), *Male Fertility and Sterility*, p. 37. Academic, New York (1974).

35 Setchell, B. P., Waites, G. M. H., The Blood-Testis Barrier, *in* Hamilton, D. W., and Greep, R. O. (eds.), *Handbook of Physiology*, Sect. 7, Vol. 5, Chapter 6. American Physiological Society, Washington, DC (1975).

36 Waites, G. M. H., Fluid Secretion, *in* Johnson, A. D., and Gomes, W. R. (eds.), *The Testis*, Vol. 4, Chapter 3. Academic, New York (1977).

37 Pande, J. K., Das Gupta, P. R., and Kar, A. B. Biochemical composition of human testicular fluid collected post mortem. *J. Clin. Endocrinol. Metab.* **27**, 892 (1967).

38 Pande, J. K., Das Gupta, P. R., and Kar, A. B., Effect of antispermatogenic agents on biochemical composition of testicular fluid of rhesus monkeys and rats. *Indian. J. Exptl. Biol.* **6**, 135 (1968).

39 Neaves, W. R., The Blood-Testis Barrier, *in* Johnson, A. D., and Gomes, W. R. (eds.), *The Testis*, Vol. 4, Chapter 4. Academic, New York (1977).

40 Voglmayr, J. K., Metabolic Changes in Spermatozoa during Epididymal Transit, *in* Hamilton, D. W., and Greep, R. O. (eds.), *Handbook of Physiology*, Sect. 7, Vol. 5, Chapter 21. American Physiological Society, Washington, DC (1975).

CHAPTER TWO

THE EPIDIDYMIS

LOURENS J. D. ZANEVELD

*Departments of Physiology and Biophysics
and Obstetrics and Gynecology
College of Medicine
University of Illinois at the Medical Center
Chicago, Illinois*

1 INTRODUCTION

The epididymis is closely apposed to the testicle and consists of a single, highly coiled tubule that fuses with the vasa efferentia at its testicular end, and the vas deferens at its distal end (Fig. 1). The lumenal surface of the epididymal tubule is lined by cells that either do or do not possess stereocilia (microvilli). A small number of cells with motile cilia are also present. By way of the rete testis and a number of vasa (ductuli) efferentia, the spermatozoa and testicular fluids pass to the head (caput) of the epididymis. From here, the contents pass through the middle (corpus) of the epididymis into its tail portion (cauda). Transport of the spermatozoa from the corpus to the cauda requires from one to three weeks and appears to occur primarily by contractions of the epididymal tubule, although fluid flow and pressure exerted by the incoming testicular fluid may also play a role.

The testicles produce a considerable amount of fluid every day (10–20 ml in rodents and rabbits; 20–40 ml in the ram, goat, and bull). Most of this fluid (approximately 98% in the boar and 73–80% in the rat) is resorbed in the caput epididymis, and some more is resorbed in the corpus. This causes a tremendous increase in the concentration of spermatozoa as the gametes pass through the epididymis (to $3–7 \times 10^9$ sperm/ml in the cauda of farm animals, and to $2–18 \times 10^8$ sperm/ml in the cauda of the rabbit and rodents). Because of the resorptive activity of the caput and corpus epididymis, only small amounts of fluid are present in the cauda epididymis. For instance, from a cannula in the cauda epididymis of the bull, ram, and boar, respectively 0.2–0.6 ml, 0.1–0.2 ml, and approximately 3.8 ml can be collected when the animals are sexually stimulated (each event). By microaspiration, only 50–300 μl can be collected from the epididymis of the rat and rabbit.

The caput and corpus epididymis resorb not only water but also certain biochemical components from the testicular fluid. Selectivity appears to be present in regard to the type and amount of compound that is resorbed. Besides being resorptive, the epididymis is also a secretory organ through the action of the epithelial cells of the lumen and thus contributes to the epididymal fluid components. Because of this selective resorption as well as secretory activity by the epididymis, the composition of the epididymal fluid changes as it passes from the caput to the cauda.

Testicular spermatozoa are immature, that is, they do not have the ability to move or can do so only poorly and cannot fertilize the ovum. During their transport through the caput and corpus epididymis, the spermatozoa gain fertilizing capacity and the ability to move. However, the gametes remain immotile as long as they reside in the male genital tract, presumably as a measure to preserve energy. Indeed, epididymal spermatozoa may remain viable for several weeks or longer. A number of morphological changes, for example, the caudal migration of the cytoplasmic droplet and a modification of the acrosomal and plasma membranes (in the guinea pig, rabbit, some species of monkey, and the

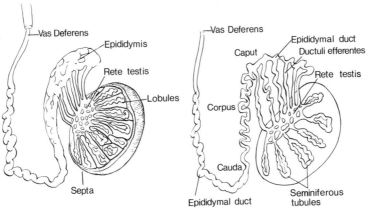

Figure 1 Schematic representation of the testis and epididymis. (Prepared by the Biocommunications Art Department, University of Illinois at the Medical Center, Chicago, Illinois.)

chinchilla, but not man), as well as biochemical alterations (see later) occur as the spermatozoa migrate through the epididymis.

The functional activity of the epididymis, whether the maturation of spermatozoa or the resorption and secretion of fluid, is highly dependent on androgen stimulation. For instance, after castration of the rat, large decreases in epididymal glycerolphosphorylcholine and carnitine occur and can be restored by testosterone treatment.

From a reproductive standpoint, the fluid of the epididymis is of primary interest because it controls or mediates epididymal function. This chapter therefore deals primarily with the biochemistry of epididymal fluid and spermatozoa, although a brief summary of the biochemistry of the cells lining the tubular lumen has been added. Although a number of studies have been reported on extracts of epididymal tissue, these have not been included because such extracts consist of a mixture of a number of different cell types as well as blood, and because one cannot be certain that any of the components measured in this fashion actually pass into the epididymal fluid. Specific references have been omitted from the text. However, a number of reviews and pertinent articles can be found in the reference list (1–22).

2 ORIGIN OF THE EPIDIDYMAL FLUID COMPONENTS

The components of the lumenal fluid of the epididymis either (1) derive from the incoming testicular fluid, (2) arrive through transudation from blood, (3) are released from the spermatozoa, or (4) are actually synthesized by the epididymis itself. Only a few of the biochemical components of epididymal fluid have been shown to be primarily synthesized by the epididymis. The best studied are the lipids: glycerylphosphorylcholine (GPC), carnitine, and acetyl carnitine. In a number of species, these compounds are almost entirely specific for the epididymis and may be useful to determine the epididymal fluid contribution to the ejaculate. Other compounds whose synthesis is known to take place in the epididymis are steroids, glycoproteins, and sialic acid, but these may also arrive there by the testicular fluid.

Synthesis of compounds occurs in the lumenar epithelium of the epididymal tubule. A number of histochemical and biochemical studies have been performed on the two primary types of epithelial cells found in the epididymis. Stereociliated cells react with the periodic acid Schiff (PAS) stain and occasionally with Alcian blue, and possess phospholipids, acid DNase, nucleoside diphosphatase, aliesterase, lactate dehydrogenase (LDH), NAD diaphorase, glucose-6-phosphate dehydrogenase, N-acetyl-β-glucosaminidase, cholinesterase, aminopeptidase, and plasminogen activator. They probably possess alkaline phosphatase, but this is argued.

The nonstereociliated cells stain with PAS, undergo osmic acid reduction, and contain sulfhydryl groups, choline, acid phosphatase, thiamine pyrophosphatase, adenosine triphosphatase, nucleoside diphosphatase, alkaline nucleoside phosphatase, succinate dehydrogenase, LDH, isocitrate dehydrogenase, NAD and NADP diaphorases, glucose-6-phosphate dehydrogenase, and possibly alkaline phosphatase, but they stain negative for RNA.

A number of steroid-metabolizing enzymes (hydroxysteroid hydrogenases) are associated with the epithelial cells, indicating that they are capable of steroid synthesis. The concentration of certain components in the epithelial cells varies between the caput, corpus, and cauda, implying that different anabolic and catabolic processes take place in the various parts of the epididymis. As will be discussed in more detail later in this chapter, these differences may be related to the various functions of the individual portions of the epididymis. The synthetic activity of the epididymis is androgen dependent.

Whether any of the other components found in epididymal fluid can be synthesized in the epididymis or originate only from other sources remains to be established. Particularly confusing is the contribution of sperm components to the overall fluid composition either naturally or during the preparative separation of the spermatozoa from epididymal fluid. This applies in particular to typical sperm glycolytic enzymes such as LDH, glutamic-oxaloacetic transaminase (GOT; aspartate aminotransferase), glutamic-pyruvic transaminse (GPT; alanine aminotransferase), and the acid and alkaline phosphatases, which are fairly readily released from the spermatozoa. The origin of each epididymal fluid component is therefore often difficult to establish.

3   COLLECTION AND PREPARATION
OF THE EPIDIDYMAL CONTENTS

Epididymal contents can be collected by a number of techniques. (1) *Excision*—after slaughter or euthanasia, the epididymis is excised, and the lumenal contents are collected by squeezing, aspiration with a micropipette, or by entering a cannula and flushing the epididymal tubule. (2) *Cannulation*—the cauda epididymis is cannulated, and the cannula is passed through the skin; or a fistula is established, and the epididymal contents are collected during sexual rest or ejaculation (the cannula remains patent for up to three months, if done carefully). (3) *Micropuncture*—the epididymal tubule is aspirated after micropuncture while the animal is under anesthesia.

Each technique has certain disadvantages and the results obtained can differ significantly with the collection method used. The disadvantages of the first method are that contamination with blood and lymph is more likely to occur and the biochemical composition of the epididymal fluid and spermatozoa may change on even short-term storage of the epididymis and exposure to different temperatures. The disadvantages of the second method are that only cauda epididymal fluid can be obtained, and bacterial contamination is difficult to avoid. The third technique is the best for obtaining uncontaminated samples, but only small volumes can be collected.

Several other, probably less reliable, methods have been used to determine the composition of epididymal fluid. One method, primarily applied to the boar, consists of surgically removing the seminal vesicles and bulbourethral glands. Atropine is administered just before ejaculation to prevent secretion by the prostate gland, and the ejaculated fluid consists mostly of epididymal fluid. A second method involves obtaining a partitioned (split) ejaculate and, with the aid of some typical epididymal fluid components such as glycerylphosphorylcholine (GPC), performing a compartmental analysis on the ejaculate. Thirdly, the vas may be sectioned or tied off immediately adjacent to the epididymis and the decrease of certain components in the ejaculate measured. "Epididymal-like" spermatozoa can be collected without surgery by the partitioned (split) ejaculation technique or by ejaculating semen in a large volume of buffer.

Separation of the epididymal fluid components from the spermatozoa is usually accomplished by centrifugation with or without a sucrose density gradient. No matter how much care is taken, some contamination of the fluid portion with sperm components will occur. The degree of contamination varies with the length of time before the separation procedures are initiated, the temperature at which the epididymal fluid is maintained and at which the separation techniques are performed, and the speed and length of the centrifugation. Therefore, reported data should be scrutinized carefully for the conditions used to collect, maintain, and treat the epididymal contents. In the tables of this chapter, the method of fluid collection is indicated after each reference quoted.

The variability in techniques makes it quite difficult to compare the results from one investigator to another. An attempt has been made to summarize a number of the reported observations in tabular form. It is suggested that the amounts listed be considered as approximations rather than as absolute values. Reported values that differed totally from others were deleted. When values differed but were in the same range, the values were averaged. These averaged values are marked with an asterisk (*).

4   BIOCHEMICAL COMPOSITION OF EPIDIDYMAL FLUID

4.1   Man and Nonhuman Primates

Very little is known regarding the fluid composition of the human epididymis except that plasminogen activator is absent and calcium concentrations are 1.5 mM. The epididymal fluid of the rhesus macaque contains significant amounts of protein, free and bound sialic acid, and lipid (mostly phospholipid). The Na^+, K^+, Mg^{2+}, and Cl^- concentrations were reported by Jones (Ref. 20, Table 1) to be, respectively, 18.2, 49.2, 4.0, and 10.5 mEq/l. The total protein, GPC, carnitine, phosphorus, and cholesterol content are, respectively, 24.0, 12.9, 3.5, 2.4, and 0.07 mg/ml. Acid and alkaline phosphatase, α-mannosidase, β-N-acetyl-glucosaminidase, acid DNase, and lactate dehydrogenase are the enzymes so far identified in rhesus monkey epididymal fluid. The protein, phospholipid, and free sialic acid concentrations are highest in the fluid from the caput epididymis, whereas the bound sialic acid and the other (nonphospholipid) lipid levels are highest in the cauda. The myoinositol levels of the caput, corpus, and cauda epididymal fluid obtained by micropuncture from the baboon were reported to be respectively 3.5, 5.9, and 1.3 mM, and those of the rhesus macaque to be respectively 25.1, 18.3, and 17.2 mM.

4.2   Farm Animals

4.2.1   Electrolytes, pH, Acids, and Carbohydrates

The biochemical composition of the epididymal fluids of the bull, boar, ram, and stallion can be found in Tables 1–3. In addition, sodium, potassium, and calcium levels have been reported for the goat. These values for the cauda and caput respectively are as follows: sodium, 95 and 339 mg/100 ml; potassium, 168 and 64 mg/100 ml; and calcium 3 and 9 mg/100 ml. Compared with blood, the epididymal fluid of farm animals contains low levels of sodium and chloride and high levels of potassium (Table 1). In spite of this, epididymal fluid is still hypertonic compared with blood plasma. In all animals, a large decrease occurs in the concentrations of sodium and chloride from the caput to the cauda epididymis, whereas the concentration of potassium tends to be highest in the corpus or caput. In the bull, ram, and goat, the concentrations of calcium tend to be highest in the caput; but in the boar, the highest levels are present in the cauda. Lactate, bicarbonate, phosphate, and magnesium concentrations are lowest in the cauda epididymis, whereas the total phosphorus and acid-soluble phosphorus concentrations are highest in the cauda epididymis. Similar to that of blood, the pH of epididymal fluid is near neutral and, at least in the bull, is essentially the same for all parts of the tract.

The sodium, potassium, calcium, phosphate, protein, and GPC levels of cauda epididymal fluid from calves aged 209–242 days (the time when spermatozoa first appear in the epididymis) were reported by Crabo to be respectively 61.8 mEq/l, 16.9 mEq/l, 1.13 mEq/l, 14.65 mmol/l, 5.2 mg N/ml, and 62.5

mmol/1. Compared with the levels in mature bulls, the levels of sodium are significantly higher and those of potassium and protein lower.

In the bull, the total concentration of carbohydrates tends to be highest in the cauda epididymis, but in the ram, the concentration is somewhat higher in the caput epididymis. Only very small amounts of fructose and other reducing sugars are present in epididymal fluid, and basically all of the reducing sugars in the ejaculate originate from the accessory sex glands. The low level of reducing sugars in the epididymis indicates that the epididymal spermatozoa do not rely on these sugars for their metabolism.

4.2.2 Lipids

High levels of GPC are typically present in epididymal fluid. The levels of GPC are much higher in the cauda epididymis than in the caput of the bull, boar, and ram. In contrast to the case with other species, the lipids of ram epididymal fluid have been studied in some detail (Table 1). With the exception of the plasmalogens and nonesterified cholesterol, which show highest concentrations in the caput epididymis, and phosphatidylcholine, which is highest in the cauda, all other lipids are approximately evenly distributed. Some discrepancy is present, however, in the reported observations. For instance, Voglmayr et al. (Ref. 28, Table 1) found that the amount of total phospholipid is 0.042 mg/ml in ram rete testis fluid but 1.34 mg/ml in the cauda epididymal fluid of this species. In contrast, others claim that the total phospholipid levels are approximately the same in the caput and cauda of the ram, approximating 6.5 mg/ml (Table 1). The prostaglandin $F_{2\alpha}$ content of ram cauda epididymal fluid was reported to vary from 1.0 to 3.2 ng/ml which is 300–1000-fold less than that found in ram seminal plasma but about the same as that found in ram rete testis fluid.

The testosterone levels are highest in the caput. Even lower levels of testosterone (2.6 ng/ml) than those given in Table 1 have been reported in ram cauda epididymal fluid. It was calculated in the bull that 943 ng of 17β-hydroxylated androgens (mostly testosterone) arrive daily in the caput epididymis but that only 5 ng leave the cauda each day. Thus although the epididymis is capable of steroid synthesis (see section 2), the testosterone in epididymal fluid derives primarily from the testis (most likely bound to androgen-binding protein, ABP) and is absorbed to a large extent in the caput epididymis, presumably exerting biological activity. The ABP levels of ram cauda epididymal fluid are 79.9 pmol/ml.

Besides testosterone, the following steroids have been found in the cauda epididymal fluid of the bull: dehydroepiandrosterone (7.9 ng/ml), 5-androstenediol (4.3 ng/ml), progesterone (7.4 ng/ml), androstenedione (1.7 ng/ml), 3α-androstanediol (4.0 ng/ml), 3β-androstanediol (6.5 ng/ml), and estrogen (26 pg/ml). Dihydrotestosterone is almost absent from bull testicular fluid (1.3 ng/ml) but is present in the cauda epididymal fluid of the bull and ram (Table 1). The pregnenolone level in ram testicular fluid is only 0.85 ng/ml but is 9.9 ng/ml in the cauda epididymal fluid of the ram. The androstenedione concentration in ram cauda epididymal fluid is 1.3 ng/ml.

Table 1 Biochemical Composition of the Epididymal Fluid from Farm Animals[a,b]

Component	Unit	Bull (1–14)			Boar (4, 5, 14–20)			Ram (19–28)		Stallion (20)
		Cauda	Corpus	Caput	Cauda	Corpus	Caput	Cauda	Caput	Cauda
Protein										
Total Protein (For enzymes and amino acids, see Tables 2 and 3)	mg/ml	32.3*	7.0	5.7	23.3*	—	—	21.9	—	12.7
Carbohydrate										
Total carbohydrate	mg/ml	3.78*	2.44	2.80	4.71	—	—	2.69*	2.93	5.53
Acid-insoluble carbohydrate	mg/ml	1.84	1.22	1.26	—	—	—	1.28	1.71	—
Acid-soluble carbohydrate	mg/ml	—	—	—	3.16	—	—	1.63*	1.46	2.65
Total reducing sugar	mg/ml	0.02	0.06	0.13	—	—	—	0.04	—	—
Total anthrone-reactive carbohydrate	mg/ml	—	—	—	1.84	—	—	—	—	—
Total amino sugar	mg/ml	—	—	—	1.28	—	—	—	—	—
Fructose	mg/ml	0.07	—	—	0.04	—	—	0.10	—	—
Inositol	mg/ml	—	—	—	0.95	—	—	—	—	—
Myoinositol	mmol/l	—	—	—	1.22	—	—	5.10	—	—
Glucose	mg/ml	Trace	—	—	—	—	—	—	—	—
Lipid										
Total Lipid	mg/ml	—	—	—	1.84	—	—	—	—	—
Glycerylphosphoryl-choline (GPC)	mg/ml	16.14*	8.6*	3.5*	22.3*	13.97	1.48	23.17*	4.6*	12.7
Carnitine	mmol/l	—	—	—	17.0	9.8	5.8	12.6	—	1.8 (mg/ml)

Phosphorylcholine	mg/ml	Trace	—	—	0	—	—	—	—	—
Phosphatidylcholine (lecithin)	mg/ml	—	—	—	—	—	—	2.11	1.39	—
Glycerylphosphorylethanolamine	mg/ml	0	—	—	Trace	—	—	—	—	—
Phosphorylethanolamine	mg/ml	0	—	—	0	—	—	—	—	—
Phosphatidylethanolamine (cephalin)	mg/ml	—	—	—	—	—	—	0.62	0.59	—
Plasmalogen	mg/ml	—	—	—	0.01	—	—	—	—	—
Choline plasmalogen	mg/ml	—	—	—	—	—	—	0.59	1.11	—
Ethanolamine plasmalogen	mg/ml	—	—	—	—	—	—	0.90	1.79	—
Sphingomyelin	mg/ml	—	—	—	—	—	—	0.63	0.43	—
Total phospholipid	mg/ml	—	—	—	—	—	—	6.45	6.66	—
Cholesterol (esterified + nonesterified)	mg/ml	—	—	—	0.7	—	—	0.43*	1.03	0.67
Volatile fatty acids	mEq/l	11.5	—	—	—	—	—	1.70	—	—
Testosterone	ng/ml	20.3	—	—	—	—	—	10.8	24.8	—
Dihydrotestosterone	ng/ml	—	—	—	—	—	—	25.8	33.8	—

Other

Sodium	mEq/l	42.7*	62.2*	90.1*	27.7*	54.7	95.8	37.6*	90.0*	41.3
Potassium	mEq/l	31.4*	50.8*	43.0*	33.5*	37.9	33.5	31.1*	48.8*	32.1
Magnesium	mEq/l	2.7	5.7	6.3	2.0*	—	—	1.1*	2.9	2.4
Calcium	mEq/l	1.1*	2.1*	1.8*	4.5*	2.4	1.7	0.9*	1.6	1.1
Zinc	μmol/l	—	—	—	136.5*	—	—	25	—	50
Strontium	μmol/l	—	—	—	8.7	—	—	—	—	—
Chloride	mEq/l	39.8*	77.2*	85.7*	15.3*	32.7	56.2	21.7*	61.9	12.0

Table 1 (continued)

Component	Unit	Bull (1-14)			Boar (4, 5, 14-20)			Ram (19-28)		Stallion (20)
		Cauda	Corpus	Caput	Cauda	Corpus	Caput	Cauda	Caput	Cauda
Bicarbonate	mEq/l	0	0.4	3.1	—	—	—	—	—	—
Phosphate	mEq/l	20.9	35.4	34.3	—	—	—	—	—	—
Citric acid	mg/ml	0	0	0	0.12	—	—	0.05	0.04	—
Lactic acid	mg/ml	0.36*	0.75	0.73	0.97	—	—	0.64*	0.99*	—
Ammonia	mg/100 ml	—	—	—	3.5	—	—	21.9	17.2	4.8
Ammonium ion	mEq/l	22.00	—	15.9	—	—	—	—	—	—
pH		6.8	6.6	6.7	6.9	—	—	7.1*	—	6.8
Total phosphorus	mg/ml	4.35	1.68	1.31	2.63*	—	—	3.98*	1.56	3.68
Inorganic phosphorus	mg/ml	—	—	—	0.37	0.36	—	0.26*	0.69	0.13
Acid-soluble phosphorus	mg/ml	3.70	1.72	1.29	3.78	—	0.34	3.20*	1.07	3.55
Acid-insoluble phosphorus	mg/ml	—	—	—	—	—	—	0.13	0.30	—
Acid-insoluble, nonlipid phosphorus	mg/ml	0	0	0.05	—	—	—	0.06	0.22	—
Total water-soluble phosphorus	mg/ml	—	—	—	1.5	—	—	3.52	0.64	1.0
Lipid phosphorus	mg/ml	0.07	0.12	0.12	—	—	—	2.3	—	—
Total nitrogen	mg/ml	—	—	—	7.2*	—	—	6.3	—	13.9
Nonprotein nitrogen	mg/ml	—	—	—	2.4*	—	—	—	—	—
Urea nitrogen	mg/ml	—	—	—	0.13*	—	—	0.2	—	0.2

Urea	mg/ml	0.29*		0.36		—		—	—	—	—	0.15	0.21	
Dry weight	mg/ml	—		—		56.8*		—	—	—	—	—	—	
Freezing-point depression	−°C	0.66		—		—		—	—	—	—	0.64	0.61	
Osmolarity	mOsm/l	281		—		324.8		—	—	—	—	—	—	
Osmotic pressure	ΔT_f, °C	—		—		0.60		—	—	—	—	0.60*	0.62	

[a] At times, different values have been reported for a single component. These values were averaged unless one of the reported values was completely different from the others, in which case it was not included. The averaged values are marked with an asterisk (*).

[b] References: (1) Crabo, Proc. 5th Int. Congr. Anim. Reprod., Trento, p. 566 (1964) [Excised]. (2) Crabo and Gustafsson, J. Reprod. Fert. 7, 337 (1964). [Excised]. (3) Wales et al., J. Reprod. Fert. 12, 139 (1966). [Excised]. (4) Salisbury and Cragle, Proc. 3rd Int. Cong. Anim. Reprod., Cambridge, p. 25 (1956). [Excised]. (5) Dawson et al., Biochem. J. 65, 627 (1957). [Excised]. (6) Seidel and Foote, Biol. Reprod. 2, 189 (1970) [Partitioned ejac.]. (7) Sorensen and Anderson, Proc. 3rd Int. Cong. Anim. Reprod. Cambridge, p. 45 (1956). [Excised]. (8) Sexton et al., J. Dairy Sci. 54, 412 (1971). [Cannula]. (9) Ganjam et al., Clin Res. 19, 771 (1971). [Cannula]. (10) Mann, in Handbook of Physiology 5, p. 461 (1975). [Not stated]. (11) Crabo, Acta Vet. Scand. 6, Suppl. 5 (1965). [Excised]. (12) Ganjam and Amann, Endocrinology 99, 1618 (1976). [Cannula]. (13) Wales et al., J. Reprod. Fert. 12, 139 (1966). [Excised]. (14) Mann, in Reproduction in Domestic Animals, Academic Press, p. 51, (1959). [Not stated]. (15) Bower et al., J. Reprod. Fert. 33, 319 (1973). [Removal of sex glands, atropine]. (16) Einarsson, Acta Vet. Scand. Suppl. 36, (1971) [Cannula]. (17) Hartree and Mann, Biochem. J. 71, 423 (1959). [Excised]. (18) Hinton et al., J. Reprod. Fert. 58, 395 (1980). [Micropuncture]. (19) Hinton et al., J. Reprod. Fert. 56, 105 (1979). [Micropuncture]. (20) Jones, Comp. Biochem. Physiol. 61B, 365 (1978). [Excised]. (21) Quinn and White, Aust. J. Biol. Sci. 20, 1205 (1967). [Excised]. (22) White and Hudson, J. Endocr. 41, 291 (1968). [Excised]. (23) Scott et al., J. Reprod. Fert. 6, 49 (1963). [Excised]. (24) White et al., Proc. 4th Int. Cong. Anim. Reprod., The Hague, p. 226, (1961). [Excised]. (25) White and Wales, J. Reprod. Fert. 2, 225 (1961). [Cannula]. (26) Murdoch and White, Aust. J. Biol. Sci. 21, 483 (1968). [Excised]. (27) Jones and Glover, in Biol. Male Gamete, Academic Press, 367 (1975). [Not stated]. (28) Voglmayr et al., J. Reprod. Fert. 49, 245 (1977). [Cannula].

4.2.3   Proteins and Amino Acids

The total concentration of proteins in the epididymal fluid of farm animals varies from 12 to 32 mg/ml. In the bull, the protein concentration is sixfold higher in the cauda epididymis than in the caput. A large number of enzymes have been identified in the epididymal fluid of farm animals; the one that is present in remarkably high amounts is N-acetylglucosaminidase (Table 2). The concentrations of several of the other hydrolases are also high, but significant species variations are present. For instance, LDH levels are high in the bull and ram but much lower in the boar. Glutamic-pyruvic transaminase (GPT) and glutamic-oxaloacetic transaminase (GOT) concentrations are high in the bull, but the enzymes are almost absent from ram epididymal fluid. The cauda epididymal fluid of the boar and ram has six to eight times higher concentrations of trypsin inhibitor than rete testis fluid. The amount of trypsin inhibitor in the ram is 45 times higher than that of blood and only slightly less than that of seminal plasma.

Besides spermatozoa, a number of different particles are present in epididymal fluid. With the exception of the work of Garbers *et al.* on the bull (Ref. 30, Table 2), none of the authors attempted to separate the particulate fraction from the fluid portion, even though many enzymes are present in these particles. Compared with the particle-free fluid, the particles have a particularly high concentration of nucleotide phosphatase. The fluid contains much higher amounts of β-N-acetylglucosaminidase, β-galactosidase, α-mannosidase, and acid phosphatase than the particles. Enzymes associated with intermediary metabolism, such as glucose-6-phosphate dehydrogenase, aldolase, lactate dehydrogenase, NADP$^+$ malic enzyme, sorbitol dehydrogenase, or GOT are either completely absent from the particles or present in only very small quantities. It is thought that these particles represent secretory granules of the epithelial cells.

The cytoplasmic droplets of the spermatozoa also possess enzymes. However, 10–50-fold higher concentrations of β-galactosidase, acid phosphatase, α-mannosidase, β-N-acetylglucosaminidase, and neutral esterase are present in particle-free bull epididymal fluid than in the cytoplasmic droplets. The cytoplasmic droplets possess twofold to threefold higher levels of β-glucuronidase, nucleotide phosphatase, and alkaline phosphatase than the fluid.

The free amino acid composition of the cauda epididymal fluid is shown in Table 3. Remarkably high levels of glutamic acid are present in all three species studied; this amino acid makes up about half of the total amount of free amino acids. In the bull (the only species in which this was measured), very high levels of hypotaurine are also present. When the epididymal fluid of the bull is compared with the testicular fluid, the concentrations of most of the amino acids, including glutamic acid, arginine, leucine, and methionine, are higher in the fluid from the epididymis than in that from the testis, but the levels of glycine and alanine are lower. Comparisons in the ram show that the concentration of each amino acid with the exception of aspartic acid and the basic amino acids (lysine, histidine, and arginine) is higher in epididymal fluid than in seminal plasma. After vasectomy of the ram, a 65-fold decrease in threonine, glutamine,

Table 2 Enzymes of the Cauda Epididymal Fluid from Farm Animals[a]

Enzyme[b]	Bull[c] (29, 30)[d]	Boar (5, 15–17, 20, 31, 33)	Ram (20, 26, 27, 31, 32)	Stallion (20, 33)
Acid phosphatase	13,049	9,336*	745*	603
Alkaline phosphatase	10,152	26,604*	11,508*	291,700
Phosphodiesterase	—	1.9	1.5	1.0
Nucleotide phosphatase	72,804	—	—	—
5′-Nucleotidase	1,250	1,572	38	596
α-Galactosidase	—	—	—	1,500
β-Galactosidase	1,777	—	—	2,813
β-N-Acetylglucosaminidase	134,032	702,500	23,145	138,750
β-Glucuronidase	349	—	—	840
α-Mannosidase	3,337	2,650	35	11,250
β-Mannosidase	—	—	—	2,700
α-D-Glucosidase	158	—	—	300
β-D-Glucosidase	45	—	—	0
Neutral esterase	2,413	—	—	—
Aryl sulfatase C	27	—	—	—
Glutamic-oxaloacetic trans-aminase (GOT)	271*	74	19	—
Glutamic-pyruvic trans-aminase (GPT)	40	—	Trace	—
Lactate dehydrogenase (LDH)	2,463*	230	8,127	1,115
Glucose-6-phosphate dehydrogenase	68	—	—	—
Aldolase	100	—	—	—
NADP+ malic enzyme	81	—	—	—
Sorbitol dehydrogenase	52	—	—	—
Acid DNase	—	4	2	11
Trypsin inhibitor	—	82	209	—

[a] The asterisk (*) indicates averaged values (see Table 1).

[b] Nanomoles substrate hydrolyzed or product formed per minute per milliter (see references for specific assay techniques). Since the investigators at times used different assay techniques, species comparisons may not be valid.

[c] Fluid + light particulate fraction combined.

[d] References: (1–28) See Table 1. (29) Stallcup et al., J. Reprod. Fert. 15, 317 (1968). [Excised]. (30) Garbers et al., Biol. Reprod. 3, 327 (1970). [Excised; values calculated based on 32.3 mg protein/ml, and the combined particle-free and particulate fractions]. (31) Suominen and Setchell, J. Reprod. Fert. 30, 235 (1972). [Cannula]. (32) Alumot and Schindler, J. Reprod. Fert. 10, 261 (1965). [Cannula]. (33) Conchie and Mann, Nature, Lond. 176, 1190 (1957). [Not stated].

Table 3 Free Amino Acids of Cauda Epididymal Fluid from Farm Animals[a]

Amino acid (μmol/100 ml)	Boar (1)	Bull (2)	Ram (3)
Glutamic acid	2547	2488	1800
Aspartic acid	51	173	42
Glycine	36	310	110
Alanine	69	419	143
Valine	4	137	59
Isoleucine	2	68	36
Leucine	4	166	95
Phenylalanine	1	Trace	Trace
Tyrosine	1	Trace	Trace
Threonine	16	146	⎫
Serine	25	260	⎬ 914
Asparagine + glutamine	96	—	⎭
Proline	—	Trace	—
Hydroxyproline	Trace	—	—
Histidine	Trace	28	13
Lysine	4	154	65
Arginine	2	35	26
Methionine	2	45	—
Cystine	—	Trace	—
Phosphoethanolamine	80	—	—
α-Amino-n-butyric acid	3	—	—
Ornithine	34	—	—
Taurine	929	—	—
Hypotaurine	4494	—	—

[a] References: (1) Sexton et al., J. Dairy Sci. 54, 412 (1971). [Cannula]. (2) Johnson et al., J. Anim. Sci. 24, 430 (1972). [Excision]. (3) Setchell et al., Biochem. J. 105, 1061 (1967). [Cannula].

asparagine, serine, aspartic acid, and arginine occurs in semen, indicating that these amino acids are mostly contributed to the ejaculate by the epididymis.

A number of electrophoretic, gel filtration, chromatographic, and immunologic studies have been performed on the epididymal fluids of the bull, boar, and ram. The molecular weight of the protein components varies from 12,000 to 150,000 daltons. By electrophoresis, 17–30 bands that stain for protein, approximately half of which are glycoproteins, can be identified in all three species. Several of these bands represent the hydrolases discussed previously, including isozymes of acid phosphatase, alkaline phosphatase, and nonspecific esterase, as well as albumin, α- and β-globulins, and transferrin. A number of protein components are completely specific for epididymal fluid. Other components are shared with rete testis fluid but are absent from blood, whereas yet others are present in blood, rete testis fluid, and epididymal fluid. Several of the epididymal proteins cannot be detected in seminal plasma by electrophoretic or immunological techniques. This does not mean that some of the epididymal fluid components are not ejaculated. Rather, the dilution of these components with seminal plasma may be so great that they become undetectable.

4.2.4 Pathological Conditions and Surgical Procedures

As was reported by Gustafsson (9), decreased spermatogenesis of the bull resulting in azoospermia or oligospermia, whether spontaneous or artificially induced, does not cause any major alterations in the biochemical composition of the fluid from the caput or corpus epididymis. Only the protein and GPC levels of the corpus fluid decrease, twofold and fourfold, respectively. Some significant changes occur in the cauda epididymal fluid components, however. In this fluid, the sodium levels increase twofold and the chloride levels threefold to fourfold, and the GPC levels decrease twofold. These results indicate that spermatozoa have a major influence on the biochemical composition of the cauda epididymis, less in the corpus and virtually none in the caput.

The formation of a high number of morphologically abnormal spermatozoa by bull testis, whether spontaneous or induced by short durations of scrotal insulation or treatment with estradiol benzoate, does not alter the composition of the epididymal fluids to any large extent. Ligation of the corpus epididymis also has little effect on the biochemical composition of the epididymal fluid from all three parts of the epididymis. However, ligation of the ductuli efferentia causes a twofold increase in the sodium levels of the corpus and cauda epididymal fluid, a fivefold decrease in GPC in corpus fluid, and a fivefold increase in chloride levels in cauda fluid. The decrease in GPC is surprising because it is supposedly synthesized by the epididymis, and not the testis, and this decrease does not occur after efferent duct ligation in the rat (see section 4.3.4 of this chapter). It is possible that in the bull some essential components for the synthesis of GPC are carried from the testis to the epididymis.

4.3 Laboratory Animals

4.3.1 Electrolytes, Acids, and pH

The biochemical compositions of the epididymal fluid of the rat, rabbit, hamster, guinea pig and dog are shown in Table 4. When such studies were performed, the general comments made for farm animals also apply to laboratory animals. A fourfold decrease in sodium occurs from the caput to the cauda in the rat and hamster, whereas the concentration of potassium increases twofold in both species. However, in contrast to the bull, a significant amount of bicarbonate is present in rat and rabbit cauda epididymal fluid. Further, the potassium levels of rat cauda epididymal fluid are approximately two times higher than those of any other species, and the magnesium concentrations of the rabbit cauda epididymal fluid are about five to six times higher. The electrolyte composition and the pH of the epididymal fluid of laboratory and farm animals is otherwise essentially identical.

Table 4 Biochemical Composition of the Epididymal Fluid from Rodents the Rabbit and the Dog

Component	Unit	Rat (1–11, 18, 19)[a,b]			Rabbit (4, 10, 12–15, 19)	Hamster (8, 10, 14–19)	Guinea Pig (18, 19)	Dog (19)
		Cauda	Corpus	Caput	Cauda	Cauda	Cauda	Cauda
Protein[c]								
Total protein	mg/ml	39.5*	—	—	42.4*	44.3*	39.4*	30.2
Glutamic oxaloacetic transaminase (GOT)		2,548*	—	—	37*	570*	260	—
Acid phosphatase		494*	—	—	150	823*	58,540	188
Alkaline phosphatase		52.7*	—	—	23,935*	75,120*	240	406,000
Lactate dehydrogenase (LDH)		7,887*	—	—	1,391*	4,060*	13,400	2,303
Acid DNase		0.28	—	—	7.19	7.36	0.22	0.95
Carnitine acetyl transferase		0	—	—	—	—	—	—
α-Mannosidase		11,330	—	—	924*	25	570	6,160
β-N-acetylglucosaminidase		336,600	—	—	51,570	250	48,100	56,200
5'-Nucleotidase		—	—	—	258	494	—	1.1
Esterase (arginine)		—	—	—	0	0	—	—
Phosphodiesterase		5.1	—	—	4.3	3.8	—	4.5
Proteinase inhibitor		—	—	—	Present	Present	Present	—

Carbohydrate

Total carbohydrate	mg/ml	2.46	—	—	4.15	6.6	6.8	10.8
Sialic acid (free + bound)	μmol/l	244*	237	99	156	2.0 (mg/ml)	2.3 (mg/ml)	2.7 (mg/ml)
Fructose	mg/ml	—	—	—	0	—	—	—
Myoinositol	mmol/l	30.4*	5.3	7.8	—	74.5	—	—

Lipid

Glycerylphosphoryl-choline (GPC)	mg/ml	10.0*	14.2	12.1	11.9*	8.4*	11.7*	9.2
Carnitine	mmol/l	56.0*	26.5	10.6	16.8	9.7	10.8 (mg/ml)	3.2 (mg/ml)
Acetylcarnitine	mmol/l	13.3	—	—	—	—	—	—
Free fatty acids (total)	μmol/l	958	—	—	—	—	—	—
Phosphocholine	mmol/l	21.4	17.9	17.7	—	—	—	—
Cholesterol	mg/ml	—	—	—	0.81	—	—	0.25

Other

Sodium	mEq/l	19.0*	44.0*	85.0*	18.4*	21.3*	16.5*	22.0
Potassium	mEq/l	46.8*	36.2*	19.5*	24.2*	31.3*	23.6	37.6
Magnesium	mEq/l	2.6	—	—	14.2*	2.5*	4.4	2.2
Calcium	mEq/l	0.22	—	—	1.13*	0	0.2	1.7
Chloride	mEq/l	21.8*	24.4*	31.0	13.8*	11.1	10.0	15.5
Bicarbonate	mEq/l	6.7	—	2.7	—	—	—	—
Zinc	mEq/l	—	—	—	0.21	0.05	0.05	0.1

Table 4 (*continued*)

Component	Unit	Rat (1–11, 18, 19)[a,b]			Rabbit (4, 10, 12–15, 19)	Hamster (8, 10, 14–19)	Guinea Pig (18, 19)	Dog (19)
		Cauda	Corpus	Caput	Cauda	Cauda	Cauda	Cauda
Inorganic phosphorus	mg/ml	0.30	—	—	0.12*	0.28*	0.17	0.22
Inorganic phosphate	mmol/l	13.6	8.8	7.2	—	—	—	—
Total phosphate	mmol/l	84.1	80.8	74.4	—	—	—	—
Total nitrogen	mg/ml	5.3	—	—	9.3	6.8	4.9	12.0
Urea nitrogen	mg/ml	0.14	—	—	0.23	0.13	0.26	0.09
Ammonia	mg/ml	1.7	—	—	0.02	0.02	0.01	0.03
Lactic acid	mg/ml	—	—	—	0.05	0.08	0.06	0.98
pH		6.9	—	6.56*	6.76*	6.70*	7.0	6.7
Osmolarity	mOsm/l	329	340	315	—	—	—	—
Osmotic pressure	ΔT_f,°C	0.59	—	—	0.58*	0.60*	0.61	0.59
Potential difference	MV	−27.0	−20.7	−5.6	—	—	—	—

[a] The asterisk (*) indicates averaged values (see Table 1).

[b] References: (1) Levine and Marsh, *J. Physiol. Lond.* **213**, 557 (1971) [Micropuncture]. (2) Gupta *et al.*, *Andrologia* **6**, 35 (1974) [Excised]. (3) Back *et al.*, *J. Reprod. Fert.* **40**, 211 (1974) [Cannula]. (4) Back *et al.*, *J. Reprod. Fert.* **45**, 117 (1975) [Cannula]. (5) Brooks *et al.*, *J. Reprod. Fert.* **36**, 141 (1974) [Excised]. (6) Marquis and Fritz, *J. Biol. Chem.* **240**, 2197 (1965) [Excised]. (7) Hinton and Setchell, *J. Reprod. Fert.* **58**, 401 (1980) [Microaspiration]. (8) Hinton *et al.*, *J. Reprod. Fert.* **58**, 395 (1980) [Micropuncture]. (9) Turner *et al.*, *Fertil. Steril.* **28**, 191 (1977) [Micropuncture]. (10). Hinton *et al.*, *J. Reprod. Fert.* **56**, 105 (1979) [Micropuncture]. (11) Levine and Kelly, *J. Reprod. Fert.* **52**, 333 (1978) [Micropuncture]. (12) Jones and Glover, *J. Reprod. Fert.* **34**, 395 & 405 (1973) [Cannula]. (13) Jones, *Fertil. Steril.* **25**, 432 (1974) [Cannula]. (14) Meizel and Mukerje, *Biol. Reprod.* **13**, 83 (1975); *Biol Reprod.* **14**, 444 (1976). (15) Morton *et al.*, *Fertil. Steril.* **29**, 695 (1978) [Excised]. (16) Jones and Glover, *Biol. Male Gamete*, Academic Press, p. 367 (1975) [Not stated]. (17) Jessee and Howards, *Biol. Reprod.* **15**, 626 (1976) [Micropuncture]. (18) Back and Shenton, *Experientia* **15**, 464 (1975) [Micropuncture]. (19) Jones, *Comp. Biochem. Physiol.* **61B**, 365 [Cannula].

[c] See references for specific enzyme assays. All enzyme activities are expressed in nanomoles substrate hydrolyzed or product formed per minute per milliliter. Due to different assay techniques, species comparisons may not be valid.

4.3.2 Lipids

In all species tested, the cauda epididymal fluid of laboratory animals possesses high levels of GPC, although these levels are about two times lower than those found in farm animals. High amounts of carnitine are also present, 5–20% being in the form of acetylcarnitine. The levels of carnitine in the cauda epididymal fluid of the rat are 2000 times higher than those of blood. Carnitine acetyltransferase (CAT) is absent from epididymal fluid but present in spermatozoa, even though carnitine is mostly associated with the fluid rather than the spermatozoa. Tissue levels of carnitine are highest in the cauda epididymis, compared with the corpus and caput.

The free fatty acid composition of rat cauda epididymal fluid differs from that of blood, and the total concentration of free fatty acids is three to four times greater. It has been suggested that these high lipid levels function in part to increase the osmotic pressure of the epididymal fluid, because the measured osmotic pressure cannot be accounted for by the concentrations of electrolytes alone. However, other potentially ionizable substances such as proteins and carbohydrates are present in epididymal fluid so that the extent to which the lipids contribute to the osmotic pressure remains to be established.

4.3.3 Carbohydrates and Proteins

High levels of sialic acid are present in rat and rabbit cauda epididymal fluid, the amount in the rat being lowest in the caput. No fructose can be detected in rabbit epididymal fluid. Much higher amounts of myoinositol are present in cauda than in caput epididymal fluid of the rat. Differences are less significant in the hamster, in which the myoinositol concentrations are 46.5 mM in the caput and 74.5 mM in the cauda. The concentration of protein in the cauda epididymal fluid of laboratory animals is generally somewhat higher than that in the fluid of farm animals. In laboratory animals, the same enzymes as in farm animals are typically present. Besides the enzymes listed in Table 4, rat epididymal fluid contains a galactosyltransferase. Proteinase (trypsin) inhibitor is present in rabbit, hamster and guinea pig cauda epididymal fluid, but an esterase that cleaves arginine bonds (such as trypsin) is absent.

A number of protein components (at least 14) can be found on electrophoretic analysis of rat, mouse, and rabbit epididymal fluid. The protein composition of the fluids from the cauda and caput of the rat differs in number (highest in the cauda) and in type. Several glycoproteins of the rabbit and hamster are secreted primarily by the caput rather than the cauda epididymis. As with farm animals, some protein components are unique to the epididymis, some are shared with rete testis fluid but not with blood, and most are present in epididymal fluid, rete testis fluid, and blood. Prealbumin and albumin bands have been identified in rat epididymal fluid, and γ-globulin has been identified in guinea pig epididymal fluid.

"Epididymal-like" fluid was obtained from the mouse, rat, and rabbit by keeping epididymal tissue in solution, and the amino acid composition was determined. The major amino acids are glutamic acid, glycine, serine, alanine, and aspartic acid. The relative concentrations vary somewhat among the various species and among different segments of the epididymis.

Hamster epididymal fluid also contains a low-molecular weight, heat-stable factor that increases the survival time of diluted spermatozoa. Rabbit and hamster epididymal fluids contain a factor ("decapacitation factor") that decreases the fertility of capacitated spermatozoa. Rat and rabbit epididymal fluid further contains ABP, which originates in the testis and carries androgens from the seminiferous tubules to the caput epididymis, where the steroids are resorbed to a large extent (see section 4.2.2 of this chapter). The rat shows a threefold decrease in ABP between the caput and cauda.

4.3.4 Surgical Procedures and Drug Treatment

Isolation of the cauda epididymis of the rabbit by placing ligatures at the vas–epididymal junction and at the corpus epididymis does not cause major changes in the fluid components or their concentrations over an eight-week period, even though a fivefold increase in fluid volume takes place. Exceptions are enzymes that originate from the spermatozoa. After five weeks, when the sperm concentration decreases drastically and almost all the spermatozoa have died, a tremendous increase in LDH occurs, as well as an increase in acid phosphatase.

Vasectomy of the rat does not alter the number of protein bands visible after electrophoresis, at least not over a 4–10-day period. Ligation of the vasa efferentia in the rat causes only a small decrease in GPC but almost completely eliminates ABP, showing that the latter derives primarily from rete testis fluid. Efferent duct ligation in the rat also causes an increase in the tissue levels of lipid, specifically free esterified cholesterol, total cholesterol, and triglycerides, indicating that these components in caput epididymal fluid derive from the epididymis itself and not the testis.

Artificial cryptorchidism of the rabbit for as long as five weeks has no effect on the concentration of sodium, potassium, chloride, protein, alkaline phosphatase, acid phosphatase, β-N-acetylglucosaminidase, or osmotic pressure and pH of the cauda epididymal fluid. Although the levels of magnesium, inorganic phosphate, and LDH initially increase, and those of GPC decrease, the levels tend to return to normal after some time, with the exception of the magnesium levels, which remain high. Artificial cryptorchidism of the rat does not affect the levels of GPC and carnitine.

α-Chlorohydrin causes infertility in the rat but has only a poor antifertility effect in the rabbit. The antifertility activity appears to be achieved by preventing sperm maturation in the epididymis. In the rabbit, α-chlorohydrin does not cause any changes in the biochemical composition of the epididymal fluid. In the rat, changes only in LDH and GOT are observed, that is, alterations in components that originate primarily from spermatozoa. This indicates that the anti–sperm-maturation effect is not mediated by changes in epididymal fluid components.

5 EPIDIDYMAL SPERMATOZOA

5.1 Introduction

The following discussion deals primarily with the biochemical changes that the spermatozoa undergo as they are transported through the epididymis and with differences .hat have been reported between epididymal and ejaculated spermatozoa. The basic biochemical composition of the spermatozoa and the characteristics of the components can be found in the chapters dealing with ejaculated spermatozoa (Chapters 5 and 6).

Spermatozoa undergo maturation changes during their epididymal transit, rendering them motile and fertile. The spermatozoa are subject to a number of biochemical alterations as they pass through the epididymis. It can be assumed that some of these changes are essential for sperm maturation but that others occur concomitant with or after the maturation process, and are not part of this process. Other biochemical events may be related to sperm storage, whereas yet others may have no physiological relevance at all. At present, it is impossible to state with certainty whether any of the reported biochemical alterations are essential for sperm function. Therefore, speculation in this regard has been omitted.

5.2 Dry Weight, Specific Gravity, Surface Charge, and Electrolytes

The dry weight of caput epididymal spermatozoa of the bull is somewhat higher than that of cauda spermatozoa: $3.2 \ mg/10^8$ sperm vs $2.6 \ mg/10^8$ sperm. Cauda epididymal spermatozoa of the boar have a dry weight of $1.9 \ mg/10^8$ sperm and those of the guinea pig, $5.3 \ mg/10^8$ sperm.

The specific gravity of cauda epididymal bull spermatozoa varies from 1.100 to 1.125, which is significantly less than that of ejaculated bull spermatozoa, but higher than that of caput sperm. The change in specific gravity is due to alterations in the sperm head, which apparently undergoes a general shrinkage and dehydration process as the spermatozoon matures. The specific gravity of the sperm tail is not altered.

Cauda epididymal spermatozoa of the rabbit, hamster, rat, guinea pig, and mongoose possess only few, if any, negative charges on the sperm head surface, even though the neck and tail areas are highly negatively charged. By contrast, bull spermatozoa are negatively charged over their entire surface. Testicular rabbit and hamster spermatozoa are essentially devoid of negative surface charges, and although caput spermatozoa are more negatively charged than testicular spermatozoa, they are much less so than cauda spermatozoa. Similarly, human spermatozoa develop an ability to bind positively charged ions during epididymal transit, that is, they gain electronegativity.

Electrophoretic studies confirm the increase in negative (anionic) charge on the sperm tail as spermatozoa migrate through the epididymis. Ejaculated and cauda epididymal rabbit spermatozoa orient themselves tail first to the positive electrode in a short period of time, whereas spermatozoa from the caput or corpus orient themselves much more slowly, not at all, or even in the opposite direction.

Bull spermatozoa from the caput epididymis have a higher electrolyte concentration than those from the cauda. The caput-vs-cauda values (in mEq/l) are respectively (1) sodium: 75 vs 29; (2) potassium: 76 vs 55; (3) calcium: 12 vs 4; and (4) chloride: 45 vs 18. The rate of exchange of both sodium and potassium through the membranes of boar spermatozoa is much more rapid in sperm from the caput than from the cauda epididymis. The concentration of zinc in rat spermatozoa also decreases during epididymal transit from 1.4 μmol/10^9 sperm in the caput to 0.75 μmole/10^9 sperm in the cauda.

5.3   Lipids, Carbohydrates, and Nucleic Acids

The most extensively studied components of epididymal spermatozoa are the lipids (Table 5). In all species tested so far, a large decrease in lipid content occurs as spermatozoa pass from the caput to the cauda epididymis. Cauda spermatozoa have a somewhat higher lipid content than ejaculated spermatozoa. There are species variations in the specific class of lipid that changes during sperm transport. For instance, rabbit and rat spermatozoa were reported to accumulate choline plasminogen as they migrate through the epididymis, but boar spermatozoa show a decrease in this lipid, and it remains approximately the same in the ram (Table 5). The primary fatty acids of the phospholipids of epididymal boar spermatozoa are docosapentanoic acid (22:5) and docosahexanoic acid (22:6) that amount to approximately 30% and 57% of the total phospholipid fatty acids. The 14:0, 16:0, and 18:0 acids are the primary fatty acids of the neutral lipids of epididymal boar spermatozoa, making up, respectively, 38%, 35%, and 15% of the total.

The sialic acid content of primate, rat, and hamster spermatozoa decreases as the spermatozoa pass from the caput to the cauda. Also, caput but not cauda rat spermatozoa are good sialyl acceptors when treated with epididymal extract containing a glycoprotein-glycosyl transferase that transfers sialyl residues from cytidine-5'-monophosphate sialic acid. Apparently it is not the sialic acid composition that causes the increase in electronegativity of the spermatozoa during epididymal transit (see section 5.2 of this chapter).

Caput and cauda bull spermatozoa possess the same amount of phosphorus and DNA, indicating that there is no change in sperm DNA content during epididymal transport, at least not in this species.

5.4   Enzymes and Other Proteins

The protein content of bull and rhesus monkey spermatozoa generally decreases as they pass from the caput to the cauda epididymis. Not only do changes occur in the number of proteins, but also in the amount of certain proteins. An exception is a 37,000-dalton glycoprotein that is present in cauda epididymal rat spermatozoa but is absent from caput sperm. Of the amino acids, a decrease

Table 5 Lipids of Epididymal Spermatozoa

Lipid[a]	Rhesus Monkey (1)[b]			Bull (2, 3, 4)[b]		Boar (5)[b]			Ram (6, 7)[b]		Rat (7–10)[b]	
	Cauda	Corpus	Caput	Cauda	Caput	Cauda	Corpus	Caput	Cauda[d]	Caput[d]	Cauda	Caput
Total lipid	5.6	20.9	157.3	3.8	5.6	—	—	—	—	—	12	30
Total neutral lipid	—	—	—	—	—	—	—	—	—	—	3.8	9.6
Total phospholipid	1.3	4.5	35.2	1.9	2.7	—	—	—	1.7	3.2	8.2	20.4
Phosphatidylcholine (lecithin)	0.5	1.8	11.0	0.44[c]	—	3.9[d]	6.1[d]	8.8[d]	0.3	0.7	1.1	1.0
Lysophosphatidylcholine (lysolecithin)	—	—	—	0.06[c]	—	Trace	0.3[d]	0.3[d]	—	—	—	—
Phosphatidylserine	—	—	—	0.01[c]	—	0.5[d]	0.6[d]	0.9[d]	—	—	—	—
Phosphatidylethanolamine (cephalin)	0.2	1.0	7.0	0.11[c]	—	1.6[d]	2.0[d]	3.6[d]	0.2*	0.4*	0.7	0.7
Ethanolamine plasmalogen	—	—	—	0.14[c]	—	1.1[d]	1.7[d]	2.4[d]	0.3	0.4	0.1	0.1
Choline plasmalogen	—	—	—	0.91[c]	—	0.5[d]	0.6[d]	0.9[d]	0.5*	0.6*	0.6	0.1
Cholesterol	—	—	—	0.4	0.6	—	—	—	0.4	0.5	1.2	2.9
Sphingomyelin	100	600	6500	0.24[c]	—	1.2[d]	1.3[d]	1.8[d]	0.2	0.3	0.1	0.3
Cardiolipin	—	—	—	0.12[c]	—	—	—	—	—	—	—	—
Proteolipid	—	—	—	0.09[c]	—	—	—	—	—	—	—	—
Glycerylphosphorylcholine	—	—	—	—	—	—	—	—	0.4	0.2	0.0	—
Carnitine	—	—	—	1.0[e]	0.1[e]	—	—	—	—	—	0.6[f]	—
Acetylcarnitine	—	—	—	0.2[e]	0[e]	—	—	—	—	—	0.4[f]	—

[a] Unless indicated otherwise, all values are expressed in mg/10⁹ sperm.

[b] References: (1) Arora et al., *Contraception* **11**, 689 (1975). (2) Lavon et al., *J. Reprod. Fert.* **23**, 215 (1970). (3) White in *Adv. in Biosc. 10; Schering Workshop on Contraception: The Masculine Gender*, Pergamon, p. 157 (1973). (4) Casillas. *J. Biol. Chem.* **248**, 827 (1973). (5) Grogan et al., *J. Reprod. Fert.* **12**, 431 (1966). (6) Quinn and White, *Aust. J. Biol. Sci.* **20**, 1205 (1967). (7) Dawson and Scott, *Nature, Lond.* **202**, 292 (1964). (8) Terner et al., *J. Reprod. Fert.* **45**, 1 (1975). (9) Brooks et al., *J. Reprod. Fert.* **36**, 141 (1974). (10) Marguis and Fritz, *J. Biol. Chem.* **240**, 2197 (1965).

[c] Indicates μg atoms P/10⁹ spermatozoa.

[d] Indicates μmole/100 mg dried spermatozoa.

[e] μmole/10⁹ spermatozoa.

[f] Indicates averaged values (see Table 1).

† The asterisk (*) indicates averaged values (see Table 1).

in glutamic acid, aspartic acid, leucine, alanine, methionine, and isoleucine occurs as the spermatozoa pass from the caput to the cauda, but an increase in cystine and arginine takes place. The polyamine content (putrescine, spermidine, spermine, and histamine) of bull spermatozoa also decreases from the caput to the cauda. The increase in cystine groups, that is, in disulfide bonding, in further illustrated by the decrease in free sulfhydryl (SH) groups during sperm transit in the rat, rabbit, and monkey. The increase in disulfide bonding occurs mostly in the sperm tails, although the nucleoproteins of the sperm heads also show an enhanced number of disulfide bonds (see further on in this section). The zinc of rat spermatozoa is localized almost entirely within the tail structures stabilized by—S—S—(disulfide) bonds, the large majority of which are the dense fibers of the tail flagellum. Actin has been localized in the postacrosomal region and cytoplasmic droplet of a variety of spermatozoa, but only minor changes appear to occur as the spermatozoa pass from the testis into the epididymis and are ejaculated.

In general, the enzymes of epididymal spermatozoa are identical to those of ejaculated spermatozoa. The type and properties of these enzymes are discussed in Chapters 5 and 6. The changes that occur in the protein content, disulfide bonding, and metabolism of the spermatozoa during epididymal transit may be caused by certain enzymes. Thus the activity and quantity of a certain enzyme may differ between caput and cauda epididymal spermatozoa, but this has not often been investigated.

Rat caput epididymal spermatozoa possess higher amounts of acid phosphatase and alkaline phosphatase than cauda spermatozoa, but have lower amounts of LDH and monoglyceride lipase, and approximately equal amounts of hexokinase, glucose-6-phosphate dehydrogenase, α-glycerophosphate dehydrogenase, and triglyceride lipase. Cauda epididymal rat spermatozoa possess lower amounts of surface ATPase, which differs somewhat in characteristics from that of caput spermatozoa. Bull caput spermatozoa possess approximately four times higher levels of adenylyl cyclase than cauda sperm. However, cauda epididymal bull spermatozoa possess higher amounts of cyclic AMP (cAMP) and cAMP-independent kinases than caput sperm, although both sperm types contain equal amounts of cAMP-dependent kinase.

The basicity of the nucleoproteins that surround sperm DNA increases somewhat during passage of the spermatozoa through the seminiferous and other testicular tubules, but this process is completed before the spermatozoa reach the epididymis. Thus even though the chromatin condenses tremendously before and/or shortly after completion of spermatogenesis, it appears that the binding between sperm DNA and its nucleoproteins is not altered significantly during epididymal transit.

The nucleoproteins and nuclear-membrane proteins do appear to change in character, however, as spermatozoa pass from the caput to the cauda, as studies by Calvin and Bedford indicate (2, 3, 5). The heads of the testicular spermatozoa of the rat, mouse, guinea pig, bull, boar, cat, ferret, shrew, and man swell readily in the presence of sodium dodecylsulfate (SDS) alone, whereas caput spermatozoa require both SDS and dithiothreitol (DTT, a disulfide reducing agent) for swelling. Cauda spermatozoa swell much less readily in the presence of SDS and DTT than caput sperm. Therefore, as with the sperm tail (see earlier in this section), an increase in disulfide-bond formation occurs in the sperm head while the spermatozoa reside in the epididymis. This does not necessarily require the epididymal surroundings, since caput spermatozoa undergo the same changes when kept *in vitro* in distilled water. In contrast to these eutherian animals, opossum spermatozoa have few disulfide bonds in their nuclei, even when mature, and SDS alone can cause the heads to swell. The opossum sperm tail possesses many disulfide bonds, however, similar to eutherian animals.

5.5 Metabolism and Motility

While the spermatozoa traverse the epididymis, they gain the ability to move actively although the sperm do not actually do so until after ejaculation. Metabolic changes must take place during sperm transit, therefore. There are only small differences in the content of metabolic enzymes between caput and cauda spermatozoa, and the alterations in metabolism are most likely at least in part due to something other than the enzyme content of the spermatozoa.

Hoskins and co-workers (10) have obtained evidence that two separate processes are involved in the acquisition of motility by spermatozoa during epididymal transit, at least in the bull. First, an increase in intrasperm cAMP occurs, followed by the binding of a specific "forward-motility protein" from the epididymal fluid to the spermatozoa. Together these cause the spermatozoa to be capable of forward progression. Forward-motility protein is present in highest concentrations in the cauda epididymal fluid but is also present in high amounts in seminal plasma. A small amount can be found in rete testis fluid. It appears to be a glycoprotein whose monomeric form has a molecular weight of 37,000 daltons. Similar factors that sustain or induce sperm motility have been found in the epididymal fluid of other species. A second increase in cAMP and stimulation by calcium ions most likely occurs at the time of ejaculation, causing the spermatozoa to be actively motile.

There are distinct differences in the metabolic activities of cauda epididymal spermatozoa and ejaculated spermatozoa. Research in this area has focused primarily on farm animals. Reported differences include the following. (1) In contrast to ejaculated spermatozoa, epididymal spermatozoa fail to show a constant rate of respiration when different concentrations of sperm suspensions are tested. (2) Epididymal spermatozoa have a lower endogenous respiration rate than ejaculated spermatozoa. (3) Glucose stimulates the respiration of epididymal spermatozoa but decreases the endogenous respiration of ejaculated spermatozoa. (4) Epididymal spermatozoa utilize maltose poorly. (5) Epididymal spermatozoa have a lower aerobic and higher anaerobic glycolysis than ejaculated spermatozoa and show a much greater Pasteur effect (inhibition of glycolysis by oxygen—this may not be the case for ram sperm). There was much argument in the older literature regarding the cause for these differences. However, it was shown that the addition of fructose to cauda epididymal spermatozoa induces many of the changes seen in ejaculated spermatozoa, and it now appears that most of the metabolic properties of cauda epididymal spermatozoa are the same as those of ejaculated spermatozoa, with the exception that the former have not been in contact with the accessory-sex-gland secretions and the available substrates therein.

REFERENCES

1 Acott, T. S., and Hoskins, D. D., Bovine sperm forward motility protein: partial purification and characterization. *J. Biol. Chem.* **253**, 6744 (1978).
2 Bedford, J. M., Calvin, H., and Cooper, G. W., The maturation of spermatozoa in the human epididymis. *J. Reprod. Fert.* Suppl. **18**, 199 (1973).
3 Bedford, J. M., Maturation, Transport and Fate of Spermatozoa in the Epididymis, *in* Green, R. O., and Astwood, E. B. (eds.), *Handbook of Physiology*, Sect. 7, Vol. 5, p. 303. American Physiological Society, Washington DC (1975).
4 Bedford, J. M., and Cooper, G. W., Membrane Fusion Events in the Fertilization of Vertebrate Eggs, *in* Poste, G., and Nicolson, G. L. (eds.), *Membrane Fusion, Cell Surface Reviews*, Vol. 5, p. 65. North Holland, New York (1978).
5 Calvin, H., Electrophoretic evidence for the identity of the major zinc-binding polypeptides in the rat sperm tail. *Biol. Reprod.* **21**, 873 (1979).
6 Crabo, B. Studies on the composition of epididymal content in bulls and boars. *Acta Vet. Scand.* **6**, Suppl. 5 (1965).
7 Einarsson, S., Studies on the composition of epididymal content and semen in the boar. *Acta Vet. Scand.*, Suppl. 36 (1971).
8 Ganjam, V. K., and Amann, R. P., Steroids in fluids and sperm entering and leaving the bovine epididymis, epididymal tissue, and accessory sex gland secretions. *Endocrinology* **99**, 1618 (1976).
9 Gustafsson, B., Luminal contents of the bovine epididymis under conditions of reduced spermatogenesis, luminal blockage and certain sperm abnormalities. *Acta Vet. Scand.* Suppl. 17 (1966).
10 Hoskins, D. D., Brandt, H., and Scott, T. S., Initiation of sperm motility in the mammalian epididymis. *Fed. Proc.* **37**, 2534 (1978).
11 Jones, R. and Glover, T. D., Interrelationship Between Spermatozoa, the Epididymis and Epididymal Plasma, *in* Duckeff, J. G., Racey, P. A. (eds.), *Biology of the Male Gamete*, p. 367. Academic, New York (1975).
12 Mann, T., Biochemistry of Semen and Secretions of Male Accessory Organs, *in* Cole, H. H., and Cupps, P. T. (eds.), *Reproduction in Domestic Animals*, Vol. 2, p. 51. Academic, New York (1959).
13 Mann, T. and Lutwak-Mann, C., Secretory function of male accessory organs of reproduction in mammals. *Physiol. Rev.* **31**, 2755 (1955).
14 Mann, T., Biochemical Aspects of Sperm Maturation and Survival, *in* Mancini, R. E., and Martini, L. (eds.), *Male Fertility and Sterility*, p. 89. Academic, New York, (1974).
15 Mann, T., Biochemistry of Semen, *in* Greep, R. O., and Astwood, E. B. (eds.), *Handbook of Physiology*, Sect. 7, Vol. 5, p. 461. American Physiological Society, Washington DC, (1975).
16 Martan, J., Epididymal histochemistry and physiology. *Biol. Reprod.* **1**, 134 (1969).
17 Orgebin-Crist, M. C., Studies on the function of the epididymis. *Biol. Reprod.* **1**:155 (1969).
18 Orgebin-Crist, M. C., Dernzo, B. J., and Davies, J., Endocrine Control of the Development and Maintenance of Sperm Fertilizing Ability in the Epididymis, *in* Greep, R. O., and Astwood, E. B. (eds.), *Handbook of Physiology*, Sect. 7, Vol. 5, p. 319. American Physiological Society, Washington DC (1975).

19 Prasad, M. R. N., Rajalakshini, M., Gupta, G., *et al.*, Epididymal Environment and Maturation of Spermatozoa, *in* Mancini, R. E., and Martini, L. (eds.), *Male Fertility and Sterility*, p. 459. Academic, New York (1974).

20 Setchell, B. P., Scott, T. W., and Voglmayr, J. K., Characteristics of testicular spermatozoa and the fluid which transports them into the epididymis. *Biol. Reprod.* **1**, 40 (1969).

21 Voglmayr, J. K., Metabolic Changes in Spermatozoa during Epididymal Transit, *in* Greep, R. O., and Astwood, E. B. (eds.), *Handbook of Physiology*, Sect. 7, Vol. 5, p. 437, American Physiological Society, Washington DC (1975).

22 Waites, G. M. H. and Setchell, B. P., Physiology of the Testis, Epididymis and Scrotum, *in* *Advances in Reproductive Physiology*, Vol. 4, p. 1. Logos, London (1969).

CHAPTER THREE

THE MALE ACCESSORY SEX GLANDS

STAN A. BEYLER

LOURENS J. D. ZANEVELD

*Departments of Physiology and Biophysics
and Obstetrics and Gynecology
College of Medicine, University of Illinois at the Medical Center
Chicago, Illinois*

1 INTRODUCTION

The function of the male accessory sex organs (the vas deferens, ampulla, seminal vesicle, prostate, and bulbourethral glands) is to provide an optimal milieu for sperm viability inside the male and female reproductive tracts by supplying metabolic substrates and agents that serve protective and nutrient functions. Additionally, from a mechanical standpoint, some structures (e.g., the vas deferens) are involved in sperm transport and storage, whereas others (the prostate gland and seminal vesicles) aid in flushing spermatozoa from the urethra, and, in many species, sequester sperm in the vagina by coagulum formation. The emphasis of this chapter is to present an overview of the salient biochemical characteristics of the secretions from these organs, and their relationships to the reproductive process in mammals.

The growth, morphology, and biochemical composition of the accessory sex organs are under androgenic control. Prior to the development of the radioimmunoassay, determinations of androgen-dependent enzymatic and other chemical components of the secretions of the accessory sex glands constituted useful bioassays for androgens. Reviews on the specific actions of testosterone and other androgens on these structures can be found in Williams-Ashman (1).

A good deal of research has been directed toward elucidating the biochemical composition of whole semen, seminal plasma, and split ejaculates from various species (see Chapter 4). However, relatively little has been learned about the composition of the fluids from the individual accessory glands. Additionally, much of the functional significance of many of these secretions and their constituents remains obscure. Part of the reason for this is the tremendous variability in the accessory sex glands among species, both morphologically and in the biochemical composition of their secretions. Indeed, some species totally lack structures common to most others (Fig. 1). For example, in carnivores such as the dog and cat, seminal vesicles are absent. Additionally, one finds seemingly homologous structures that differ greatly among species in the composition of their secretions. For instance, fructose in semen originates mainly from the seminal vesicles and ampullae of man and the bull, whereas in the rat, fructose is a product of the prostate. Some structures can even be excised (e.g., boar and bull seminal vesicles) without a concomitant loss in fertility. This variability increases the complexity of comparative reproductive biochemistry and physiology and makes studies regarding the functional significance of the accessory-sex-gland secretions difficult.

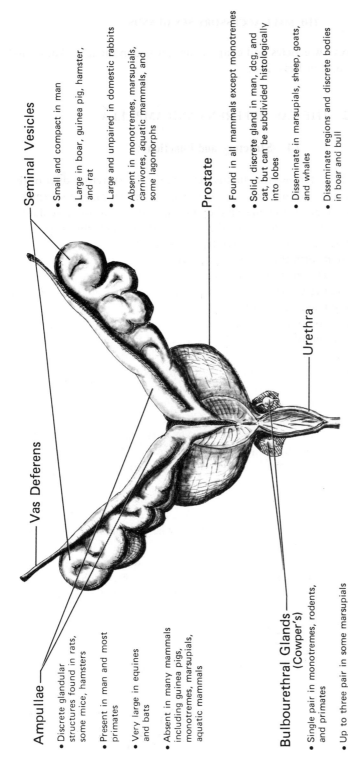

Ampullae
- Discrete glandular structures found in rats, some mice, hamsters
- Present in man and most primates
- Very large in equines and bats
- Absent in many mammals including guinea pigs, monotremes, marsupials, aquatic mammals

Vas Deferens

Seminal Vesicles
- Small and compact in man
- Large in boar, guinea pig, hamster, and rat
- Large and unpaired in domestic rabbits
- Absent in monotremes, marsupials, carnivores, aquatic mammals, and some lagomorphs

Prostate
- Found in all mammals except monotremes
- Solid, discrete gland in man, dcg, and cat, but can be subdivided histologically into lobes
- Disseminate in marsupials, sheep, goats, and whales
- Disseminate regions and discrete bodies in boar and bull
- Several lobes in rodents and lagomorphs, with anterior lobes termed coagulating glands
- Only accessory gland in weasel, dog, bear, and aquatic mammals

Urethra

Bulbourethral Glands (Cowper's)
- Single pair in monotremes, rodents, and primates
- Up to three pair in some marsupials
- Very large and complex in squirrels
- Large and cylindrical in boars
- Absent in aquatic mammals, bears, and dogs

Figure 1 Accessory sex organs of man with comparison to other mammals.

A number of review articles and chapters are quoted in the text. Additional reviews can be found in Refs. 2–6.

2 THE VAS DEFERENS AND AMPULLA

2.1 Structure and Function

The vas deferens (also termed the ductus deferens) is the excretory duct of the testis and epididymis, and functions in the transport and storage of spermatozoa. The vas arises from the cauda epididymis and extends to the ejaculatory duct, which is formed by the fusion of the vas with the ducts from the seminal vesicles (Fig. 1). The vas deferens is a relatively straight tube with distal glandular enlargements that form the ampullary glands. These glands may be part of the wall of the vas deferens (e.g., man, stallion), or may form discrete glands that surround the entire vas (hamster, some strains of mice, rats). Guinea pigs and boars lack ampullae, whereas those of the stallion are quite large.

The primary function of the vas deferens is the rapid transport of spermatozoa from the cauda epididymis to the urethra during the ejaculatory process. During the early excitatory process the vas begins to fill with spermatozoa (emission), which are subsequently extruded by strong contractile movements. The vas and ampulla can also store spermatozoa. For instance, after vasectomy in man, a number of ejaculations (often over a one to two month period) are necessary to obtain ejaculates free of spermatozoa. Data further indicate that spermatozoa are also transported through the vas during prolonged sexual rest, presumably as one of the mechanisms to rid the cauda epididymis of excess spermatozoa. Not only are spermatozoa transported to the urethra, but immediately after ejaculation, material in the vas deferens is moved toward and into the epididymis, at least in the rabbit (7). Finally, some maturation changes may occur as the spermatozoa traverse the vas deferens. For example, epididymal spermatozoa often possess cytoplasmic droplets, whereas ejaculated spermatozoa do not.

The vas deferens itself is composed of circular and longitudinal smooth muscle layers, a well-developed lamina propria, and a convoluted epithelium consisting mainly of stereociliated columnar and underlying cuboidal cells. The epithelium of the proximal vas is quite similar to that found in the epididymis, whereas the epithelium of the terminal ampullary portion resembles that of the seminal vesicles (8). Such differences in histology and ultrastructure may have functional significance.

2.2 Biochemistry

Although the vas deferens is usually considered to be simply a conduit for sperm passage, this organ has both an absorptive and secretory capacity, thereby enabling it to modify the biochemical content within its lumen. The epithelium of the rat vas deferens allows transport of protein (by means of pinocytosis) from the lumen to intracellular lysosomes, which possess high levels of acid hydrolase (9). Other than protein, little transport of solutes appears to occur across the luminal epithelium. Alkaline phosphatase and ATPase activities that are normally associated with active absorption of solutes in the kidney and intestine are absent in the stereocilia of the vas. Many researchers have theorized that the vas deferens plays a role in the reabsorption of excess spermatozoa. However, there is yet no strong evidence for either direct phagocytosis of intact spermatozoa or secretion of hydrolytic enzymes that digest the gametes, followed by absorption of the products of hydrolysis.

The secretory activity of the vas deferens is somewhat better established than its absorptive activity. The presence of rough endoplasmic reticulum in the principal cells of the vas epithelium indicates that they are capable of protein synthesis and secretion. Additionally, holocrine cells have been found in the luminal epithelium. Although little is known about the biochemistry or physiology of these secretory proteins, they may attach to the sperm surface. Also, at least one proteinase inhibitor has been identified in the epithelium of the boar vas deferens. This protein inhibits the sperm enzyme acrosin, and it may have a protective function toward the spermatozoon and may be involved in capacitation (see Chapters 5 and 8). Other components that are secreted by the vas deferens include sialic acid, water, sodium ions, and potassium ions.

Information concerning the composition of the secretions from the ampullary portion of the vas deferens from different species is sparse, with the exception of the stallion and bull. These species produce a sufficiently large volume of ampullary secretions to allow biochemical analysis (Table 1). In general, ampullary secretions tend to be similar to seminal vesicle secretions. This is not surprising, considering the common embryonic origin of the two glands. However, some notable differences exist. The ampullary secretions of the stallion contain high amounts of ergothioneine and glycerylphosphorylcholine (GPC), whereas stallion vesicular secretions contain only small amounts of these compounds. Likewise, bull ampullary fluid has a much higher concentration of GPC than does bull vesicular fluid. Fructose has been detected in the ampullary secretions from man, bull, and rat but not in those of the stallion, although other carbohydrates (mainly sialic acid) are present in stallion ampullary secretions. High concentrations of citric acid have been found in the ampullary secretions of the bull and rabbit; much smaller amounts are present in the stallion, whereas none is found in the jackass or rat.

A large variety of glycosidases are present in stallion ampullary secretions. In particular, a large amount of β-N-acetylglucosaminidase and α-mannosidase activities are present. It is interesting to note that stallion seminal vesicle secretions have lower amounts of these enzymes compared with secretions from other species (Section 3). It appears that in stallion semen these enzymes originate mainly from the ampulla.

Table 1 Components of Ampullary Secretions[a]

Component[b,c]	Bull	Goat	Jackass	Stallion	Rabbit	Rat
Total phosphate	328 (1)	—	>33 (4)	68 (3)	—	—
Ergothioneine	—	—	50 (4)	36 (3)	—	—
Citric acid	550 (6)	—	0 (4)	1 (3)	280 (7)	0 (6)
Lactic acid	—	—	53 (4)	—	—	—
Fructose	Present (8)	—	—	0 (8)	23 (7)	186 (9)
Inositol	—	—	—	20 (10)	—	—
Glyceryl-phosphoryl-choline	94 (5)	—	—	120 (5)	598 (7)	—
Phosphoryl-choline	trace (5)	—	—	0 (5)	—	—
Glycerylphos-phorylethan-olamine	0 (5)	—	—	0 (5)	—	—
Calcium	32 (1)/21 (2)/38 (11)	6 (2)	—	—	—	—
Sodium	137 (1)/161 (2)/249 (11)	112 (2)	—	—	—	—
Potassium	240 (2)/109 (11)	120 (2)	—	—	—	—
Chloride	—	—	—	128 (3)	—	—

[a] Analyses done on ampullary fluid from excised glands except for Ref. 4 (split ejaculate).
[b] In mg/100 ml except for ref. 7 (mg/100 gm tissue).
[c] References: (1) Cragle et al., J. Dairy Sci. **41**, 1273 (1958). (2) Salisbury and Cragle, Proc. 3rd Int. Cong. Anim. Reprod. Cambridge (1956), p. 25. (3) Mann et al., J. Endocrin. **13**, 279 (1956). (4) Mann et al., J. Reprod. Fert. **5**, 109 (1963). (5) Dawson et al., Biochem. J. **65**, 627 (1957). (6) Humphrey and Mann, Biochem. J. **44**, 97 (1949). (7) Holtz and Foote, Biol. Reprod. **18**, 286 (1978). (8) Mann, Ref. 13 at end of chapter, p. 53. (9) Fouquet, Comp. Biochem. Physiol. **40**, 305 (1971). (10) Hartree, Biochem J. **66**, 131 (1957). (11) Quinn et al., J. Reprod. Fert. **10**, 379 (1965).

3 THE SEMINAL VESICLE

3.1 Structure and Function

The seminal vesicles are relatively large, paired glands that originate embryologically from the mesonephric duct. In most species, the seminal vesicles empty into the ejaculatory duct, although in some species, the contents are released directly into the urethra. These glands are large in the guinea pig, hamster, rat, boar, bull, and ram, relatively small in man, and entirely absent in the cat, dog, monotremes, marsupials, and some species of primates and lagomorphs. The rabbit has two organs, the glandula seminalis and the glandula vesicularis, which are homologous to the seminal vesicles of other species. Because of the variability in size of the seminal vesicles, secretions from this gland constitute different percentages of the ejaculate in different species. In man, about 70% of the ejaculate is estimated to be of vesicular origin, averaging approximately 2.5 ml.

The epithelium of the seminal vesicles is typical of a secretory organ, and is either simple (boar, stallion, hamster, rat) or pseudostratified. The ultrastructure of the seminal vesicle and other accessory sex organs has been reviewed by Cavazos (10).

The seminal vesicles were originally thought to be sperm reservoirs, but this is a fallacy. Spermatozoa have been found in the seminal vesicles of some species, including man and the nonhuman primate, but these spermatozoa are often immotile and/or degenerated. Even though the seminal vesicles have several functions related to sperm maintenance, they do not appear to be essential for fertility in many species. For instance, in the boar and bull, excision of the seminal vesicles has no effect on fertility. In rats, removal of the seminal vesicles and coagulating glands (anterior prostate) renders the animals infertile, but excision of the seminal vesicles alone greatly reduces, but does not eliminate, fertility. By contrast, removal of the seminal vesicles from the hamster renders the animal sterile. These findings in rodents may result from the lack of copulatory plug formation after excision of the accessory sex glands. This plug prevents escape of spermatozoa from the vagina, and is thought to initiate cervical contractions that aid in sperm transport. In most species, the seminal coagulum has a less important role, as illustrated by the fact that washed spermatozoa, that is, spermatozoa from which the seminal plasma has been removed, are transported and subsequently fertilize ova when placed in the vagina. In man, the seminal coagulum entraps spermatozoa, preventing their transport into the cervix until the coagulum lyses. The coagulum and its formation is discussed more thoroughly later in this chapter in relation to the individual proteins involved in this process.

Although the seminal vesicles are not entirely necessary for fertility in many species, the secretions from these glands appear to protect the spermatozoa, and enhance their lifespan and fertilizing capacity. Fructose and other substances can be utilized by the spermatozoon for its metabolic processes and thus aid in the maintenance of sperm motility. The buffering power of the relatively alkaline vesicular secretions helps neutralize the natural acidity of the vagina, which can be detrimental to sperm viability. Ergothioneine, a sulfur-containing base found in boar vesicular and stallion ampullary secretions, can protect spermatozoa from the poisonous action of oxidizing agents. Vesicular fluid contains protein moieties such as decapacitation factor (a glycoprotein) and proteinase inhibitors that may serve a protective function by stabilizing the sperm membranes and preventing the premature release of hydrolytic acrosomal enzymes. Seminal vesicle proteins are also a major component of the seminal coagulum or plug (see above). Additionally, since the seminal vesicle fluids are ejaculated last and represent the largest portion of the total ejaculate, at least in species such as man, they help to flush spermatozoa from the urethra.

Studies comparing the characteristics of spermatozoa from different portions of the bull ejaculate have indicated that vesicular secretions may aid in the removal of cytoplasmic droplets. Incubation of epididymal spermatozoa with seminal vesicle fluid results in a greater dispersion of cytoplasmic droplets compared to incubation with buffered saline or prostatic secretions. Eliasson and co-workers (11) have shown that the last portion of the human ejaculate (representing mainly vesicular fluid) decreases the viability of spermatozoa, whereas prostatic fluid has the opposite effect.

There are still a large number of secretory components from the seminal vesicle (e.g., choline derivatives, inositol, organic acids, a number of proteins) whose functional activity is unknown. Speculation on the possible roles of these constituents in the reproductive processes is presented later in this section.

3.2 Biochemistry

3.2.1 Fructose

Reducing and yeast fermentable sugars have been detected in mammalian semen as early as 1928, although the main sugar component was not identified until 1945, when Thaddeus Mann showed that the semen of man and many other mammals contains D-fructose and little or no glucose (Table 2). The seminal vesicles are the main source of fructose in many mammals, although a small percentage is contributed by the ampullary glands. This is not surprising, considering the common embryological origin of these organs. The most striking species variation occurs in the rat, where the seminal vesicles do not produce fructose at all; rather, fructose is produced in the dorsolateral prostate and coagulating glands (anterior prostate) of this species. In the rabbit, fructose is produced in the glandula vesicularis, prostate, and ampulla of the vas deferens.

Fructose constitutes approximately 25% of the total dry weight of the seminal vesicle secretion in man. Thus the amount of fructose in whole semen is a common criterion used clinically to evaluate seminal vesicle function.

Considerable effort has been devoted to elucidating the mechanism by which fructose is synthesized in the seminal vesicles and the other fructose-producing structures. Incubation of tissue from these structures with glucose results in formation of fructose. *In vivo*, the amount of fructose in semen is directly proportional to blood glucose levels. The exact manner by which the conversion of glucose to fructose occurs has been subject to some controversy. The most convincing evidence suggests that glucose is first reduced to sorbitol, which is in turn oxidized to form fructose (12). NADPH and NAD$^+$ serve as hydrogen donors and receptors respectively, and the corresponding reduction and oxidation of these cofactors is coupled to citric acid synthesis. The proposed pathway for fructogenesis in the seminal vesicle is shown in Fig. 2. Additionally, some fructose appears to arise from dephosphorylation of fructose-6-phosphate by fructose-6-phosphatase and/or alkaline phosphatase. Small amounts of other carbohydrates including glucose, ribose, fucose, sorbitol, inositol, sialic acid, and glycoproteins, have been found in seminal vesicle secretions. Metabolism of carbohydrates and other nutritive components of seminal plasma by spermatozoa is discussed in Chapter 6.

Table 2 Biochemical Constituents of Seminal Vesicle Secretions[a]

Constituent[b]	Man[c]	Boar[c]	Bull[c]	Stallion[c]	Guinea Pig[c]	Rat[c]
Fructose	315 (7)	59 (1)/52 (7)	970 (7)/840 (16)	3–15 (5)	100 (6)/630 (17)	0 (7)
Glucose	390 (11)	—	40 (9)	—	125 (17)	—
Inositol (total)	—	2150 (1)	—	—	26 (9)	666 (16)
Citric Acid	125 (7)	635 (1)/580 (4)	670 (4)	127 (5)/72 (19)	320 (6)/192 (17)	115 (7)
Lactic Acid	—	21 (20)	—	319 (19)	9 (17)	—
Ascorbic Acid	5 (13)	4 (7)	14 (7)	—	9 (13)	4 (7)
Uric Acid	—	—	10–30 (10)	—	—	—
Ergothioneine	<1 (7)	57 (1)/79 (8)	Trace (2)	—	—	Trace (2)
Phosphorylcholine	—	0 (2)	Trace (2)	—	—	Trace (2)
Glycerylphosphorylcholine	—	190 (2)	Trace (21)	Trace (14)	275 (21)	654 (2)
Total Protein	9,000 (11)	11,000 (12)	—	—	—	29,000 (18)
Dry Weight	—	14,200 (12)	—	—	—	25,900 (15)

[a] Analyses done on vesicular fluid from excised glands except for ref 7 in man (split ejaculate).

[b] All values are expressed in mg/100 ml except for ref. 16 (mg/100 gm tissue).

[c] References: (1) Mann, Ref. 13 at end of chapter, p. 104. (2) Dawson et al., Biochem. J. **65**, 627 (1967). (3) Huggins and Neal. J. Exp. Med. **76**, 527 (1942). (4) Humphrey and Mann, Biochem. J. **44**, 97 (1949). (5) Mann et al., J. Endocrin. **13**, 279 (1956). (6) Ortiz et al., Endocrinology **59**, 479 (1956). (7) Mann, Ref. 13 at end of chapter, p. 49. (8) Mann and Leone, Biochem. J. **53**, 140 (1953). (9) Hartree, Biochem. J. **66**, 131 (1957). (10) Leone, Bull. Soc. Ital. Biol. Sper. **29**, 513 (1953). (11) Huggins et al., Am. J. Physiol. **136**, 467 (1942). (12) Einarsson, Acta Vet. Scand. **36**, 1 (1971). (13) Berg et al., Am. J. Physiol. **133**, 82 (1941). (14) Dawson and Rowlands, Quart. J. Exp. Physiol. **44**, 26 (1959). (15) Porter and Melampy, Endocrinology **51**, 412 (1952). (16) Eisenberg and Bolden, Nature, **202**, 599, (1964). (17) Prendergast and Veneziale, Ref. 12 at end of chapter. (18) Geiger et al., Immunology **27**, 729 (1974). (19) Mann et al., Am. J. Physiol. **133**, 82 (1941). (20) Mann, Ref. 13 at end of chapter, p. 106. (21) Mann, Ref. 13 at end of chapter, p. 204.

73

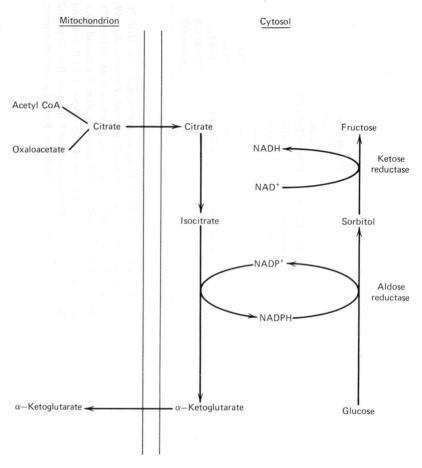

Figure 2   Relationship between citrate synthesis and fructogenesis. From Prendergast and Veneziale (12).

3.2.2   *Polyols*

Two polyols (sometimes called sugar alcohols), sorbitol (D-glucitol) and inositol (mesoinositol) have been detected in the semen of most species and are synthesized in the accessory sex glands. Sorbitol has been found in the seminal vesicles of the ram and guinea pig and in the coagulating glands of the rat. It appears that the concentration of sorbitol in semen tends to be largest in those species that produce high levels of fructose. The stallion is an exception: sorbitol is present in stallion semen even though this animal produces very little fructose. The relationship between fructose and sorbitol content in semen can be explained in terms of activity of the enzymes aldose reductase and ketose reductase (sorbitol dehydrogenase), which are involved in the conversion of glucose to fructose (Fig. 2).

Inositol was first discovered in the seminal vesicle secretions of the boar by Mann (13), who found that this substance constitutes between 40 and 70% of the total dialyzable material from this gland. Although the highest levels of inositol are found in the boar, it has also been found in the vesicular secretions of the bull, guinea pig, hedgehog, and monkey, and in the seminal plasma of the stallion, ram, and man (Table 2). Although the mechanism of inositol synthesis is still unknown, it is believed to derive from glucose. The functional significance of inositol is still obscure, since spermatozoa have not been shown to directly metabolize inositol, as they do sorbitol. Possible functions may be that (1) inositol helps maintain osmotic equilibrium (the boar seminal vesicle secretes very little sodium chloride), and (2) inositol may serve as a carbohydrate reserve for hexoses (13).

3.2.3 Organic Acids

Citric acid, another frequently cited compound, is produced chiefly by the seminal vesicles of the bull, ram, boar, jackass, stallion, goat, guinea pig, and rat (Table 2). It is also produced by the glandula vesicularis of the rabbit and the ventral prostate of the rat. In man, it is only secreted by the prostate gland. This organic acid is both synthesized and catabolized by the tricarboxylic acid cycle, at least in the rat prostate, and the same is probably true in the seminal vesicle. The citric acid content—like the fructose content—of the accessory-sex-gland secretions is dependent upon androgen stimulation. The relationship between fructogenesis and citrate synthesis and catabolism is illustrated in Fig. 2. The physiological role of citric acid in semen is not yet fully understood. Although it has been reported to have a beneficial effect upon sperm motility, there is no direct evidence that spermatozoa utilize citrate aerobically or anaerobically. It has been theorized that the calcium-binding properties of citric acid may be related to the calcium-dependent coagulation process of seminal plasma.

Other organic acids present in semen that originate mainly from the seminal vesicle secretions are ascorbic acid, uric acid, lactic acid, and pyruvic acid (Table 2). Ascorbic acid has been found in the seminal vesicle secretions of man, boar, bull, guinea pig, and rat. Uric acid is present in high concentrations in human seminal vesicle secretions and is found in ram and bull seminal plasma. The major reducing agent in boar seminal vesicle fluid is ergothioneine, a sulfur-containing derivative of histidine. Ergothioneine does not appear to be synthesized, but after ingestion is concentrated in the male reproductive tract by way of the blood. The putative function of ergothioneine and other reducing agents that arise mainly from seminal vesicle secretions is to protect spermatozoa from the toxic effects of oxidation. Lactic acid and pyruvic acid are found in the seminal plasma of most species, and are mainly derived from the seminal vesicles. These compounds are a product of the highly active metabolic processes that occur in the vesicular secretory epithelium, and spermatozoa can use the compounds as metabolic substrates.

3.2.4　Lipids, Phospholipids, and Choline Derivatives

The majority of the lipids found in seminal plasma originate from the prostate gland. However, in most species, prostaglandins and phospholipids arise from the seminal vesicles. Although a considerable amount of research on prostaglandins and their derivatives has been done, their role in reproduction remains unclear (see section 4.2.3 of this chapter).

The seminal plasma of the rat and several other species is the richest source of choline of any tissue or fluid in these animals. Addition of iodine to semen results in the formation of brown crystals of choline derivatives, and is the basis for the medicolegal test for human semen stains. Among the choline-containing compounds present in semen are phosphorylcholine, glycerylphosphorylethanolamine, and glycerylphosphorylcholine (GPC), with the latter generally being prevalent. Bull seminal vesicle secretions contain phosphatidylcholine and very small amounts of phosphorylcholine, but no GPC. By contrast, boar, guinea pig, rabbit, and rat seminal vesicles secrete much larger amounts of GPC than any other choline derivative (Table 2). Lecithin (phosphatidylcholine) is also present in rat seminal vesicle fluid; cytidine diphosphocholine is the key intermediate in its production. Choline and its derivatives are not metabolized by spermatozoa, have no effect upon sperm motility, and have not been shown to be functionally significant. These compounds may be products of membrane fragments resulting from apocrine secretions of the glandular epithelia.

3.2.5　Proteins and Enzymes

Flickinger (14) studied the formation of protein components and their secretion in the rat seminal vesicle epithelium and found that these are relatively rapid processes compared with those of other tissues. Seminal vesicles produce the protein components of the coagulum (coagulinogen) and in rodents, the enzyme procoagulase. Procoagulase is activated to coagulase by the enzyme vesiculase, which is secreted by the coagulating gland of rodents. Upon activation, coagulase converts seminal vesicle coagulinogen to the coagulated protein, forming the vaginal plug of the rodent ejaculate. Further detail on the coagulation of semen can be found in Tauber and Zaneveld (15) and in Section 4.2.5. Rat seminal vesicle fluid has a surprisingly high concentration of protein (Table 2) and is the major source of protein in the semen from this species. Immunological studies have shown that unlike most other body fluids, rat seminal vesicle secretions contain no free serum proteins. A number of interesting functional characteristics of seminal vesicle proteins have been observed, including high-affinity calcium-binding proteins in the bull, and basic proteins that have the ability to agglutinate spermatozoa in the boar. These proteins are not involved in seminal coagulum formation, since no coagulum is found in the bull and only "tapioca-like" material is found in the boar (15).

Although the enzymes of seminal plasma have been studied extensively, the accessory sex glands from which they originate are not always known. Seminal vesicle (and prostate) secretions are produced by an apocrine secretory mechanism; during this process, cytoplasm from the glandular epithelial cells is also secreted. Therefore, fluids produced by these glands possess a great variety of enzymes, most of which were associated with intracellular processes of the glandular epithelium.

Thus, although seminal vesicle (and prostate) secretions are rich in enzymes, most of these probably have little physiological significance. The seminal vesicle fluid of the bull has been the most thoroughly characterized with respect to enzymatic constituents; the following enzymes were identified: alkaline phosphatase, xanthine oxidase, diphosphopyridine and triphosphopyridine nucleosidases, 5'-nucleotidase, DNAse, and a chymotrypsin-like enzyme. Boar seminal vesicle fluid possesses acid and alkaline phosphatase and an esterase. The guinea pig seminal vesicles and rabbit glandula vesicularis also produce alkaline phosphatase, but this is not the case for the hamster, mouse, and rat. A number of different glycosidase activities have been identified in accessory-sex-gland secretions (16), and those of the seminal vesicles are summarized in Table 3. Proteinases (seminin, coagulase, fibrinolysin, fibrinogenase, aminopeptidase, plasminogen activator) found in semen originate mainly from the prostate although a pepsin-like enzyme is associated with seminal vesicle secretions. The seminal vesicle is a rich source of proteinase inhibitors that are proteins or glycoproteins. The two major inhibitors found in human seminal plasma are a trypsin-chymotrypsin inhibitor, and a trypsin-acrosin inhibitor, which have molecular weights of 12,500 and 5400 respectively (17). Proteinase inhibitors have also been found in seminal vesicle secretions of the bull, boar, rat, mouse, hamster, and guinea pig. These inhibitors become associated with spermatozoa and presumably function to protect the reproductive tract against digestion by the sperm acrosomal proteinase, acrosin, which can be released from dead sperm. These inhibitors are removed during the capacitation process in the uterus, which enables acrosin to function in the sperm–ovum fusion process (Chapter 8).

Table 3 Relative Glycosidase Activity in Seminal Vesicle Secretions[a]

Enzyme	Boar	Bull	Monkey	Stallion
α-Mannosidase	+++	++	++	0
β-Mannosidase	0	0	0	0
α-Glucosidase	0	−	0	+
β-Glucosidase	0	−	0	+
α-Galactosidase	0	−	0	+
β-Galactosidase	+	++	0	+
β-N-Acetylglucosaminidase	++++	+++	+++	+
β-Glucuronidase	−	+	+	+

[a] Derived from Conchie, J., and Mann, T., *Nature* **179**, 1190 (1957).

Table 4 Inorganic Constituents of Seminal Vesicle Secretions[a]

Constituent	Unit	Man[b]	Bull[b]	Boar[b]	Goat[b]	Ram[b]
Hydrogen	pH	7.29 (2)	—	—	—	—
Sodium	mEq/l	103 (2)	109 (34)/108 (7)	27 (6)	83 (3)	93 (7)
Potassium	mEq/l	18 (2)	74 (3)/27 (7)	24 (5)/54 (6)	—	—
Calcium	mEq/l	—	21 (3)/35 (4)/25 (7)	6 (5)/6 (6)	6 (3)	4 (7)
Magnesium	mEq/l	—	14 (4)/8 (7)	56 (5)	—	6 (7)
Manganese	mEq/l	—	<0.02 (4)	—	—	—
Copper	mEq/l	—	0.3 (4)	—	—	—
Zinc	mEq/l	—	6 (4)	—	—	—
Chloride	mEq/l	16–36 (1)	—	20 (5)/3 (6)	—	—
Inorganic phosphate	mg/100 ml	30–60 (1)	—	7 (6)	—	—

[a] Method of collection is indicated after each reference.
[b] References: (1) Huggins and Johnson, *Am. J. Physiol.* **103**, 514 (1933) [Massage]. (2) Huggins et al., *Am. J. Physiol.* **136**, 467 (1942) [Massage]. (3) Mann, Ref. 13 at end of chapter, p. 102 [Excised]. (4) Mann, Ref. 13 at end of chapter, p. 98 [Excised]. (5) Einarsson, *Acta Vet. Scand.* **36**, 1 (1971) [Excised]. (6) Mann, Ref. 13 at end of chapter, p. 104 [Excised]. (7) Quinn et al., *J. Reprod. Fert.* **10**, 379 (1965) [Excised].

3.2.6 Ions and Electrolytes

Seminal vesicle secretion, like prostatic fluid, is approximately iso-osmotic with blood serum. In general, vesicular fluid tends to be more alkaline and has a higher dry weight than fluid from the prostate, because of a higher protein content. Sodium tends to be the major cation in vesicular fluid with the exception of the boar, in which as much or more potassium is present. As with most other constituents, there appears to be a good deal of variability in ion concentration among different species (Table 4).

4 THE PROSTATE

4.1 Structure and Function

The prostate is a compound tubuloalveolar gland that is present in all mammals with the exception of monotremes. It is the only accessory sex gland in many carnivores such as the bear, dog, ferret, and weasel, and in cetaceans (e.g., dolphins, porpoises, and whales).

Structurally, the prostate varies among species (see Fig. 1) and may be of three types: (1) disseminate or diffuse, with glandular acini that are located within the lamina propria between the lumenal basement membrane and the muscle of the urethra; (2) a more discrete gland that is sometimes lobed and remains outside the urethral muscle; that is, between the muscle and urethral surface; and (3) a combination of types 1 and 2. For example, the sheep and goat have a disseminate prostate, whereas the bull and boar prostates possess both a disseminate region and a discrete body. Dog and man have solid, compact prostates, whereas the rodent and lagomorph have lobular prostates that are distinctly separate. The rodent has three pairs of prostatic lobes. One pair is termed the coagulating glands, otherwise known as the anterior (or cranial) lobes. The other two pairs are termed the dorsolateral (or middle) lobes, and the posterior (or ventral) lobes. In man, the prostate (although a single, compact structure) can be subdivided into anterior, posterior, middle, and two lateral lobes. There are also functional differences between these lobes, as illustrated by the fact that the posterior lobe is a common site for prostatic carcinoma, whereas the anterior lobe is more susceptible to benign hyperplasia. The middle lobe of the human prostate appears to be homologous to the coagulating gland of rodents. Considerable research has been performed on the prostatic secretion of the dog, since unlike the prostate of other species, the canine prostate has the propensity to develop prostatitis, benign hyperplasia, and adenocarcinoma similar to the human prostate.

A prostate gland homolog is present in females of many species. In women, these are termed the paraurethral glands of Skene. The prostatic cells in females are vestigial and are apparently not functionally significant. In many other species, these glands have properties similar to their male counterparts subsequent to androgen stimulation. A review of the female prostate can be found in Price and Williams-Ashman (18).

The secretory epithelium of the prostate is simple columnar and is capable of both apocrine and merocrine secretions, explaining the presence in prostatic fluid of both lytic and cytoplasmic enzymes that are associated with metabolism. Reviews of the ultrastructure and histochemistry of the prostatic epithelium can be found in Cavazos (10).

Functionally, the necessity of a prostate for fertility is questionable. During ejaculation, prostatic secretions aid in sperm transport through the urethra. However, excision of the coagulating glands and/or the other lobes of the guinea pig prostate does not prevent ejaculation or fertilization. Removal of both prostate and seminal vesicles in rats abolishes fertility. In rodents and many other species, enzymes in prostatic fluid catalyze the coagulum (plug) formation of semen. In man, a proteinase called seminin originates from the prostate gland and is involved in the liquefaction of the seminal coagulum. The physiological significance of many of the biochemical constituents of prostatic secretions is discussed individually later in this section. Because the prostate gland is not a homogeneous structure and different lobes and their secretions have different compositions, distinctions between prostatic fluid obtained from the entire gland (i.e., by induced secretion) and fluid from excised lobes will be made wherever possible.

Table 5   Organic Constituents of Dog and Human Prostatic Fluid[a]

Constituent[b]	Dog[c]	Man[c]
Glucose	15 (1)	16.4 (2)
Fructose	1 (4)	—
Inositol	—	148 (3)
Citric acid	1 (4)/2.6 (5)	1580 (5)
Ascorbic acid	0.8 (6)	0.5 (6)
Spermine	4 (12)	243 (11)
Total protein	800 (1)/2800 (4)	2550 (2)
Total lipid carbon	95 (8)	1220 (8)
Total lipid nitrogen	0 (8)	0 (8)
Total lipid phosphorus	0.2 (4)/1.7 (8)	2.1 (8)
Phosphatidylcholine (lecithin)	0 (9)	0 (10)
Phosphatidylethanolamine (cephalin)	—	107 (10)
Cholesterol	17 (8)	322 (8)/80 (10)/78 (11)
Total lipid	17–142 (7)	286 (10)

[a] Method of collection is indicated after each reference.
[b] In mg/100 ml.
[c] References: (1) Huggins et al., *J. Exp. Med.* **70**, 543 (1939) [Cannulation]. (2) Huggins et al., *Am. J. Physiol.* **136**, 467 (1942) [Massage]. (3) Lewin and Beer, *Fertil. Steril.* **24**, 666 (1973) [Massage]. (4) Wales and White, *J. Reprod. Fert.* **9**, 69 (1965) [Massage]. (5) Huggins and Neal, *J. Exp. Med.* **76**, 527 (1942) [Cannulation/massage]. (6) Berg et al., *Am. J. Physiol.* **133**, 82 (1941) [Cannulation/massage]. (7) Rosenkrantz, et al., *Am. J. Vet. Res.* **22**, 1057 (1961) [Pilocarpine-induced secretion]. (8) Moore et al., *J. Urol.* **46**, 132 (1941) [Cannulation/massage]. (9) Seaman, *J. Urol.* **75**, 324 (1956) [Tissue]. (10) Scott, *J. Urol.* **53**, 712 (1945) [Massage]. (11) Anderson and Fair, *Invest. Urol.* **14**, 137 (1976) [Massage]. (12) Tabor and Tabor, *Pharmacol. Rev.* **16**, 245 (1965) [Tissue].

4.2   Biochemistry

4.2.1   Collection of Fluid

Essentially four methods are employed for the collection of prostatic fluid for biochemical analysis: (1) excision of the gland, (2) cannulation, (3) digital massage of the prostate through the rectum, and (4) collection of a split ejaculate (with prostatic fluid primarily constituting the first portion of the ejaculate). Although the final two are useful clinically, contamination from other accessory-sex-gland fluids may interfere with accurate biochemical analyses. Methods of prostatic fluid collection and the mechanisms involved in formation of these secretion are reviewed by Smith (19). Prostatic fluid is best characterized in man and dog (Tables 5 and 6.)

4.2.2   Fructose and Citric Acid

In most species the prostate gland does not produce fructose or polyols, but does secrete citric acid. However, in others the structures designated as prostate

Table 6 Inorganic Constituents of Prostatic Fluid from Dog and Man[a]

Constituent	Unit[c]	Dog[b]	Man[b]
Water	gm/100 ml	98.1 (1)	93.2 (2)
Hydrogen	pH	5.5–6.2 (1)/6.3 (3)	6.4 (2)/7.6 (7)
Sodium	mEq/l	159 (1)/156 (3)	153 (2)
Potassium	mEq/l	5.1 (1)/7.0 (3)	48.3 (2)
Calcium	mEq/l	0.3 (1)/0.2 (3)	60.4 (2)/128 (6)
Magnesium	mEq/l	0.3 (4)	40 (6)
Zinc	mEq/l	2.3 (4)	10.2 (6)/10.8 (7)
Chloride	mEq/l	160 (1)/148 (3)	38 (2)/99 (5)
Carbon dioxide	mEq/l	0.8 (1)	4.2 (2)
Total nitrogen	mg/100 ml	154 (1)	416 (2)
Inorganic phosphorus	mg/100 ml	—	3.4 (5)

[a] Method of collection is indicated after each reference.
[b] References: (1) Huggins et al., J. Exp. Med. 70, 543 (1939). [Cannulation]. (2) Huggins et al., Am. J. Physiol. 136, 467 (1942) [Massage]. (3) Wales and White, J. Reprod. Fert. 9, 69 (1965) [Massage]. (4) Bartlett, J. Reprod. Fert. 3, 190 (1962) [Split ejaculate]. (5) Huggins and Johnson, Am. J. Physiol. 103, 574 (1933) [Massage]. (6) Hommonai et al., Fertil. Steril. 29, 539 (1978) [Massage]. (7) Anderson and Fair, Invest. Urol. 14, 137 (1976) [Massage].

gland may secrete some fructose or citric acid, or both. For instance, in rats, fructose is secreted by the dorsolateral and anterior prostate (coagulating gland), whereas citric acid is produced mainly in the seminal vesicles and ventral prostate. In the rabbit, the prostate gland and glandular seminalis and vesicularis contribute equal amounts of fructose to the ejaculate. The prostate glands of the rabbit and guinea pig secrete less citric acid than the seminal vesicles, whereas in man the prostate is the sole source of citric acid. The dog, which has no seminal vesicles, ejaculates only small amounts of fructose and citric acid. In the ram and bull, the prostate secretes approximately 20–100 times less fructose than the seminal vesicles, and approximately 100 times less citric acid. The boar prostate also produces much less citrate (1 mg/100 ml) than the seminal vesicles (635 mg/100 ml).

4.2.3 Lipids and Choline Derivatives

In man, prostatic fluid is rich in lipid, containing refractile fat droplets that occasionally give this fluid a yellowish color. Cephalin (phosphatidylethanolamine) is the major ether-soluble lipid present. Phospholipids constitute approximately 60% of the total lipid, and the remainder is mostly cholesterol. Little if any neutral fat and no lecithin (phosphatidylcholine) are present in human or dog prostatic secretions (Table 5).

The lipid composition of dog prostatic fluid is roughly similar to that of man (approximately 2.5% lipid). The rat, rabbit, and guinea pig prostatic secretions possess phosphorylcholine and GPC, but in concentrations two to five times lower than those found in seminal vesicle secretions. Phosphorylcholine and a large amount of glycerylphosphorylethanolamine have been found in human prostatic fluid.

One interesting group of fatty acids that deserves mention are the prostaglandins, which were originally found in the vesicular secretions of man and prostatic secretions of sheep. In the 1930s, researchers noted that strips of human uterus relaxed and contracted when exposed to human semen. Similar effects on smooth muscle along with vasopressor properties were also found to be produced by the seminal fluid and accessory-sex-gland fluids of other species. The active ingredients were later found to be prostaglandins. These findings have led to the hypothesis that prostaglandins facilitate conception by enhancing motility of the cervix, uterus, and oviducts (Fallopian tubes), thereby aiding sperm transport. However, current data in support of this theory are still controversial. Because of the great diversity and broad spectrum of action of these compounds upon practically every tissue, the true physiological action(s) of prostaglandins upon the reproductive process remains obscure.

4.2.4   Nitrogenous Bases

Spermine and spermidine (polyamines) are present in the prostate fluid of many species; the former is more prevalent. Spermine (α, δ -bis[γ'aminopropylamino]-butane) is the oldest known organic constituent of human semen. In the seventeenth century, Van Leeuwenhoek discovered that three-sided crystals appeared in semen over time. These crystals were later found to be composed of spermine phosphate. The formation of observable crystals is not surprising, since human semen may contain over 300 mg/100 ml spermine. The rat prostate similarly secretes relatively high amounts of polyamines, whereas the dog prostate secretes almost none. Bull semen contains only trace amounts of polyamines, which is not surprising, since the prostate of this species is essentially non-functional. Oxidation of spermine and spermidine by diamine oxidase of seminal plasma yields products that are highly toxic to spermatozoa and give semen its characteristic odor. Polyamines have been postulated to serve as cross-linking agents that participate in the coagulation of semen (see section 4.2.5 of this chapter).

4.2.5   Amino Acids and Proteins

A number of free amino acids have been found in prostatic secretions of both man and rat, and their presence depends on androgen stimulation. The free amino acid content of ejaculated semen increases with time, because of the action of endogenous proteolytic enzymes on seminal proteins. Most of the proteins from prostatic fluid of man and dog are of low molecular weight (less than 10,000 daltons). Upon electrophoresis, seven different protein components (mainly glycoproteins) have been found in human prostatic fluid. These migrate similar to serum proteins, but are immunologically distinct.

A number of enzymes are present in mammalian prostatic secretions, including acid and alkaline phosphatase, 5′ nucleotidase, pyrophosphatase, ATPase, nucleases, nucleosidases, nucleotide pyrophosphatase, glycosidases, choline esterase, aldolase, and dehydrogenases for lactate, malate, glucose-6-phosphate, 6-phosphogluconate, isocitrate, and α-glycerophosphate. Enzyme activities identified in human prostatic fluid include esterase, aminopeptidase, succinate dehydrogenase, β-glucuronidase (also identified in the dog), diastase, α-amylase, and lysozyme.

One of the best known enzymes found in prostatic fluid is acid phosphatase. High levels are present in the prostatic secretions of man, monkey, and dog. Lower amounts of this enzyme are found in rabbit, rat, and bull prostate fluid, and in guinea pig seminal vesicle secretions. The amount of acid phosphatase activity is under androgen control. Acid phosphatase from human semen has a pH optimum between 5 and 6, hydrolyzes phosphate monoesters, catalyzes the transfer of phosphate from various donors to alcoholic compounds such as glucose, fructose, and methanol, and is inhibited by calcium. High levels of serum acid phosphatase are found in cases of metastatic prostate carcinoma, and measurement of serum levels of this enzyme constitutes a diagnostic test for this disease. Alkaline phosphatase is also present in the prostatic secretions of many mammals. In man, this enzyme is stimulated by magnesium and has a pH optimum of 9. Bull prostatic fluid possesses more alkaline phosphatase activity than acid phosphatase activity. Alkaline phosphatase has also been found in prostatic and vesicular fluid from the rat, and its presence is likewise under androgenic control.

Seminal plasma and male-accessory-sex-gland secretions also contain phospholipase, deaminase, transaminase, and proteolytic activity (20). The prostate gland of the guinea pig and the coagulating gland (part of the prostate complex) of other rodents are among the richest sources of kallikreins. Kallikreins (also known as kininogenases) are proteolytic enzymes that cause the release of kinins from the precursor kininogen. Kinins have been shown to stimulate human sperm motility and the concentration of sperm in the ejaculate (apparently through stimulation of smooth muscle contractility of the vas deferens). The human prostate also secretes at least one plasminogen activator, which has urokinase-like activity, but has an unknown function.

It was shown that the acid phosphatase and fibrinogenase of the dog represent a fixed fraction of the total protein content over a wide range of protein output. This may indicate that a single subcellular system is responsible for the synthesis, packaging, and secretion of all protein (18).

The ejaculates of a number of animal species form a coagulum or plug (see Chapter 4). In man, this coagulum liquifies normally within 5–20 minutes. Partial liquifaction of the primate coagulum takes place, but this never occurs in the rodent. In rodents, the coagulating proteins originate from the seminal vesicles, and the enzyme that catalyzes coagulation is called vesiculase. This enzyme originates from a specific area of the prostate gland (the cranial or anterior lobe, also known as the coagulating gland). Vesiculase is active at a pH of 6.0–8.2 with a pH optimum of 7.4. Its activity is inhibited by EDTA. Recently, Williams-Ashman and others (21) presented evidence that the guinea pig seminal plug is formed through a transamination reaction of the coagulating protein, leading to one of the highest degrees of cross-linking ever described. These intermolecular peptide linkages are similar to those formed in the activated factor XIII reaction of blood coagulation, based on the occurrence of γ-glutamyl-ϵ-lysine dipeptides isolated from proteolytic digests of the clotted seminal protein. Therefore, vesiculase can be regarded as a transaminase. Polyamines (spermine and spermidine) may also be involved in this cross-linking process (21) as well as a glycoprotein from the bulbourethral gland (see Section 5.1).

The mechanism of coagulation of human semen remains to be elucidated. However, it is fairly well established that the liquefying agent originates from the prostate gland and is a proteolytic enzyme, called seminin. The plasminogen activator does not appear to have a role in the liquefaction process, nor do α-amylase or lysozyme. Seminin has a molecular weight of approximately 33,000. It digests gelatin very effectively, but is less active with hemoglobin and casein substrates. Seminin does not hydrolyze γ-globulin, albumin, transferrin, fibrin, or fibrinogen, and is not inhibited by any of the usual proteinase inhibitors.

Removal of the cranial prostate of the primate prevents the coagulation of semen. In the boar, the formation of the gel-like material in semen appears to be due to the interaction between the sialomucoprotein present in the secretions from the bulbourethral glands and at least two specific proteins secreted by the seminal vesicles (15).

4.2.6   Ions and Electrolytes

Prostatic secretions, like those from the seminal vesicle, vary considerably among species with respect to inorganic constituents (Table 6). Prostatic secretions are approximately iso-osmotic with blood serum, as are seminal vesicle secretions, although vesicular secretions have a higher dry weight because of a greater protein concentration. The pH of prostatic fluid is neutral to acidic, providing a more optimal condition for acid phosphatase activity, which abounds in prostatic fluid. Sodium is the major electrolyte present, although the dog prostate secretes the same amount of chloride as sodium (approximately fourfold more than that in man). Potassium is also present in dog prostatic secretions, but in much smaller concentrations than in man and boar. The concentrations of Na^+, K^+, and Cl^- in resting dog prostatic fluid are lower than their corresponding blood plasma levels.

Human prostatic secretions also contain high levels of calcium. Although human prostatic fluid has high levels of inorganic cations, it has relatively low levels of anions such as chloride, bicarbonate, and phosphate. Organic anions, that is, amino acids, proteins, and citrate, balance the cations such that the pH remains slightly acidic.

One of the most unique features of prostatic secretions is the high levels of zinc. Indeed, the rat dorsolateral prostate is one of the richest sources of zinc in nature, with an ability to concentrate zinc to a level that is more than 20 times greater than that of any other organ. The capacity for the prostate to take up and concentrate zinc is under androgen control. The physiological function of zinc is still somewhat of a mystery, although it may be involved in modulating enzyme activity, in particular carbonic anhydrase. This enzyme is found in the mammalian prostate and requires zinc as a cofactor. Zinc (and to a lesser extent, magnesium) is associated with ejaculated spermatozoa from the bull and boar. Most of the zinc is lost upon washing the spermatozoa, which causes an increase in oxygen consumption by the gametes and a concomitant increase in motility. The negative correlation between zinc concentration and sperm motility has caused some persons to suggest that the removal of zinc is associated with the capacitation process (19). Nevertheless, the necessity of zinc for fertilization is questionable, at least in the rat, since in this species excision of the lateral lobe of the dorsolateral prostate (which essentially eliminates zinc from the semen) has no effect upon fertility.

5 BULBOURETHRAL (COWPER'S) GLANDS
AND URETHRAL (LITTRÉ'S) GLANDS

5.1 Bulbourethral Glands

The bulbourethral glands are compound tubuloalveolar glands that are, for the most part, typical mucous glands. As with most other accessory glands, there is extensive variation in number and structure among species. For example, in monotremes, primates, and rodents, there is only a single pair of glands, whereas marsupials may possess three or more pairs. Most marine mammals completely lack bulbourethral glands, as do the bear and dog (Fig. 1). Structurally, similar variability occurs. For example, the bulbourethral glands of man are small and compact, whereas in the squirrel they are relatively voluminous and complex. In the boar, these glands are large and cylindrical, contributing quite extensively to the volume of the ejaculate.

The bulbourethral glands function in supplying lubricant to the urethra and the tip of the penis, facilitating ejaculation and intromission. In some species, secretions from these glands also play a role in seminal coagulum formation. In the rat, mouse, and hamster, but not the guinea pig, an interrelationship between the coagulating gland (anterior prostate) secretion, vesicular secretion, and bulbourethral secretions exists in this regard. A heat-labile, nondialyzable glycoprotein from the bulbourethral glands causes clotting of vesicular basic

Table 7   Composition of Bulbourethral (Cowper's) Gland
Secretions from the Bull and Boar[a]

Component	Bull[b]		Boar[b]	
	Concentration	Unit	Concentration	Unit
Sialic acid	—		4250 (2)	mg/100 ml
Fructose	2 (1)	mg/100 ml	Trace (3)	mg/100 gm
Citric acid	20 (1)	mg/100 ml	0 (4)	mg/100 gm
Lactic acid	0 (1)	mg/100 ml	—	
Ergothioneine	—		Trace (3)	mg/100 gm
Total nitrogen	14.8 (1)	mg/100 ml	1379 (3)	mg/100 gm
Phosphorus	1 (1)	mg/100 ml	7 (3)	mg/100 gm
Sulfur	—		18 (3)	mg/100 gm
Hydrogen	7.8 (1)	pH	—	
Sodium	—		207 (3)	mg/100 gm
Potassium	—		109 (3)	mg/100 gm
Chloride	132 (1)	mEq/l	372 (3)	mg/100 gm

[a] Method of collection is indicated after each reference.
[b] References: (1) Lutwak-Mann and Rowson, *J. Agric. Sci.* **43**, 131 (1953) [Analysis performed on initial portion of ejaculates]. (2) Hartree, *Nature* **196**, 483 (1962) [Initial portion of ejaculates]. (3) Mann, Ref. 13 at end of chapter, p. 105 [Fresh tissue]. (4) Humphrey and Mann, *Biochem. J.* **44**, 97 (1949) [Fresh tissue].

proteins by means of a nonenzymatic, noncovalent mechanism. Vesiculase from the coagulating gland (anterior prostate) potentiates this clot formation by formation of covalent cross-linkages. The vesiculase activity is stimulated by calcium and inhibited by EDTA, a calcium-chelating agent (22). In the boar, it appears that one or more protein factors from the seminal vesicles promote swelling of the sialic-acid-rich bulbourethral gland mucin, forming a rigid, elastic gel (23). Since cross-linking of the mucin occurs, at least one of the protein factors of vesicular origin may be an enzyme.

Information concerning the biochemical components of bulbourethral gland secretions is quite sparse. Investigations into the composition of this fluid have been limited to the bull and boar (Table 7). The high concentration of sialic acid in the secretions from the boar is noteworthy. The rat bulbourethral gland similarly secretes high amounts of sialic acid. As with all other accessory-sex-gland secretions, bulbourethral gland activity is androgen dependent.

5.2   Urethral (Littré's) Glands

Urethral glands, or glands of Littré, are located in the wall of the urethra and occur along its entire length. These glands are found in man and most other mammals. They secrete a clear, watery fluid that is rich in mucoproteins similar to those produced by the bulbourethral glands. These glands presumably also function in the lubrication of the urethra and penis, thereby facilitating intromission and preparing the urethra for the transport of the ejaculate.

REFERENCES

1 Williams-Ashman, H. G., Metabolic Effects of Testicular Androgens, *in* Hamilton, D. W., and Greep, R. O. (eds.), *Handbook of Physiology*, Sect. 7, Vol. 5, p. 473. Williams and Wilkins, Baltimore (1975).

2 Huggins, E., The physiology of the prostate gland. *Physiol. Rev.* **25**, 281 (1945).

3 Risley, P. L., Physiology of the Male Accessory Organs, *in* Hartman, C. G. (ed.), *Mechanisms Concerned with Conception*, p. 73. Macmillan, New York (1963).

4 Burgos, M. H., Biochemical and Functional Properties Related to Sperm Metabolism and Fertility, *in* Brandes, D. (ed.), *Male Accessory Sex Organs*, p. 151. Academic Press, New York (1974).

5 Neaves, W. B., Biological Aspects of Vasectomy, *in* Hamilton, D. W., and Greep, R. O. (eds.), *Handbook of Endocrinology*, Sect. 7, Vol. 5, p. 383, Williams and Wilkins, Baltimore (1975).

6 White, I. G., Accessory Sex Organs and Fluids of the Male Reproductive Tract, in Alexander, N. J. (ed.), *Animal Models for Research in Contraception and Fertility*, p. 105. Harper & Row, Hagerstown (1979).

7 Prins, G. S. and Zaneveld, L. J. D., Contractions of the rabbit vas deferens following sexual activity: a mechanism for proximal transport of spermatozoa. *Biol. Reprod.* **23**, 904 (1980).

8 Hamilton, D. W., The Mammalian Epididymis, *in* Balin, H., and Glasser, S. (eds.), *Reproductive Biology*, Excerpta Med. Found., p. 268. Amsterdam (1972).

9 Friend, D. S. and Farquhar, M. G., Functions of coated vesicles during protein absorption in the rat vas deferens. *J. Cell Biol.* **35**, 357 (1967).

10 Cavazos, L. F., Fine Structure and Function Correlates of Male Accessory Sex Glands of Rodents, *in* Hamilton, D. W., and Greep, R. O. (eds.), *Handbook of Physiology*, Sect. 7, Vol. 5, p. 353. Williams and Wilkins, Baltimore (1975).

11 Eliasson, R.; Johnson, O., and Lindhomer, C., Effects of Seminal Plasma on Some Functional Properties of Human Spermatozoa, *in* Mancini, R. E., Martini, L., (eds.), *Male Fertility and Sterility*, p. 107. Academic Press New York (1974).

12 Prendergast, F. G. and Veneziale, C. M., Control of fructose and citrate synthesis in guinea pig seminal vesicle epithelia. *J. Biol. Chem.* **250**, 1282 (1975).

13 Mann, T., *The Biochemistry of Semen and of the Male Reproductive Tract*. Wiley, New York (1964).

14 Flickinger, C. J., Synthesis, intracellular transport, and release of secretory protein in the seminal vesicle in the rat, as studied by electron radiography. *Anat. Rec.* **180**, 407 (1974).

15 Tauber, P. F. and Zaneveld, L. J. D., Coagulation and Liquefaction of Human Semen, *in* Hafez, E. S. E. (ed.), *Human Semen and Fertility Regulation in Men*, p. 153. C. V. Mosby, St. Louis (1974).

16 Conchie, J. and Mann, T., Glycosidases in mammalian sperm and seminal plasma. *Nature*, **179**, 1190 (1957).

17 Fritz, H., Schiessler, H., Schill, W. B., *et al.*, Low Molecular Weight (Acrosin) Inhibitors From Human and Boar Seminal Plasma and Spermatozoa and Human Cervical Mucus—Isolation, Properties, and Biological Aspects, *in* Reich, E., Rifkin, D. B., and Shaw, E. (eds.), *Proteases and Biological Control*, p. 767. Cold Spring Harbor Laboratory, New York (1975).

18 Price, D. and Williams-Ashman, H. G., The Accessory Reproductive Glands of Mammals, *in* Young, W. C. (ed.), *Sex and Internal Secretions*, Vol. 1, p. 366. Williams and Wilkins, Baltimore (1961).

19 Smith, E. R., The Canine Prostate and Its Secretion, *in* Thomas, J. A., and Singhal, R. L. (eds.), *Molecular Mechanisms in Gonadal Hormone Action, Advances in Sex Hormone Research*, Vol. 1, p. 167. University Park Press, Baltimore (1975).

20 Zaneveld, L. J. D., Polakoski, K. L., and Schumacher, G. F. B., The Proteolytic Enzyme Systems of Mammalian Genital Tract Secretions and Spermatozoa, *in* Reich, E., Rifkin, D. B., and Shaw, E. (eds.), *Proteases and Biological Control*, p. 683. Cold Spring Harbor Laboratory, New York (1975).

21 Williams-Ashman, H. G., Wilson, J., Beil, R., *et al.* Transglutaminase reactions associated with the rat semen clotting system: modulation by polyanions. *Biochem. Biophys. Res. Comm.* **79**, 1192 (1977).

22 Beil, R. E. and Hart, R. G., Cowper's gland secretion in rat semen coagulation. *Biol. Reprod.* **8**, 613 (1973).

23 Boursnell, J. C., Hartree, E. F. and Briggs, P. A., Studies of the bulbo-urethral (Cowper's) gland mucin and seminal gel of the boar. *Biochem. J.* **117**, 981 (1970).

CHAPTER FOUR

SEMINAL PLASMA

KENNETH L. POLAKOSKI

MAZIE KOPTA

Department of Obstetrics and Gynecology
Washington University
St. Louis, Missouri

1   INTRODUCTION

Mammalian seminal plasma is the nongamete portion of an ejaculate and is composed of cells, cellular particles, and fluids from the testis, excretory ducts, and accessory sex glands of the male. In most species, including man, the accessory-sex-gland secretions are not ejaculated simultaneously. There first appears a "presperm" fraction that originates in Cowper's and Littré's glands and is used to lubricate the urethra. The second fraction is a "sperm-rich" portion from the prostate, ampulla, and epididymis. The "postsperm" fraction is the final and largest fraction originating from the seminal vesicles. Like most generalities, this one has notable exceptions. For instance, the horse occasionally ejaculates a fourth fraction known as the "tail-end" or "postejaculation drip," whereas the boar has no ampulla, and the spermatozoa are ejaculated in a number of sperm-rich waves. There is no ampulla or seminal vesicles in the dog, and the "postsperm" fraction is derived from the prostate. The bull appears to have a simultaneous ejaculation; however, it can be split by electro-ejaculation procedures.

These are a few examples of the numerous species-specific male reproductive peculiarities that have confused our understanding of the possible physiological significance of many aspects concerning mammalian semen. Nonetheless, the biochemistry of mammalian seminal plasma has received considerable attention for several reasons. First, although it is a subject for serious argument (see section 2 of this chapter), many investigators believe that seminal plasma can influence the fertility potential of the spermatozoa. Secondly, some of the seminal constituents are organ specific, and their concentrations can be useful for assessing the secretory capacity of the various sex glands whose development and secretion are controlled by androgens produced in the testis, namely testosterone. Finally, from a biochemical standpoint, mammalian seminal plasma is a fascinating secretion, for it contains many interesting and unusual substances, some of which are not measurable in any other body fluid or tissue.

A review of the literature on the biochemistry of seminal plasma shows a patchiness of knowledge. The earlier studies were basically descriptive and were performed on species that were readily available to the particular investigator. The majority of more recent biochemical studies have been focused on semen from man to find possible correlations between various seminal constituents and male infertility. This is because approximately 6 to 7% of the male population is infertile, and semen from these individuals is readily available in clinics.

The emphasis on studies concerned with bull seminal plasma has been to determine fertility indexes and to increase the quality of frozen semen to be used for artificial insemination. Semen preservation is also a major goal in other species of agricultural interest such as the boar and buffalo, in which recovery of viable spermatozoa after freezing is much more difficult. Much less is known about the seminal plasma of the ram and stallion since large individual variations result from these animals' being seasonal breeders, as are the jackass, goat, camel, hare, and free-ranging monkeys. Research on the seminal plasma from nonhuman primates has been hampered by the high cost of the animals and the small volume of ejaculate that is produced (up to 70% or more of the total ejaculate consists of a nonliquefying seminal gel).

The semen from laboratory animals that has received the most attention is that from dogs and rabbits. Since the majority of dog seminal plasma is of prostatic origin, this animal has been used mainly for studies on diseases of the prostate. Rabbit semen is readily available and has been used in many *in vivo* fertilization experiments, especially for investigating the effects of antifertility compounds. A few investigations on cat seminal plasma have been reported, but the cat ejaculates a very small volume, and the semen is somewhat difficult to collect. A number of studies have been performed on rodent semen, with emphasis on elucidating the mechanism of seminal plug formation; however, little is known about the composition and biochemistry of the other constituents.

Because of space limitations, the following review primarily deals with those species about which the greatest amount of information is available. For more detailed information and additional references, the reader is referred to the reviews listed at the end of this chapter, particularly the excellent monographs written by Thaddeus Mann (1) and edited by E. S. E. Hafez (2).

2 PHYSIOLOGICAL SIGNIFICANCE

Seminal plasma acts as a fluid for the transport of the male gametes into the female reproductive tract. It accomplishes this task by acting as a buffered medium that contains nutrients for the spermatozoa in a species-specific volume and sperm density. If these conditions are appreciably altered, the fertilizing potential of the spermatozoa is potentially hampered (3).

An argument can be made that seminal plasma is nothing more than a physiological buffer functioning as a diluent. For example, it has been known since before the turn of the century that not all the secretions from the male accessory sex glands are required for fertility, as shown by experiments with rats in which either the prostate or seminal vesicles had been excised (Chapter 3). Similar experiments were performed with boars, which remained fertile even after their seminal vesicles and Cowper's glands were removed (a process that reduced the total volume of the ejaculate by about 50%). Not only were the boars still able to impregnate sows, but the semen from these animals had a reduced number of abnormal spermatozoa and an increase in the duration of sperm motility (4). Additionally, epididymal spermatozoa that have not had contact with accessory sex gland secretions are capable of becoming capacitated (Chapter 8) and able to fertilize eggs. It therefore appears that seminal plasma and, consequently, its constituents are not an absolute requirement for fertilization.

Not only may seminal plasma be unnecessary for conception but it may even have a detrimental effect on fertility. For instance, the chance for conception with artificial insemination may decrease when human semen from proven fertile men is mixed with semen from infertile men. Moreover, bull seminal plasma tends to reduce the respiration and motility of spermatozoa and makes the gametes more susceptible to cold shock. Human seminal vesicle secretions are also detrimental to sperm motility and are the source of the coagulum (clot)-forming constituents (Chapter 3). The majority of the human spermatozoa are trapped in the coagulum and before the sperm become motile, the clot must liquefy. Absence of liquefaction has been associated with human male infertility. Additionally, in the woman, seminal plasma does not pass into the cervical mucus but some spermatozoa penetrate this mucus within 1½ to 3 minutes post ejaculation. These spermatozoa would have had only minimal contact with the seminal plasma making it very unlikely that the seminal components were able to influence the fertility of these gametes.

On a molecular level, several constituents in seminal plasma have been shown to be "decapacitation factors," which are entities that inhibit the sperm's fertilization potential. Other compounds that are present in high concentrations, such as pepsinogen in human semen, are never subjected to conditions in which they can be utilized. There are also a large number of seminal compounds like the prostaglandins whose levels are extremely high in some species but that are apparently lacking in others, indicating that these are not generally required for fertility. Furthermore, although more than 100 compounds have been found in the seminal plasma of numerous species, there is no concrete evidence that any one of the seminal constituents is required for the spermatozoon to fertilize an egg. In conclusion, the devil's advocate proposes that the basic premise for the seminal biochemist must be that "Components in seminal plasma should be considered to have no function until proven otherwise."

In support for the thesis that mammalian seminal plasma acts as more than just a transfer medium for spermatozoa, it has been noted that several species of epididymal spermatozoa have a nondirectional type of movement and obtain forward progression only when mixed with seminal plasma or buffers that mimic this fluid. Furthermore, seminal plasma has a protective effect when used as a thawing diluent for frozen boar sperm. Although seminal plasma has a tendency to decrease sperm motility, the addition of bull or buffalo seminal plasma from normally motile ejaculates to semen samples containing poorly motile spermatozoa was found to enhance sperm motility and the conception rate of the poor quality samples.

Although it is true that the seminal vesicle secretion may be detrimental to the motility of spermatozoa in most species, it is secreted last and usually contains the coagulum-forming constituents, which may assist in preventing the loss of semen from the female reproductive tract following penis withdrawal. Thus, the coagulum may aid as well as hinder fertility (see above). It was noted that semen from boars who had had their seminal vesicles and Cowper's glands removed did not contain the seminal gel and had a "great loss of semen from the vagina during service." Although no coagulum is found in dog semen, in this species the bulbourethral gland enlarges during intercourse and prevents the penis from withdrawing immediately after ejaculation. Such species' idiosyncrasies are extremely important in defining or proposing roles for the various seminal constituents since differences in reproductive physiology could require different mechanisms of control and, consequently, differences in the requirement for certain constituents.

Several physical parameters of semen from various species are apparently related to the site of insemination, that is, either vaginal or uterine. Furthermore, even though the spermatozoa of some species may be in contact with the seminal plasma for only several minutes before passing into the cervix, this is more than adequate time for the spermatozoa to adhere to a number of seminal components that may be beneficial to the sperm's fertilization potential and survival. Contact of the fertilizing spermatozoon with seminal plasma may actually be longer because the first spermatozoa to enter the female reproductive tract may not be the ones that actually fertilize the eggs.

Spermatozoa can metabolize several of the seminal plasma constituents, and some seminal compounds have been shown to have other beneficial effects toward spermatozoa. An "antiagglutinin" protein that is derived from the prostate prevents the spermatozoa from autoagglutination. This is particularly important for species in which the spermatozoa must penetrate through cervical mucus (bovine, human), since agglutinated spermatozoa cannot do so. In spite of the fact that seminal plasma as a whole tends to decrease sperm motility, it contains a number of substances that stimulate the motility of spermatozoa *in vitro*, namely, forward-motility protein, arginine, and possibly insulin. In the rat, a heat-labile, nondialyzable factor required for maintenance of sperm motility is found in the seminal plasma. Several spasmogens (factors causing stimulation of a spasmodic response in isolated uterine tissue) have been found in prostatic fluid and semen of both rats and stallions. These may aid sperm transport in the female reproductive tract, a function that has also been suggested for the prostaglandins found in human and ram semen. The decapacitation factors (see above) are removed from spermatozoa during capacitation in the female reproductive tract and may be used to regulate the sperm membrane or enzyme systems so that the sperm's ability to penetrate cells is functional only when it is in the vicinity of the ovum.

Seminal compounds may have other functions besides being involved in the fertilization process. For instance, human semen contains material that supresses lymphocyte activation; these seminal immunosuppressive substances could assist in preventing the development of antibodies against spermatozoa. Consequently, one must look at the overall effect that seminal plasma has on both the male and female reproductive tracts before its constituents are dismissed as nonfunctional. Semen also contains a number of constituents with bacteriostatic properties, that is, zinc, spermine, lysozyme, glucosidases, and secretory IgA. These may be important because some types of bacteria have been shown to immobilize and/or agglutinate spermatozoa.

Finally, the biochemical study of seminal plasma is important because it has given insights into the secretory activity of the various accessory sex glands. This could have ramifications in understanding the general process of glandular secretion.

As outlined above, many factors must be considered when discussing the functional biochemistry of mammalian seminal plasma. One group of andrologists claims that no compound in seminal plasma has ever been shown to be required for fertilization, whereas another group argues that some components are of potential importance but their function remains to be defined. However, Williams-Ashman has cautioned about "reading in" functions of the individual seminal constituents, for even though they may have effects *in vitro*, they may not be significant *in vivo*. Mann has also noted that much confusion has resulted from a lack of knowledge of the importance of seminal components *in vivo*. Perhaps, similar to what has occurred in the study of blood diseases, major insights will be obtained through the systematic investigations of the abnormal (infertile) rather than the normal (fertile) males.

3   VARIATION

A major stumbling block to much of our understanding of the functional biochemistry of mammalian seminal plasma is the extreme variability in semen parameters that occurs between animals of different species, breeds, individuals within a breed, and even in different ejaculates from the same animal.

It is extremely difficult to compare the functional biochemistry of seminal plasma among different species, for the volume of an ejaculate can be as small as 0.01 ml for the cat and greater than 500 ml for the boar. Even the concentration of spermatozoa can affect the semen samples, since the spermatozoa can account for as much as one-third the total volume of a ram ejaculate and only a few percent of the ejaculates of boar and man. Although the secretion of many of the seminal plasma components is often confined to one of the accessory glands and consequently can be used as a chemical indicator of the secretory functions of that gland, there are species differences in the origin of the particular components. For example, the main source of seminal citric acid in man is the prostate, whereas in the boar, bull, mouse, ram, and stallion, it is the seminal vesicles. In the rabbit, both the prostate and vesicular glands secrete citric acid.

Large differences in seminal values may also exist within a breed of animal, for instance, domesticated rabbits and monkeys have less of a seasonal breeding cycle than their wild counterparts. Ejaculates obtained during different times of the breeding cycle can have marked variations. For example, the volume of semen from a stallion in December may be less than 25 ml, whereas the same animal in March may ejaculate greater than 50 ml.

Besides breeding cycles, there are many other factors that affect individual ejaculations (5). Psychic factors and the degree of sexual excitement influence semen samples. For example, it has been shown that estrous cows can cause bulls to have up to an 18% increase in ejaculate volume and sperm concentration. Endocrine and genetic agents can also affect the animal's libido. There are numerous other factors that can influence the quantity and quality of semen and consequently the biochemistry of the seminal plasma. Some of these have been related to climate and include such things as photoperiodism, temperature, and relative humidity. Nutrition, body weight, exercise, age, and sexual maturity are important in influencing the composition of semen. Also important is the frequency of ejaculation, as demonstrated in depletion trials during which animals are continuously ejaculated. Long periods of rest can also have an influence on seminal composition since they generally result in an increased number of degenerating spermatozoa, whose constituents can affect the composition of the seminal plasma.

An often overlooked but significant variation that confuses our understanding of the functional biochemistry of mammalian seminal plasma is the actual measurement of the constituents that are present. The collection and storage conditions are extremely important because seminal plasma contains a large number of hydrolytic enzymes that affect the constituents. The most noticeable of these are the proteolytic enzymes. However, nonprotein constituents are also affected with storage, as exemplified by the formation of spermine phosphate crystals and changes in the various prostaglandins in human semen on standing. While refrigerated or frozen, many enzymes remain active, whereas others are irreversibly denatured. So, it is safest to assume changes are occurring during storage unless proven otherwise.

Another factor that may interfere with accurate measurements is that seminal plasma contains a diverse array of compounds, including inorganic, organic, carbohydrate, protein, and lipid constituents that may either inhibit or modify many chemical and enzymatic tests. Well-known examples of this are the numerous seminal proteolytic enzyme and prostaglandin inhibitors; however, there are also compounds that bind to and prevent the detection of other compounds. For instance, several older papers claim a lack of testosterone in seminal plasma. It has now been shown by several investigators that testosterone is present, but the majority is bound to a type of androgen-binding protein. Much confusion has also arisen from the fact that many of the enzymatic activities (for instance, benzoyl-arginine methylesterase, leucine aminopeptidase, pepsin, plasminogen activators, acid and alkaline phosphatase) that have been measured are the result of not one but several enzymes. This makes their evaluation difficult, for the different forms may have different requirements for optimal activity (pH, salt concentrations, etc.).

Finally, an area that is often overlooked concerns the spermatozoa that are present. As already noted, spermatozoa can account for a significant portion of the total ejaculate, and these cells are actively incorporating and metabolizing compounds as well as releasing constituents into the seminal plasma. This release is particularly active in the case of dead or dying spermatozoa.

In conclusion, there are many reasons to expect variations in the composition of mammalian seminal plasma. Consequently, the published values are often general estimates and should be considered only as guidelines. The data presented in the Tables should be treated as such. They represent the most commonly reported values in the literature. Due to the very large number of articles used to derive these data, specific references have been omitted. Review chapters and articles can be found listed at the end of this chapter.

4 PHYSICOCHEMICAL CHARACTERISTICS

The major physicochemical characteristics of semen from several mammalian species are summarized in Table 1. These can be influenced by the length of postejaculate time prior to the analysis, the sperm cell concentration, the non-spermatozoa particulate bodies, and the presence or absence of a gel or coagulum.

The physicochemical measurements of mammalian seminal plasma should be performed as soon after ejaculation as possible. This should be within minutes of coagulum liquefaction with human semen; in animals whose clot remains intact or dissolves slowly, the gel should immediately be removed by centrifugation or filtration. Semen is a relatively weak buffer between pHs of 6.5 to 8.5 and the pH of human semen increases upon storage at ambient temperatures. By contrast, the pH of bull semen decreases upon storage because of the lactic acid produced by the spermatozoa. The freezing point of human semen rises from 0.58 to 0.70 during a four-hour incubation at 37° C. The seminal hydrolases are particularly potent in altering seminal constituents, and large variations in protein content occur in human semen within 15 minutes of ejaculation, even at 20° C. Large spermine phosphate crystals also appear in human seminal plasma on standing and affect the physicochemical composition.

The appearance of most mammalian semen samples is usually a light yellow to creamy white, depending to a great extent on the sperm concentration. The deep yellow sometimes seen is an indication of the presence of urine (however, the semen of some bulls has yellow pigmentation from high concentrations of riboflavin). Semen samples with a red hue are usually associated with blood contamination.

An area that has received very little attention is the nonspermatozoa particulate matter that is present in semen. In some species (for example, the rabbit), this material can be present in concentrations greater than the concentration of spermatozoa. The particulate matter includes exfoliated cells, cytoplasmic droplets that originate from spermatozoa, lipid bodies, colostrum corpuscles, corpora amylacea, and the firm, stonelike calcium phosphate structures known as prostatic calculi. Many of the exfoliated cells are the result of a sloughing of the germinal epithelium. These are rich in intracellular material and therefore can affect the biochemical analysis of semen. The majority of this so-called cellular debris originates from the prostate gland and, as Matthew Freund has recently shown, can possibly be used in the diagnosis of prostatic cancer and possibly other disease states. The white blood cells, bacteria, viruses, yeasts, and mycoplasma that are occasionally found in semen are considered to be abnormal and, as discussed earlier, may be a cause of infertility.

The factor that has perhaps the greatest influence on the physical appearance of semen is the gel, or coagulum, that can form shortly after ejaculation. In the case of bull and dog semen, it remains a liquid. However, in most other species, the seminal plasma coagulates when the fluids from the accessory glands are

Table 1 Physicochemical Properties of Mammalian Semen[a]

Parameter	Man	Boar	Bull	Ram	Stallion	Jackass	Dog	Rabbit
Volume (ml)	3.5(1.5–6.0)	250(150–>500)	4(2–10)	1(0.7–2)	70(30–300)	69(36–120)	9(2–15)	1(0.4–6)
Spermatozoa (millions/ml)	100 (50–150)	100 (25–300)	1000 (300–2000)	3000 (2000–5000)	120 (30–800)	229 (100–475)	300 (100–900)	150 (100–200)
Dry weight (mg/ml)	—	46(22–62)	95	148	31(22–38)	49(10–80)	2.3–2.7(%)	—
pH	(7.0–7.5)	7.5(6.8–7.9)	6.9(6.4–7.8)	6.9(5.9–7.3)	7.4(6.2–7.8)	7.5(7.2–7.7)	(6.1–7.0)	(6.6–7.5)
Freezing point depression (°C)	(−0.55 to −0.58)	(−0.54 to −0.63)	(−0.54 to −0.73)	(−0.55 to −0.70)	(−0.58 to −0.62)	(−0.56 to −0.62)	(−0.58 to −0.60)	(−0.55 to −0.59)
Total nitrogen (mg/100 ml)	913(560–1225)	613(334–765)	897(441–1169)	875	310	—	307–480	—
Nonprotein nitrogen (mg/100 ml)	90(53–137)	22(15–31)	48	57	55	—	23–34	—
Protein (gm/100 ml)	4.5(3.3–6.8)	3.7	6.8	5.0	1.0–2.0	—	1.74–2.61	6.3–6.0
Ash (gm/100 ml)	0.9	—	—	—	(0.91–0.93)	—	—	—
Water (gm/100 ml)	91.8(89.1–99.4)	95(94–98)	90(87–95)	85	98	—	97.6	—
Specific gravity	1.035 (1.031–1.039)	1.035	1.034 (1.015–1.053)	—	1.0116	1.027 (1.015–1.034)	1.011	—
Conductivity (mho × 10^{-4})	(88–107)	—	—	—	—	—	(129–138)	94(85–100)

[a] Mean values. The range of reported values is presented in parentheses. If too few data are available to firmly establish a mean value, only the range is given.

mixed. The degree of solidification can range from an increase in viscosity to the hard *bouchon vaginal* or vaginal plug seen in the semen of some rodents. As described in Chapter 3, this plug is formed by transglutaminases from the anterior prostate (coagulating gland) which calalyze ϵ-γ-glutamyl cross-linkages between seminal vesicle proteins, a process that is accelerated by polyanions from the bulbourethral gland secretions (6). Some cross-coagulations have been observed between mixed semen samples of the rat and some monkeys but not with the semen from rat and man. The material that forms the clot in human semen originates from the seminal vesicles while the liquefication agents (hydrolytic enzymes) are found in the prostate secretions (7).

Much less is known about seminal coagulation in other species. Semen from nonhuman primates contains a coagulum that constitutes more than 50% of the volume. If it is left in contact with the spermatozoa, many will become incorporated into the coagulum, which does not liquify *in vitro*. Boar semen is noted for its "tapioca-like" gelatinous material that is derived from an interaction of sialomucins from the bulbourethral glands and several proteins from the seminal vesicles. The *in vivo* coagulation that occurs in opossum and bat semen depends on mixing of the seminal constituents with the female reproductive tract secretions, whereas the appearance of a seminal coagulum is seasonal in horses and jackasses, and only occasionally seen in elephant semen. Again, there can be large variations, since the semen from a given rabbit may have up to 6 ml of gelatinous material or as little as 0.5 to 1 ml of fluid devoid of gel.

5   LOW-MOLECULAR-WEIGHT CONSTITUENTS

There is a wide range of interesting and unusual components present in mammalian seminal plasma (1, 8). Some of these either are not found in any other body fluids or are present in much higher concentrations in the seminal plasma. The levels of many of the nonprotein constituents that are normally found in seminal plasma are presented in Tables 2–6. It should be noted, however, that the actual concentration of all of the constituents is subject to large variations as explained in section 3. The assumption was made that the reported values were obtained from normal fertile males but this was rarely indicated. It was also not always clear whether the values were for whole semen or sperm-free seminal plasma. This can be extremely important for species like the ram, which have small semen volumes and high sperm concentrations. For example, the total plasmalogen concentration in whole ram semen is about 380 mg/100 ml, whereas it is only about 60 mg/100 ml in seminal plasma.

Numerous attempts have been made to correlate the concentration of various seminal constituents with the fertility of the male. However, these have generally met with little success and there is presently no single biochemical criterion for differentiating between a fertile and an infertile male. Nevertheless, in some species, some of the components are tissue specific, and the biochemical analysis of seminal plasma can give a very convenient and quantitative

Table 2 Low-Molecular-Weight Organic Seminal Constituents[a]

Constituent	Man	Boar	Bull	Ram	Stallion	Dog	Rabbit
Ascorbic acid	(2.6–14.4)	3.5(2–5)	(3–24)	5	—	—	—
Adenosine triphosphate (µg/ml)	1.1	—	(0.6–1.5)	—	—	(0.1–0.6)	0.3
Citric acid	376(96–1430)	130(30–330)	720(340–1150)	140(110–260)	26(8–53)	0–30	50–600
Glycerylphosphoryl-choline	70(54–90)	(110–240)	350(110–496)	1650(1101–2040)	38–113	180(110–240)	280(215–370)
Fructose	235(40–640)	13(3–50)	530(26–981)	(250–372)	2(0–6)	0.5–0.6	(40–400)
Galactose	—	(4–20)	—	—	0	—	—
Glucose	7(0–99)	—	300	—	82	116	Trace
Glycerol	—	(10–20)	2	—	23	—	—
Glycerylphosphoryl-inositol (nmol/ml)	—	260	1400	1520	250	—	—
Glycerylphosphoryl-ethanolamine	—	Trace	Trace	Trace	Trace	—	—
Glycogen	Trace	Trace	—	—	—	—	—
Inositol	(50–70)	530(380–630)	35(25–46)	(7–41)	30(11–47)	—	(21–30)
Lactic acid	35(20–50)	27	35(20–50)	36	12(9–15)	44(22–77)	(11–30)
Pyruvic acid	29(11–56)	—	(4–5)	—	—	—	—
Sorbitol	10	12(6–18)	(10–140)	72(26–120)	40(20–60)	<1	80

[a] Values are expressed in mg/100 ml unless otherwise indicated. The numbers without parentheses represent mean values; those within parentheses show the range of reported values. If two few data are available to firmly establish a mean value, only the range is given.

Table 3 Inorganic Constituents in Mammalian Seminal Plasma[a]

Constituent	Man	Boar	Bull	Ram	Stallion	Dog	Rabbit	Goat
Bicarbonate	54(43–74)	50	16	16	24	(1–6)[b]	—	—
Boron	—	—	1.73	—	—	—	—	—
Calcium	30(14–62)	5(2–6)	44(35–60)	11(6–15)	26	(0.4–0.9)[b]	—	11.3(5–15)
Chloride	155(100–203)	330(260–430)	174(110–290)	86	270(90–450)	151	99[b]	125(82–215)
Copper	(0.006–0.024)	—	Present	0.05[b]	—	0.51	—	—
Hematin iron	—	—	—	0.01	—	—	—	—
Iron	—	—	2(1–4)	0.16	—	(0.02–0.09)	—	—
Magnesium	14	11(5–14)	9(7–12)	8(2–13)	(3–9)	(0.3–0.7)[b]	—	3(1–4)
Manganese	—	—	<0.04	—	—	—	—	—
Phosphorus								
Total	112(90–120)	357	82	142	17(12–28)	(12.7–13.2)	—	—
Inorganic	11	17	(9–26)	12	—	1	—	—
Acid soluble	(27–110)	171	(33–46)	132	14(11–22)	—	—	—
Lipid	6	6	9	3	—	—	—	—
Potassium	89(56–107)	240(80–380)	172(50–387)	90(150–190)	(60–103)	(7.9–8.2)[b]	29[b]	158(76–255)
Sodium	281(240–319)	650(290–850)	258(152–370)	190(120–250)	(70–275)	(72–180)[b]	82[b]	103(60–183)
Sulfite	3% of ash	16(12–22)	10	—	—	—	—	—
Sulfur	—	—	—	—	8	(7.1–8.7)	—	—
Zinc	14(5–23)	2.2	2.8(2.6–3.7)	0.28	—	—	—	—

[a] Values are expressed in mg/100 ml unless otherwise indicated. The numbers without parentheses represent mean values; those within parentheses show the range of reported values. If too few data are available to firmly establish a mean value, only the range is given.
[b] Values are expressed in mEq/l.

Table 4 Nitrogenous Substances in Mammalian Seminal Plasma[a]

Substance	Man	Boar	Bull	Ram	Stallion
Adrenalin (μg/ml)	(1.0–2.1)	—	0.1	—	—
Ammonia	2	1.5(0.5–2)	2	2	1.3(0.3–2.4)
Carnitine	(11.5–53.5)	—	—	—	—
Creatine	17	—	3	21	3
Creatinine	—	0.3	12	—	12
Flavin	—	—	1.89	—	—
Ergothoneine	Trace	15(16–30)	Trace	Trace	7.6(3.5–14.7)
Histamine (μg/ml)	2	—	—	—	—
Niacin	—	—	(0.1–0.55)	—	—
Pantothonic acid	—	—	(0.232–0.466)	—	—
Riboflavin	—	—	(0.15–0.3)	—	—
Spermine	(20–250)	0	Trace	—	0
Spermidine	Trace	—	—	—	—
Thiamine	—	—	(0.12–0.20)	—	—
Urea	72	5	4	44	3
Uric acid	(2–6)	3	2.5(0.82–4.4)	(4–23)	—
Serotonin (ng/ml)	150	50	1000	500	—
Putrescine (nmol/ml)	230	—	—	—	—

[a] Values are expressed in mg/100 ml unless otherwise indicated. The numbers without parentheses represent mean values; those within parentheses show the range of reported values. If too few data are available to firmly establish a mean value, only the range is given.

indication of the contribution of the individual accessory genital glands to the entire ejaculate. Since the maintenance and secretion of these glands is androgen dependent, many researchers and clinicians have attempted to use various components as chemical indexes to assess testicular androgen secretions. However, there is little quantitative correlation between the plasma level of testosterone and the concentration of most of the constituents in seminal plasma. The reason for this is that subnormal levels of plasma testosterone can maintain normal concentrations of seminal constituents and exogenous testosterone does not increase their concentrations.

One of the principal seminal constituents in which such androgen correlations have been attempted is citric acid. It is present in high concentrations in the semen of most mammalian species and is an important regulatory agent in many biochemical systems including the tricarboxylic acid cycle. Citric acid also stimulates hexosediphosphatase, which is a key enzyme involved in the regulation of glucose synthesis from pyruvate, and it inhibits phosphofructokinase, the primary regulatory mechanism responsible for the Pasteur effect. Citric acid is believed to be beneficial to sperm motility; however, the actual mechanism is not known, for there is no evidence that it influences sperm metabolism. Since the prostate is the sole source of human citric acid and citric

Table 5 Free Amino Acids in Seminal Plasma

Amino Acid	Man (mg/100 ml)	Boar (μmole/100 ml)	Bull (μmole/100 ml)
Neutral			
Alanine	29.1	23.8	49
β-Aminobutyric acid	31.7	Trace	—
Cystine	3.4	—	Trace
Glycine	58.9	147.7	85
Isoleucine	62.4	4.7	—
Leucine	96.7	10.7	15
Methionine	3.8	2.4	—
Phenylalanine	28.2	6.4	16
Proline	26.9	3.8	Trace
Threonine	47.7	8.9	13
Tyrosine	51.4	5.9	13
Serine	111.6	14.3	62
Valine	49.8	17.1	15
Acidic			
Aspartic acid	99.8	18.0	32
Glutamic acid	197.7	204.7	279
Basic			
Arginine	79–90	1.2	11
Histidine	109.1	3.2	16
Lysine	152.1	Trace	13

acid is a strong chelator, Mann has proposed that by binding calcium ions, citric acid may function in preventing prostatic calculi and stones. Low levels of citric acid have been used to indicate chronic prostatitis or hypogonadism. However, one should remember that in other species seminal citric acid is derived from several accessory sex glands (see section 3 of this chapter).

As only low levels of cysteine, sulphite, uric acid, and glutathione are present in seminal plasma, the main dialyzable reducing agents are fructose and ascorbic acid. Ascorbic acid is essential for a large variety of biological oxidation processes throughout the body. A lack of ascorbic acid not only results in scurvy but also in alterations in connective tissue and a decreased resistance to some infections. Seminal ascorbic acid originates in the seminal vesicle; however, a specific reason for its high concentration compared with human blood plasma (0.7–1.2 mg/100 ml) is not known. Neither is a specific function known for the other water-soluble B vitamins that are present in seminal plasma: B_{12}, choline, inositol, niacin, pantothenic acid, riboflavin, and thiamine. Lactate and pyruvate are the only other seminal organic acids present in appreciable amounts. The concentration of lactate rapidly increases in the presence of viable spermatozoa since it is the end product of fructolysis. Conversely, pyruvate concentration is rapidly decreased, as it is metabolized by spermatozoa.

Table 6 Lipids in Mammalian Seminal Plasma[a]

Lipid	Man	Boar	Bull	Ram	Goat	Buffalo
Cardiolipin	0.3^b	—	7.5	—	1.1(0.2–2.8)	4.4
Cholesterol	47(16–94)	—	20(11–31)	6.2	—	—
Dehydroepiandrosterone (free)	0.032	—	0.039	0.086	—	59
Phospholipid	84(48–133)	(1.9–2.6)	149	127	57	—
Choline plasmalogen	0.31^b	—	49	22	10.9(8.9–12.8)	10.3
Ethanolamine plasmalogen	4.8^b	—	13.4	38	1.7(0.9–2.4)	2.4
Phosphatidylserine	4.4^b	(0.24–0.33)	1.9	—	2.3(0.6–3.3)	1.7
Phosphatidylethanolamine	3.3^b	(0.55–0.77)	8.3	37	5.1(0.6–8.9)	6.9
Phosphatidylinositol	0.7^b	—	1.1	—	2.0(0.8–3.4)	1.7
Phosphatidylcholine (lecithin)	3.0^b	(0.19–0.28)	36	13	10(9–12)	13
Prostaglandins						
A+B group	5.0	0.00002	0.00002	0.8–1.0	—	—
E group	5.3	0.00005	0.00004	0.75–3.6	—	—
F group	0.8	0.00002	0.00005	0.036–0.112	—	—
19-OH A+B group	20	—	—	0	—	—
Sphingomyelin	17.16^b	(0.6–0.89)	17.2	11	8.6(6–11.3)	7.8
Testosterone (free)	0.068	0.102	0.077	—	—	—
Total lipid	185(166–206)	(4.7–6.1)	104(60–142)	—	136	150
Fatty acids (volatile) (mEq/l)	0.73(0.5–0.95)	—	0.8(0.2–1.4)	1.6(0.7–2.7)	—	—

[a] Values are expressed in mg/100 ml unless otherwise indicated. The numbers without parentheses represent mean values; those within parentheses show the range of reported values. If too few data are available to firmly establish a mean value, only the range is given.
[b] In μg atoms lipid P/100 ml.

103

Numerous electrolytes help to make seminal plasma an isotonic solution (Table 3). The importance of the various ions has been extensively studied in reference to their effects on sperm motility, respiration, and viability during sperm dilution and washing experiments. A relatively narrow range of concentrations of potassium, magnesium, and phosphate are required for optimal bull and ram sperm motility. Calcium decreases sperm motility and respiration; however, the calcium that is in semen is probably present in a chelated form. Chloride ions are probably used for maintaining the osmotic pressure in most species except the boar, where the concentrations are minimal. Copper, iron, and phosphorus are secreted into the semen by the ampullary glands of the bull, and the majority is either bound to various enzymes or in the case of phosphorus, is mainly found as phosphorylcholine or glycerylphosphorylcholine with a trace in adenosine phosphate. Zinc is found in very high concentrations in human and dog seminal plasma and is of prostatic origin. Although a small amount of zinc is associated with seminal carbonic anhydrase, the majority is bound to low-molecular-weight ligands, which are probably peptides (Parrish, unpublished observations).

The presence of low concentrations of bicarbonate (approximately 4% of the osmotically active anions in human prostatic fluid) can function to stimulate oxygen consumption and fructolysis in spermatozoa, whereas larger concentrations cause a temporary quiescence of the spermatozoa by decreasing their metabolism. This latter effect has been used in diluents for preserving bull and boar spermatozoa. The ammonia content of seminal plasma is higher than that found in blood but much less than that found in urine (67 mg/100 ml). Some of it is obtained as a by-product from sperm oxidation of seminal amines and amino acids.

6 NONPROTEIN NITROGEN

Low-molecular-weight nitrogenous substances are rare in most other body fluids but are present in high concentrations in mammalian semen (Table 4). The polyamines (spermine and spermidine) originate in the prostate, and the largest concentrations are found in human and dog seminal plasma. Spermine relaxes smooth muscle and can bind to ribosomes, RNA, DNA, cell walls, and membranes. The "musk" odor of human semen is believed to result from the oxidation of spermine by seminal diamine oxidase. Oxidized spermine and spermidine are bacteriostatic and at high concentrations are capable of immobilizing and killing spermatozoa.

With time, spermine phosphate crystals form in human semen. This results from the accumulation of inorganic phosphate that is released from phosphate esters by the action of seminal phosphatases. These crystals were first observed by Leeuwenhoek in 1677 and have been used in forensic medicine for the identification of seminal stains. Choline is another nitrogenous base that has been used in forensic medicine and is detected by the Florence reaction. Phosphorylcholine and glycerolphosphorylcholine are bound forms of choline present in the semen of most species, and the majority originates from the seminal vesicles and epididymis, respectively.

Ergothioneine is a reducing substance that is present in high concentrations in semen from boars, stallions, jackasses, zebras, moles, hedgehogs, and opossums but is not found in appreciable quantities in semen from bulls, men, or rams. It is a chemical marker for boar seminal vesicle secretions, but in stallions and jackasses the majority originates from the large ampullae present in these species. It is possible that ergothioneine functions in protecting spermatozoa from the various thio-agents that may have a detrimental effect on the sperm whereas reduced glutathione could serve a similar function in human semen.

There are many other small peptides in seminal plasma. One has been shown to be a strong inhibitor of the angiotensin-converting enzyme in human seminal plasma. However, the function of the other peptides is presently not known, and many are probably degradation products resulting from the presence of active proteinases. C-peptide, the inactive conversion product of proinsulin to insulin, has recently been demonstrated by immunoassay in human, boar, and bull seminal plasma in concentrations of 4.9–5.1, 5.5, and 3.0 ng/ml, respectively (Polakoski and Clark, unpublished observations). The actual concentrations in boar and bull semen are probably higher than the values given here because there is rather poor cross-species immunoreactivity in the C-peptide because of differences in amino acid composition and sequence.

Almost all the common amino acids are found in seminal plasma (Table 5). Their concentrations in human seminal plasma greatly exceed those found in blood plasma. When expressed as milligrams per 100 milliliters, the comparative amounts in seminal plasma and serum are as follows: neutral amino acids, 638 vs 19; acidic amino acids, 280 vs 10; and basic amino acids, 340 vs 6. Glutamic acid is the most common free amino acid in boar, bull, camel, human, and ram seminal plasma. It has been shown to originate from the epididymal secretions in bulls and may play a role in maintaining osmotic pressure and pH. Alanine originates from the bull ampulla and seminal vesicle fluid. Both alanine and serine have been shown to increase dog sperm motility, whereas arginine stimulates the motility of human and rabbit spermatozoa.

Carnitine has been used as an indicator of epididymal secretions in the bull and rat, although it is also secreted by the seminal vesicles in man. Its possible seminal function is unknown; however, it is a growth factor for some organisms and is used as a carrier molecule for transferring acyl groups across membranes. Creatine and creatinine are both found in seminal vesicle secretions of most species. In other body systems, creatine is normally found intracellularly (only 0.2–0.9 mg/100 ml in human blood) and is used in the storage and transmission of phosphate-bond energy. It is excreted as creatinine, which is present in concentrations of 1–2 mg/100 ml in human blood plasma but is about 70 times higher in urine.

Other low-molecular-weight nitrogenous substances in seminal plasma include urea, uric acid, ammonia, and the pharmacological substances adrenaline, noradrenaline, serotonin, and histamine.

7 CARBOHYDRATES

Mammalian seminal plasma contains many carbohydrates, which are present in either bound or free forms. These include arabinose, fructose, fucose, galactose, glucose, glucuronic acid, hexosamine, inositol, mannose, ribose, sialic acid, sorbitol, and uronic acid (Table 2). Carbohydrate is present in bound form as polysaccharide or in glycopeptides and glycoproteins, most of which originate from the seminal vesicles. The concentrations of bound carbohydrates in milligrams per milliliter found in boar, bull, human, and stallion semen are 1–1.3, 0.3–1.5, 0.7, and 0.4, respectively. After acid hydrolysis, the main constituents are galactose, hexosamines and sialic acid. Small quantities of fucose, glycerol, mannose, and uronic acid have also been observed. Contrary to several reports, it appears that only trace quantities of glycogen (less than 0.2 mg/ml for human semen) are present in mammalian semen. Glycosaminoglycans are also found in human and boar seminal plasma; however, their concentrations have not been quantified (Polakoski and Wincek, unpublished observations).

The majority of the carbohydrate in semen is found in the free form (Table 2). Over 35 years ago, Mann identified the major seminal carbohydrate as fructose. Fructose is believed to be derived from blood glucose and is found in concentrations as high as 1.2 g/100 ml in semen of diabetics but returns to normal levels upon insulin treatment. The majority of the fructose is usually of seminal vesicle origin. However, small amounts can originate from the ampullary glands, whereas in some species like the rabbit and rat, the prostate may also secrete fructose, as can the coagulating gland of the mouse. Dog and stallion semen contain little or no fructose. The spermatozoa from all mammalian species so far tested, including the dog, can readily metabolize fructose. Because of the large concentrations of fructose in most seminal plasma, it is generally believed that this is the major source of exogenous glycolytic energy available to the spermatozoa. However, glucose (resulting from the hexokinase preference by spermatozoa) accounts for as much as 40% of the seminal sugar utilized by human spermatozoa *in vitro* (9). Moreover, the spermatozoon's glucose utilization may be particularly important during its residence in the female reproductive tract, where fructose is normally not detectable. Seminal fructolysis by spermatozoa is correlated with the number of motile sperm cells, and there are reports that either substantiate or refute its correlation to fertility. However, normal levels of fructose in a human ejaculate indicate that the seminal vesicles are actively secreting fluids. Numerous case reports document the absence of fructose in semen samples from patients with congenital bilateral absence of the vas deferens and seminal vesicle or obstruction of the ejaculatory ducts.

Inositol is another carbohydrate that is produced in the seminal vesicles and is found in both free and bound forms in the seminal plasma of most mammalian species. Although inositol is a growth factor for certain microorganisms and several animals, it is not believed to be metabolized by spermatozoa. However, it may function in the regulation of the osmotic pressure of seminal plasma, especially in species such as the pig that have high levels of inositol but almost no sodium chloride.

8 LIPIDS

Almost all the major classes of lipids are present in mammalian semen (Table 6). Human semen contains the highest concentration of total lipid (10), most of which is of prostatic origin. Seminal plasma from the boar contains the least, with relatively comparable levels in most of the other species. However, there are rather large species differences in specific lipids. The reported lipid values are rather variable because of differences in extraction procedures, the amount of contamination by nonsperm bodies (cytoplasmic droplets have been estimated to account for up to two-thirds of the extracellular phospholipid in bull semen), and even in the length of abstinence prior to collection of the ejaculate.

The free fatty acids are metabolically the most active and may serve as an important energy source for spermatozoa. The most prevalent of the saturated free fatty acids is palmitic acid, which is rapidly incorporated into spermatozoa. The long-chain polyunsaturated fatty acids are abundant in semen of all the species investigated and contain more 20- and 22-carbon acids than the polyunsaturated fatty acids of any other mammalian fluid. Linoleic acid, an "essential fatty acid," is present in large concentrations and is required for the biosynthesis of arachidonic acid, a precursor to the prostaglandins. However, bull seminal plasma contains large quantities of both linoleic and arachidonic acid but only barely detectable amounts of prostaglandins.

The most widely studied unsaturated fatty acids in semen are the prostaglandins. These pharmacological agents have been associated with almost every organ; however, the richest known source is the secretion of the human seminal vesicles. Thirteen different forms of prostaglandins have been isolated from, and eight additional forms identified in, human seminal plasma, all of which can be grouped into four series: prostaglandin A, B, E, and F (PGA, PGB, PGE, and PGF). The prostaglandins affect smooth muscle contraction, and the members of each group generally express similar biological properties. Some have also been shown to influence blood pressure and to regulate cellular metabolism, including the adenylyl clycase-cAMP system. Most of the prostaglandins have a notable lack of tissue specificity and have a very short half-life *in vivo*; more than 96% of PGE is inactivated within 90 seconds. It has been postulated that the seminal prostaglandins are involved in the smooth muscle contractions that occur during ejaculation, or that they may aid sperm transport in the female reproductive tract. Furthermore, *in vitro* studies have shown that the PGFs can influence sperm penetration into cervical mucus. Available evidence is highly controversial, however, for any physiological effect *in vivo*. Numerous attempts have been made to obtain correlations between the concentration of prostaglandins in semen and male fertility or infertility, but direct relationships are still argued. If a relationship is present in man, it may be possible to regulate the concentration of seminal prostaglandins, since it has recently been shown that the levels of PGE's and PGF's can be lowered by administering aspirin, a prostaglandin synthetase inhibitor.

The ram is the only species besides man that has significant levels of prostaglandins in its semen. A significant increase in seminal PGE's occurs during the breeding season. The addition of physiological concentrations of PGE_2 and $PGF_{2\alpha}$ to diluted ram semen has been reported to increase the conception rate following artificial insemination.

Sphingomyelins are usually associated with brain and nerve tissue. The phosphoglycerides are characteristically major components of cell walls and normally are not present in appreciable concentrations elsewhere. However, these phospholipids are present in reasonable concentrations in mammalian semen. The major phospholipids in boar and human seminal plasma are sphingomyelin, phosphatidylethanolamine, and phosphatidylserine; choline plasmalogen is the most prevalent in buffalo, bull, and goat, with ethanolamine plasmalogen the major phospholipid in ram and rhesus monkey seminal plasma. Cardiolipin, which is normally associated with mitochondria, has been found in human, bull, goat, and buffalo seminal plasma.

Another class of seminal lipids is the gangliosides, which contain sialic acid and can function to detoxify certain toxins. They are present in buffalo, bull, and goat seminal plasma in concentrations of 0.92, 1.08, and 0.90 mg/100 ml, respectively. The gangliosides may be the spasmogens previously described in horse and rat semen. The very insoluble wax esters are also present and account for approximately 5–7% of the total lipid.

Human semen has the largest concentration of cholesterol, which is the most prevalent steroid in seminal plasma. Since cholesterol is almost totally insoluble in aqueous media, it would be interesting to determine if it is transported by means of micelles or possibly through lipoprotein complexes. Lipoproteins are present and, if they are similar to those in blood serum, they may form hydrophilic lipoprotein complexes that could be used to transport large quantities of hydrophobic lipid in seminal fluid (possibly the lipid bodies).

Several other steroids have also been found in bull, human, rabbit, and rat seminal plasma. The most notable are testosterone and dihydrotestosterone. However, measurable quantities of androstenedione, cortisol, estrone, estriol, estradiol-17B, pregnenolone, and dehydroepiandrosterone have also been found in the seminal plasma of various species. It should be noted that the majority of seminal testosterone and several other steroids are found in conjugated forms, presumably bound to a type of androgen-binding protein; consequently, many of the previously reported values need to be reexamined.

9   PROTEINS

9.1   Nonenzymatic

Seminal plasma is a rich source of a wide variety of proteins (1, 3, 7), and the concentration of protein in most species rivals that in human blood plasma (7.2

g/100 ml). The majority of the protein originates from the seminal vesicles, thus explaining the low concentration of protein in dog semen. Similar to the blood proteins, the major classes have been separated and studied by paper, starch, immuno- and cellulose acetate electrophoresis as well as by salt precipitation, gel filtration, ion-exchange chromotography, ultracentrifugation, and isoelectric focusing. Species-specific differences in the relative amounts of the major protein bands and slight differences among the same species are present. A number of seminal proteins are similar to those in blood plasma. All animal species appear to have prealbumin, albumin, and the α-, β-, and γ-globulins, transferrin, and IgG immunoglobulins. IgA and IgM are not present in boar seminal plasma but human semen contains IgE, IgA, secretory IgA, and occasionally IgM. Other serum proteins present in human seminal plasma include α_1-antitrypsin, α_{1x}-antichymotrypsin, α_1-glycoprotein, B_1A/C-globulin, follicle-stimulating hormone (FSH), luteinizing hormone (LH), inhibin, insulin, kininogen, β-lipoprotein, orosomucoid, prolactin, relaxin, steroid-binding proteins, and two proteins with the electrophoretic mobility of β_1- and α_2-globulins. The Lewis blood group substances (Le^a and Le^b) and the ABO-group antigens are present in semen of "secretors" (about 80% of the population secrete these antigens into various body fluids), and the antigens can be matched to those found in saliva for forensic purposes. Serum proteins not detectable by immunochemical methods in human semen include antithrombin III, B-globulin, C/B-globulin, A-globulin (C_3 component of complement), C_{1s} inactivator, factor XIII, fibrinogen, and plasminogen. Consequently, the complement system is not functional. Also, the mechanism of coagulation of semen must differ from that of blood even though they both undergo a cross-linking of γ-carboxyl groups of glutamic acid (at least during the formation of the rat coagulum).

A number of proteins are found in seminal plasma but not blood plasma. Several forms of lactoferrin are present in human, bull, and boar semen, and its heterogeneity results from variations in the amount of sialic acid as treatment with neuraminidase converts the mixture to one electrophoretically homogeneous form. Neuraminidase also changes the electrophoretic patterns of five to seven other specific seminal glycoproteins. Boursnell and his colleagues have demonstrated a glycoprotein from boar seminal plasma that forms a copious precipitate when cooled or dialyzed at an ionic strength of less than 0.3. They have also isolated and partially characterized a hemagglutinin that originates from boar seminal vesicles. It has a molecular weight of 68,000 and an isoelectric point of 9.4. This hemagglutinin is believed to be capable of binding to and causing a release of enzymes from spermatozoa. Another sperm-binding protein that is present in semen from the boar, bull, man, rabbit, and stallion prevents sperm head-to-head agglutination. This antiagglutinin is a nondialyzable mucoprotein that contains sulphate residues and must be in its reduced form (pI = 3.0) to bind to spermatozoa. However, the active portion of the macromolecule is not protein. A heat-labile hemolytic protein with a molecular weight of 80,000 and an isolective point of 4.5 has been isolated from bull semen and probably is derived from the seminal vesicles. This protein may have phospholipase activity. Bull semen also contains a nondialyzable, heat-labile, toxic protein that depresses sperm viability. Cysteine protects the sperm from this toxin, which may be the causative factor of poor sperm storage of some animal species (for instance, the boar) but its presence in the semen of these species remains to be established.

Several other inhibitory proteins are present. Bull seminal plasma has a protein ("inhibin") that inhibits gonadotropin action. A protein that inhibits the sperm associated estradiol-17β dehydrogenase has been found in the seminal plasma of the bull, rabbit, dog, man, and stallion (order of decreasing concentration). Seminal plasmin is an antimicrobial protein in bull seminal plasma that inhibits RNA synthesis in *Escherichia coli*. Several low-molecular-weight, acid-stable proteinase inhibitors have been isolated from the seminal plasma; they bind to spermatozoa and are removed during the capacitation process in the female reproductive tract (Chapter 8).

The majority of antibodies prepared against seminal plasma antigens (11) do not react with serum proteins; however, before the antigens are considered specific, it must be determined whether they are absent from other tissues and fluids. Antibodies to several antigens believed to be specific for semen cross-react with fluids from the female reproductive tract, indicating that both fluids may contain identical proteins. This may be a reason why the female reproductive tract only rarely develops antibodies against seminal plasma. In addition, several of the seminal proteins poorly stimulate the production of precipitating antibodies. The antigens from the same species of animal but different breeds (for instance, the ram) may have different degrees of antigenicity.

The possible functions of some of the seminal proteins have been alluded to; however, most proteins have no known biological significance. Albumin (3.7–4.7 g/100 ml serum) is known to function in the maintenance of osmotic equilibrium in blood and to transfer various constituents, most notably fatty acids. However, the levels of albumin in human seminal plasma are only 30–97 mg/100 ml, indicating that the concentration is probably too low for albumin to have the same functions as it has in serum. Conversely, other proteins normally associated with blood plasma are present in higher concentrations in seminal plasma. For example, immunoreactive insulin occurs in two to four times higher amounts in human semen than in serum, and high levels are also present in boar and bull seminal plasma. One report claims that insulin regulates sperm metabolism and stimulates sperm penetration of cervical mucus; however, others, using different methodology, have not been able to substantiate these results.

Several complications have led to anomalous results in the study of proteins in seminal plasma. The first concerns the presence of the seminal gel and debris that are seen in most species. A common practice is to remove these by passing the semen through glass wool or a mucin, both of which can bind proteins and give misleading results. The second problem concerns the presence of several proteolytic enzymes, which are present in especially high concentrations in dog and human semen. Rapid proteolysis occurs in human seminal plasma even during refrigeration. A twofold increase in the nonprotein nitrogen (peptides and amino acids) occurs if whole semen is incubated for 15 min at 20°C or 3 hr at 1°C. Consequently, care must be taken to obtain reproducible results. A smaller amount of hydrolysis occurs in each fraction of a split (compartmentalized) ejaculate, as the hydrolytic enzymes originate in the prostatic fluid, whereas the majority of the other proteins are secreted by the seminal vesicle.

9.2 Enzymatic

A few of the many enzymes present in mammalian seminal plasma (1, 5) are listed in Table 7 with their approximate activities. The values given are for comparison purposes only. Occasionally different units are given for the same enzyme, as it is not unusual to find an enzyme assayed by a number of different procedures.

Animals that have a seasonal breeding period are particularly subject to large differences in their seminal enzyme concentrations. Although seminal catalase is not found in measurable amounts in most mammals, it is present in ram seminal plasma at a concentration of approximately 13 μg/ml in the winter and 6–7 μg/ml in the spring. This is surprising because one would expect the greatest activity of catalase when conception is desired since spermatozoa are very susceptible to hydrogen peroxide damage, a by-product of their oxidative deamination of amino acids.

A number of enzymes found in human seminal plasma are present in extremely high concentrations (3) when compared with other tissues and blood plasma. Present in concentrations that exceed those in sera by at least 50-fold are γ-glutamyltranspeptidase, acetylcholinesterase, creatine phosphokinase, nitrocatecholsulfatase, and acid phosphatase. The presence of acid phosphatase (more than 50 King-Armstrong units/ml) in the vaginal fluid of suspected rape cases is considered evidence of recent intercourse by many courts. In fact, acid phosphatase and creatine phosphatase have been used to detect seminal stains on clothing six months after they have dried. Acid phosphatase is used as a marker for human prostatic function and, since it is immunochemically different from the serum enzyme, blood levels of this enzyme are being used to detect prostatic carcinoma and to monitor various treatments of this disease. In other species in which the prostate contributes only a small amount to the semen, as in the bull, the acid phosphatase levels are much smaller than the alkaline phosphatase levels. Not all of the seminal enzymes are as stable as the phosphatases noted above. For example, human seminal adenosine triphosphatase activity is decreased by 60% after a 45-minute incubation at 37°C. This is probably the result of the high concentrations of proteinases which both human and dog seminal plasma are noted for.

Seminal proteolytic activity is the result of two plasminogen activators, a chymotrypsin-like enzyme called seminin, a collagenase-like proteinase, a kinase, and several aminopeptidases. The pepsinogen-pepsin system is another group of proteinases present in human seminal plasma. The pepsin activity of normal seminal plasma approximates that in gastric juice, although the pH of

Table 7 Common Enzymatic Activities Found in Mammalian Seminal Plasma[a]

Enzyme	Man	Boar	Bull	Stallion	Dog	Rabbit	Buffalo	Goat
Acid phosphatase	66(49–72)[c]	—	25[c]	0.07–0.86	—	0.85[d]	12(5–26)[e]	—
Alkaline phosphatase	1.3–12[c]	2149[e]	106[c]	0.25–2.55	—	1.80[d]	8(4–22)[e]	—
Aldolase	—	—	—	—	—	—	70[f]	—
Amylase	9(3–25)[g]	—	—	—	—	—	—	—
Glutamic oxaloacetic transaminase	14(7–25)[h]	11.1	1751(1200–2900)[b]	—	—	329(40–1120)[b]	167[i]	—
Glutamic pyruvic transaminase	2.5(1.0–4.0)[h]	8.1	—	—	—	—	34[i]	—
Lactate dehydrogenase	1.87(0.68–4.75)	219[k]	1689(1560–1805)[k]	—	—	—	1671[k]	—
DNase I (units × 10⁻⁴/ml)	248	—	49	—	3.8	58	—	—
DNase II (units × 10⁻⁴/ml)	376	—	100	—	34	98	—	—
β-N-Acetylgalactosaminidase	11[l]	—	44[l]	—	—	—	2.4[i]	115[l]
β-N-Acetylglucosaminidase	7288[m]	101,500[m]	15,200[m]	468[m]	3250[m]	15,680[m]	32[l]	1500[l]
α-Galactosidase	0.2[l]	0[m]	0.1[l]	0[m]	0[m]	257[m]	0[l]	0.4[l]
β-Galactosidase	68[d]	48[d]	320[d]	38[d]	68[d]	837[d]	1.6[l]	28[l]
α-Glucosidase	82[d]	16[d]	0.7[l]	0[d]	20[d]	54[d]	0.2[l]	0.6[l]
β-Glucosidase	60[d]	40[d]	0.2[l]	0[d]	0[d]	0[d]	0.1[l]	0.4[l]
β-Glucuronidase	232[n]	0[n]	1028[n]	31[n]	12[n]	70[n]	3.2[l]	2.8[l]

α-Mannosidase	218[d]	1112[d]	628[d]	228[d]	405[d]	1320[d]	2.2[l]	5.4[l]
β-Mannosidase	10[l]	0[m]	263[m]	0[m]	15[m]	1022[m]	2.4[l]	14[l]
Hyaluronidase	1.8[l]	—	5.2[l]	—	—	—	16.8[l]	20[l]
α-L-Fucosidase	4.3[l]	—	4.5[l]	—	—	—	3.6[l]	3.8[l]
β-D-Fucosidase	0.2[l]	—	0.5[l]	—	—	—	0.2[l]	1.3[l]
5'-Nucleotidase	—	—	29	—	—	—	—	—

[a] Values are expressed as international units/ml unless otherwise indicated. The numbers without parentheses represent mean values; those within parentheses show the range of reported values. If too few data are available to firmly establish a mean value, only the range is given.

[b] Decrease in optical density at 340 min 40.001/min.

[c] Sigma units.

[d] In μg nitrophenol liberated per hour at 37°C.

[e] King-Armstrong units.

[f] Sibley-Lehninger units.

[g] Street-Close units per milliliter per 15 min at 37°C.

[h] King units.

[i] In μg pyruvate.

[j] Reitman-Frankel international units.

[k] Berger-Broida units.

[l] In μmoles aglycone per N-acetylglucoseamine liberated per hour at 37°C.

[m] In μmoles phenol liberated per hour at 37°C.

[n] In μgrams phenolpthalein liberated per hour at 37°C.

semen does not become acidic enough (pH less than 3.0) for the enzyme to be of physiological importance. The plasminogen activators have been highly purified and shown to be immunologically identical to urokinase even though they have slightly different chemical properties. Partially purified seminin has a molecular weight of 33,000, is unstable below pH 4.0, and is partially inhibited by α_1-antichymotypsin and hydrocinnamic acid but not by the other natural or synthetic proteinase inhibitors. Recent data indicate that seminin, but not the plasminogen activators, is involved in the liquefaction of the seminal coagulum of man. Seminin has also been implicated in sperm penetration of cervical mucus but the evidence is circumstantial. The proposed function of other seminal enzymes is also insufficiently substantiated. For example, *in vitro*, amylase stimulates lactate production by rabbit spermatozoa which is beneficial for the maintenance of sperm motility. Exogenous β-amylase and β-glucuronidase have been reported to increase the conception rate of artificially inseminated bull semen either when added separately or combined. β-Glucuronidase may also aid in the *in vitro* capacitation of epididymal hamster spermatozoa.

Many of the enzymes found in seminal plasma have a number of interesting biochemical properties not shared by similar enzymes from other tissues or fluids. For example, the 5'-nucleotidase present in high concentrations (2900 units/ml) in bull seminal plasma is found in three active forms that have pH optima at both 8.6 and 9.6 in the presence of magnesium chloride but only one optimum at pH 8.6 in its absence. The enzyme is believed to be a coating antigen for bull spermatozoa. Bull seminal ribonuclease (RNase BS1) has a molecular weight of 29,000 and is the only ribonuclease that is present as a dimer. The seminal nicotinamide adenine dinucleotide nucleosidase (NADase) from bull semen differs from the brain enzymes by exhibiting first-order kinetics. There is also evidence to suggest that bull alkaline endonuclease, but not the acid endonuclease, can activate chromatin for DNA synthesis *in vitro*.

Since mammalian seminal plasma contains proteinases, lipases, phospholipases, nucleases, and glycosidases, it is capable of degrading almost all classes of macromolecules. Although spermatozoa usually do not significantly contribute to the physiochemical composition of semen (with the possible exception of rams), membrane damage of spermatozoa caused by standing at room temperature, by freezing, by refrigeration, or by centrifugation may result in a loss of intracellular enzymes from spermatozoa. Indeed, up to 50% of the lactate dehydrogenase activity found in human seminal plasma is LDH-X, which originates from the spermatozoa. The spermatozoon also releases all of the hyaluronidase and cytochrome C that is found in seminal plasma. The assay of these enzymes has been used to advantage for monitoring cell damage that results from various types of storage and handling procedures.

10   FACTORS

Several of the seminal constituents that have various biological effects on different systems are referred to as factors until their composition is known. An example is the oxytocic (uterus-stimulating) factor (not a prostaglandin) that is present in ram, human, and rat semen, and was shown to be a type of ganglioside. The material in boar seminal vesicle fluid and seminal plasma that lowers the frequency and amplitude of the spontaneous contractions of uterine strips remains a factor because its chemical composition is unknown. Hamster seminal fluid contains a factor that prevents damage to spermatozoa by dilution and another that is nondialyzable, partially heat-labile, and decreases the aerobic metabolism and motility of epididymal spermatozoa. A factor is present in ram and human semen that inhibits the action of prostaglandins. However, the factors that have received the most attention are known as decapacitation factors (DFs). Spermatozoa can be considered to be capacitated when they have the ability to penetrate the outer layers of the egg and can fertilize the female gametes (Chapter 8). Under normal conditions, spermatozoa become capacitated in the female reproductive tract, a process that in most species requires several hours or more. Several seminal factors when added to capacitated spermatozoa can prevent them from penetrating the outer layers of an ovum. Such a compound is called decapacitation factor (DF). Such factors include a large-molecular-weight glycoprotein as well as a number of low-molecular-weight components that have been obtained from hydrolytic digestion of larger macromolecules. Several low-molecular-weight acid-stable protein inhibitors that inhibit acrosin have been isolated from seminal plasma and are types of DF. Membrane vesicles obtained from semen demonstrate DF activity and, under certain conditions, cholesterol can be considered to be a DF. It is hoped that future research will determine the mechanisms of the various decapacitation factors so that new avenues of fertility regulation can become available.

11 CONCLUSIONS

A relatively large amount of descriptive knowledge on the biochemistry of seminal plasma is available. This information has been extremely useful for developing methods of sperm preservation, diagnosing pathological conditions, and aiding in the management of fertility and infertility. However, with the exception of a few isolated instances, very few of the mechanistic aspects are known.

Several significant areas that require further investigation have been alluded to in this chapter. One example is the mechanisms of coagulation formation and liquefaction of human semen; these mechanisms are virtually unknown. A thorough understanding of this process has important implications both in the treatment of some cases of infertility and in possible contraceptive approaches. More information on seminal plasma may also allow better methods of preserving semen in many important species. This is required to ensure a maintenance of reproductive integrity in humans and could be extremely beneficial in a number of agriculturally important animals such as the buffalo (the majority of milk in India is from buffalo cows; semen extenders are not very efficient nor can fertility be routinely maintained when buffalo semen is frozen).

An area of tremendous importance that needs to be investigated concerns the transport of constituents from the accessory sex glands to the blood plasma and vice versa. It is known that prostatic acid phosphatase can be found in serum and is eliminated after prostatic massage or during certain pathological conditions. However, it appears that many of the other constituents of the accessory sex glands are probably not secreted into serum, that is, prostaglandins, hemolytic proteins, and so forth. In one study, all the individuals injected with tetanus toxoid were found to have antibodies to the toxoid in their sera; however, some also had antibodies to the toxoid in their semen. One wonders whether the antibodies were transported to the accessory sex glands or if the toxoid entered the tissue and local antibodies were produced toward it.

Answers to these problems may also be beneficial for the study of autoantibodies directed to spermatozoa, because there are indications that some cases of male infertility are associated with measurable levels of sperm antibodies in the sera of the men. If these antibodies could also enter into the semen, they could have adverse effects on the sperm. A number of mutagenic and toxic compounds are present in the environment and, if these compounds enter into the semen, they could result in infertility or worse situations, such as the reproductive problems that have been associated with males employed in various chemical industries. A somewhat related problem concerns the inhibition of seminal prostaglandin synthesis by aspirin. Conversely, it is likely that other compounds may stimulate prostaglandin synthesis; if so, these could be used in some cases of infertility. Alternately, stimulation of the synthesis of seminal factors that are detrimental to spermatozoa may be useful as male contraceptives.

Various hormones are currently believed to be synthesized only in specific organs, then transported to various target tissues through the bloodstream. This probably occurs in the male reproductive system in some instances, since testosterone injections can maintain the secretory activity of the accessory sex glands in castrates. However, other hormones that are present in much higher concentrations in semen than in serum (insulin, FSH, and LH) either come from serum and are concentrated and repackaged for resecretion into semen by an interesting but yet unknown mechanism or are synthesized in the accessory sex glands themselves. If local synthesis is not occurring, one wonders about the presence in semen of C-peptide, a degradation product of proinsulin that has no known function or transport system. Investigations into these processes could give insight into various questions in endocrinology.

More investigations into the nonsperm particles and cells in seminal plasma could lead to better diagnosis and monitoring of treatments in a number of pathological conditions. For example, it may eventually be possible to have a type of "pap smear" for the detection of prostatic cancer.

REFERENCES

1   Mann, T., *The Biochemistry of Semen and of the Male Reproductive Tract*. Wiley, New York (1964).

2   Hafez, E. S. E. (ed.), *Human Semen and Fertility Regulation in Men*. Mosby, St. Louis (1976).

3 Polakoski, K. L., Syner, F. N., and Zaneveld, L. J. D., Biochemistry of Human Seminal Plasma, *in* Hafez, E. S. E. (ed.), *Human Semen and Fertility Regulation in Men*, p. 133. Mosby Co., St. Louis (1976).

4 McKenzie, F. F., Miller, J. C., and Bauguess, L. C., *The Reproductive Organs and Semen of the Boar*, Bulletin No. 279. University of Missouri Experimental Station (1938).

5 Perry, E. J., Factors Influencing the Quality and Quantity of Semen, *in* Perry, E. J., (ed.), *The Artificial Insemination of Farm Animals*, 4th ed., p. 76, Rutgers University Press, New Brunswick (1968).

6 Williams-Ashman, H. G., Wilson, J., Beil, R. E., and Lorand, L., Transglutaminase reactions associated with the rat semen clotting system: modulation by macromolecular polyanions. *Biochem. Biophys. Res. Comm.* **79**, 1192 (1977).

7 Tauber, P. F., and Zaneveld, L. J. D., Coagulation and Liquefaction of Human Semen, *in* Hafez, E. S. E. (ed.), *Human Semen and Fertility Regulation in Men*, p. 153, Mosby, St. Louis (1976).

8 Mann, T., Physiology of Semen and of the Male Reproductive Tract, *in* Cole, H. H., and Cupps, P. T. (eds.), *Reproduction in Domestic Animals*, 2nd ed, p. 277, Academic Press, New York and London (1969).

9 Peterson, R. N., and Freund, M. Metabolism of Human Spermatozoa, *in* Hafez, E. S. E. (ed.), *Human Semen and Fertility Regulation in Men*. p. 176, Mosby, St. Louis (1976).

10 White, I. G., Darin-Bennett, A., and Poules, A., Lipids of Human Semen, *in* Hafez, E. S. E. (ed.), *Human Semen and Fertility Regulation in Men*, p. 144, Mosby, St. Louis (1976).

11 Hekman, A., and Rumke, P., Seminal Antigens and Autoimmunity, *in* Hafez, E. S. E. (ed.), *Human Semen and Fertility Regulation in Men*, p. 245, Mosby, St. Louis (1976).

CHAPTER FIVE

THE SPERM HEAD

ASOK K. BHATTACHARYYA

LOURENS J. D. ZANEVELD

*Departments of Physiology and Biophysics
and Obstetrics and Gynecology, College of Medicine
University of Illinois at the Medical Center
Chicago, Illinois*

1 INTRODUCTION

The head of a spermatozoon varies greatly in shape and size among different mammals, for example, it is oval in the rabbit, boar, and bull, resembles a long cylinder in fowl, and is curved in the mouse, hamster, and rat. In some species, for instance the guinea pig and squirrel, the sperm head has a more bizarre shape. Size variations also exist. For instance, the human sperm head is about 5.0 μm long, 3.5 μm wide, and 2.0 μm thick, whereas that of the bull is approximately 8.5 μm long, 4.0 μm wide, and 0.3–0.5 μm thick. Whether the diversity in shape and size of the mammalian sperm head is of significance in the fertilization process remains to be established. It could be of importance in selective sperm penetration through cervical mucus and the layers surrounding the oocyte.

The sperm head can be divided into two parts: the nucleus and the surrounding membrane structures (Fig. 1). The nucleus contains the haploid number of chromosomes and the nucleoproteins. These are so densely packed that individual chromosomes are not visible even on electron microscopy. Small, clear regions of irregular shape are often present in the sperm nucleus. These are referred to as the nuclear vacuoles. In the human sperm nucleus, these vacuoles are very common and vary significantly in shape and size, averaging 0.5 μm in diameter.

The membrane structures of the sperm head consist of (1) the plasma membrane that covers the entire surface of the sperm head; (2) the acrosome, a baglike structure that surrounds the anterior portion of the nucleus; (3) the postnuclear cap, a cytoplasmic sheath that covers the posterior part of the nucleus; and (4) the equatorial segment, which represents the area of overlap between the posterior cap and acrosome. The acrosome consists of an inner acrosomal membrane that is in close contact with the nuclear membrane and an outer acrosomal membrane that is closely apposed to the plasma membrane. The acrosome proper is located in between these two membranes.

The sperm head is connected to the tail by a narrow region called the neck, which is the most fragile portion of the spermatozoon. The neck contains a centriole from which the tail fibers and microtubules originate.

The head of the spermatozoon has two important functions. First of all, within its nucleus, it carries paternally derived, genetic information to the egg. Secondly, its membranes and associated enzymes are essential for the penetration of the spermatozoon through the layers surrounding the egg. As described in more detail later in this chapter and in chapter 8, when spermatozoa that have been capacitated (activated) in the female genital tract come in contact with the egg, the plasma membrane and outer acrosomal membranes fuse, vesiculate and disappear. This is called the acrosome reaction without which the spermatozoon is unable to pass through the zona pellucida of the egg. Biochemical changes occur in the sperm head surface during the capacitation process. Certain enzymes are associated with the acrosome that appear to be essential for fusion of the spermatozoon with a specific ovum investment,

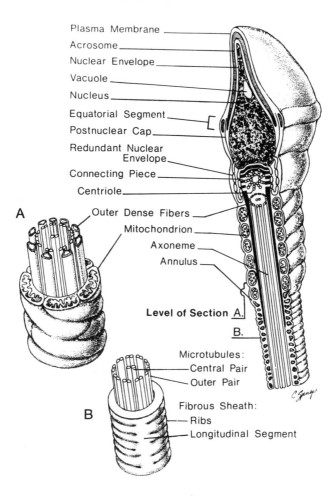

Plasma Membrane
Acrosome
Nuclear Envelope
Vacuole
Nucleus
Equatorial Segment
Postnuclear Cap
Redundant Nuclear
        Envelope
Connecting Piece
Centriole

A

Outer Dense Fibers
Mitochondrion
Axoneme
Annulus

Level of Section A.
                 B.

Microtubules:
  Central Pair
  Outer Pair

Fibrous Sheath:
  Ribs
  Longitudinal Segment

B

Figure 1 Diagram of the human sperm head. From Zaneveld, *in* Wynn (ed.), *Obstetrics and Gynecology Annual* (Vol. 7), p. 15. Appleton-Century-Crofts, New York (1978).

for instance, hyaluronidase—the cumulus oophorus, and acrosin—the zona pellucida. Acrosomal enzymes are also involved in the induction of the acrosome reaction and possibly in sperm-egg recognition and binding, in vitelline membrane penetration by the spermatozoon and in the decondensation of the sperm nucleus in the egg cytoplasm. Clearly, the sperm head is an important physiological entity that warrants detailed study. In spite of this, with the exception of some enzymes, very little is known about its biochemistry. A number of books reviewing the subject matter are listed in the references (1–6). A graphic overview of the biochemical constituents of the human sperm head is presented in Fig. 2.

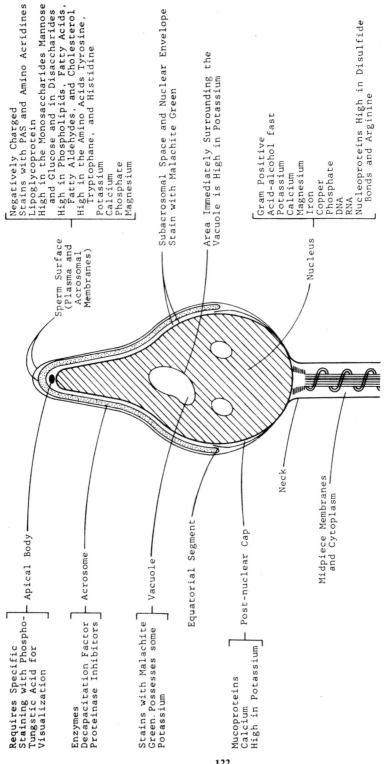

Requires Specific
Staining with Phospho- ─ Apical Body
Tungstic Acid for
Visualization

Enzymes
Decapacitation Factor ─ Acrosome
Proteinase Inhibitors

Stains with Malachite
Green. Possesses some ─ Vacuole
Potassium

Equatorial Segment

Mucoproteins
Calcium ─ Post-nuclear Cap
High in Potassium

Negatively Charged
Stains with PAS and Amino Acridines
Lipoglycoprotein
High in the Monosaccharides Mannose
    and Glucose and in Disaccharides
High in Phospholipids, Fatty Acids,
    Fatty Aldehydes, and Cholesterol
High in the Amino Acids Tyrosine,
    Tryptophane, and Histidine
Potassium
Calcium
Phosphate
Magnesium

Subacrosomal Space and Nuclear Envelope
Stain with Malachite Green

Area Immediately Surrounding the
Vacuole is High in Potassium

Gram Positive
Acid-alcohol fast
Potassium
Calcium
Magnesium
Iron
Copper
Phosphate
DNA
RNA
Nucleoproteins High in Disulfide
    Bonds and Arginine

Sperm Surface
(Plasma and
Acrosomal
Membranes)

Nucleus

Neck

Midpiece Membranes
and Cytoplasm

Figure 2   Some biochemical components of the human sperm head. From Zaneveld and Polakoski, *in* Hafez (ed.), *Human Semen and Fertility Regulation in Men*, p. 167. Mosby, St. Louis (1976).

2 PLASMA MEMBRANE

The plasma membrane of the sperm head can be divided into several morphologically distinct parts, and significant differences appear to exist in the functional characteristics of these regions. For instance, the anterior portion of the plasma membrane that covers most of the acrosome is specifically involved in the acrosome reaction (Chapter 8; Fig. 1). The biochemical characteristics of each region remain to be established, however.

It is difficult to isolate the plasma membrane so that it is uncontaminated with other sperm components. Commonly used procedures include extraction with detergents, sonication, hypotonic shock, or Potter homogenization, followed by sucrose density gradient centrifugation. Recently, the technique of nitrogen cavitation has been applied with success (7). Even though completely pure plasma membrane fractions can be obtained by some of these techniques, it is always possible that compounds are released from the other sperm membranes during the extraction process and bind to the plasma membranes. Therefore, one has to be very cautious in the interpretation of the data. Enzyme markers that are typically associated with cell membranes, such as 5′-nucleotidase, Na⁺-K⁺-activated ATPase, or acetylcholinesterase, have not yet been shown to be entirely specific for the sperm plasma membrane.

The plasma membrane consists of protein, lipid, and carbohydrate (8–11). Very little is known about the exact nature of each of these components. Because of the carbohydrate moiety, several plant lectins adhere to the plasma membrane of mammalian sperm. Some of these lectins, for example, concanavalin A, predominantly bind to the plasma membrane of the sperm head rather than to the entire sperm surface. Although all species of spermatozoa tested bind the same lectins, significant species differences exist in the amount of lectin binding (Table 1). The plasma membrane is also high in sialic acid residues, which are in part responsible for the surface to be more or less negatively charged.

A number of phospholipids and hydrocarbons are associated with the outer membranes of the sperm head. These include sphingomyelin, phosphatidylcho-

Table 1 Lectin Binding to Mammalian Spermatozoa[a]

Lectin[b]	Sites per Cell		
	Rabbit	Hamster	Mouse
Concanavalin A	1.00×10^7	1.90×10^7	4.90×10^7
Ricin	0.22×10^7	0.55×10^7	—
Wheat germ	1.20×10^7	3.30×10^7	0.70×10^7

[a] From Nicolson and Yanagimachi, *Science*, **177**, 276 (1972).
[b] Concanavalin A binds to terminal residues of α-D-glucose, α-D-mannose and their derivatives; ricin to residues of D-galactose and L-arabinose; and wheat germ lectin to residues of *N*-acetyl-D-glucosamine.

line, phosphatidylethanolamine, phosphatidylserine, choline plasmalogen, cardiolipin, and several others. Relatively high concentrations of fatty acids and aldehydes are also present as well as low amounts of glycolipids. The studies were not done with isolated plasma membranes, however, and it is not known what portion of the lipids is associated with the acrosomal rather than the plasma membranes. A more general discussion of sperm lipids and a table summarizing the lipids found to be associated with spermatozoa from various animal species can be found in Chapter 2.

Species differences exist in the protein composition of the plasma membranes. For instance, isolated plasma membranes from boar and human spermatozoa show different profiles on sodium dodecyl sulfate (SDS) gel electrophoresis. In both species, approximately 20 polypeptides can be detected having molecular weights varying from 12,000 to 230,000. Keratin-like proteins have been demonstrated in the ram, bull, boar, and human sperm head. These insoluble and sulfur-rich proteins that contain 4–6% sulfur and 11.5% cystine are apparently derived from the outer membranes of the sperm head. The keratin-like proteins are thought to aid in stabilizing the plasma membrane. This is important because damage to this membrane may cause sterility.

A number of seminal plasma proteins, for example, lactoferrin, proteinase inhibitors, and decapacitation factor (Chapters 4 and 8), bind to the surface of the spermatozoon. These proteins cannot be removed by washing the gametes. Because of their antigenic nature, one or more of these proteins is occasionally referred to as sperm-coating antigen (SCA).

The plasma membrane appears to have an essential role in the maintenance of sperm motility (Chapter 6). Mammalian spermatozoa maintain ionic gradients across the plasma membrane. In dog, ram, bull, rabbit, and human spermatozoa, the concentration of sodium ions is lower inside than outside the plasma membrane, whereas the concentrations of potassium, magnesium, and calcium ions are higher inside the plasma membrane. Of all the mammalian species studied, dog and guinea pig spermatozoa have the highest gradient of potassium and sodium ions across the plasma membrane. When dog spermatozoa are stored at a low temperature, potassium ions are lost from the spermatozoa and sodium ions accumulate in the sperm cell. On incubation at higher temperatures, the uptake of potassium ions increases rapidly. The extrusion of sodium ions and the uptake of potassium ions by dog spermatozoa is inhibited by fluoride, iodoacetate, 2,4-dinitrophenol, cetyltrimethylammonium bromide, and ouabain.

Fluorescent dyes have been used to determine the plasma membrane potential. In a physiological solution, the resting potential of the guinea pig sperm membrane is +13 mV. The potential remains unaltered when sodium ions are replaced by potassium ions in the suspending medium. At least in boar spermatozoa, the membrane potential appears to be dependent on the metabolic activity of the gametes. Rapid depolarization can be induced by drugs with anesthetic-like properties.

The plasma membrane of mammalian spermatozoa binds radioactive calcium with high affinity. In a purified plasma membrane preparation from boar spermatozoa, binding of calcium is markedly stimulated in the presence of sodium and potassium ions and by low concentrations of lanthanum (La^{3+}), and holmium (Ho^{3+}) ions, but not by other large electrophilic cations. This ability of mammalian sperm membranes to bind high concentrations of calcium ions suggests that the plasma membrane may play an important role in the storage and release of calcium. Calcium ions appear to be involved in the induction of the acrosome reaction (Chapter 8).

The sperm plasma membrane possesses one or more specific receptors for the zona pellucida (12). Ejaculated human spermatozoa or cauda epididymal rat or guinea pig spermatozoa fail to bind to the porcine zona even if the spermatozoa are added in high concentrations or if the motility of the spermatozoa is stimulated by caffeine. Although epididymal golden hamster spermatozoa and ejaculated bull spermatozoa do bind to the porcine zona, the numbers are very low in comparison to boar spermatozoa.

3 ACROSOME AND POSTACROSOMAL REGION

3.1 Introduction

The acrosome of mammalian spermatozoa is an intracellular, membrane-limited organelle that surrounds the anterior part of the sperm head (see Section 1 of this chapter). Although the size, shape, and internal structure of the acrosome vary greatly from one species to another, all mammalian spermatozoa that have been examined possess an acrosome. Some mammals, such as the guinea pig and golden mantled squirrel, have very large acrosomes, whereas those of others are much smaller. Using monoamines and the fluorescent amine, 9-aminoacridine, the internal pH of the acrosome, at least that of hamster sperm, was estimated to be 5.0 or less. The volume of the hamster acrosome was approximated as 0.4 μ^3.

The acrosome of mammalian spermatozoa contains various hydrolytic enzymes (13, 14) at least some of which appear to be required by the spermatozoon for fusion with the different layers surrounding the egg (Chapter 8). Morphological evidence also points to the acrosome as being directly involved in sperm–ovum fusion. Since the acrosome is produced by the Golgi apparatus during spermiogenesis and has other similarities to a lysosome, it is not surprising that its enzymes are similar to those found associated with lysosomes. They are all hydrolases of one type or another. Certain biochemical differences are usually present between the lysosomal and acrosomal enzymes, however.

A number of methods can be used to disrupt the acrosome and obtain its enzymatic content as well as portions of its membranes. These include the use of detergents (e.g., Hyamine and Triton X-100), sonication, shaking with glass beads, treatment with $MgCl_2$ or acid (HCl or acetic acid), hypotonic shock, or freeze-thawing. Further purification of the acrosomal contents requires conventional biochemical techniques.

The postacrosomal region of the sperm head consists of the postnuclear cap and the posterior portion of plasma membrane of the sperm head. This region is of interest because the first contact of spermatozoon with the cytoplasmic membrane of the oocyte (the vitelline membrane) occurs at this site. The plasma membrane covering the postacrosomal region of the golden hamster sperm head contains a dense population of finely granular, intramembranous particles. Very little is known about the biochemical composition and functional characteristics of the postacrosomal region. An actin-like substance has been detected as a major component in the postacrosomal region of man, dog, rabbit, bull, hamster, guinea pig, and mouse, but not in the rat. It has been postulated that these actin-like contractile proteins are important in sperm–egg interaction during fertilization and that the actin filaments, in conjugation with egg cytoskeletal systems, are involved in the migration of the sperm nucleus into the egg cytoplasm.

3.2 Proteinases and Proteinase Inhibitors

3.2.1 Introduction

The spermatozoon possesses different types of proteinases; several of them are known to be associated with the sperm acrosome (15, 16). An exopeptidase, arylamidase, has been found in bull spermatozoa. The enzyme has a molecular weight of 112,000, is particularly active against L-methionine-β-naphthylamide with a pH optimum of 8.0, is activated by dithiothreitol and inhibited by diisopropylfluorophosphate (DFP), 1,10-phenanthroline, and tosyllysine chloromethyl ketone (TLCK), but not by ethylenediaminetetraacetate (EDTA) or tosylamide-2-phenylethyl chloromethyl ketone (TPCK).

All other acrosomal proteinases studied so far are endopeptidases, that is, cleave internal peptide bonds. These enzymes can be divided into proteinases that have their optimal activity at acidic pH and those with optimal activity at basic pH. Several proteinases have been found in each category. The most widely studied proteinase is acrosin, which has its highest activity at pH 7.6–8.2.

A number of proteinase inhibitors are also associated with the sperm acrosome. They are peptides varying in molecular weight from 5000 to 15,000, and appear to be identical to the low-molecular-weight proteinase inhibitors in seminal plasma (Chapter 4). At least some of these sperm proteinase inhibitors inhibit acrosin, so that after extraction and activation of acrosin (see Section 3.2.2 of this chapter), between 70 and 95% of the acrosin is inhibited at basic pH. In man, two proteinase inhibitors have been found on the spermatozoon, one with a molecular weight of 10,100 that inhibits trypsin but not acrosin, and another with a molecular weight of 6200 that inhibits both acrosin and trypsin.

The exact location of the proteinase inhibitors is not well established, but most investigators consider them to be associated with the outer sperm head membranes. However, a certain portion of the inhibitors is probably also located within the acrosome proper or on the inner acrosomal membrane. It is not known if any of the inhibitors are present in association with acrosin or if they are localized at a different site. In some species such as the rabbit and stallion, only a small amount of acrosin inhibitory activity has been found in epididymal spermatozoa. In such species, the spermatozoa apparently accrue the inhibitor(s) during contact with seminal plasma. In other species, both epididymal and ejaculated spermatozoa possess high levels of acrosin inhibitor. It has been postulated that the physiological function of the inhibitor is to bind to acrosin when it is released from dead or dying spermatozoa and to inactivate this highly lytic enzyme so that it cannot damage the reproductive tract.

3.2.2 Acrosin and Proacrosin

Acrosin (EC 3.4.21.10) is a unique proteinase with a histidine and a serine at its active site. It is a glycoprotein and has many properties in common with pancreatic trypsin. The enzyme has been found in the spermatozoa from all mammalian species studied to date, including the boar, bull, ram, stallion, guinea pig, hamster, mouse, rabbit, monkey, and man. Some species differences exist in the total amount of acrosin present. Most of the enzyme is present on the spermatozoon in an inactive, zymogen form called proacrosin (17). The amount of acrosin in proacrosin form on the spermatozoa of man is approximately 93%; boar, 99%; ram, 95%; dog, 79%; guinea pig, 89%; mouse, 73%; and hamster, 92%.

The properties of the acrosins from different species are similar, although quantitative differences are present, particularly in regard to the hydrolysis of certain synthetic substrates and the interaction with inhibitors (Table 2). The acrosins from different species of mammals differ somewhat in their amino acid composition but show major differences from the trypsins of the same species (Table 3). The peptide portion of rabbit proacrosin makes up approximately 92% of the molecule, the rest being carbohydrate consisting of glucosamine, mannose, galactose, fucose, and sialic acid, respectively, amounting to 9.63, 9.43, 3.38, 0.49, and 2.34 residues per mole acrosin. Boar and ram acrosin, respectively, consist of 10% and 3% of carbohydrate. Immunological cross-reactions occur at times between species. For instance, ram acrosin cross-reacts with bull acrosin but not with that of the rabbit, boar, or man. Acrosin does not cross-react with trypsin.

Acrosin only cleaves arginine and lysine bonds and, in general, hydrolyzes arginine bonds faster than lysine bonds. The enzyme digests the proteins azocasein, ribonuclease, lysozyme, and insulin B chain, but not elastin and collagen. Acrosin hydrolyzes ester substrates typical for those of trypsin such as α-N-benzoyl-L-arginine ethyl ester, α-N-benzoyl-D,L-arginine-β-naphthylamide, α-N-benzoyl-D,L-arginine-p-nitroanilide, and p-tosyl-L-arginine methyl ester (Ta-

Table 2 Some Kinetic Properties and Inhibitors of Mammalian Acrosin

Species	Activity (nmoles BAEE hydrolyzed/10^6 min^-1 sperm)[a]	Enzyme Activity with Different Substrates[a]					Interaction with Inhibitors[a]	
		Substrate[b]	pH Optimum	$K_m(M)$	$K_{cat}(min^{-1})$	V_{max}(BAEE)/V_{max}(TAME)	Inhibitor[b]	$K_i(M)$ or other constants[c,d,e]
Boar	133.8 (1)	BAEE	8.5 (3)	$4.8\text{–}5.0 \times 10^{-5}$ (2)	8900 (4, 7)		L-Arginine	3×10^{-3} (2)
		BAPNA	8.0 (5)	2.7×10^{-4} (5)			Leupeptin	$5.9\text{–}8.6 \times 10^{-8}$ } (22)
			7.5 (5)	3.9×10^{-4} (6)	780 (4, 7)		Antipain	7.1×10^{-8}
		BLPNA	7.8 (6)				DFP	630^c
			8.0 (6)	6.5×10^{-4} (6)	54 (4)		TLCK	23^c (6)
		Azo-casein						
Bull	15.9 (1)	BAPNA }	8.7 (3)					
		BANA }	8.0–8.7 (8)					
		TAME }						
Hamster		BAEE	—	3×10^{-5} (13)			p-Aminobenzamidine	1×10^{-5}
		TAME	8.2 (13)	2×10^{-5} (13)		0.21 (13)	Soybean trypsin inhibitor	7×10^{-10} } (13)
Man	10.6 (10)	BAEE	8.0 (24)	3.6×10^{-5} (24)			NPGB	3.0×10^{-9} (25)
	4.7 (11)	BANA	8.2 (12)	2.8×10^{-5} (12)			TLCK	1.5×10^{-8} (9)
	6.5 (20)	BAPNA		1.8×10^{-4} (24)			Soybean trypsin inhibitor	2.6×10^{-8} (24)
		TAME		2.8×10^{-4} (24)			Antipain	2.5×10^{-8} (25)
		BAME		3.2×10^{-4} (24)			Benzamidine	3.0×10^{-5} (25)
				2.0×10^{-5} (24)			p-Aminobenzamidine	4.0×10^{-6} (25)
							Seminal inhibitor II	2.0×10^{-9} (23) } [d]

Species		Substrate	pH opt. (ref)		K_m (ref)	Inhibitor	K_i (ref)
Mouse	12.6 (21)	BAEE	8.0 (17)		8.2×10^{-6} (17)	Benzamidine	5.1×10^{-4}
						p-Aminobenzamidine	2.2×10^{-5}
						NPGB	9.8×10^{-10d} } (17)
Rabbit		BAEE	8.2 (16)		$5.1-5.2 \times 10^{-6}$ (16)	TLCK	1.95–2.00 (15)
			8.0 (19)		6.4×10^{-5} (18)	TLCK	0.032 (18)^e
					$4.0-4.3 \times 10^{-3}$ (15)	Benzamidine	$2.3-2.6 \times 10^{-5}$
		BAPNA	8.2 (16)		$10.0-10.9 \times 10^{-3}$ (18)	p-Aminobenzamidine	$6.0-8.5 \times 10^{-6}$
		BANA	8.2 (15)		$7.8-10.5 \times 10^{-5}$ (15)	Lima bean inhibitor	$1.8-2.0 \times 10^{-5}$ } (16)
		TAME			$2.8-3.3 \times 10^{-5}$ (15)	Soyban inhibitor	1.0×10^{-10}
Ram	26.2 (1)	TAME			2.1×10^{-5}		2940
		BAEE	8.1 (14)		2.8×10^{-4}		6900
		BAEE (+ Ca^{+2})			1.2×10^{-4} } (14)		7200 } (14)
		BAEE (+ Ca^{+2}, + Na$^+$)			1.6×10^{-4}		8040

^{a} References: (1) Brown and Harrison, Biochim. Biophys. Acta, **526**. 202 (1978). (2) Polakoski and McRorie. J. Biol. Chem. **248**. 8183 (1973). (3) Polakoski et al., J. Biol. Chem. **248**. 8178 (1973). (4) Schiessler et al., Hoppe Seyler's Z. Physiol. Chem. **356**. 1931 (1975). (5) Schleuning and Fritz. Hoppe Seyler's Z. Physiol. Chem. **355**. 125 (1974). (6) Schiessler et al., Hoppe Seyler's Z. Physiol. Chem. **353**. 1638 (1972). (7) Schleuning et al., Hoppe Seyler's Z. Physiol. Chem. **356**. 1915 (1975). (8) Multamaki and Niemi. Int. J. Fert. **17**. 43 (1972). (9) Bhattacharyya et al., J. Reprod. Fert. **47**. 97 (1976). (10) Goodpasture et al., J. Androl. **1**. 16 (1980). (11) Bhattachayya and Zaneveld, Fertil. Steril. **30**. 70 (1978). (12) Gilboa et al., Eur. J. Biochem. **39**. 85 (1973). (13) Meizel and Mukerji, Biol. Reprod. **14**. 444 (1976). (14) Brown et al., Biochem. J. **149**. 133 (1975). (15) Meizel and Mukerji, Biol. Reprod. **13**. 83 (1975). (16) Stambaugh and Buckley, Biochim. Biophys. Acta **284**. 473 (1972). (17) Beyler and Zaneveld. Unpublished. (18) Conners et al., Biol. Reprod. **9**. 57 (1973). (19) Zaneveld et al., Biol. Reprod. **6**. 30 (1972). (20) Polakoski et al., Fertil. Steril. **28**. 668 (1979). (21) Bhattacharyya et al., Am. J. Physiol. **237**. E40 (1979). (22) Fritz et al., Hoppe Seyler's Z. Physiol. Chem. **354**. 1304 (1973). (23) Zaneveld et al., J. Reprod. Fert. **32**. 525 (1973). (24) Anderson et al., Biochem. J. **199**. 307 (1981); (25) Anderson et al., unpublished.

^{b} Abbreviations: BAEE = α-N-benzoyl-L-arginine ethyl ester; TAME = p-tosyl-L-arginine methyl ester; BANA = α-N-benzoyl-D.L-arginine-β-naphthalamide; BAME = α-N-benzoyl-L-arginine methyl ester; BAPNA = α-N-benzoyl-D.L-arginine-p-nitroamalide; BLPNA = α-N-benzoyl-D.L-lysine-p-nitroanilide; DFP = diisopropylfluorophosphate; TLCK = N-α-tosyl-L-lysine chloromethyl ketone; NPGB = p-nitrophenyl-p'-guanidinobenzoate.

^{c} Bimolecular velocity constant for irreversible inhibition.

^{d} Concentration causing 50% inhibition.

^{e} Limiting inactivation constant (min^{-1}) for irreversible inhibition.

Table 3 Amino Acid Composition of Mammalian Acrosins and Some Trypsins[a]

Amino Acid	Acrosin				Proacrosin	Trypsin			
	Boar (1)	Ram (2)	Man (1)[c]	Rabbit (4)	Rabbit (8)	Boar (5)	Ram (7)	Man (3)	Bull (6)
Lysine	16	16	18	10	34	10	12	11	14
Arginine	25	25	19	7	21	4	4	6	2
Histidine	6	9	6–7	4	17	4	3	3	3
Aspartic acid and asparagine	23	24	31	22	47	24	20	21	22
Glutamic acid and glutamine	27	29	30–32	27	59	17	14	21	14
Threonine	20	19	19	12	29	10	15	10	10
Serine	18	25	25–26	22	42	27	26	24	33
Proline	40	36	31–34	8	37	9	9	9	9
Methionine	6	4	4	2	6	2	2	1	2
Glycine	34	31	38	20	36	25	19	20	25
Alanine	19	25	27	14	38	15	17	13	14
Valine	24	26	22	10	30	16	17	16	17
Isoleucine	17	18	21	8	17	16	10	12	15
Leucine	21	21	22	13	41	16	14	12	14
Tyrosine	10	10	10	5	19	8	6	7	10
Phenylalamine	11	10	10–11	6	18	4	5	4	3
Cystine (half)	8	12	8	ND[b]	18	12	12	8	12
Tryptophan	3	3	ND	ND	5	4	ND	3	4
Total	328	343	341–349	190	514	223	205	201	223

[a] References: (1) Muller-Esterl et al., Hoppe Seyler's Z. Physiol. Chem. **361**, 1811 (1980). (2) Brown and Hartree, Biochem. J. **175**, 227 (1978). (3) Travis and Roberts, Biochem. **8**, 2884 (1969). (4) Stambaugh and Smith, Science **186**, 745 (1974). (5) Hermodson et al., Biochem. **12**, 3146 (1973). (6) Walsh and Neurath, Proc. Natl. Acad. Sci. **52**, 884 (1964). (7) Travis, Biochem. Biophys. Res. Commun. **30**, 730 (1968). (8) Mukerji and Meizel, J. Biol. Chem. **254**, 11,721 (1979).
[b] Not determined.
[c] Tentative.

ble 2). With these substrates significant differences often exist between the
Michaelis constants of acrosin compared with those of trypsin. Acrosin does
not react with substrates for chymotrypsin and exopeptidases.

Acrosin is inhibited by all the synthetic and natural inhibitors of trypsin,
although large kinetic differences are occasionally present. One inhibitor appears to be an exception: the 10,100-dalton peptide inhibitor from human
seminal plasma and spermatozoa that inhibits trypsin but not acrosin. Typical
synthetic inhibitors of acrosin include DFP, TLCK, and p-nitrophenyl-p'-guan-
idinobenzoate (NPGB). The enzyme is also inhibited by a large number of
amidino and diamidino compounds and, at least boar acrosin, by L-arginine.
Acrosin is not inhibited by chymotrypsin inhibitors such as TPCK, or by
iodoacetamide, cysteine, 1,10-phenanthroline, or EDTA. The enzyme is effectively inhibited by high- and low-molecular-weight, naturally occurring trypsin
inhibitors such as kunitz pancreatic trypsin inhibitor (Trasylol), α_1-antitrypsin
(although species differences may be present), inter-α-trypsin inhibitor, soybean
trypsin inhibitor, and others. Particularly effective are the microbial inhibitors
leupeptin and antipain. In contrast to the activity of human trypsin, the activity
of human acrosin is inhibited by certain sugars (monosaccharides), such as
D-fructose, α-methyl mannoside, D-mannose, D-fucose, D-ribose, and D-arab-
inose. Certain cations such as Fe^{2+}, Fe^{3+}, Zn^{2+}, and Hg^{2+} inhibit acrosin, whereas Ca^{2+} stabilizes acrosin and, in some situations, stimulates enzyme activity.
The activity of boar acrosin, but not that of human acrosin, is also stimulated
by spermine and other polyamines.

A number of different molecular weight forms of acrosin are present in the
same species (15). These different molecular weight forms may be due to aggregation of the enzyme or to the formation of dimers, and so forth. However, in
most cases, the different forms result from the removal of peptides from the
original molecule while retaining catalytic activity. For instance, boar proacrosin occurs in two molecular weight forms, 55,000 and 53,000, both of which, on
activation, form active acrosin at a molecular weight (MW) of 49,000 (m_α-
acrosin). This, in turn, produces a 34,000-mw form (m_β-acrosin) and finally a
25,000-MW form (m_γ-acrosin). Two major fractions of ram acrosin have been
obtained. Ram m_β-acrosin was isolated in homogeneous form, is quite stable at
low pH, and has a molecular weight of 38,000. Ram m_ψ-acrosin is unstable and
contains at least three components. At least three different molecular weight
forms of human acrosin are present, the highest one (m_α-acrosin) measuring
approximately 70,000 daltons. The molecular weight of rabbit proacrosin is
68,000. On activation, it forms acrosin, with a molecular weight of 34,000, and
it was suggested that rabbit proacrosin consists of two identical polypeptide
(acrosin) chains. The various forms of acrosin may have different kinetic interactions with substrates and inhibitors (Table 4).

The activation of proacrosin to acrosin follows a typical sigmoidally shaped
curve and occurs maximally at a pH of 7–8. At present, little is known regarding the conversion of proacrosin to acrosin. Synthetic inhibitors of acrosin
inhibit activation, which would indicate that the process depends on the activity

Table 4   Some Properties of the Different Molecular Forms of Boar Acrosin[a]

Property	m_α-acrosin	m_β-acrosin	m_γ-acrosin
Molecular weight	49,000	34,000	25,000
Relative migration (R_f) in the pH 4.3 electrophoresis system	0.39	0.54	0.65
K_m (BAEE) (mM)	0.053	0.050	
K_m (BAME) (mM)	0.068	0.027	
K_m (TAME) (mM)	0.048	0.077	
V_{max} (BAEE) (μmole/min/ unit acrosin)	1.34	3.31	
V_{max} (BAME) (μmole/min/ unit acrosin)	1.23	2.21	
V_{max} (TAME) (μmole/min/ unit acrosin)	0.95	3.01	
Stimulation of specific activity (%) by			
Calcium	240	240	
Spermine	215	230	
Inhibitors			
L-arginine (K_i) (mM)	6.1 (competitive)	3.0	
Benzamidine (K_i) (μM)	4.4 (competitive)	4.0	
p-aminobenzamidine (K_i) (μM)	0.61 (competitive)	1.2	
Diisopropylflurophosphate	Irreversible inhibition	Irreversible inhibition	
Tosyllysine chloromethyl ketone	Irreversible inhibition	Irreversible inhibition	
Tosylphenylalanine chloromethyl ketone	No inhibition	No inhibition	
Ovomucoid trypsin inhibitor	Inhibition	Inhibition	
Lima bean trypsin inhibitor	Inhibition	Inhibition	
Seminal plasma trypsin inhibitors	Inhibition	Inhibition	

[a] Data taken from Paulson *et al.*, *Int. J. Biochem.* **10**, 247 (1979), Parrish and Polakoski, *Int. J. Biochem*, **10**, 391 (1979), and Parrish and Polakoski, *J. Biol. Chem.* **253**, 8424 (1978).

of a proteinase (18). This proteinase may be acrosin itself, although activation of human proacrosin is delayed in the presence of spermine and calcium, which do not inhibit acrosin, at least not in man. Alternately, proacrosin may activate spontaneously, or the activation may involve a thermolysin-like enzyme, called acrolysin. Extracts of rabbit, cat, hamster, and boar sperm acrosomes appear to contain this thermolysin-like enzyme, which was proposed to convert proacrosin to acrosin by cleaving the aminopeptide bonds of hydrophobic aminoacyl residues. Acrolysin is inhibited by chelating agents and Zn^{2+}, and is activated by Ca^{2+} (19).

The exact location(s) of acrosin within the sperm acrosome is not completely known. At least half the acrosin (proacrosin) appears to be bound to the inner acrosomal membrane and, in the guinea pig spermatozoon, remains bound even after removal of the plasma and outer acrosomal membranes. Some acrosin and/or proacrosin also appears to be associated with the outer acrosomal membrane and/or the acrosome proper, but the relative amount at each site is not known.

Treatment of spermatozoa with natural or synthetic inhibitors of acrosin prevents fertilization, and acrosin appears to have an esssential role in the fusion of the spermatozoon with oocyte, specifically by enabling spermatozoa to pass through the zona pellucida, a mucopolysaccharide layer surrounding the oocyte (Chapter 8). It does so either by inducing the acrosome reaction (this reaction is prevented on addition of acrosin inhibitors to spermatozoa) and/or by lysing a passage for the spermatozoon through the zona pellucida (20–30). A role for acrosin has also been suggested in the initial attachment of the spermatozoon to the zona pellucida, in the passage of the sperm through the vitelline membrane, and in the release of sperm chromatin in the egg cytoplasm by hydrolysis of sperm nuclear proteins. Although it had also been proposed that acrosin is involved in sperm penetration through cervical mucus, this was shown not to be the case.

3.2.3 Other Proteinases

Several proteinases with optimal activity at acid pH against casein, hemoglobin, and other protein substrates have been found in a number of mammalian spermatozoa (15, 16). An acid proteinase from boar spermatozoa was shown to have pH optima of 2.8 and 3.5, to have a molecular weight of more than 150,000, and to be inhibited by iodoacetate, but not by DFP or EDTA. Mouse spermatozoa possess an acid proteinase with pH optimum between 3.5 and 5.0 that is inhibited by pepstatin but not by typical trypsin inhibitors. The molecular weight of the enzyme is approximately 40,000 and it may be cathepsin D.

An enzyme from bull, rat, and human spermatozoa that hydrolyzes the peptide Pz-Pro-Leu-Gly-Pro-D-Arg at a pH optimum of 7.5 has also been found. It has a molecular weight of approximately 110,000 and was called a "collagenase-like peptidase." It is not active against a number of protein substrates such as hemoglobin and serum albumin, but it digests gelatin. A chymotrypsin-like enzyme may also be present in rabbit spermatozoa; it reacts with typical synthetic substrates of chymotrypsin. Finally, proteolytic activity at a very basic pH (9 or above) was shown to be present in rabbit, human, bull, ram, and mouse spermatozoa. At least in the rabbit and ram, this enzyme activity is not inhibited by typical trypsin inhibitors, and in the mouse, a molecular weight of approximately 150,000 or more was found. Whether there is more than one enzyme that is active in this high pH region remains to be established.

Table 5 Properties of Hyaluronidase of Mammalian Spermatozoa[a]

Species	Acrosomal Extracts Specific Activity[b] (units/mg protein)	Method of Extraction	pH Optimum	Salt Requirement	Activators	Inhibitors	MW (method)
Bull	0.221 (ejac.) (1) 90–250[c] (ejac.) (9)[d] 1.2[c] (ejac.) (10)	Detergent Sucrose Barbituric acid	3.75 (1) 3.6 (3)[e] 3.8 (11)[f]	0.15 M NaCl 0.15 M KCl	BSA Histones Protamine sulfate Hyamine 2389 Spermine Spermidine Cations	Fe^{2+} Fe^{3+} Heparin Phosphorylated hesperidin	110,000 (1) (Sephadex G-100) 62,000 (11) (SDS gel electrophoresis)[f]
Buffalo	11.35 (ejac.) (5)	Detergent					
Rabbit	0.56 (ejac.) (1) 0.60 (epid.) (1) 16.4[c] (epid.) (7) (Nuclei—11.4; Acrosomes—48.5) 0.17[c] (ejac.) (10)	Detergent Detergent Sonication Barbituric acid	3.75 (1)	0.15 M NaCl	Histones	Fe^{2+} Fe^{3+} Heparin	11,000 (7)[g] (Sucrose density gradient centrifugation)

Species	Activity[b]		pH optimum		Inhibitors	MW
Ram	1.54 (ejac.) (1)	Detergent	4.3 (2)	Unknown	Heparin	62,000 (2) (SDS gel electrophoresis)
	1.22 (epid.) (1)	Detergent			Chondroitin sulfates A, B, and C	
	5.3–8.5 (ejac.) (4)	Freeze-thaw				
	8.9–11.3 (epid.) (4)	Freeze-thaw				
Man	0.067 (ejac.) (1)	Detergent	3.5–6.5 (8)	Unknown	Not stated; inactivated at pH < 3	Not stated
Boar	0.045 (ejac.) (1)	Detergent				
	0.029 (epid.) (1)	Detergent				
Rat	0.016 (epid.) (1)	Detergent			Phosphorylated hesperidin (6)	

[a] References: (1) Zaneveld et al., J. Biol. Chem. 248, 564 (1973). (2) Yang and Srivastava, J. Reprod. Fert. 37, 17 (1974). (3) Doak and Zahler, Biochim. Biophys. Acta 570, 303 (1979). (4) Brown, J. Reprod. Fert. 45, 537 (1975). (5) Kaur et al., Experientia 32, 436 (1976). (6) Martin and Beiler, Science 115, 402 (1952). (7) Stambaugh and Buckley, J. Reprod. Fert. 19, 423 (1969). (8) Salegui et al., Arch. Biochem. Biophys. 121, 548 (1967). (9) Masaki and Hartree, Biochem. J. 84, 347 (1962). (10) Bernstein and Teichman, J. Reprod. Fert. 33, 239 (1973). (11) Yang and Srivastava, J. Biol. Chem. 250, 79 (1975).

[b] Unless otherwise indicated, the activity was measured colorimetrically with hyaluronic acid as substrate.

[c] Turbidimetric method.

[d] Activity per $0.35–0.89 \times 10^6$ sperm heads per milliliter.

[e] pH optimum in the presence of serum albumin (1 mg/ml) or histone (0.5 mg/ml). Shows optimum at pH 6.0 with none.

[f] Enzyme was isolated from seminal plasma, K_m (hyaluronic acid) = 37 mg/ml and V_{max} = 2.4 μmole/min.

[g] Reported MW of 59,000 for hyaluronidase-acrosin complex.

3.3   Hyaluronidase

Large variations exist among species in regard to the hyaluronidase content of spermatozoa, at least that obtained on extraction: it is highest in buffalo, ram, rabbit, and bull spermatozoa; less is present in boar, rat, mouse, and human spermatozoa; very little can be found in dog spermatozoa; and the enzyme is almost absent from stallion spermatozoa. Hyaluronidase appears to be associated with the outer membranes of the acrosome and is readily released from intact, viable spermatozoa. Even though it has been suggested that hyaluronidase exists in complex with acrosin in the sperm acrosome, this is unlikely, since the release rate of acrosin from sperm is much slower than that of hyaluronidase.

Physiologically, much of the hyaluronidase is liberated from the spermatozoa during capacitation and the acrosome reaction (Chapter 8). On addition to eggs, hyaluronidase causes the dispersion of the follicle-cell layer. In some species such as the rabbit, the follicle cells surrounding the egg can be divided into two layers: an outer cumulus oophorus and an inner corona radiata. In such cases, hyaluronidase disperses only the cumulus cells, not the corona cells. It is thought that at the time of contact of the spermatozoon with the egg, the enzyme lyses a path through the follicle-cell layer, thus allowing the spermatozoon to pass through this egg investment. Addition of hyaluronidase inhibitors or hyaluronidase antibodies to spermatozoa prevents fertilization and, at least in the case of mouse eggs, specifically penetration through the cumulus oophorus.

Bull, rabbit, and ram sperm hyaluronidase have been studied quite extensively (Table 5). The properties of the sperm enzyme appear to be identical to those of testicular hyaluronidase but are different from those of the lysosomal hyaluronidase of other tissues. The pH optima of bull and rabbit sperm hyaluronidase is approximately 3.75, with a pH range of activity varying from 3 to 7. Ram sperm hyaluronidase has a pH optimum of 4.3, shows no activity at pH 3.0, and only 10% of the maximal activity at pH 8.0. Neither bull nor rabbit sperm hyaluronidase are degraded significantly by either freezing and thawing or refrigeration. By contrast, the hyaluronidase of human spermatozoa is destroyed when spermatozoa are kept at $-20°C$. Rabbit sperm hyaluronidase activity is lost completely after 1 hr at pH 3.0; whereas bull acrosomal hyaluronidase is more stable and loses activity only after 12 hr at pH 3.0. Partially purified ram acrosomal hyaluronidase loses 36% of its activity within a week at $4°C$, and 70% of the activity is lost when it is kept frozen for a year. Both bull and rabbit sperm hyaluronidase are denatured at temperatures above 50°, and require salt (NaCl) for activity. These enzymes are inhibited by Fe^{2+}, Fe^{3+}, and heparin, but not by Mn^{2+}, Mg^{2+}, Ca^{2+}, or EDTA. Ram sperm hyaluronidase is also inhibited by chondroitin sulfate B. The molecular weight of ram and bull sperm hyaluronidase when assessed by SDS gel electrophoresis is 62,000. However, the enzyme may occur in its native form as a dimer because a molecular weight of 110,000 was found for bull sperm hyaluronidase by Sephadex gel filtration.

Table 6 Lytic Enzymes of the Mammalian Sperm Head Other Than Proteinases or Hyaluronidase

Enzyme	Species[a]	Probable Subcellular Localization	Properties
ATPase	Man (1), Rabbit (1)	Outer acrosomal membrane	pH optimum 9.0; inactive at pH 7.0; requires Ca^{2+} with ATP as substrate; not inhibited by ouabain.
	Guinea pig (2)	Outer acrosomal membrane and plasma membrane.	Requires ATP and Ca^{2+}; pH optimum 9.0; no inhibition with PCMB[b] and NEM[c].
	Rabbit (1)	Plasmalemma	Dephosphorylates ATP at pH 7.0; requires no divalent ions; no effect with EDTA; inhibited by ouabain.
Aryl sulfatase	Rabbit (8)	Acrosome	Present as isoenzymes (A and B forms); pH optima 4.8, 5.6, 6.0; temperature optimum 50°C; inhibited by phosphate, sulfite, decapacitation factor (DF); shows higher affinity for p-nitrocatechol sulfate than p-nitrophenyl sulfate.
	Ram (6)	Acrosome	Activity: 0.6–1.0[d], pH optimum 4.9; both A and B forms are present.
	Rat (10)	Acrosome	pH optimum 5.2.
	Bull (3, 4)	Acrosome and post-acrosomal segment	Activity: 0.03–3.2[e]; inhibited by NaF.
Acid phosphatase	Bull (9)	Head	Activity: 0.13[f]
	Man (3, 7, 18)	Acrosome	Activity: 3.7–13.8[e] activity 1.2–68.1; pH optimum 3.8–5.6; MW 50,000–80,000; inhibited by PCMB.

Table 6 (*continued*)

Enzyme	Species[a]	Probable Subcellular Localization	Properties
	Rabbit (3, 5)	Acrosome and post-acrosomal segment	Activity: 0.03–0.47[c]; two forms are present (S$_4$ and S$_{zn}$); inhibited by NaF and creatine phosphate.
	Ram (6)	Acrosome	Activity: 7.8–9.2[d], pH optimum 4.5.
	Ram (9)	Surface membrane	Activity: 0.73[g]
β-Glucuronidase	Bull (3)	Acrosome	Activity: 2.3[e]
	Man (3)	Acrosome	Activity: 0.15–1.2[e]
	Rabbit (3)	Acrosome	Activity: 0.05[e]
β-Aspartyl-N-acetylgluco-samine amido hydrolase (aspartyl amidase)	Boar (11)	Acrosome	Activity: 73[h]
	Man (11)	Acrosome	Activity: 47.8[h]; pH optimum 7.8; temperature optimum 70°C.
	Ram (11)	Acrosome	Activity: 145[h]
	Squirrel monkey (11)	Acrosome	Activity: 89.1[h]
Corona-penetrating enzyme (CPE)	Man (12)	Acrosome	Activity: 4.0[i]; unstable at −20°C; stable at −70°C; inhibited by decapacitation factor (DF).
	Rabbit (12)	Acrosome	Activity: 3.7–3.8[i]; unstable at −20°C; stable at −70°C; inhibited by decapacitation factor (DF). Appears to be an esterase (14).
β-N-Acetylglucosaminidase	Buffalo (13)	Acrosome	51% soluble; 49% bound.
	Bull (15)	Acrosome	pI 7.96; pH optimum 4.5; MW 190,000 (Sephadex G-200); K_m 0.52 mM; V_{max} 0.088 μmole/min/mg protein; inhibited by Hg^{2+}, lactones, PCMB, and iodoacetamide.

138

	Goat (13)	Acrosome
	Ram (6)	Acrosome
	Rabbit (16)	Inner acrosomal membrane
Neuraminidase		45% soluble; 55% bound; solubilized at pH 3.0. Activity: 14.6–16.8.[d] Activity: 0.37 μmoles sialic acid released/min/mg protein; K_m 1.72 $\times$ 10^{-6}M with mucin; K_m 1.17 $\times$ 10^{-5}M with fetuin; K_m 8.8 $\times$ 10^{-4}M with sialyllactose; V_{max} 0.112 (mucin), 0.071 (fetuin) and 0.038 (sialyllactose) μmole/min/mg; pH optimum 5.0 on mucin and 4.3 on sialyllactose; inhibited by Ca^{2+}, Mg^{2+}, Mn^{2+}, Co^{2+}, and Cu^{2+}.
Cyclic AMP-dependent protein kinase	Rat (17)	Outer surface	Cyclic AMP (2.5 μM) activates the kinase and causes the release of the enzyme from the sperm surface; approximately 80% of the kinase is released after 30 sec of incubation.

[a] References: (1) Gordon, *J. Exp. Zool.* **185**, 111 (1973). (2) Gordon and Barnett, *Exp. Cell Res.* **48**, 395 (1967). (3) Bernstein and Teichman, *J. Reprod. Fert.* **33**, 239 (1973). (4) Teichman and Bernstein, *J. Reprod. Fert.* **27**, 243 (1971). (5) Gonzales and Meizel, *Biochim. Biophys. Acta* **320**, 166 (1973). (6) Allison and Hartree, *J. Reprod. Fert.* **21**, 501 (1970). (7) Singer *et al.*, *Arch. Androl.* **5**, 195 (1980). (8) Yang *et al.*, *Proc. Soc. Expl. Biol. Med.* **145**, 721 (1974). (9) Dott and Dingle, *Exp. Cell Res.* **52**, 523 (1968). (10) Seiguer and Castro, *Biol. Reprod.* **7**, 31 (1972). (11) Bhalla *et al.*, *J. Reprod. Fert.* **34**, 137 (1973). (12) Zaneveld and Williams, *Biol. Reprod.* **2**, 363 (1970). (13) Anand et al., *Hoppe Seyler's Z. Physiol. Chem.* **358**, 685 (1977). (14) Bradford *et al.*, *Biol. Reprod.* **15**, 102 (1976). (15) Khar and Anand, *Biochim. Biophys. Acta* **483**, 141 (1977). (16) Srivastava and Abou-issa, *Biochem. J.* **161**, 193 (1977). (17) Majumder, *Biochim. Biophys. Res. Comm.* **83**, 829 (1978). (18) Kipping, *Wochenschr.* **48**, 1127 (1970).

[b] Parachloromercuribenzoic acid.

[c] *N*-ethylmaleimide.

[d] Micromoles of substrate hydrolyzed per 30 minutes calculated back to 1 ml semen.

[e] Activity is expressed as micrograms of product formed per microgram protein per hour.

[f] Activity is expressed as millimoles of nitrophenol formed by 10^6 sperm per hour.

[g] Activity is expressed as micrograms of nitrophenol formed by 10^6 sperm per hour.

[h] Activity is expressed as nanomoles *N*-acetylglucosamine per milligram protein per hour.

[i] Egg denudation index: from 0 (lowest) to 4 (highest).

3.4   Other Enzymes

A number of other enzymes have been found associated with the sperm acrosome or plasma membrane. Those that have been studied in some detail are listed in Table 6, and their properties are summarized. Of these, only the acid phosphatase from rabbit spermatozoa has been obtained in a highly purified form. This enzyme has also been detected in ram, bull, rat, and human spermatozoa. The acid phosphatase of rabbit ejaculated spermatozoa exists in at least five multiple forms; only two of these forms (S_4 and S_{zn}) are associated with the acrosomes. The different forms are distinguished by the effects of NaF, $ZnCl_2$, L-tartaric acid, and creatine phosphate on the hydrolysis by the enzyme of α-naphthyl phosphate at pH 5.0.

Besides the ones listed in the table, evidence is present for the association of the following enzymes with the sperm head: several esterases, alkaline phosphatase, and phospholipase A. The proposed role of some of the listed enzymes in the fertilization process can be found in Chapter 8. An enzyme (corona-penetrating enzyme, CPE) that disperses the corona radiata layer of rabbit eggs has been extracted from rabbit and human sperm acrosomes and was proposed to have a role in the penetration of the spermatozoon through the corona radiata. This enzyme appears to be an esterase. It is inhibited by a high-molecular-weight glycoprotein from seminal plasma, often termed the decapacitation factor (DF) (Chapters 4 and 8). This glycoprotein is present on ejaculated spermatozoa and, when isolated, prevents fertilization on addition to capacitated spermatozoa. Sperm neuraminidase may aid in the induction of the zona reaction (hardening of the zona pellucida, preventing more than one spermatozoon from passing through this layer, that is, blocking polyspermy) but material released from the cortical granules during fertilization appears to be the major inducer of this reaction. Phospholipase and ATPase have been proposed to be intimately involved with acrosin in the removal of the outer sperm-head membranes during the acrosome reaction.

4   THE SPERM NUCLEUS

4.1   Introduction

Generally, more than half the mature mammalian sperm head is occupied by the extremely dense nucleus. It is filled with tightly packed chromatin material consisting almost entirely of deoxyribonucleoproteins (DNPs), which are composed of deoxyribonucleic acid (DNA) conjugated to certain highly basic nuclear proteins (24). A packing ratio of 55 equivalent lengths of DNA to each unit length of DNP can be found in the human and bull chromatin fibers.

Chromatin fibers of varying sizes have been demonstrated by ultrastructural techniques. At least in the human, bull, and rabbit, the fibers are composed of discrete spherical units with diameters of 150 and 400 Å. Biochemically, human sperm nuclear chromatin appears to be composed almost entirely of double helical DNA strands that aggregate into thick bundles, which in turn condense into tightly packed supercoiled structures. The DNA molecules are complexed to basic proteins, giving rise to a tertiary configuration. This supercoiled, tertiary configuration of the chromatin molecule confers a high degree of stability of the genetic materials toward environmental changes and to denaturation.

The DNA of spermatozoa is so tightly complexed to the nuclear proteins that transcription is almost impossible. Although it could be expected that the amount of DNA would be directly proportional to the size of the sperm nucleus, this is not true, since different sperm species undergo different degrees of condensation of the nuclear material during spermiogenesis (Chapter 1).

These typical properties of the mammalian sperm nucleus, that is, a high degree of stability and inertness, a high density, tight packing of DNA and protein, supercoiling, replacement of somatic histones by a new class of highly basic proteins (see further), lack of DNA transcription, and others, make the study of the DNA and nucleoproteins from spermatozoa very interesting.

The content of ash, nitrogen, phosphorus, and sulfur in the sperm head of the bull is approximately 2.0, 18.5, 4.0, and 1.5%, respectively. Besides the nucleic acids and nucleoproteins, which will be discussed in detail further, the nucleus also contains small amounts of magnesium, iron, copper, potassium, and phosphates (Fig. 2). A layer of high potassium concentration appears to exist around the nuclear vacuoles, even though the concentration of this ion within the vacuoles is very low.

4.2 Deoxyribonucleic Acid

4.2.1 Extraction and Isolation

After penetration of sperm nucleus into the mammalian egg, the decondensation of the chromatin into DNA and nucleoprotein takes place almost immediately. However, the artificial dissolution of DNA and nucleoproteins *in vitro* is very difficult because of the tight binding of the DNA with the basic proteins and the large number of —S—S— (disulfide) bonds in the proteins (see later). Early attempts to extract nuclear contents, that is, to induce nuclear decondensation, included treatment with strong acids, alkalis, urea, reducing agents not active toward disulfide bonds, and physical disruption with fine glass beads, but these were mostly unsuccessful except at very high concentrations. Treatment of spermatozoa (bull) with lipase, chymotrypsin, lysozyme, or even exposure to 9 M urea is also ineffective. However, sulfhydryl-reducing agents, for instance, mercaptoethanol, alkaline thioglycolate, 2,3-dimercaptopropane, dithiothreitol (DTT), either alone or in combination with certain other reducing agents such as cysteine, urea, guanidine hydrochloride, and sodium dodecyl sulfate (SDS), are effective in extracting the nuclear contents from mammalian spermatozoa by breaking down the nuclear membrane and subsequently the nucleoproteins. The effect of some agents in decondensing sperm nuclei of several mammalian species can be found in Table 7.

Table 7   Reagents Effective in Decondensing Sperm Nuclei
of Five Different Mammalian Species[a]

Reagents	Effective Molar Concentration of Reagents[c]				
	Hamster	Rat	Guinea Pig	Rabbit	Dog
SH-containing compounds					
Dithiothreitol in 0.5% SDS	0.002	0.2	0.002	0.02	0.002
2-Mercaptoethanol in 0.5% SDS	0.1	0.1	0.1	0.1	0.1
L-Cysteine in 0.5% SDS	1.0	1.0	1.0	1.0	1.0
Thioglycolate	0.8	0.8	0.8	0.8	0.8
Sulfides					
$(NH_4)_2S$	0.5	0.5	0.5	0.5	0.5
K_2S	0.1	0.5	0.25	0.25	0.5
Na_2S	0.1	0.1	0.01	0.05	0.01
Bases					
KOH	1.0	1.0	0.3	0.8	1.0
NaOH	1.0	0.8	0.2	0.8	1.0
NH_4OH[b]	7.0	—	—	4.0	—
KCN	2.0	2.0	2.0	1.0	1.0

[a] From Mahi and Yanagimachi, *J. Reprod. Fert.* **44**, 293 (1975).
[b] The NH_4OH produced only slight swelling of hamster and rabbit spermatozoa and none in the other species.
[c] Period of incubation was 30 min.

Species differences exist in the resistance of sperm nuclei to DTT mixed with SDS. Rat sperm nuclei are the most resistant, and hamster, guinea pig, and dog spermatozoa are least so. The chromatin fibers of squirrel spermatozoa, unlike those of most mammals, are not completely compact, and decondensation of the sperm nucleus occurs relatively readily. Rabbit sperm nuclei are rather resistant to DTT and SDS unless the concentration of SDS is doubled. In the rat and mouse, guanidinium chloride, mercaptoethanol, and urea are all highly effective in dissociating sperm nucleoproteins.

Once the nuclear material is extracted, the proteins are hydrolyzed by treatment with trypsin, thermolysin, or proteinase K before further purification of the DNA by standard techniques (25). Alternately, if isolation of the nuclear proteins is of interest, the DNA is first degraded by treatment of the nuclear contents with DNase.

4.2.2   Content

The DNA content is approximately 1.36×10^{-9} mg in a single nucleus of a human spermatozoon. In bull, rabbit, and ram, the DNA measures 3.22×10^{-9} mg, 3.1×10^{-9} mg, and 2.93–3.2×10^{-9} mg per sperm nucleus, respectively.

Although fertile men and domestic mammals have a fairly constant DNA content per nucleus, sterile men and infertile animals occasionally show discrepancies and variable amounts of DNA in their spermatozoa. In an extensive study on infertile men, large abnormalities in the DNA content were noted in most of the spermatozoa, in spite of the apparently normal sperm morphology, normal size of the sperm nuclei, and normal motility. By contrast, no significant differences were observed in the DNA content of the spermatogonia from fertile and infertile men. The changes in DNA in the spermatozoa from infertile men apparently occurs during spermatogenesis. Variability in the DNA content of nuclei of round spermatids and epididymal spermatozoa from F_1 hybrids of the laboratory mouse and the tobacco mouse have also been found. In these hybrid mice, abnormal meiotic segregation causes an increased variability of 6% in the amount of DNA in the sperm nuclei.

4.2.3 Characteristics (25–27)

In man, the absorptions at wavelengths of 260 and 280 nm of a DNA preparation have a ratio of 1.93. Purified human sperm DNA shows a double-strand size of approximately 100,000 base pairs (a Mw of about 70×10^6) and a single-strand size of about 17,000 bases (25). Besides double-strandedness, a high degree of stability of mammalian sperm nuclear DNA is indicated by the relatively higher melting temperature (Tm) and increased hyperchromicity in comparison to DNAs from other cells (Table 8). This nativeness is further supported by the failure of human sperm nuclear DNA to react with formaldehyde.

Although the composition of the DNA in respect to the purine and pyrimidine sequence as well as their ratio may be expected to vary somewhat from one species to another, in both species examined so far (man and bull), the ratio of adenine to thymine or guanine to cytosine and of purine to pyrimidine is one. Trace amounts of a methylated base, 5-methylcytosine, have also been found in the DNA of human, bull, and ram spermatozoa. The molar ratio of cytosine to methylcytosine of ram sperm DNA is 1:0.05.

The buoyant density of human and bull sperm DNA is identical, measuring 1.701 g/cm³. The presence of satellite human sperm DNA, with a buoyant density value of 1.704 g/cm³, has also been detected. Although the function of this satellite DNA still remains unclear, the molecule may serve to maintain nuclear organization and structural entity.

Five DNA fractions with varying molecular sizes were obtained from bull spermatozoa. The major fraction had a sedimentation coefficient of 17.7 S. A DNA preparation from human sperm heads showed an intrinsic viscosity (η) of 40 dl/gm and a sedimentation coefficient value of 16.5 S. The intrinsic viscosities of the DNA from bull and rabbit spermatozoa are 45 to 53 dl/gm and 19 dl/gm, respectively (Table 8).

Reassociation kinetic studies with sheared, purified human sperm DNA showed a sharp change in the curve for C_0t values of approximately 10^{-2}, which may indicate the presence of highly repetitive DNA sequences (28). In man, the number of repetitive sequences is significantly higher in the DNA of sperm than

Table 8 Some Properties of DNA from Mammalian Sperm Nuclei

Property	Reference[a]	Man	Bull	Rabbit
Content ($\times$ 10^{-9} mg/ nucleus)	1, 2	1.35–1.37	3.22	3.1
	1	2.45–2.86 (oligospermic semen samples)		
Molecular weight ($\times$ 10^{-6})	3[b]	70		
	5	4.0–4.4	4.5–8.2	1.8–2.3
Melting temperature (T_m) in $^\circ$C (medium)	5	89.0 (0.2 M NaCl)	87–90 (0.2 M NaCl)	88 (0.2 M NaCl)
	4	85.8 $\pm$ 0.5		
Sedimentation velocity (s)	5[c]	16.5	18.1–23.4	12.8
	4[d]	4.8 $\pm$ 0.5		
Intrinsic viscosity (n) mdl/gm	5	40	45–53	19
Hyperchromic effect (Δ260 nm; in %)	5	30	33–37	33
	4	40.6 $\pm$ 2.6		
Buoyant density (g/cm³)	5	1.701	1.701	
	4[e]	1.697 (87%)		
Base composition (moles/100 moles)	5[e]			
Adenine		29.8–30.5	25.9–28.9	
Thymine		28.9–32.5	26.3–29.4	
Guanine		18.6–19.9	21.6–22.9	
Cytosine		19.3–20.6	21.6–22.1	
Purine/pyrimidine	5	0.94–1.01	0.96–1.06	
Adenine + thymine / Guanine + cytosine	5	1.47–1.64	1.23–1.31	

[a] References: (1) Frajese et al., Fertil. Steril. 27, 14 (1976). (2) Silverstroni et al., Arch. Androl. 2, 257 (1979). (3) Wallace and Salser, Gene 7, 343 (1979). (4) Hernández et al., Biochim. Biophys. Acta 521, 557 (1978). (5) Borenfreund et al., Nature 191, 1375 (1961).
[b] DNA averaged approximately 100,000 base pairs.
[c] Average sedimentation coefficients, $S_{50\%}$, corrected to 20°C and water.
[d] After fragmentation, both alkaline sucrose gradient and sedimentation-velocity methods were used. Results correspond to a single strand chain of 200–250 nucleotides (MW approximately 75,000).
[e] Two DNA forms were found, the major one constituting 87% of the total. Buoyant density (g/cm³) of the satellite DNA (13%) was 1.704. Guanine + cytosine content, calculated from the density values, were 37.7% for the main DNA form and 44.8% for the satellite form.

in that of leukocytes. The increased number of repetitive sequences in sperm DNA may aid in giving the sperm nucleus its characteristic properties and may be the recognition sites for the sperm-specific, highly basic proteins.

4.3 Ribonucleic Acid

The mature mammalian sperm nucleus contains an extremely low amount of RNA. High levels of RNA synthesis occur in spermatogonia and primary spermatocytes of the rat, mouse, hamster, ram, and bull. Although a small burst in postmeiotic nuclear RNA synthesis occurs in mouse, rat, and bull spermatids, RNA synthesis decreases sharply upon shedding of the spermatozoa from the germinal epithelium, and does not occur, or only to a very small extent, in mature spermatozoa. One group of investigators claims that mature spermatozoa are capable of transcription and translation but that these activities are associated with the mitochondria of the sperm rather than the nucleus. This remains to be confirmed.

The majority of the RNA in the mouse and human sperm nucleus has a sedimentation coefficient of 28 S and 18 S. Most (85%) of the RNA is synthesized in primary spermatocytes, that is, during meiosis. Rat spermatocytes synthesize RNA with sedimentation coefficients of 32 S, 28 S, and 18 S. In this species, most of the RNA is labile. It appears that the primary gene product is actually a 45 S RNA that cleaves into 32 S and 18 S units. The 32 S RNA is probably not stable and dissociates into the 28 S RNA. RNA synthesis, associated with the developing germ cells, is maximal at 37°C in the presence of magnesium ions, all four ribonucleoside triphosphates, mercaptoethanol, and glycine–sodium hydroxide buffer (pH 9.0).

4.4 Nuclear Polymerases

Recently, several reports have appeared that claim that human, bull, and mouse sperm nuclei are capable of a very low level of nucleic acid synthesis and contain high-molecular-weight RNA polymerase and DNA-dependent DNA polymerase associated with particulate fractions. This DNA polymerase, while in association with the particulate complex, can synthesize DNA even in the absence of an exogenous template (polynucleotides). When dissociated from the complex, the enzyme activity can only be demonstrated in the presence of polynucleotides.

Unscheduled DNA synthesis, which is considered to be DNA repair, occurs during spermatogenesis in the mouse, hamster, rat, and rabbit. After x-ray- or UV-induced damage, the maximum rate of unscheduled DNA synthesis is observed in the nuclei of pachytene primary spermatocytes. It is markedly lower in spermatogonia, and essentially no DNA synthesis is observed in nuclei of spermatids and testicular spermatozoa. The replacement of histones by protamine in late spermatids (see next section) apparently marks the end of any unscheduled DNA synthesis in the mammalian sperm cell, regardless of the nature of the chemical or physical agents used to damage the nuclear DNA.

Appreciably high activity of RNA polymerase has been detected in bull, rat, and hamster spermatozoa. The enzyme was solubilized and isolated from the heads of mature bovine spermatozoa. Two different RNA polymerases were obtained. Both enzymes were only active in the presence of exogenous DNA, required all four ribonucleoside triphosphates for maximal activity, and were stimulated by β-mercaptoethanol. Quantitation of reaction parameters and inhibition studies with α-amanitin and rifampicin indicated a strong similarity between eukaryotic nuclear RNA polymerases 1 and 2 and the two RNA polymerases associated with the sperm head. The physiological role, if any, of the enzymes in the mature spermatozoon remains to be established, because RNA synthesis does not occur, or only to a very small extent, beyond the spermatid stage.

4.5   Nuclear Proteins

4.5.1   Introduction

Several complicated intracellular changes occur during spermiogenesis in mammals (24). Concomitant with these changes, most or all of the histones, basic nuclear proteins of molecular weight between 11,000 and 20,000, are replaced by protamines, arginine-rich proteins of lower molecular weight (about 5000–7000). In eutherian mammals, these protamines, contain a high number of disulfide bonds, and are at times referred to as basic keratins. The concentration of —S—S— bonds varies from 8 to 16% depending on the species. In mammalian sperm nuclei, the basic proteins are associated with the DNA in approximately a 1:1 ratio by weight. The DNA–basic-protein complex becomes somewhat more compact during maturation of the spermatozoon in the epididymis, and an increase in the number disulfide bonds occurs. Although some significant species differences exist in the reactive—SH (sulfhydryl) groups before the spermatozoa are mature, these differences mostly disappear after epididymal transit.

The chromatin fibers of the nucleus consist of DNA that is folded in a supercoiled manner and is tightly complexed to nucleoproteins. The high arginine content of sperm nuclear proteins not only influences the supercoiling of DNA, but also increases the thermal stability of the DNP complex. A high degree of hydrogen bonding in the DNA molecule itself and the formation of stable disulfide cross-links in the basic proteins result in the stabilization of the nuclear contents of the mammalian spermatozoa.

Extraction techniques for the purification of the nuclear proteins were presented in Section 4.2.1.

4.5.2   Properties

Considerable heterogeneity and variation in the composition of the nuclear proteins are present among species (29). On fractionation of ram nuclear proteins, eight different molecular weight forms are obtained. Except for one, they all contain cysteine. Bull and rat sperm nuclear proteins demonstrate several fractions with different electrophoretic mobility. It is, of course, not certain that this many proteins are also naturally present within the sperm nucleus, since treatment with disulfide-reducing agents with or without SDS may break down the natural protamine to different molecular weight forms. For instance, on incubation with mercaptoethanol, rabbit, hamster, and guinea pig sperm protamines produce components that migrate electrophoretically faster than the original, untreated protamines.

Treatment of rabbit, hamster, guinea pig, and bull sperm nuclear extract with mercaptoethanol produces low-molecular-weight components that disappear upon further incubation, indicating that proteolysis takes place. By contrast, if rat sperm nuclear extract is incubated for 2 hr under similar conditions, no proteolysis occurs. The proteinase that is responsible for the degradation of rabbit and bull sperm chromatin has properties similar to those of acrosin. The enzyme may be a small amount of acrosin that is associated with the nuclear membrane.

The physicochemical properties of the nucleoproteins from several species are shown in Table 9. One nuclear protein has been studied in great detail, and the amino acid sequence is known (Fig. 3). It was isolated from bull sperm

Table 9 Physicochemical Properties of Mammalian Sperm Nucleoprotein

Species[a]	Melting Temp. (°C) (medium)	Hyperchromicity (%)	Enthalpy (Kcal/mole)	Entropy (Cal/mole-°K)
Man (1)	89.1 ± 0.1 (0.15 M NaCl + 0.015 M sodium citrate)	30.5 ± 2.6		
Man (2)	80–100 (Ethylene glycol)		43.2 ± 11.9	54 ± 33
Bull (2)	90–110 (Ethylene glycol)	90–110	55.3 ± 5.3	83 ± 14
	75–85 (Ethylene glycol + β-mercaptoethanol		101.4 ± 66.5	222 ± 118
Boar (3)	79 (0.15 M NaCl + 0.015 M sodium citrate + 52% sucrose)			
	72 (0.15 M NaCl + 0.015 M sodium citrate + 70% glycerol)			

[a] References: (1) Hernandez et al., Biochim. Biophys. Acta **521**, 557 (1978). (2) Bearden and Bendet, J. Cell Biol. **55**, 489 (1972). (3) Chamberlain and Walker, J. Mol. Biol. **11**, 1 (1961).

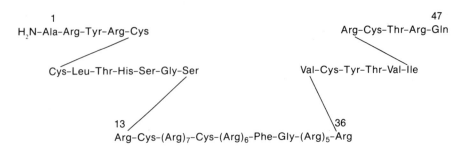

Figure 3   Amino acid sequence of bull protamine. From Coelingh *et al.*, *Biochim. Biophys. Acta* **322**, 173 (1973).

nuclei and has a molecular weight of 6190, being composed of 47 amino acids (30). The molecule possesses a highly symmetric primary structure with a central zone of 24 amino acid residues (numbers 13–36). This central portion is highly basic and contains 20 arginine residues divided into three clusters. Altogether, 24 arginine residues are present in the molecule. The N-terminal and C-terminal ends of the molecule contain 12 and 11 arginine residues, respectively, each terminal containing 2 arginines. Among the 6 half-cystine residues, 2 are found in the central portion and 2 in each terminal portion. After cleavage of the central portion with proteinase, a heterogeneous mixture of peptides is formed, varying in molecular weight from 400 to 1500.

Partial C-terminal sequences have been determined for three other mammalian sperm protamines (Fig. 4). Also, the amino acid compositions of the major fraction of protamines from a number of mammalian species are known and are shown in Table 10. Although all the mammalian sperm protamines are rich in arginine (22–30 residues) and poor in lysine (0–2 residues), their amino acid composition differs considerably from species to species. Other than arginine, the amino acids of the nuclear proteins number between 20 and 27, and the amount of half-cysteine residues varies from 4 to 8. Human sperm protamines, unlike those of other mammals, contain a high amount of histidine and glutamic acid. Electrophoretic analyses of human sperm protamines show two major

	NH$_2$ terminal	COOH terminal
Boar:	Ala–(----------------------)–Cys–(Arg)$_4$–Cys	
Stallion:	Ala–(----------------------)–(Arg)$_2$–Cys–(Arg)$_3$	
Ram:	Ala–(-------------------)–(Cys, Thr)–(Arg)$_2$–Gln	

Figure 4   The C-terminal sequence of mammalian sperm protamines. From Monfoort *et al.*, *Biochim. Biophys. Acta* **322**, 173 (1973).

Table 10 Amino Acid Composition of Sperm Protamines[a]

Amino Acid	Man Prota-mine 1	Man Prota-mine 2	Bull	Ram	Boar	Stal-lion
Arginine	22	24	24	27	24	27
Lysine	0	2	0	0	0	0
Histidine	1	8	1	1	1	0
Glutamic acid and glutamine	4	2	1	1	0	3
Serine	5	4	2	2	3	4
Threonine	1	2	3	3	1	1
Cystine (half)	5	4	6	6	8	6
Tyrosine	4	1	2	2	2	2
Phenylalanine	0	0	1	1	0	0
Alanine	2	0	1	1	2	1
Proline	2	0	0	0	2	0
Glycine	0	2	2	1	0	0
Valine	0	0	2	3	2	3
Isoleucine	0	1	1	0	1	0
Leucine	0	1	1	1	0	1
Methionine	1	0	0	0	0	0
Aspartic acid and asparagine	0	0	0	0	0	0
Tryptophan	—	—	0	0	0	0
Number of non-arginine residues	25	27	23	22	22	21
Number of different residues	10	11	13	12	10	9
N-terminal residue	Ala	prob-ably Gly	Ala	Ala	Ala	Ala
C-terminal residue	?	?	Gln	Gln	Cys	Arg

[a] Data taken from Monfoort *et al.*, *Biochim. Biophys. Acta* **322**, 173 (1973); Coelingh *et al.*, *Biochim, Biophys. Acta* **285**, 1 (1972); Kolk and Samuel, *Biochim. Biophys. Acta* **393**, 307 (1975).

bands and four to five distinct minor bands. Analyses of the two major human protamine fractions indicate significant differences in amino acid composition. Protamine 1 resembles the other mammalian protamines, whereas protamine 2, with its lysine and high histidine content, is quite different.

REFERENCES

1 Duckett, J. G., and Racey, P. A., *The Biology of the Male Gamete*. Academic Press, New York (1975).

2   Baccetti, B., and Afzelius B. A., *The Biology of the Sperm Cell*. Karger, Basel (1976).

3   Fawcett, D. W., and Bedford J. M., *The Spermatozoon—Maturation, Motility, Surface Properties and Comparative Aspects*. Urban & Schwarzenberg, Baltimore (1979).

4   Hafez, E. S. E., *Human Semen and Fertility Regulation in Men*. Mosby, St. Louis (1976).

5   Hafez, E. S. E., and Thibault, C. G., *The Biology of Spermatozoa*. Karger, Basel (1975).

6   Hubinont, P. O., L'Hermite, M., and Schwers, J., *Sperm Action*. Karger, Basel (1976).

7   Gillis, G., Peterson, R., and Russell, L., Isolation and characterization of membrane vesicles from human and boar spermatozoa: Methods using nitrogen cavitation and ionophore induced vesiculation. *Prep. Biochem.* **8**, 363 (1978).

8   Calvin, H. I., Keratinoid Proteins in the Heads and Tails of Mammalian Spermatozoa, *in* Duckett, J. G., Racey, P. A. (eds.), *The Biology of the Male Gamete*, p. 257, Academic Press, New York (1975).

9   Clarke, G. N., Boyd, R. L., and Muller, H. K., Actin-like Protein in Human Sperm Heads, *in* Boettcher, B. (ed.), *Immunological Influence on Human Fertility*, p. 211, Academic Press, New York (1977).

10 Millette, C. F., Cell Surface Antigens During Mammalian Spermatogenesis, *in* Moscona, A. A., Monroy, A. (eds.), *Current Topics in Developmental Biology*, Vol. 13, p. 1. Academic Press, New York (1979).

11 White, I. G., Darin-Bennett, A., and Poulos, A., Lipids of Human Semen, *in* Hafez, E. S. E. (ed.), *Human Semen and Fertility Regulation in Men*, p. 144, Mosby, St. Louis (1976).

12 Peterson, R. N., Russell, L., Bundman, D., and Freund, M., Sperm-egg interaction: Evidence for boar sperm plasma membrane receptors for porcine zona pellucida. *Science* **207**, 73 (1980).

13 Harrison, R. A. P., Aspects of the Enzymology of Mammalian Spermatozoa, *in* Duckett, J. G., and Racey, P. A. (eds.), *The Biology of the Male Gamete*, p. 301, Academic Press, New York (1975).

14 Zaneveld, L. J. D., The Acrosomal Enzymes of Mammalian Spermatozoa, *in* Hafez, E. S. E., Thibault, C. G. (eds.), *The Biology of Spermatozoa*, p. 192, Karger, Basel (1975).

15 Morton, D. B., The Occurrence and Function of Proteolytic Enzymes in the Reproductive Tract of Mammals, *in* Barrett, A. J. (ed.), *Proteinases in Mammalian Cells and Tissues*, p. 445, North Holland, Amsterdam (1977).

16 Zaneveld, L. J. D., Polakoski, K. L., and Schumacher, G. F. B., The Proteolytic Enzyme Systems of Mammalian Genital Tract Secretions and Spermatozoa, *in* Reich, E., Rifkin, D. B., and Shaw, E. (eds.), *Proteases and Biological Control*, p. 683, Cold Spring Harbor Laboratory, New York (1975).

17 Parrish, R. F., Polakoski, K. L., Mammalian sperm proacrosin-acrosin system. *Int. J. Biochem.* **10**, 391 (1979).

18 Parrish, R. F. and Polakoski, K. L., Boar m_α-acrosin: Purification and characterization of the initial active enzyme resulting from the conversion of boar proacrosin to acrosin. *J. Biol. Chem.* **253**, 8428 (1978).

19 McRorie, R. A., Turner, R. B., Bradford, M. M., and Williams, W. L., Acrolysin, the aminoproteinase catalyzing the initial conversion of proacrosin to acrosin in mammalian fertilization. *Biochim. Biophys. Res. Comm.* **71**, 492 (1976).

20 McRorie, R. A., and Williams, W. L., Biochemistry of mammalian fertilization. *Ann. Rev. Biochem.* **43**, 777 (1974).

21 Meizel, S., The Mammalian Sperm Acrosome Reaction—a Biochemical Approach, *in* Johnson, M. H. (ed.), *Development in Mammals*, Vol. 3, p. 1, North Holland, Amsterdam (1978).

22 Yanagimachi, R., Specificity of Sperm-Egg Interaction, *in* Edidin, M., and Johnson, M. H. (eds.), *Immunobiology of Gametes*, p. 225, Cambridge University Press, Cambridge (1977).

23 Stambaugh, R. and Smith, M., Sperm Enzymes and Their Role in Fertilization, *in* Hubinont, P. O., L'Hermite, M., and Schwers, J. (eds.), *Sperm Action*, p. 222, Karger, Basel (1976).

24 Gledhill, B. L., Nuclear Changes during Mammalian Spermiogenesis, *in* Duckett, J. G., and Racey, P. A. (eds.), *The Biology of the Male Gamete*, p. 215, Academic Press, New York (1975).

25 Wallace, B., and Salser, W., Isolation of human germ-line DNA suitable for recombinant DNA studies. *Gene* **7**, 343 (1979).

26 Zaneveld, L. J. D., and Polakoski, K. L., Biochemistry of Human Spermatozoa, *in* Hafez, E. S. E. (ed.), *Human Semen and Fertility Regulation in Men*, p. 167, Mosby, St. Louis (1976).

27 Monk, M., Biochemical Studies on Mammalian X-chromosome Activity, *in* Johnson, M. H. (ed.), *Development in Mammals*, Vol. 3, p. 189, North Holland, Amsterdam (1978).

28 Hernández, O., Bello M. D. L. A., and Rosado, A., The human spermatozoa genome— Analysis by DNA reassociation kinetics. *Biochim. Biophys. Acta* **521**, 557 (1978).

29 Coelingh, J. P., and Rozijn, T. H., Comparative Studies on the Basic Nuclear Proteins of Mammalian and Other Spermatozoa, *in* Duckett, J. G., and Racey, P. A. (eds.), *The Biology of the Male Gamete*, p. 245, The Linnean Society of London. Academic Press, New York (1975).

30 Coelingh, J. P., Monfoort, C. H., Rozijn, T. H., *et al.*, The complete amino acid sequence of the basic nuclear protein of bull spermatozoa. *Biochim. Biophys. Acta* **285**, 1 (1972).

CHAPTER SIX

THE SPERM TAIL AND MIDPIECE

R. NICHOLAS PETERSON

*Departments of Medical Physiology
and Pharmacology
School of Medicine
Southern Illinois University
Carbondale, Illinois*

1 INTRODUCTION

The midpiece and tail of the spermatozoon contain the apparatus for coordinated movement and the power pack that generates the chemical energy re-

The author expresses appreciation to Dr. Lonnie Russell for preparation of the electron micrographs.

quired for motility. Yet, despite intensive study, the biochemical mechanisms involved in motility and, more basically, the function of motility in fertilization and sperm survival are still not completely understood. In most and perhaps all mammalian species, spermatozoa are rapidly transported to the site of fertilization by the contraction of the female reproductive tract. It is possible, therefore, that motility plays its major role as spermatozoa pass through the outer investments of the egg to gain access to the egg surface. Indeed there is evidence that indicates that the motility of spermatozoa increases in the presence of oviductal fluid and as spermatozoa contact egg investments. Motility may also serve another function in that it allows spermatozoa to circulate rapidly within fluids of the reproductive tract and thus avoid depletion of nutrients and oxygen in the microscopic environments that surround them. This may account in part for the ability of some species of spermatozoa to survive for several days in the female tract. The major aspects of the structure of the flagellum of the mature spermatozoon have been described in detail and leave less to speculation; yet here also the mechanisms involved in producing the coordinated forward motion of this cell are poorly understood. But much has also been learned, and this chapter presents a review of our current knowledge of the biochemistry of the sperm midpiece and tail. Some emphasis will be placed on energy metabolism and its control, and on the role of the plasma membrane in the control of motility. Specific references have been omitted from the text. However, a number of reviews and pertinent articles are listed in the references (1–23).

2   FUNCTIONAL ANATOMY OF THE FLAGELLUM

The major structural features of the flagellum of the mammalian spermatozoon are illustrated in Fig. 1. The organization of the tailpiece is quite complex when viewed in cross section as shown. At the center of the flagellum is the axoneme with the characteristic 9 + 2 microtubular structure typical of virtually all cilia and flagella. The outer doublets contain an ATPase activity that is thought to generate the chemical energy that powers motility by inducing a sliding motion between these doublets. This mechanism will be described in more detail in a later section of this chapter. The structures that surround the axoneme vary throughout the length of the flagellum. Perhaps they are most complex in the midpiece region, where a ring of nine large (coarse) fibers and mitochondria surround the axoneme. The nine coarse fibers (sometimes called gamma fibers) extend from the anterior portion of the midpiece and taper in size, ultimately disappearing in the principal piece of the flagellum. These coarse fibers, which differ in size and thickness, are found in most vertebrate species, some invertebrates and even in certain insects. They are best developed, however, in mammals. The function of these coarse fibers is not entirely known, although it is thought that they play at least a supportive role in motility. The size of these fibers varies among mammalian species, and spermatozoa of those species with thick, long fibers move more slowly than spermatozoa of species such as the man, in which the fibers are comparatively thin and short.

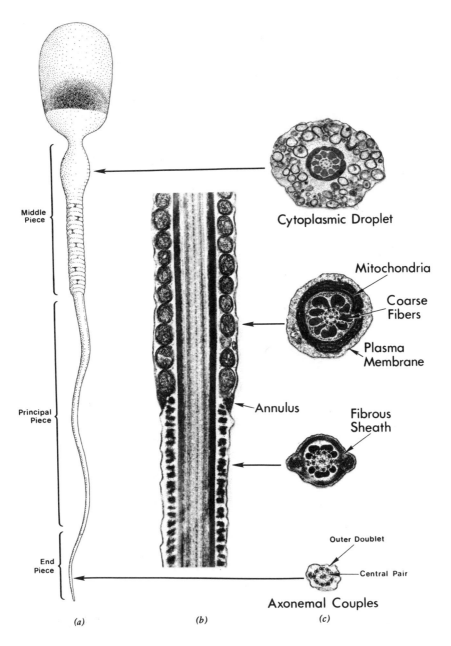

Middle
Piece

Principal
Piece

End
Piece

Cytoplasmic Droplet

Mitochondria

Coarse
Fibers

Plasma
Membrane

Annulus

Fibrous
Sheath

Outer Doublet

Central Pair

Axonemal Couples

(a)   (b)   (c)

Figure 1 Anatomical representations of the mature mammalian spermatozoon. *a*) Idealized normal spermatozoon. *b*) Electron micrographs of longitudinal sections of the midpiece and tail of the boar spermatozoon. *c*) Electron micrographs of cross sections of a boar spermatozoon.

155

The coarse fibers are surrounded in the midpiece by mitochondria, which, as in other cells, provide a substantial portion of the chemical energy used by the spermatozoon under aerobic conditions. In mammals the mitochondrial sheath consists of many elongated mitochondria encircling, in spiral form, the coarse fibers of the midpiece. But the structure of this sheath as well as the morphology of the midpiece in general, varies among species. In the sea urchin, for example, the midpiece contains a single mitochondrion wrapped in spiral form around the anterior portion of the tail.

The enzymes of the glycolytic apparatus are thought to exist within the midpiece, although their exact location in this region has not been determined. Indeed, the possibility that glycolytic enzymes may exist in regions of the flagellum other than the midpiece has not been rigorously excluded. It is improbable that all glycolytic enzymes exist in free solution in the limited volume of the cytosol of the flagellum. Hexokinase, for example, is not easily released from sperm suspensions by various methods of cell disruption, and at least a portion of this enzyme is bound to particulate matter. Studies by Storey and his associates of the glycolytic activity of rabbit spermatozoa that had been repeatedly washed in hypotonic medium also suggest that several glycolytic enzymes are bound to structural components of spermatozoa, such as the axonemal complex or the plasma membrane. There is ample evidence for such plasma membrane associations in other cell types. Aldolase and glyceraldehyde-3-phosphate dehydrogenase, for example, bind to specific sites on the red blood cell membrane.

The plasma membrane undoubtedly plays a special role in the initiation and control of motility by virtue of its function in nutrient and ion transport. However, until recently, very little was known about the structure and composition of this membrane. Seen in cross section, the plasma membrane lies tightly apposed to the mitochondria and perhaps in contact with them. Freeze-fracture studies of guinea pig spermatozoa by Friend have shown rows of particles in the plasma membrane of this region that may be associated with the mitochondria. The particles appear to change in their distribution with a change in motility or after disruption of mitochondrial function by poisons such as cyanide. The relationship of these changes to energy metabolism is of obvious interest but remains to be defined. There is evidence that a glycoprotein-polysaccharide layer, or glycocalyx, is associated with the plasma membrane of spermatozoa, since the spermatozoa surface reacts strongly with ruthenium red and other polysaccharide-staining reagents. The function of this material and of other proteins that become bound to spermatozoa during epididymal transit or during contact with the seminal plasma is beginning to be clarified, and recent studies implicate the glycocalyx and peripheral-surface glycoproteins in the control of ion transport and motility.

Another major anatomical segment of the tail of a mammalian spermatozoon is the principal piece, which extends caudally from the midpiece to the endpiece. The principal piece is characterized by the presence of a fibrous sheath that encircles the axoneme and that, in some species such as the human, shows connections to some of the outer doublets of the axoneme. The structure of this fibrous material has not been investigated in any great detail, and its role in motility is mostly speculative. It is generally thought to have an elastic or supportive function. Recent studies by Friend with guinea pig spermatozoa show that the plasma membrane over the principal piece displays some unusual features, including a thickening in the region opposite the coarse fibers. Extending farther down the principal piece, these dense regions appear as a "zipper of particles" in freeze fractures and surface replicas. Such structural specializations, which are enigmas from a functional point of view, are manifest in the flagella of a wide variety of species of spermatozoa.

The last segment of the flagellum, the endpiece, lacks any structural detail other than the axoneme.

3 METABOLISM

3.1 Glycolysis

The major metabolic reactions occurring in mammalian spermatozoa involve pathways for the utilization of simple sugars to form the chemical energy to drive motile mechanisms and to maintain cellular osmotic balance. Very little sugar is converted to glycogen in mammalian spermatozoa, although glycogen storage occurs in many other vertebrate and invertebrate species. Hexose catabolism involves the conversion of hexose to pyruvic acid or lactic acid by the Embden-Myerhoff pathway and by oxidation of either or both of these prod-

Table 1   Oxygen Consumption by Various Species of Spermatozoa

Species[a]	Oxygen uptake (37°C) $\mu l/10^8$ sperm/hr
Bull (1)	21
Cock (1)	7
Rabbit (1)	11
Ram (4)	22
Boar (2)	30
Guinea Pig (3)	42
Human (4)	3

[a] References: (1) Lardy and Phillips, *Am. J. Physiol.* **138**, 741 (1948). (2) Unpublished observations from the authors' laboratory. (3) Frenkel et al., *Fertil. Steril.* **26**, 144 (1975). (4) Peterson and Freund, *J. Reprod. Fert.* **17**, 357 (1968).

ucts in the mitochondria (Fig. 2). Glycolysis in mammalian spermatozoa is characterized by unusually high aerobic rates, which in some species such as man proceed at a rate almost as rapid as in the absence of air. Indeed potent mitochondrial poisons such as azide or dinitrophenol have, in contrast to their effect on most other cell types, only a small effect on glycolysis in human spermatozoa, and these cells maintain high motility in the presence of these poisons or in the absence of air. Oxygen consumption by human spermatozoa is unusually small but nevertheless is also capable of supporting motility for short periods in the absence of glycolysis. There is a striking difference among species of mammalian spermatozoa in their ability to consume oxygen and in their dependence on oxygen for survival (Table 1). In guinea pig spermatozoa, for example, oxygen uptake is about 10-fold greater than it is in human spermatozoa, and guinea pig spermatozoa cannot survive in the absence of air, regardless of the amount of metabolizable sugars present. Boar spermatozoa are also highly dependent upon oxygen for survival, whereas spermatozoa of other species, such as the bull, are less dependent on an aerobic environment. The physiological function of these metabolic peculiarities is not known.

The high rate of aerobic glycolysis of many species of mammalian spermatozoa has its parallel in similar metabolic patterns in many tumor cells. High aerobic glycolytic rates reflect, in part, an absence of significant feedback on the glycolytic pathway by the metabolites of mitochondrial metabolism (the so-called Pasteur effect). Although the reason for this has not been adequately explained, it does not appear to be due to any usual properties of key glycolytic enzymes, which respond to the allosteric* effects of mitochondrial metabolites much as do these enzymes in other cells. One plausible explanation is that a significant portion of the energy generated by mitochondrial electron transport

---

* Allosteric effects on enzymes are responses (either stimulation or inhibition) of enzyme activity to ligands that bind to sites on the enzymes other than the active site. The ligands may be metabolites or cofactors in metabolism. Thus the enzyme phosphofructokinase is inhibited by citric acid, which is generated by Krebs cycle oxidations, and by ATP, which is generated both by glycolysis and Krebs cycle oxidations. Note that ATP is also a substrate for phosphofructokinase (Fig. 2).

Figure 2 Pathways of energy metabolism in spermatozoa. As in other cells in spermatozoa, the ATP required for mechanical and chemical work is generated by glycolysis and oxidative phosphorylation. In the abbreviated pathways shown in the figure, the following points are noted. The energy-consuming reactions occur at steps 1 and 2, catalyzed by the enzymes hexokinase and phosphofructokinase; steps 4 and 5 generate ATP. Control of the glycolytic rate in human and other primate sperm appears to involve primarily the enzymes phosphofructokinase and glyceraldehyde phosphate dehydrogenase (steps 2 and 3); the latter enzyme catalyzes the conversion of triose phosphate to diphosphoglyceric acid. Acetyl coenzyme A, formed from the end product of glycolysis, is oxidized by Krebs cycle enzymes (6) in the mitochondria. This series of cyclical reactions generates the reducing equivalents (NADH, FADH) that are oxidized by the components of the respiratory chain (7) to produce additional ATP. Marked quantitative differences in the rates of glycolytic metabolism and oxidative metabolism exist among species of spermatozoa.

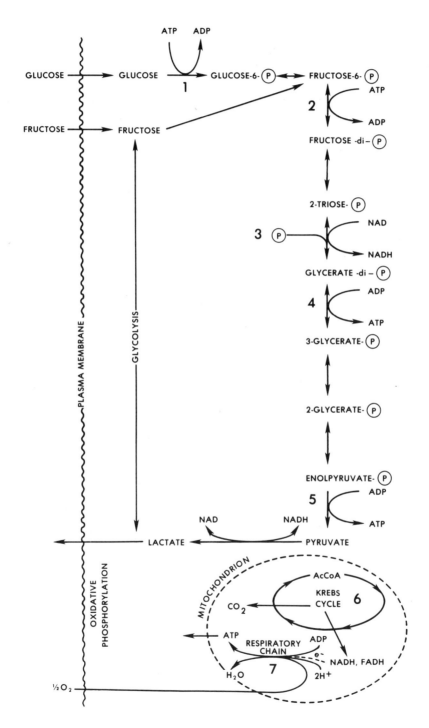

is converted into the energy needed to carry out non–ATP-dependent ion transport, which may be very active in motile cells.

Glucose and fructose serve as sources of metabolic fuel in most species of spermatozoa, although mannose and possibly galactose can also be used. Fructose abounds in the seminal plasma of some species such as man, bull, and monkey, and glucose is found in ample concentration in the fluids of the female reproductive tract. Attempts have been made to rationalize the high concentrations of seminal plasma fructose in terms of sperm survival in the female tract, but spermatozoa uptake mechanisms show a considerably higher affinity for glucose than for fructose, and since seminal plasma must be diluted to some extent in the female reproductive tract (in many species, seminal plasma does not pass through the cervix), glucose probably is the major source of energy for most species of mammalian sperm.

Transport of glucose into the spermatozoon may be considered the first step in metabolism, since it occurs by a specific transport mechanism that is inhibited by low concentrations of the antibiotic cytochalasin B and other specific inhibitors of hexose transport. Fructose uptake is unimpeded by cytochalasin B and appears to be transported into the cell by a separate mechanism. Very little is known about sugar transport in spermatozoa, primarily because of technical difficulties in measuring a rapid sugar uptake into a small metabolizing intracellular volume of only 16–20 μm^3. However isotopic measurements with human spermatozoa suggest that the initial rate of uptake of glucose exceeds the rate of lactate production at physiological glucose concentrations, indicating that transport is probably not a limiting factor in glucose utilization in this species.

The Embden-Myerhoff pathway is the predominant and probably the only significant pathway for glucose degradation in mammalian spermatozoa. Table 2 shows the concentration of glycolytic enzymes found in homogenates of human spermatozoa and steady-state concentrations of glycolytic intermediates in motile spermatozoa provided with glucose as a source of energy. The comparative concentrations of these enzymes as well as the steady-state concentrations of the metabolites indicate probable sites of glycolytic control, as will be discussed below. Note that several enzymes of the hexose monophosphate shunt are present in mature sperm, but it is likely that these enzymes are associated with residual unshed cytoplasm (cytoplasmic droplets) present on some ejaculated spermatozoa, since there is little evidence for pentose synthesis or metabolism in these cells. Moreover a need for this pathway, which ordinarily generates the nucleotide NADPH needed for biosynthesis and ribose for the synthesis of nucleic acids and their precursors in other cells, has not been demonstrated in the mature mammalian spermatozoon.

The control of glycolysis in mammalian spermatozoa has been analyzed by several investigators. Hoskins has postulated that in bull sperm (and very probably in other spermatozoa including human, as we suggest) glycolytic rates operate near capacity and are controlled primarily by the energy charge of the cell (i.e., by the fraction of high-energy phosphoanhydride bonds contained in the total nucleotide pool). This argument rests in part on the observation that

Table 2 Glycolytic Enzymes and Metabolites in Washed
Suspensions of Human Spermatozoa[a]

Enzyme	Activity μmol/ hr/10^8/ sperm	Metabolite	Intracellular Steady-State Concentration nmol/10^8/ sperm
Hexokinase	1.4	Glucose	—
Phosphohexosisomerase	22.5	Glucose-6-phosphate	1.5
Phosphofructokinase	1.8	Fructose-6-phosphate	0.5
Aldolase	5.5	Fructose diphosphate	5.0
Glyceraldehyde-3-phosphate dehydrogenase	5.7	Triose phosphate	8.2
Phosphoglycerate kinase	40.5	3-Phosphoglycerate	1.1
Enolase	7.1	2-Phosphoglycerate	—
Pyruvate kinase	26.0	Phosphoenolpyruvate	0.6
Lactate dehydrogenase	49.1	Pyruvate	16.4
		ATP	12.6
		ADP	9.1

[a] From Peterson and Freund, *Fertil. Steril.* **21**, 151 (1970).

when motility is increased by elevation of intracellular cyclic AMP (cAMP) the glycolytic rate in spermatozoa increases, but at the expense of a fall in the ATP:ADP ratio in the cell. This can be explained by the inability of glycolysis to increase the rate of ATP production sufficiently when the energy demands of increased motility exceed a certain limit. A lower steady state of ATP:ADP is established under these conditions of near-maximum glycolysis. High levels of ATP have also been postulated to account for the inhibition of glycolysis during bovine and guinea pig sperm passage through the epididymis. The induction of motility at ejaculation very probably also contributes to a reduction in ATP levels in epididymal sperm and thereby contributes to a needed increase in the glycolytic flux.

By measuring the changes in the steady-state concentration of the substrates and products of each glycolytic enzyme before and after an induced change in the glycolytic rate (for example, a change from an aerobic to anaerobic environment), it has been possible to identify those spermatozoal enzymes primarily involved in controlling glycolytic rates. These analyses have indicated that phosphofructokinase and glyceraldehyde-3-phosphate dehydrogenase play key roles in the regulation of glycolysis in bovine, monkey, and human spermatozoa. The activities of these enzymes are inhibited by ATP in most cell types (allosteric effect) and therefore their activities and the rate of glycolysis will decrease when energy reserves in the form of ATP are high.

The observation that energy metabolism in spermatozoa operates near capacity is perhaps not surprising in view of the short life span of these cells and the absence of significant biosynthesis that could impress tighter regulation of the utilization of high-energy metabolites such as that occurring in the more complex metabolic pathways of the somatic cell.

3.2   Oxidative Metabolism

Mitochondrial oxidative metabolism is capable of maintaining spermatozoal motility in the absence of exogenous substrates, indicating that endogenous oxidizable reserves, presumably lipids, are present in spermatozoa. Early studies by Terner showed that labeled acetate (and therefore acetyl-CoA) was incorporated into the fatty acids of human and bull spermatozoa, and labeled glucose was incorporated into the glycerol portion of triglycerides. However, since exchange reactions probably account for some of this incorporation, it is not clear how much these reactions contribute to net lipid synthesis in addition to their obvious effect of maintaining lipid reserves.

A considerable species variation exists in the ability of spermatozoa to oxidize various substrates. For example, pyruvate, lactate, and α-glycerophosphate are rapidly oxidized by boar and guinea pig spermatozoa, but not by human spermatozoa. Variability in the oxidative capacity of the respiratory chain among these species does not appear to be involved because, given an appropriate substrate (e.g., succinate for human spermatozoa), the maximum rate of oxygen uptake by each species is comparable. Undoubtedly significant differences in the regulation of Krebs cycle oxidations exist among different species, and more experimental work is needed to determine the mechanism of this regulation.

Lactate dehydrogenase (LDH), particularly the X isozyme (LDH-X), appears to play an important role in mitochondrial metabolism in spermatozoa. The enzyme, which is located in the cytosol of most other cells, is present almost exclusively in the mitochondria of mammalian spermatozoa. LDH competes effectively for reducing equivalents with respiratory-chain enzymes and permits the direct oxidation of the lactic acid produced by glycolysis. The advantage of this specialization is not obvious, although it has been suggested that mitochondrial LDH may participate in shuttling reducing equivalents from the cytosol to the mitochondria, where energy yields are higher.

An important role in mitochondrial metabolism for L-carnitine and L-carnitine acetyl transferase in spermatozoa has recently been demonstrated by work from several laboratories. Carnitine and its derivatives are present in high concentration in epididymal spermatozoa, and it is thought that "active acetate" in the form of acetylcarnitine may provide fuel for spermatozoa during epididymal transit. Acetylcarnitine is also thought to function in carbohydrate metabolism and in the utilization of pyruvate. Carnitine levels in monkey spermatozoa are some 2000 times greater than CoA levels, and since pyruvate and lactate are converted to acetylcarnitine, this intermediate can serve as a reserve supply of active acetate. This would buffer against otherwise rapid changes in acetyl-CoA levels and thus ensure a steady flow of intermediates into the citric acid cycle. These metabolic pathways seem particularly well suited for the maximum utilization of glucose and pyruvate by spermatozoa under aerobic conditions, particularly since these compounds have been implicated as major sources of energy during capacitation.

It is quite difficult to determine the relative contributions of glycolysis and respiration to the overall energy demands of the spermatozoon, since such a calculation depends on very precise knowledge of metabolic rates and parameters of hydrodynamic work that are not easily measured. In one important study with bull sperm, however, Rickmenspoel has estimated that approximately 25% of the free energy of energy metabolism was generated by fructolysis. Of the free energy generated by mitochondrial metabolism, only 30–40% was available to directly support motility (ATP generation). The remaining free energy presumably was used to drive ion transport, which suggests the importance of this process in motility.

3.3 Role of Cyclic AMP in Metabolism and Motility

Increased levels of intracellular cAMP have marked effects on the motility and metabolism of epididymal and washed ejaculated spermatozoa. Garbers and his associates carried out the first comprehensive study of the effects of pharmacologic agents that alter internal levels of cAMP on the motility of bovine epididymal spermatozoa. They showed that the increased levels of cAMP induced by caffeine and papaverine, both potent cAMP phosphodiesterase inhibitors, increased fructolysis, respiration, and motility (Table 3), whereas just the

Table 3 Respiration of Bovine Epididymal Spermatozoa With Various Substrates at 10 mM in the Presence and Absence of Caffeine[a]

| Substrate (10 mM) | n | μg-Atoms of O_2/10^8 Cells/hr | | |
		Substrate Only	+2 mM Caffeine	Increase
Endogenous	8	0.80	1.20	0.40
Glucose	1	2.08	2.69	0.61
Fructose	4	1.68	2.88	1.20
Lactate	3	1.18	3.00	1.82
Pyruvate	12	1.48	4.90	3.42
Acetate	7	1.45	4.21	2.76
Citrate	4	0.90	1.31	0.41
β-Ketoglutarate	4	0.91	1.27	0.36
Succinate	1	1.46	1.69	0.23
Fumarate	3	0.82	1.11	0.29
Malate	2	1.12	2.53	1.41
Oxaloacetate	3	1.63	4.88	3.25
β-Hydroxybutryate	2	1.17	4.04	2.87

[a] From Garbers et al., Biochemistry 10, 1825 (1971).

opposite effects occurred in the presence of imidazole, an agent that stimulates phosphodiesterase (see Fig. 3 for a general scheme of the relationships among cAMP, phosphodiesterase, and protein kinases). Later, Hoskins and his colleagues were able to show that cAMP-dependent protein kinases were present in abundance in bovine spermatozoa, and other studies by these workers showed that cAMP plays an important role in the development of motility in the epididymis. Hoskins and co-workers have recently shown a relationship

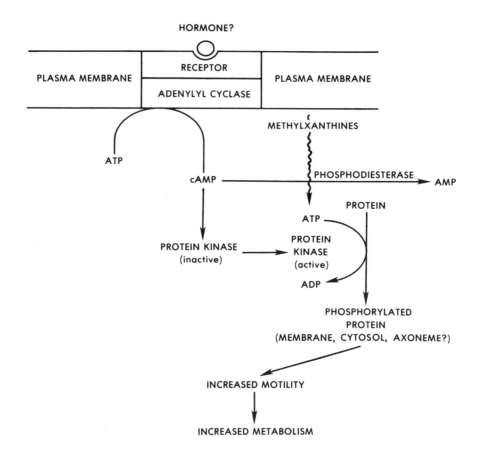

Figure 3   Role of cyclic AMP in cellular metabolism. The diagram depicts the general scheme for the mechanism of action of cAMP. Membrane-bound adenylyl cyclase, possibly activated by an as-yet-unidentified hormone, converts ATP to cAMP. The cyclase and the putative hormone receptor are very likely free to move with the fluid lipid bilayer. This reaction is controlled in many cells by calcium and a calcium-binding protein, calmodulin, which recent experiments suggest may also exist in spermatozoa. Cyclic AMP is known to activate protein kinases, which in turn may act to phosphorylate a variety of proteins throughout the cell. These phosphorylated proteins enhance motility by mechanisms that are not known. Caffeine, theophylline, and other phosphodiesterase inhibitors inhibit the breakdown of cAMP, and thereby potentiate its actions.

between the cAMP-dependent phosphorylation of soluble proteins and motility. Considerable interest is now focused on the locus of action of these proteins.

As mentioned above, it is thought that increases in metabolism arise secondarily to cAMP-induced increases in motility, but the mechanism by which motility is increased is poorly understood. Both the plasma membrane and axonemal proteins have been implicated as sites of action of cAMP-activated protein kinases. Studies by Herman and studies in our laboratory indicate that in bull, human, and boar spermatozoa, adenylyl cyclase activity is concentrated in plasma membranes, as it is in most cell types. But this may not be the only cellular locus of this enzyme; Herman also found that a significant portion of adenylyl cyclase in bull sperm could be isolated in soluble form, and cytochemical studies at the electron microscopic level in this laboratory suggest that a portion of the cellular adenylyl cyclase activity in human spermatozoa may be located along the axonemal complex. However, the localization of adenylyl cyclase at sites other than the plasma membrane has been reported only for a few cell types, and these cytochemical observations must be confirmed by direct enzyme assay using isolated axonemal preparations. So far this has not been accomplished.

It seems reasonable to expect that alteration in the rate of ion flow across the plasma membrane accompanies a change in motility, and indirect evidence for cAMP-mediated changes in ion flow across plasma membranes of boar and human spermatozoa have been observed. However, it must be stressed again that too little is known about the motile apparatus itself (see below) to permit all but guesses on the effects of phosphorylation on proteins associated with the apparatus.

4   THE PLASMA MEMBRANE, ION TRANSPORT, AND MOTILITY

Aside from the structural specializations noted above, little is known about the composition of the plasma membrane of the flagellum. Regional differences in the charge density of the surface membrane have been noted by several investigators using colloidal iron as a cytochemical stain. Lectin binding to the surface membrane of mouse spermatozoa also shows regional differences in the specificity and the amount of lectin bound, and certain surface antigens appear to be restricted to discrete regions of the plasma membrane. Studies in this laboratory, in which plasma membranes from different regions of boar spermatozoa were isolated by sucrose density centrifugation after selective vesiculation by gas cavitation and treatment with antibiotics, also suggest regional differences in protein composition along the surface membrane of this species. Differences between the proteins of the head and flagellar plasma membranes are not marked, however, and appear to be due to differences in the concentration of specific proteins as well as the presence or absence of certain proteins. Those proteins present in high concentration appear to be present throughout the plasma membrane, whereas those proteins restricted to particular regions may correspond to only a small fraction of the total membrane protein.

Studies with membrane-active drugs have indicated that the plasma membrane plays an active role in the control of metabolism and motility. Experiments in the author's laboratory and those carried out by Webster and Foote have shown that ouabain, which specifically interacts with the membrane-bound Na^+-K^+-activated ATPase, inhibits metabolism and motility in human and bull spermatozoa. The studies by Webster and Foote are particularly interesting because they observed that testosterone, which stimulates motility, blocked that action of ouabain and vice versa, suggesting that both agents were acting on the same enzyme. Although these effects conceivably could result from changes in intracellular ATP levels induced by inhibition or acceleration of enzyme activity, a direct effect of a change in ion flow on motility is an attractive alternative. Indeed, there is substantial evidence that motility is highly sensitive to the ionic composition of supporting media. For example, high concentrations of potassium ions are known to depress sperm motility. In a detailed study of ionic effects on the motility of bull and chimpanzee spermatozoa, which measured the velocity, frequency, and amplitude of flagellar motion, McGrady and Nelson noted the following. In bull spermatozoa, motility was depressed in a potassium-free medium or in media containing potassium in excess of 6 mM; motility was also depressed in media containing low concentrations of chloride or sodium ions and when the calcium concentration exceeded 1 mM. The effects of changes in the concentrations of these ions in chimpanzee sperm were slightly different, and it was thought that this was related to the known differences in the ionic composition of the seminal plasma from both species. Unfortunately, the mechanistic basis for these changes in motile patterns must await more-detailed knowledge of the contractile processes in the flagellum. It is not unreasonable to expect that ion fluxes, perhaps by their effects on surface- or intracellular-propagated action potentials, play key roles in developing progressive forward flageller motion.

Recent studies on the role of ions in motility have focused on calcium, since this ion is known to be indispensable for the acrosome reaction, and since there is some evidence that calcium may increase motility during capacitation. Reports on the effects of calcium on spermatozoa motility vary considerably and may reflect species differences. Calcium ions have been reported to stimulate motility in some species of quiescent epididymal sperm but not in others, and in other studies, both inhibition and stimulation of motility, dependent on the calcium concentration used, have been observed. There also appear to be distinct species differences in response to the influx of calcium ions mediated by the divalent cation ionophore A23187.* The motility of quiescent epididymal bull sperm is increased by the ionophore plus calcium, whereas these agents rapidly abolish the motility of rat, guinea pig, and boar spermatozoa. Human spermatozoa, on the other hand, are not affected by the ionophore and calcium unless high concentrations are employed. Some of these species differences

* Ionophores are lipid-soluble compounds that contain polar regions highly specific in their ability to bind ions. The compounds, which are often antibiotics, can act as ion carriers rapidly transporting highly charged ions across hydrophobic barriers such as cell membranes.

seem clearly related to effects on metabolism; the rapid uptake of calcium into the mitochondria of spermatozoa that are highly dependent on oxidative metabolism may lead to an uncoupling of phosphorylation and cell death. Still, this leaves unexplained the observations that both ionophore-mediated calcium influx and calcium alone can initiate motility in quiescent epididymal sperm, and that calcium enhances motility in some species of capacitated spermatozoa. Recent reports of an increase in cAMP levels in capacitated guinea pig spermatozoa in the presence of calcium suggest that activation of adenylyl cyclase by calcium may play a role in this change of motility.

Regarding epididymal spermatozoa, it is quite possible, as some recent experiments by Morton suggest, that calcium may be involved in the initiation of motility, but that once started, external calcium is not required to maintain motility. Regarding capacitated spermatozoa, Singh and co-workers have observed changes in the pattern of calcium influx during *in vitro* capacitation in the guinea pig. In the early phase of capacitation, calcium is apparently loosely bound to the surface of the spermatozoa, then a secondary phase of calcium uptake occurs that involves uptake and binding to stable sites deep within the membrane and parallels the time course of the acrosome reaction. This leads Babcock to postulate a common mechanism that allows more-rapid entry of calcium into the cell for direct stimulation of the contractile apparatus and for induction of the acrosome reaction. However, direct effects of calcium on the contractile apparatus of spermatozoa have still to be convincingly demonstrated. A very recent study by Brokaw has shown that the flagellar beat of demembranated, reactivated sea urchin spermatozoa is asymmetrical unless calcium is added to the reactivating medium; however, again the functional significance and mechanistic explanation of these effects remain to be provided.

5   THE ROLE OF SURFACE PROTEINS IN MOTILITY

The capacity for forward motility develops in spermatozoa during epididymal transit. Although this has been known for more than 60 years, it is only recently that the biochemical basis for this change has been determined. The characteristic rapid and directed movements of mature spermatozoa appear to be due to several major changes in the biochemical properties of epididymal spermatozoa. An increase in intracellular cAMP is an important component of the motility increase, and as demonstrated by Hoskins and co-workers for bull sperm, the levels of the cyclic nucleotide almost double during epididymal transit. Caput epididymal spermatozoa of the guinea pig, rat, bull, and other species are either immobile or move slowly in circular paths when resuspended in buffers *in vitro*. The addition of cAMP phosphodiesterase inhibitors or dibutyryl-cAMP enhances motility, but this motile change is characterized by a twitching motion without forward direction. Several studies using ligation of the caudal epididymis to retain sperm for long periods in the caput epididymis demonstrated that immature spermatozoa under these circumstances develop normal motility with enhanced forward progression. Controversy soon arose regarding the role of the epididymis in developing the motile patterns of mature spermatozoa. One investigator noted that in blocked epididymides in man, rapid forwardly moving spermatozoa could be recovered, and suggested that these motility patterns developed as a result of time-dependent properties intrinsic to the spermatozoon. But other studies have implicated extrinsic epididymal factors in the development of normal motile patterns in other species. Evidence has been presented that the development of normal patterns of motility in the rat requires androgen stimulation of the epididymis, and this supports the idea that a secretory product of the epididymis is involved in changing motility patterns.

The first suggestion that proteins secreted by the male reproductive tract were important in motility came from the work of Lindholmer, who showed that human seminal plasma induced progressive forward motility in sperm trapped by occlusion in the caput epididymis. Recently, Hoskins and his colleagues have isolated and partially purified a glycoprotein from bovine seminal plasma that induces progressive forward motility in caput spermatozoa under conditions of elevated levels of cAMP. The experiments indicate that this protein originates from the secretions of the cauda epididymis and may act together with cAMP in developing the motile patterns of the mature spermatozoon. The protein is not an enzyme, nor does it appear to operate through a simple enzymatic mechanism. We note, again, however, that although these experiments have provided new information regarding the contribution of epididymal secretions and cAMP on motility, much remains to be learned about how these agents interact with the contractile apparatus. The involvement of glycoproteins in this interaction, and thus presumably the surface of the spermatozoon, again suggests that coordinated motile movements in spermatozoa may be controlled in part by signals originating from the plasma membrane.

6   OTHER MECHANISMS OF MOTILITY CONTROL

A neuronal theory for the control of motility has been suggested by Nelson. This idea arises from the effects of cholinergic drugs on spermatozoal motility. After establishing the presence of a cholinesterase in spermatozoa, further studies showed that inhibitors of cholinesterase (e.g., eserine) stimulated motility. Acetylcholine also stimulated motility, and, provided that the sperm membrane was made permeable to allow access to intracellular binding sites, anticholinergic compounds inhibited motility. Nelson proposes the existence of a cholinergic receptor oriented in the plasma membrane such that its active site is located on the cytoplasmic surface. Activation of the receptor (by binding to acetylcholine or its analogs) induces changes in ion transport that increase motility. It is speculated that calcium may play a key role in the changes as a stimulus for an excitation wave that is responsible for the propagation of contractile events (see below).

7 DRUG EFFECTS ON MOTILITY AND METABOLISM

In contrast to the studies with cholinergic agents (see section 6 of this chapter), other studies have shown that a variety of common pharmacologic agents inhibit motility, very likely as the result of their interaction with the plasma membrane. Agents such as propranolol, lidocaine, and diphenydramine depolarize the membrane of spermatozoa. Some of these agents (chlorpromazine, benztropin mesylate) are particularly potent and inhibit the motility of human spermatozoa at concentrations as low as 50 μM. An initial membrane depolarization is induced by these drugs and can, in part, be reversed by the potassium-specific ionophore valinomycin; a second and slower depolarization is not reversible. Part of the effects of these agents is, therefore, probably related to changes in plasma membrane ion flow, and it has been established that propranolol and certain other local-anesthetic-like compounds displace divalent cations from plasma membranes in other cells. These experiments have also suggested that the plasma membrane interacts directly with the glycolytic apparatus. Glycolysis is inhibited by propranolol in intact spermatozoa, and triose phosphates and fructose diphosphates accumulate. This has suggested that a plasma membrane interaction with one or more glycolytic enzymes may be disturbed when local-anesthetic-like drugs enter the membrane. Perhaps related to this are the electron microscopic observations, described earlier, for metabolically induced organizational changes in the plasma membrane of the midpiece. Conceivably, tight association between the plasma membrane and the apparatus for energy metabolism may be required to integrate the closely related processes of ion transport, motility, and energy metabolism.

A particularly interesting drug is tetraphenylboron, which, unlike many other active drugs, reversibly inhibits spermatozoal motility during short periods of contact. This lipophilic organic anion has several effects on membrane structure, but an important interaction involves binding to exoplasmic, postively charged, membrane phospholipids. One effect of the drug interaction is to cause a decline in sperm cAMP levels. Interestingly, the enzyme adenylyl cyclase in isolated membranes is not affected by the drug if the membranes are treated with a detergent such as Triton X-100. This suggests that membrane phospholipids probably play an important role in adenylyl cyclase activity in the intact spermatozoon. The effects of tetraphenylboron can be reversed by albumin or potassium ions, which insolubilize the drug, and motility can be reinitiated by these antagonists. This drug holds promise as a probe for a more detailed analysis of the ionic events that initiate motility in spermatozoa.

8 THE MOTILITY APPARATUS

Although spermatozoal motility has been the concern of biologists for many years, these scientists had to be content with the fact that little was known about the motile apparatus itself. Even today, the mechanism of motility in molecular terms is only poorly understood, but the rapid growth of knowledge of contractile events in nonmuscle cells promises a short life for this ignorance.

The sliding microtubule model originally proposed by Satir for cilia and flagella (Fig. 4) satisfies many of the conditions for movement. Electron micrographs had established that neighboring outer microtubule doublets of the axonemal complex slide past each other without contracting. This sliding induces a localized bending in the flagellum. Sliding is apparently due to the cyclic attachment and detachment of the outer radial arms of the complete (13-subunit) A tubule with an adjacent B tubule. These extensions contain the ATPase dynein, thought to be involved in these cyclic reactions. It is postulated that the sliding of the outer doublets and the constraints imposed on sliding by (1) the connections between doublets (nexin fibers) and (2) the radial spokes that connect the doublets to the sheath surrounding the central doublets give rise to characteristic bending patterns. The length of the tubules does not

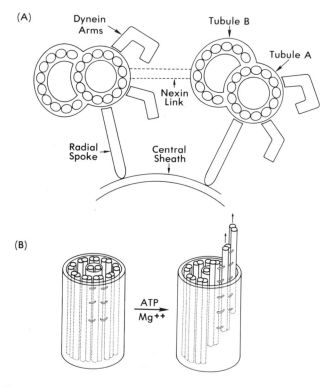

Figure 4 The axoneme complex. (*A*) Outer doublets. Note dynein arms in the A tubules. These are believed to form cross bridges with a neighboring B tubule, the making and breaking of which causes sliding of adjacent tubules past one another. This is depicted in simplified form in *B*. For clarity, only a single connection is used to represent tubule A–tubule B reactions, and not all dynein links are shown. The nexin links between subunits and the radial spokes are believed to provide the resistance to tubule sliding that gives rise to bending in the flagellum.

change during bending. Elegant experiments by Summer and Gibbons on isolated demembranated axonemes of sea urchin spermatozoa clearly showed that ATP induces sliding between the doublets. Using trypsin to destroy supporting structures (nexin and radial spokes) these investigators showed that in the presence of ATP the outer tubules rapidly slid past one another, and as one tubule of the doublet became exposed, it was destroyed by the trypsin present in solution. In another study, Kincaid and co-workers were able to remove by salt extraction one of the two (dynein) arms that extend out from the A tubule of the doublet. The effect was to slow the flagellar beat by a factor of two.

Dynein is a magnesium-activated ATPase, but the similarity of the action of this ATPase with the Ca^{2+}-activated ATPase of muscle (myosin) is striking. The ATPase activity is concentrated in the headpiece of muscle myosin, and it is this region that is involved in sliding interactions with actin.

Other experiments relate dynein ATPase activity with cross-bridging reactions between doublets. Studies with demembranated sea urchin spermatozoa have shown that in the absence of ATP, rigor wave forms (i.e., flagella "frozen" in a given oscillatory state) occur and are maintained until ATP is provided. This has suggested that fixed cross bridges within the axoneme structure prevent tubules from sliding past one another and thus force the axoneme to remain in a particular pattern against the pressure of elastic forces that tend to relax it. Since ATP and other nucleotides induce sliding and loss of rigor, and since the specificity of dynein ATPase for different nucleotides is the same, it has been concluded that cross bridges between doublets are controlled by dynein ATPase.

9 PROPAGATION OF WAVE MOTION

Although the sliding of microtubule doublets and the constraints imposed on sliding by the central doublets and radial spokes explain flagellar bending, this does not explain how the wave is propagated. Since wave motion occurs in demembranated spermatozoa (so-called sperm models), wave propagation is apparently an intrinsic property of the axoneme. Brokaw proposed a mechanism for propagation that suggests that portions of the axoneme that are not actively sliding resist the sliding of neighboring doublets. This may activate the sliding mechanism in that adjacent section and thus propagate bending along the flagellum. This mechanism requires that the ATPase involved in active sliding exists along the entire axoneme.

The mechanism by which the doublets undergo coordinated sliding to provide the various types of wave motion observed in flagella, including variations in amplitude and direction (i.e., helical rotation), remains to be resolved. Similarly unresolved are the factors that control beat frequency and the induction of spontaneous beating. Protein phosphorylation mediated by cAMP appears to play a role in the control of beat frequency in mammalian spermatozoa, and perhaps the cholinergic mechanism described above also plays a role. The responsiveness of the flagellum to external stimuli and the coordinated directed motion induced by surface glycoproteins also indicate the importance of the plasma membrane in motility. A hypothesis integrating the biochemical and biomechanical mechanisms involved in motility and its control seems still to be far off. But like so many areas in modern biology, interesting surprises are probably near at hand.

REFERENCES

1   Afzelius, B. A. (ed.), *The Functional Anatomy of the Spermatozoon*, Proceedings of the 2nd International Symposium, Wenner-Gren Center, Vol. 23. Pergamon, New York (1975).

2   Acott, T. S., Hoskins, D. D., Bovine sperm forward motility protein: partial purification and characterization. *J. Biol. Chem.* **253**, 6744 (1978).

3   Baccetti, B., Afzelius, B. A., The biology of the sperm cell. *Monograph Develop. Biol.* **10**, 1 (1976).

4   Bedford, J. M., Cooper, G. W., Membrane Fusion Events in the Fertilization of Vertebrate **Eggs,** *in* Poste, F., Nicolson, G. L. (eds.), *Cell Surface Reviews, Membrane Fusion*, Vol. 5, p. 65. North Holland, New York (1978).

5   Brokaw, C. J., Flagellar movement: A sliding filament model. *Science* **178**, 455 (1972).

6   Fawcett, D. W., and Bedford, J. M. (eds.), *The Spermatozoon, Maturation, Motility, Surface Properties and Comparative Aspects*. Urban Schwartzenberg, Baltimore (1979).

7   Friend, D. S., The Organization of the Spermatozoal Membrane, in Ededin, M., and Johnson, M. H. (eds.), *Immunobiology of Gametes*, p. 5, Cambridge University Press, Cambridge (1977).

8   Garbers, D. L., Lust, W. D., First, N. L., and Lardy, H. A., Effect of phosphodiesterase inhibitors and cyclic nucleotides on sperm respiration and motility. *Biochemistry* **10**, 1825 (1971).

9   Gibbon, I. A., Structure and Function of Flagellar Microtubules, *in* Brindley, B., and Porter, K. R. (eds.), p. 348, *International Cell Biology*, Rockefeller University Press, New York (1977).

10   Hoskins, D. D., and Casillas, E. R., Hormones, second messengers, and the mammalian spermatozoon. *Adv. Sex Horm. Res.* **1**, 283 (1975).

11   Hoskins, D. D., Adenine nucleotide control of fructolysis and motility in bovine epididymal spermatozoa. *J. Biol. Chem.* **248**, 1135 (1977).

12   Hoskins, D. D., Brandt, H., and Acott, T. S., Initiation of sperm motility in the mammalian epididymis. *Fed. Proc.* **37**, 2534 (1978).

13   Lindemann, C. B., and Gibbons, I. R., Adenosine triphosphate-induced motility and sliding of filaments in mammalian sperm extracted with Triton X-100. *J. Cell Biol.* **65**, 147 (1975).

14   Lindemann, C. B., A cAMP-induced increase in the motility of demembranated bull sperm models. *Cell* **13**(1), 9 (1978).

15   Mann, T., *The Biochemistry of Semen and of the Male Reproductive Tract*. Metheun, London (1964).

16   Nelson, L., Young, M. J., and Gardner, M. E., Sperm motility and calcium transport: A neurochemically controlled process. *Life Sci.* **26**, 1739 (1980).

17   O'Day, P. M., and Rickmenspoel, R., Electrical control of flagellar activity of impaled bull spermatozoa. *J. Cell Sci.* **35**, 123 (1979).

18   Pedersen, H., Ultrastructure of the Sperm Tail, *in* Continho, E., and Fuchs, F. (eds.), *Physiology and Genetics of Reproduction*, Part A, p. 227, Plenum, New York (1974).

19 Peterson, R. N., and Freund, M., The inhibition of the motility of human spermatozoa by various pharmacological agents. *Biol. Reprod.* **13**, 552 (1975).

20 Satir, P., How cilia move. *Scientific American* **231**, 44 (1974).

21 Singh, J. P., Babcock, D. F., and Lardy, H. A., Increased calcium ion influx is a component of capacitation of spermatozoa. *J. Biol. Chem.* **172**, 549 (1978).

22 Storey, B. T., and Kayne, F. J., Energy metabolism of spermatozoa. VI. Direct intramito-chondrial lactate oxidation by rabbit sperm mitochondria. *Biol. Reprod.* **16**, 549 (1977).

23 Van Dop, C., Hutson, S. M., and Lardy, H. A., Pyruvate metabolism in bovine epididymal spermatozoa. *J. Biol. Chem.* **252** (4), 1303 (1977).

CHAPTER SEVEN

THE TUBULAR ORGANS
OF THE FEMALE
GENITAL TRACT

PANAYIOTIS M. ZAVOS

Department of Animal Science
University of Kentucky
Lexington, Kentucky

PETER F. TAUBER

Department of Obstetrics and Gynecology
University of Essen
Essen, West Germany

GEBHARD F. B. SCHUMACHER

Department of Obstetrics and Gynecology
University of Chicago
Chicago, Illinois

1 INTRODUCTION

The internal female genital tract derives during ontogenesis from the Müllerian duct and develops into three different compartments: the two oviducts (fallopian tubes), the two uterine horns (or one uterus simplex, as in the woman and nonhuman primate), and the cervix (or two cervixes in some species such as the rabbit). The fourth tubular organ, the vagina, develops from the urogenital sinus.

The mammalian female reproductive tract has received considerably more attention than the male reproductive tract. In spite of this, biochemical information is rather limited. A number of physiological processes associated with the female reproductive tract require the correct milieu in order to take place. These processes include sperm and ovum transport, sperm capacitation, fertilization, and blastocyst development and implantation. The secretory activity of the female genital tract and the concentration of the secretory components are under estrogen and progesterone control. Hormonal stimulation varies during the menstrual cycle (human, primates) or the estrous cycle. As such, the composition of the secretions also varies, resulting in environments that optimally accommodate the different physiological processes at the times that they are to take place. Since the internal milieu of the genital tract is primarily determined by the lumenal secretions, the following is a summary of our knowledge regarding these secretions. Due to space limitations, not all available data can be included, and often examples, that is, certain species, have been selected for more-detailed discussion rather than making an attempt to cover all information on every species.

2 THE VAGINA

2.1 Introduction

The cytodynamics of the vaginal squamous epithelium during the reproductive period are influenced by cyclicly released ovarian sex steroid hormones and are geared for protective, metabolic, and reproductive functions. Such cyclic changes can be followed to some extent by taking vaginal smears and observing cellular alterations. This is most accurate in animals with short estrous cycles, for example, rats and mice, but it is also used in man and nonhuman primates.

Vaginal fluid is composed of cervical mucus, transudate from vaginal mucosa, desquamated cells, polymorphonuclear leukocytes and bacteria. During sexual excitation, the production of vaginal fluid is increased. There are no glands in the vaginal mucosa. The Skene's and Bartholin's glands are located in the vestibule near the urethral meatus. The vagina is kept moist predominantly by the transudation of fluid through the vaginal epithelium and by cervical secretions. Most of our knowledge regarding vaginal secretions is of the woman, and the following relates to this species unless indicated otherwise. A summary of the components of cow vaginal fluid can be found in Table 8 (see Section 4.3 of this chapter).

2.2 Secretory Activity During Sexual Stimulation

In the woman, a transudate appears on the walls of the vagina during sexual stimulation. This fluid is a part of the vaginal milieu, but little is known concerning its composition or role in the reproductive processes other than providing lubrication prior to coitus. Qualitative studies of this transudate indicate that only little difference in the type of compounds can be found in this "stimulated" fluid compared with "unstimulated" vaginal fluid. Quantitative changes occur, however, with lactic acid and possibly glycerol and higher-molecular-weight lipids increasing during stimulation (1). The higher concentrations of these compounds suggest that the glycolysis and the hydrolysis of glycerides increases in the vaginal mucosal cells during sexual stimulation.

In the presence of the uterus and without sexual arousal, the amount of vaginal fluid secreted by women was reported to be 2–7.5 g/24 hours. During sexual arousal, a two to fourfold increase in the rate of transudation takes place. In hysterectomized women who also had an oophorectomy (excision of the ovaries), the mean amount of vaginal fluid secreted is 1.56 g/24 hours. This is increased if ovarian function is maintained or estrogen supplementation is provided (2). Because of the low amount of fluid produced in the basal state, collection of the secretions is most often accomplished by entering a tampon in the vagina, allowing it to remain for some time, and then extracting the fluid components.

2.3   Electrolytes

Vaginal fluid contains organic and inorganic substances. Inorganic substances include Na^+, K^+, and Cl^-. The concentrations of these ions vary markedly from those of blood plasma. K^+ is present in 6.6 times greater amounts than in plasma, whereas the Na^+ and Cl^- levels are approximately 46 and 61% of those of plasma. It has been postulated that in the basal (sexually unstimulated) state, both Na^+ and Cl^- continuously enter the vaginal lumen from the cervical secretions and by transudation from the blood through the vaginal epithelium. Further, Na^+ ions are actively reabsorbed through the vaginal epithelium, and it is presumed that the Cl^- ions follow passively. The reabsorption process tends to create a transvaginal electrical potential difference, which in turn creates an electrical force for the movement of plasma K^+ into the lumen. The hemodynamic alterations during sexual stimulation cause an increase in the formation of NaCl-rich vaginal fluid from the plasma. This stimulated vaginal fluid has a pH of 7.3, which aids in the partial neutralization of the acidity (pH is approximately 4.0) of the basal vaginal fluid and in turn protects the ejaculated spermatozoa from the acid environment of the unstimulated vagina.

2.4   Small Organic Constituents

A large number of small organic compounds have been identified in human vaginal secretions (Table 1). Aliphatic acids are secreted in relatively large amounts; the most consistently found aliphatic acids are acetic, propanoic, isobutyric, N-butyric, isovaleric, and 2-methylbutyric acid (3). Total aliphatic acid concentrations and the concentrations of acetic, propanoic, and butanoic acid show cyclic patterns, being highest during the proliferative phase and at midcycle (4). These acids are probably breakdown products from carbohydrates. The lactic acid content (0.5–1.0%) of vaginal fluid is high. Human vaginal fluid contains carbohydrates, which are derived from epithelial glycogen. The aerobic lactobacilli, Döderlein's bacilli, ferment the carbohydrates to aliphatic acids. The relatively high concentration of organic acids causes the fluid to be acidic, with a pH of approximately 4.0.

Several of the aliphatic acids in the human vagina are volatile. The variation in concentration of these acids during the cycle causes vaginal fluid to smell differently, apparently being more pleasant during the ovulatory period. Whether such olfactory cues, called pheromones, play a role in human reproduction remains to be established. In many other animals, the part played by olfaction in the reproductive process (the male being sexually attracted to the female during her ovulation period) is well established. The substances responsible for this effect are volatile, two-to-six-carbon aliphatic acids.

Urea levels of vaginal fluid also vary during the cycle, being highest at midcycle and the midluteal phase (1).

Table 1 Small Organic Constituents of Human Vaginal Secretions[a]

Compound	MW		MW
Aliphatic Acids		*Aromatic Compounds*	
Acetic	60	1. Aldehydes	
Propanoic	74	Benzaldehyde	106
N-butyric	88	Phenylacetaldehyde	120
Isobutyric	88	Furfural	96
Isovaleric	102	2. Alcohols	
2-Methylbutyric	102	Phenol	94
4-Methylvaleric	116	p-Cresol	108
Myristic	228	Furfuryl alcohol	98
Isomyristic	228	3. Acids	
Pentadecanoic	242	Benzoic	122
Isopentadecanoic[b]	242	p-Hydroxyphenyl-	
Palmitic	256	propionic	156
Palmitoleic	254	Salicylic[b]	138
Steric	284	Phenylpropionic[b]	150
Oleic	282	Phenylacetic[b]	136
Linoleic[b]	280	4. Nitrogen-Containing	
		Pyridine	79
		Indole	117
Alcohols		Uracil	102
N-Dodecanol	186	*Lactams*	
N-Tetradecanol	214		
N-Hexadecanol	242	2-Piperidone	99
N-Octadecanol[b]	270		
		Sulfur-Containing	
Glycols		Dimethylsulfone	94
Propylene glycol	76		
Glycerol	92	*Miscellaneous*	
		Cholesterol	386
Hydroxy Acids		Squalene	410
Lactic	90	Urea	60
Hydroxy Ketones			
3-Hydroxy-2-butonone	88		
2-Hydroxypropanone[b]	74		

[a] Adapted from Huggins and Preti, *Am. J. Obstet. Gynecol.* 1?6, 129 (1976).
[b] Tentative.

2.5 Proteins and Amino Acids

The results of immunoelectrophoretic and immunodiffusion analyses using vaginal secretions obtained from 29 women are summarized in Table 2. Similar analyses of vaginal secretions from hysterectomized women show the complete absence of α_2-haptoglobin, α_2-macroglobulin, β-lipoprotein, orosomucoid, and IgM. The absence of these proteins from the vaginal fluid of hysterectomized women indicates that they originate from cervical mucus.

Fourteen free amino acids have been identified in human vaginal fluid. These are alanine, arginine, aspartic acid, glycine, histidine, isoleucine, proline, serine, taurine, threonine, tryptophan, and valine. The histidine content ranges from 2.2 to 9.9 mg%. The histidine levels are higher during the luteal phase compared with the other phases of the menstrual cycle. At present it is not known whether the amino acids found in the vaginal fluid of women originate from the vagina or are derived from cervical mucus, which is known to contain free amino acids.

Table 2   Protein Content of Human Vaginal Secretions[a]

Protein	Immunological Cross-Reaction[b]			Positive (%)
	$-$[c]	$+$[d]	$++$[e]	
Albumin	0	4	25	100
Immunoglobulin (IgG, IgA, IgM)	1	15	13	96.5
γ Chain	2	17	10	93.1
Transferrin	2	21	6	93.1
γG	3	16	10	89.6
α_2-Haptoglobin	3	23	3	89.6
α_1-Antitrypsin	3	21	5	89.6
γA	4	18	7	86.2
α_2-Macroglobin	8	15	6	72.4
β-Lipoprotein	25	4	0	13.7
γG(K) (Bence Jones)	26	3	0	10.3
Orosomucoid	26	3	0	10.3
Ceruloplasmin	28	1	0	3.4
γM	28	1	0	3.4
C-reactive protein	29	0	0	0
Fibrin	29	0	0	0

[a] From Raffi et al. Fertil. Steril. 28:1345, 1977.
[b] Values listed indicate the number of women who do or do not show a cross-reaction. A total of 29 women were used for the study.
[c] Negative ($-$) precipitin reaction (no precipitation bands present).
[d] Positive ($+$) precipitin reaction.
[e] Strongly positive ($++$) reaction (according to the density of the precipitin bands).

3 THE CERVIX

3.1 Introduction

The cervix and the transition zone between the vagina and the cervix vary greatly among mammalian species, and so does the physiological function of the cervix (5). For instance, rats, mice, guinea pigs, pigs, and horses ejaculate directly into the cervix or the uterus, whereas vaginal deposition of semen occurs in rabbits, ruminants, dogs, primates, and women. The cervix is filled with mucus that prevents the entry of foreign substances and bacteria into the sterile uterus. In those animals in which spermatozoa are placed vaginally, the cervix also appears to have a role as sperm reservoir. The characteristics of the mucus and its secretion vary from one species to another, and during the estrous (and menstrual) cycle. However, considerable similarities in the biophysical and biochemical properties of cervical mucus have been established between the woman, cow, and nonhuman primate. Human cervical mucus has been studied biochemically in greatest detail and is presented in this section.

Estrogen stimulates the secretion of cervical mucus, whereas progestins inhibit the secretion. The daily production of cervical mucus in the woman varies from approximately 0–0.1 ml per day at the beginning and the end of the menstrual cycle to 0.2–1.0 ml per day at midcycle (6).

3.2 Physical Properties

The physical properties of cervical mucus, including quantity, *spinnbarkeit* (stretchability), ferning patterns formed by the crystals of dried mucus, white-blood-cell content, and viscosity, show a typical cyclic pattern (Table 3), which has been used in the woman to estimate the time of ovulation. Similar changes occur in other animal species and may be used as a rough index of the hormonal profile and optimal time for insemination.

According to Odeblad (7), several different types of mucus may be produced by the cervical epithelium. Based on nuclear magnetic resonance (NMR) studies, this investigator claims the presence of a network of interconnected micelles, best characterized as a "tricot-like macromolecular gel arrangement." Under estrogen stimulation, this network of filaments is structured in such a way that the intermicellar spaces (mesh size = 0.5–5.0 μm), containing the aqueous phase of the cervical mucus, allow for free movement of spermatozoa and fast forward progression (type E mucus). Under the influence of progesterone or progestational agents, a more dense network (type G) develops with much smaller aqueous cavities (mesh size = 0.1–0.5 μm); it does not allow passage of spermatozoa.

Type E mucus is thin and consists of 95–98% water, 1% protein, and 0.5–1.5% mucin. The type G mucus is thick and sticky. It contains 85–92% water, 2–4% protein, and 2–10% mucin. During the late follicular and ovula-

Table 3   Cyclic Changes of Human Cervical Mucus (Wet)[a]

Estrogen Effects (midcycle)	Gestagen Effects (luteal phase)
Profuse secretion (0.2–1.0 ml/day)	Scanty secretion (0.0–0.1 ml/day)
Low consistency (viscosity)	High consistency (viscosity)
NaCl concentration 0.8%	NaCl concentration 0.8%
High ratio of salt per organic compound	Low ratio of salt per organic compound
Low albumin content (0.1–1.0 mg/ml)	High albumin content (2.0–24.0 mg/ml)
Low enzyme content, e.g., lysozyme (0.01–0.1 mg/ml)	High enzyme content, e.g., lysozyme (0.2–2.0 mg/ml)
Low trypsin-inhibitor content, e.g., α_1-antitrypsin (0.01–0.05 ml/mg)	High trypsin-inhibitor content, e.g., α_1-antitrypsin (0.1–1.0 mg/ml)
Low immunoglobulin content, e.g., IgG (0.03–0.3 mg/ml)	High immunoglobulin content, e.g., IgG (0.2–3.0 mg/ml)
Low lipid content	High lipid content
Low electrolyte concentration	High electrolytic concentration

[a] From Prins et al., in Hafez (ed.), Human Ovulation, p. 313, North Holland, New York (1979).

tory phases of the cycle, a normal cervix produces about 95% type E mucus and only about 5% type G. During the luteal phase, 95% of the bulk cervical secretion is composed of type G mucus.

Several investigators have confirmed the presence of fibrous filaments in cervical mucus with the scanning electron microscope. These fibers run parallel at midcycle but form crossing patterns during the luteal phase. However, the theory of the network structure of cervical mucus has recently been challenged by others (8), who base their observations on data obtained by laser light-scattering spectroscopy on cow and human midcycle cervical mucus. These investigators claim that midcycle mucus is composed of an ensemble of entangled, randomly coiled macromolecules instead of a fibrillar cross-linked network. These macromolecules are glycoproteins with molecular weights estimated at 10 million and linear dimensions of approximately 0.7 μm. The cyclic changes in the rheological properties of cervical mucus during the cycle may be due to changes in the degree of hydration of the entangled, randomly coiled macromolecules.

3.3   Low-Viscosity Components

Cervical mucus is believed to be a mixture of the secretions from the cervical epithelium, blood transudate, and fluids from the uterus, fallopian tubes (oviducts), and perhaps the follicle. Leukocytes and other cellular elements may also contribute. Cervical mucus is a hydrogel composed of a matrix of high-molecular-weight glycoproteins (high-viscosity component) and an aqueous phase named cervical plasma (low-viscosity component).

Many soluble substances of lower and higher molecular weight are present in cervical secretions, which vary in quantity during the cycle (Table 3). The functions of these components include the support, preservation, and aid in migration of the spermatozoa as they transverse the cervix, and immunological defense against microbial invaders.

The water portion (85–98%) acts as a hydrating agent for both the low- and high-viscosity components of the mucus. In the woman, the water content (95–98%) is higher at midcycle and during the late follicular phase, whereas the water content is 85–92% early in the cycle or during the luteal phase. It is believed that spermatozoa will not penetrate cervical mucus with less than 94.5% water content, that is, if it contains more than 5.5% dry matter.

Distinct cyclic variations occur in the concentrations of the electrolytes in cervical mucus. This variation most likely has a regulatory, biological function to maintain isotonicity. Since the levels of mucins and other proteins decrease in midcycle cervical mucus, the electrolyte content per dry weight of organic material increases proportionally. Sodium chloride represents the major part of the electrolyte system, and, together with potassium ions, is responsible for the crystal formation when cervical mucus is dried. During the midcycle period, the crystals form patterns that appear as ferns ("ferning pattern"). Cervical mucus also changes in pH during the cycle. The role of the electrolytes in inducing this change remains to be established. An abnormal pH may render cervical mucus impassable to spermatozoa.

Calcium, magnesium, and zinc increase, whereas potassium, copper, and iron appear to decrease at midcycle in dry mucus. Sulfate and bicarbonate ions are also found in cervical mucus. Glucose, maltose, and mannose are the reducing sugars in cervical mucus, and some authors report elevated levels of these sugars at midcycle. The glycogen content of cervical mucus is rather constant, and neither cyclic changes nor treatment with steroids causes any fluctuations.

Phospholipids, cholesterol and cholesterylesters, monoglycerides, diglycerides, glycerides, fatty acids, and hydrocarbons are also present in cervical mucus. The total lipid content shows a midcycle minimum, with an average of 2 mg/gm wet weight mucus in contrast to 6 mg/gm in postmenstrual or luteal-phase mucus. The most noticeable fluctuation of the various lipids is the high level of free fatty acids and phospholipids at midcycle and the high level of palmitoleic acid found during the luteal phase. These fluctuations may be linked to the cyclic leucocyte distribution in the cervical mucus.

Cervical mucus contains cAMP, which increases significantly in human periovulatory cervical mucus (9). The mucus also contains a fairly high amount of free amino acids. The concentration of free amino acids tends to decrease during midcycle.

Soluble proteins represent approximately 30% of the nondialyzable material of cervical mucus (10). The greater portion of these proteins are immunochemically identical to certain serum proteins (Table 4). However, a considerable number appear to be locally produced. Labeled amino acids are incorporated into IgA and IgG by cultures of human cervical tissue. IgG- and IgA-producing

Table 4   Proteins, Immunoglobulins, and Enzymes of Human Cervical Mucus[b]

Proteins that arrive by transudation from serum
   Prealbumin
   Haptoglobin
   Ceruloplasmin
   Orosomucoid
   α_1-Antitrypsin
   α_{1x}-Antichymotrypsin
   Antithrombin III
   Inter-α-trypsin inhibitor
   C'_3 = Complement component
   C'_1 = Complement component
   Transferrin
   C-Reactive protein
   IgG (Immunoglobulin)
   IgA (Immunoglobulin) (part may be locally produced)
   α_2-Macroglobulin ⎫
   IgM (Immunoglobulin) ⎬ ⟶ Macroglobulins found only in traces
   α_2/β-Lipoprotein ⎭
Enzymes
   (May derive partially from local production or selective transudation[a])
   Alkaline phosphatase
   Transaminase
   Amylase
   Ribonuclease
   Kallikrein
Proteins locally produced
   Secretory piece and secretory IgA
   Lactoferrin
   Lysozyme
   Plasminogen activator
   Carbonic anhydrase
   Fucosidase
   Sialyltransferase
   Transpeptidase
   Glucose-6-phosphate isomerase
   Acid stable inhibitor of WBC proteinases (antileucoproteases)
   Nonspecific proteinases

[a] The enzyme levels in cervical mucus are often considerably higher than those in serum; this is most likely due to local production or perhaps to selective secretion.
[b] Prepared from Schumacher, *in* Insler and Bettendorf (eds.), *The Uterine Cervix in Reproduction*, p. 187, Thieme, Stuttgart (1977).

plasma cells have also been identified in the cervix. The presence in cervical mucus of these immunoglobulins that either derive from plasma or are locally produced shows the presence of an active immune system. If the antibodies are sperm directed, they can impair or prevent sperm migration and thus fertility.

It is neither clear nor to be expected that all enzymes, proteins, and inhibitors found in cervical mucus are of physiological significance. Some of the protein-aceous substances are most likely involved in nonspecific and specific defense mechanisms and other forms of protective and reparative functions, or aid in sperm transport. All the soluble proteins of the cervical mucus are subject to cyclic variations and show a pronounced decrease at midcycle. The degree of change varies considerably from substance to substance. This information may be of value in the detection of ovulation.

3.4 High-Viscosity Components

Biochemically, the core of the cervical mucus is mucin, a macromolecule that consists of long glycoprotein molecules. More than 70% of the glycoprotein consists of carbohydrates that are distributed along a polypeptide core (back-bone). These carbohydrates form numerous heterosaccharide chains consisting of galactosamine, fucose, and sialic acid that connect one polypeptide to the other. The polypeptide core contains a subunit of 19 amino acids each, includ-ing cysteine residues. Subunits from different polypeptide cores are connected by disulfide bridges. These subunits do not contain any carbohydrate. The carbohydrate-bearing part of the polypeptide (MW 30,000) contains 45 amino acids in its backbone structure and is rich in threonine and serine residues. The chain of subunits is probably held together in a linear fashion by hydrophobic interaction between the respective N- and C-terminal segments of the peptide backbone.

The secondary structure of the glycoproteins is that of a loose random coil, leaving the individual molecules rather flexible. The tertiary structure is as yet unknown.

The sialic acid content of the mucin is significantly higher during the luteal phase. The midcycle decrease in viscosity appears to be accompanied by an increase in fucose, decrease in α-L-fucosidase, and a decrease in N-acetylneur-aminic (sialic) acid residues and sialyl transferase. The lower content of sialic acid residues decreases the rigidity of the molecule by diminishing electrostatic re-pulsion of the carbohydrate side chains. At the same time the cross-linkages between the chains are reduced, which probably contributes to the lowering of the viscosity. The glycoproteins can be digested by proteinases such as pancreat-ic trypsin and the sperm enzyme acrosin (Chapter 5) (11). However, recent results indicate that acrosin is not involved in sperm migration through cervical mucus (12). Seminal plasma contains a chymotrypsin-like enzyme, called sem-inin, that also hydrolyzes the cervical glycoproteins and has been postulated to aid the entry of spermatozoa into the mucus.

4 THE UTERUS

4.1 Introduction

The uterus is the site of a number of important physiological processes. Its contractibility and the movement of the lumenal cilia aid in gamete and blastocyst transport. Uterine fluids help to maintain sperm viability. During the passage of the spermatozoa through the uterine cavity and into the oviduct, uterine fluid and oviductal fluid cause spermatozoa to undergo capacitation, without which they cannot fuse with the egg (Chapter 8). After fertilization and prior to implantation, the developing blastocyst is maintained by oviductal and uterine secretions. Following implantation of the blastocyst, embryonic and fetal development takes place, which is supported by the fluid in the uterus.

The fluid content of the endometrial cavity is very low (0.05–0.25 ml) and is believed to be made up of a combination of certain blood plasma components (through selective transudation) and the secretions from uterine glands. These processes can occur independently from each other. Transudation and the secretory activity of the uterus are highly dependent on estrogen and progesterone stimulation, and the composition of uterine fluid varies during the estrous (menstrual) cycle.

4.2 Human Uterine Fluid

A number of constituents of human uterine fluid are shown in Table 5. Using filter paper (blotting) for the collection of endometrial secretions, Schumacher *et al.* (13) showed that human uterine secretions vary significantly in the content of a particular protein during the menstrual cycle (Table 6). Immunoglobulins and proteinase inhibitors are present in considerable amounts in human uterine fluid. One of the immunoglobulins, IgM, cannot be demonstrated under physiological conditions in cervical mucus. Besides the components listed, lactoferrin and lysozyme were also found.

In a similar study, Tauber (14) analyzed endometrial-tissue cylinders obtained from hysterectomy specimens of 36 women (20–49 years of age, Table 7). All the components found in both the cervix and uterus, that is, IgG, IgA, α_1-antitrypsin, α_1-antichymotrypsin, complement component C_3, lysozyme, plasminogen activator, and albumin, were present in much higher amounts in the cervix than in the uterus with the exception of the plasminogen activator which revealed much higher activities in the uterus. The secretory piece of IgA, lactoferrin, and the complement component $C'q$ were detectable in the cervix, but were virtually absent from the endometrium. Total complement activity was extremely low in the cervix and uterus (<5% of that present in serum). No neutral proteinase activity was found in the uterus.

Both immunoglobulins IgG and IgA, plasminogen activator, and proteinase inhibitors showed their highest values per milligram wet weight of tissue during the ovulatory (midcycle) phase of the menstrual cycle. All values obtained

Table 5 Some Constituents of Human Uterine Fluid[a]

Component[b]	Follicular Phase	Luteal Phase
pH	7.85	7.83
Protein nitrogen (mg/100 ml)	269.25	283.03
Nonprotein nitrogen (mg/100 ml)	64.30	72.22
Alkaline phosphatase (mg P/100 ml/hr)	7.58	10.56
Acid phosphatase (mg P/100 ml/hr)	4.62	6.08
Hyaluronidase (units/100 ml)	0	0
Lactic dehydrogenase (units/mg protein/min)	0.105	0.102
Glucose-6-phosphate dehydrogenase (units/ml)	0 to trace	0 to trace
Glucose-6-phosphatase (Racker unit/100 ml)	0	0
Glycogen (mg/100 ml)	5.86	5.38
Glucose (mg/100 ml)	47.68	50.33
Lactic acid (mg/100 ml)	11.97	11.30
Fructose (mg/100 ml)	0	0
Ascorbic acid (mg/100 ml)	2.50	3.00
Sodium (mEq/L)	105.91	129.88
Potassium (mEq/L)	3.74	3.43
Chloride (mEq/L)	99.97	100.87
Calcium (mEq/L)	2.43	3.31
Bicarbonate (mM/L)	36.67	36.33
Total lipids (mg/100 ml)	427.66	397.00
Free sterols (mg/100 ml)	53.19	42.93
Sterol esters (mg/100 ml)	115.14	100.04
Phospholipids (mg/100 ml)	217.36	204.67
Choline (mg/100 ml)	45.98	48.54
Serine (mg/100 ml)	9.35	8.43
Ethanolamine (mg/100 ml)	33.65	32.80

[a] From Kar et al., Am. J. Obstet. Gynecol. **101**, 966 (1968).
[b] Fluid obtained by aspiration with a catheter.

during the luteal phase showed lower levels but did not reach the very low postmenstrual concentrations. In contrast to the immunoglobulins and the proteinase inhibitors, lysozyme showed its highest concentrations during the proliferative phase of the cycle. The ratio of IgG:IgA during the proliferative and secretory phases is relatively high (approximately 20:1) in the cervix and the uterus. The ratio decreases to about 10:1 around midcycle, indicating a relative increase in diffusible IgA.

4.3 Cow and Sow Uterine Fluid

Significant differences are present in the composition of the fluids from different portions of genital tracts recovered from cows following slaughter (Table 8).

Table 6   Protein Components of Human Endometrial Fluid (EF) and Serum(S)[a]

Component	Postmenstrual		Late Proliferative		2–3 Days Postovulatory		4–5 Days Postovulatory		10–11 Days Postovulatory		13–14 Days Postovulatory	
	EF	S	EF	S	EF	S	EF	S	EF	S	EF	S
Immunoglobulin G (mg/dl)	388	1100	435	1320	600	1280	324	1370	278	1460	336	1020
Immunoglobulin A (mg/dl)	176	890	110	340	76	130	20	200	60	365	60	220
Immunoglobulin M (mg/dl)	18	89	—[b]	36	71	142	—	—	18	149	19	71
Immunoglobulin D (mg/dl)	47	18	—	18	47	10	—	—	43	10	Trace	18
Secretory piece	Trace	0	—	0	Trace	0	—	—	+	0	Trace	0
C'_{1q} (in % of N-Ser. pool)	1	50	—	—	25	125	10	125	1	125	5	25
C'_3 Complement (mg/dl)	28	94	28	84	28	70	16	102	18	70	16	84

C'$_3$ Activator (mg/dl)	6	27	—	—	11	22	—	—	—	24	11	33
C'$_4$ (mg/dl)	11	27	2	27	4	27	1	27	4	27	4	16
Plasminogen (mg/dl)	4	20	—	11	3	15	5	20	1	15	4	15
Plasminogen activator (Plough units/ml)	12	0	0	0	0	0	0	0	0	0	0	0
Neutral Proteinase	0	0	0	0	0	0	0	0	0	0	0	0
α$_1$-Antitrypsin (mg/dl)	72	320	171	320	240	320	12	345	80	345	138	430
Inter-α-trypsin inhibitor (mg/dl)	12	26	—	—	22	32	2	36	16	32	28	43
α$_{1x}$-Antichymotrypsin (mg/dl)	8	69	—	—	32	56	4	64	8	69	10	69
Antithrombin III (mg/dl)	9	23	—	—	19	22	—	—	—	23	11	32
α$_2$-Macroglobulin (mg/dl)	12	190	8	140	131	322	8	215	8	215	59	248

[a] From Schumacher et al., in Beller and Schumacher (eds.), *Biology of the Fluids of the Female Genital Tract*, p. 115; Elsevier/North Holland, New York (1979).

[b] Dash indicates that a value was not determined.

189

Table 7   Diffusible Protein Components in the Human Uterine Mucosa;
Overall Concentration Differences—Cyclic and Topographic Variations[a]

Component	Pre-ovula-tory (%)	Peri-ovula-tory	Post-ovula-tory (%)	Endo-cervix	Endome-trium (%)
Immunoglobulin G	−40	4.80[b]	−25	4.26[b]	−20
Immunoglobulin A	−47	0.53[b]	−43	0.42[b]	−20
α_1-Antitrypsin	−50	1.14[b]	−25	1.10[b]	−20
α_{1x}-Antichymotrypsin	−5	0.06[b]	+5	0.07[b]	−20
Lysozyme	+8	0.05[c]	−30	0.07[c]	−50
Plasminogen activator	−40	36[d]	−50	<10	+300
C_3 complement component	—	—	—	0.03[a]	−12
Total complement activity	—	—	—	4.0[e]	−10

[a] From Tauber, Dissertation, University of Essen, West Germany, 1979.
[b] Concentrations in μg per mg wet weight tissue.
[c] In μg per mg wet weight tissue egg white lysozyme equivalents.
[d] In milli-international units per milligram wet tissue.
[e] In CH_{100} milli-units per milligram wet tissue.

Uterine fluid has a much higher percentage dry matter, total nitrogen, and reducing sugars than vaginal fluid, but the content of these components is lower than that found in the oviductal fluid (15). By contrast, the ash content of the uterine and oviductal fluid is much less than that of vaginal fluid. Cow uterine fluid contains higher concentrations of reducing substances, potassium, inorganic phosphate, and alkaline and acid phosphatase activity than blood. However, blood contains higher levels of calcium and sodium. The concentration of these components varies during the cycle.

Significant variations in the free amino acid content of cow uterine fluid were noted during the different stages of the estrous cycle (Table 9) (16). Comparative studies with the free amino acids of blood during the estrous cycle suggest that the cyclic changes of the uterine free amino acids occur independently of those in blood. Ethanolamine and two amino acids, β-alanine, and cystine, were identified in uterine fluid but could not be detected in blood. The free amino acids may be used by the spermatozoa for their metabolism, for example, bovine spermatozoa can utilize free glycine.

These results show that the uterine cavity, although to some extent dependent on the chemical composition of blood, can act independently from it to facilitate the various biological functions during the estrous cycle. Customarily, blood serum has been used as the principal medium for the *in vitro* storage and handling of ova from the cow. In view of the differences in the biochemistry of the two fluids, more-suitable media should be considered.

The protein content of the uterine fluid from cows treated for two to three months with progesterone averages 582 mg per uterus (17). No relationship

Table 8 Composition of Bovine Female Genital Tract Secretions[a,b]

	Fluid From			
	Vagina	Uterus	Oviduct	Follicle
Dry matter (%)	2.4	8.4	13.6	7.6
Ash (% of dry matter)	42.6	19.6	7.52	10.46
Total nitrogen (g/100 ml)	0.18	0.74	1.95	1.06
Reducing sugars (mg/100 ml)	16.1	78.4	89.2	35.3
Ether extract (% of dry matter)	18.0	14.0	13.5	4.7
Sodium (mg/100 ml)	170	220	208	304
Potassium (mg/100 ml)	166	183	223	36
Calcium (mg/100 ml)	11.3	15.2	11.8	12.3
Inorganic phosphorus (mg/100 ml)	1.5	7.4	9.7	2.6
Chloride (as NaCl) (mg/100 ml)	526	362	400	530
pH	7.8	7.1	6.4	7.1
Eh (volts)	−0.40	−0.25	−0.22	−0.09
Milliosmols	349	353	350	287

[a] Averaged values.
[b] From Olds and VanDemark, *Fertil. Steril.* **8**, 345 (1957).

appears to exist between the period of progesterone treatment and the amount of protein present. After ion-exchange chromatography, five major fractions are obtained (17). Two of these fractions contain serum proteins. In the other three fractions, at least nine nonserum proteins are present. Two of the nonserum proteins are an acid phosphatase and lactoferrin. The latter is the major nonserum protein present. Lactoferrin, an iron binding protein, is also present in human and sow uterine fluid. In the sow, exogenous progesterone does not stimulate either the synthesis or the secretion of lactoferrin but it does stimulate the secretion of acid phosphatase, lysozyme, leucine aminopeptidase, and cathepsin (B_1, D, and E). Lysozyme is present in sow uterine fluid, but is absent from bovine uterine secretions. Lysozyme and lactoferrin are thought to have a bacteriocidal function in the porcine uterus.

A purple metalloprotein that has acid phosphatase activity associated with it has been found in porcine uterine secretions. The secretion of this purple protein is also stimulated by progesterone, and it has been postulated to have an important role in the transport of iron to the fetus. Lactoferrin, which can also bind metal ions, may have a similar function.

4.4 Rabbit Uterine Fluid

Nine different protein fractions have been identified in the endometrial secretions of the rabbit by immunoelectrophoresis. Of these, seven are glycoproteins. The prealbumin fraction contains a high amount of *N*-acetylneuraminic acid, whereas the postalbumin and β-glycoprotein are high in glucose and galactose, respectively. Most rabbit uterine fluid proteins are relatively small molecules

Table 9 Content of Total Free Amino Acids and Amino Compounds in Bovine Uterine Fluid (UF) and blood serum (S) Collected at Different Periods of the Estrous Cycle (μM/ml)[a]

	Days of Estrous Cycle																			
	1		1-RB[b]		2		3-4		5-7		8-10		11-13		14-16		17-18		19-20	
Amino Acid	UF	S	UF	S	UF	S	UF	S	UF	S	UF	S	UF	S	UF	S	UF	S	UF	S
Aspartic acid	0.19	0.01	0.22	0.04	0.40	0.01	0.32	0.02	0.37	0.02	0.60	0.02	0.38	0.01	0.56	Trace	0.70	Trace	0.55	0.02
Threonine	0.02	0.07	0.03	0.05	0.07	0.10	N	0.12	0.04	0.10	0.15	0.05	0.07	0.07	0.17	0.07	0.10	0.04	0.13	0.05
Serine	0.10	0.09	0.10	0.07	0.17	0.15	0.03	0.18	0.16	0.14	0.47	0.10	0.23	0.13	0.43	0.10	0.31	0.05	0.47	0.08
Asparagine and/or glutamine	0.23	0.30	0.25	0.20	0.25	0.46	0.07	0.18	0.07	0.31	0.67	0.03	0.08	0.31	0.27	0.25	0.12	0.11	0.08	0.14
Proline	0.12	0.02	0.04	Trace	0.16	0.08	0.12	0.25	0.20	0.10	0.26	0.24	0.22	0.06	0.26	0.07	0.46	0.01	0.62	0.07
Glutamic acid	0.73	0.08	0.70	0.06	1.49	0.11	1.34	0.26	1.38	0.14	1.34	0.45	1.31	0.12	1.74	0.07	1.70	0.09	1.56	0.11
Glycine	0.89	0.40	0.80	0.21	1.49	0.67	1.77	0.77	3.27	0.64	4.42	0.38	1.94	0.52	1.95	0.43	1.81	0.41	1.56	0.50
Alanine	0.51	0.22	0.48	0.17	0.66	0.29	0.72	0.40	0.98	0.30	1.41	0.23	1.01	0.33	1.10	0.21	1.21	0.21	1.03	0.28
Valine	0.12	0.16	0.12	0.15	0.20	0.32	0.20	0.23	0.30	0.22	0.41	0.18	0.34	0.31	0.37	0.23	0.41	0.20	0.14	0.34
Cystine	0.12	N[c]	0.26	N	0.04	N	0.38	N	Trace	N	N	N	N	N	N	N	N	N	N	N
Methionine	0.02	0.02	0.01	0.01	0.01	0.03	Trace	Trace	0.08	0.02	0.07	0.02	0.05	0.02	0.04	0.02	0.02	0.02	0.02	Trace
Isoleucine	0.07	0.06	0.06	0.08	0.10	0.17	0.08	0.17	0.15	0.16	0.22	0.11	0.17	0.15	0.19	0.13	0.20	0.11	0.11	0.11
Leucine	0.11	0.13	0.10	0.11	0.18	0.19	0.15	0.28	0.24	0.18	0.54	0.15	0.36	0.21	0.32	0.15	0.46	0.14	0.14	0.12

Tyrosine	0.07	0.05	0.05	0.07	0.09	0.09	0.11	0.10	0.12	0.09	0.22	0.06	0.17	0.07	0.21	0.06	0.22	0.05	0.05	0.10
Phenylalanine	0.05	0.04	0.04	0.05	0.08	0.08	0.09	0.06	0.10	0.06	0.20	0.04	0.13	0.06	0.16	0.04	0.19	0.04	0.04	Trace
Lysine	0.07	0.08	0.10	0.13	0.03	0.49	0.23	N	0.06	N	Trace	N	Trace	N	0.37	N	N	Trace	Trace	Trace
Histidine	0.04	0.04	0.06	0.05	0.10	0.08	0.08	N	N	0.06	0.11	0.05	0.09	0.06	0.10	0.04	0.13	0.04	0.04	0.04
Arginine	0.02	0.13	0.13	0.01	0.04	0.21	Trace	0.12	1.69	0.18	0.04	0.16	0.06	0.16	0.07	0.11	0.05	0.10	0.10	0.10
Taurine	0.72	0.05	0.04	0.70	1.36	0.07	1.24	0.14	N	N	1.57	0.06	0.62	N	0.77	0.04	0.53	N	0.66	0.04
Citrulline	N[c]	0.09	0.08	N	N	0.16	N	0.11	N	0.14	N	0.08	N	0.16	N	0.11	N	0.11	N	0.07
Cystathionine	0.03	Trace	Trace	0.02	0.02	Trace	0.02	Trace	N	Trace	0.26	Trace	0.07	Trace	0.09	Trace	0.02	Trace	0.04	N
Ornithine	0.09	0.04	0.06	0.08	0.17	0.07	0.14	0.02	N	N	N	Trace	N	0.07	0.25	N	N	N	N	N
Ethanolamine	0.14	N	N	0.21	0.20	N	0.19	N	1.61	N	0.06	N	N	N	0.32	N	N	N	0.40	N
β-Alanine	N	N	N	N	Trace	N	N	N	0.05	N	Trace	N	N	N	N	N	N	N	Trace	N
α-Aminoadipic and/or α-amino-n-butyric acid	0.09	Trace	N	0.08	0.11	0.03	0.14	N	0.22	0.05	0.27	Trace	0.20	0.03	0.24	0.02	0.29	0.02	0.14	N
γ-Aminobutyric acid	N	N	N	N	N	N	N	N	0.22	0.06	0.40	0.10	0.22	N	0.04	0.06	0.60	0.05	0.28	0.04
Total	4.78	2.08	2.71	4.41	7.40	3.86	7.23	3.41	11.31	2.97	13.89	2.51	7.72	2.85	10.02	2.21	10.41	1.80	8.12	2.21

[a] From Fahning et al., J. Reprod. Fert. 13, 229 (1967).

[b] a "Repeat-breeder" cows.

[c] None detected.

(MW 15,800–70,000). Several large proteins are also present, for example, α_2-macroglobulin and β-M-macroglobulin, with molecular weights varying between 10^5 and 10^6. The predominant component of the postalbumin fraction is the pregnancy-specific rabbit uterine protein called uteroglobin (18). Uteroglobin is a small globular macromolecule that is absent from follicular fluid, but present of oviductal fluid, and has been suggested to be involved in the implantation of the blastocyst.

Uteroglobin makes up 50–70% of all the proteins found in rabbit uterine secretions. Its presence has not been confirmed in other species including man. Progesterone stimulates the production of uteroglobin in the rabbit. Some uteroglobin can be found in the endometrial epithelium at any time, but it reaches maximum levels three to four days postcoitally. One day earlier, maximal concentrations of this protein are found in the uterine secretions. Since the uteroglobin pattern observed during pseudopregnancy suggests that implantation itself terminates uteroglobin release or synthesis, a specific "message" appears to be delivered from the blastocyst or from the decidual tissue to terminate uteroglobin synthesis. Perfusion experiments on isolated uteri indicate that the "switch off" *in vivo* is a termination of the synthesis rather than release, since these studies show that long-term accumulation of uteroglobin occurs within the cytosol or the endometrial cells.

5   THE OVIDUCT

5.1   Introduction

The length of the oviduct (fallopian tubes) varies significantly among mammals (Table 10). The oviduct consists of three distinct regions. The first portion, the proximal end, is the infundibulum, with its fimbriae and a series of ciliated folds. It is a funnel-shaped abdominal opening near the ovary. The second portion is the ampulla, which constitutes almost half of the oviductal length. The most unique feature of this region is a series of long fingerlike folds that protrude in the lumen. It is the site where fertilization takes place. The third section is the isthmus, which is a rather tapered section of the oviduct. The connection between the two sections forms the ampullary-isthmus junction. The connection between the uterus and the oviduct is called the uterotubal junction.

Oviductal fluid provides a suitable, chemically balanced environment for the capacitation of spermatozoa, fertilization, cleavage of fertilized eggs, and blastocyst development. This fluid is the product of the secretory activity of the endosalpinx (the secretory epithelium of the oviduct) as well as transudation from blood, and its volume and composition vary, depending on the type of steroid stimulation (19). In the woman, the volume is low during the proliferative phase, increases at the onset of estrus (estrogen dominance), reaches a maximum a day later, and then declines to characteristic luteal-phase levels.

Table 10 Comparative Dimensions of the Oviduct in the Adult,
Nonpregnant Female Mammal[a]

	Cow	Ewe	Sow	Mare	Human
Oviduct length (cm)	25	15–19	15–30	20–30	8–15[b]

[a] From Hafez, *Reproduction in Farm Animals*, 3rd Ed., p. 37, Lea and Febiger (1974).
[b] From Pauerstein and Eddy, in Beller and Schumacher (eds.), *Biology of the Fluids of the Female Genital Tract*, Elsevier/North Holland, New York (1979).

5.2 Human Oviductal Fluid

Oviductal fluid from women has been collected by cannulation of the oviducts of patients who were scheduled for hysterectomy or tubal ligation. On electrophoresis, the major protein fractions consist of α-1, α-2, β-1, β-2, and γ-globulins with approximately the same distribution as those in serum. Fifteen proteins that are also present in serum were found; among these were agglutinins and immunoglobulins (with the exception of IgM). One glycoprotein is absent from blood but is immunologically identical with an ovarian-tissue protein. Oviductal immunoglobulins may have an antifertility effect. They can bind to sperm membranes and cause immobilization and agglutination of the spermatozoa, or inhibit the activity of enzymes important for fertilization. Elevated levels of secretory IgA are correlated with infertility in women.

A number of trace elements are present in human tubal fluid. Several of these are known to act as enzyme cofactors and may participate in the function of such oviductal enzymes as carbonic anhydrase and alkaline phosphatase. Zn^{2+} and Mn^{2+} appear at fairly constant levels in the human endosalpinx (layers surrounding the lumen) during each phase of the menstrual cycle, whereas Mg^{2+} reaches its highest level during the proliferative phase and decreases at ovulation. Prostaglandin $F_{2\alpha}$, which stimulates muscular activity, has also been identified in human tubal fluid. Its concentration was high before and after ovulation and exceeded serum levels substantially, suggesting synthesis by the oviduct.

5.3 Cow and Ewe Oviductal Fluid

Olds and VanDemark (15) have collected oviductal fluids from the cow continuously by catheter during the four stages of the cycle. The mean $\pm$ standard error of the volumes (ml/24 hr) were (1) estrus, 5.31 ± 1.57; (2) metestrus, 3.60 ± 0.55; (3) diestrus, 1.49 ± 0.11; and (4) proestrus, 1.36 ± 0.34. The volume collected followed a cyclic pattern, being larger during estrogen stimulation. Even so, no significant differences in protein concentration are present during the different stages of the estrous cycle (Table 11). Of the serum proteins found in oviductal fluid, albumin is present in highest concentrations, followed by γ-globulin. Similar results were obtained with rabbit oviductal fluid (20).

Table 11 Total Protein, Albumin, α-Globulin, β-Globulin, and γ-Globulin Concentrations in Bovine Oviductal Fluid[a,b]

Stage of Cycle	Total Protein	Albumin	α-Globulin	β-Globulin	γ-Globulin	Albumin/Globulin Ratio
Estrus	5.20 ± 0.80	2.18 ± 0.27	0.67 ± 0.09	0.64 ± 0.12	1.71 ± 0.33	0.73 ± 0.05
Metestrus	4.04 ± 0.32	1.84 ± 0.12	0.28 ± 0.04	0.50 ± 0.05	1.42 ± 0.16	0.82 ± 0.06
Diestrus	4.76 ± 0.20	2.06 ± 0.08	0.56 ± 0.04	0.63 ± 0.05	1.51 ± 0.09	0.78 ± 0.03
Proestrus	5.01 ± 0.60	2.17 ± 0.18	0.54 ± 0.10	0.64 ± 0.06	1.56 ± 0.27	0.78 ± 0.06

[a] Values are given in gm/100 ml and are expressed as means ± standard error.
[b] From Stanke et. al., J. Reprod. Fert. **38**, 493 (1974).

Cow oviductal fluid possesses a higher percentage dry matter, total nitrogen, reducing sugars, potassium, and inorganic phosphorus than any of the other genital tract secretions (Table 8). It has the lowest ash content and the lowest pH. Of the ions, sodium and potassium are present in highest concentrations. The sodium levels are somewhat lower than those of blood, whereas potassium levels are much higher. The calcium and chloride levels of cow oviductal fluid are approximately the same as those of blood.

A total of 20 free amino acids were found by Stanke *et al.* (20) in bovine oviductal fluid (Table 12). Other investigators have found either slightly lower or higher numbers of amino acids. Glycine, glutamic acid, alanine, lysine and arginine are present in highest amounts. An increase occurs in the amino acid levels of the rat blastocyst as it develops, possibly indicating a metabolic function of uterine amino acids during this early stage of development.

The production of oviductal fluid in the ewe follows a pattern similar to that of the cow. As measured after cannulation of the oviduct, fluid formation is greatest around estrus and diminishes during the luteal phase of the cycle. The increase in fluid begins on the last day of the estrous cycle and reaches a maximum at about day 2. Fluid flow from the oviduct into the uterus increases markedly 3.9 days after the onset of estrus and peaks at the time during which the morula or blastocyst enters the uterus.

Oviductal fluid collected continuously from ewes throughout the estrous cycle shows marked differences in composition during estrus, metestrus, and

Table 12   Concentrations of Free Amino Acids in Bovine Oviductal Fluid Collected During Various Stages of the Estrous Cycle[a,b]

Amino Acid	Estrus	Metestrus	Diestrus	Proestrus
Ornithine	20.3	20.0	21.3	18.7
Lysine	50.4	50.4	52.4	28.5
Histidine	15.8	14.7	15.4	15.1
Arginine	52.9	56.6	36.7	43.2
Aspartic acid	6.9	6.7	6.8	5.8
Serine	4.7	3.9	4.7	4.9
Glutamic acid	52.2	54.8	55.8	4.6
Glycine	58.3	53.8	81.7	69.7
Alanine	45.6	42.7	51.9	53.5
Valine	21.6	21.7	24.8	21.8
Methionine	5.6	4.9	4.6	4.2
Isoleucine	14.1	12.9	13.0	14.3
Leucine	19.7	24.9	27.5	25.3
Tyrosine	9.6	8.4	10.9	10.1
Phenylalanine	11.0	12.1	12.9	10.2
Cystine	2.8	2.1	2.6	Trace
α-Aminoadipic acid	7.2	6.7	6.2	7.5

[a] Values are given in $\mu g/ml$.
[b] From Stanke *et al.*, *J. Reprod. Fert.* **38**, 493 (1974).

Table 13   Mean Values of Some Components of Sheep Oviductal Fluid[a]

Constituent	Stage of Estrous Cycle		
	Estrus	Metestrus	Diestrus
Sodium (mEq/l)	136–138	123–128	136–146
Chloride (mEq/l)	119–126	112–139	127–133
Potassium (mEq/l)	8.24–9.90	7.49–8.38	7.84–8.99
Bicarbonate (mEq/l)	21.8–22.6	15.6–17.0	17.6–21.8
Calcium (mEq/l)	2.72–3.09	2.74–3.81	3.04–3.32
Magnesium (mEq/l)	0.58–0.71	0.67–0.74	0.88–1.04
Phosphate (mEq/l)	0.98–1.02	0.75–1.38	1.38–1.59
Total phosphorus (mg/100 ml)	2.28–4.05	1.50–21.3	3.04–4.63
Acid-insoluble phosphorus (mg/100 ml)	0.77–0.87	0.46–0.80	0.84–2.07
Total carbohydrate (mg/100 ml)	61.2–69.4	43.7–55.0	67.0–69.1
Protein (g/100 ml)	0.93–1.45	0.48–2.80	1.59–3.20

[a] From Restall and Wales, *Austr. J. Biol. Sci.*, **19**, 687 (1966).

diestrus (Table 13). For instance, lowest levels of sodium and bicarbonate are present during the metestrous stage. Concentrations of magnesium are lowest during estrus and highest during diestrus, whereas the total carbohydrate concentration is lowest at metestrus. In another study, it was found that bicarbonate concentrations are higher in the isthmus than in the ampulla, and that, in contrast to the data reported in Table 13, higher concentrations of bicarbonate are present during diestrus than during any other stage of the cycle.

5.4   Rabbit Oviductal Fluid

Rabbit oviductal fluid can be collected continuously by inserting a cannula at the fimbriated end of the oviduct and leading it to a chamber in the abdomen from which the secreted fluid can be collected. The rate of fluid secretion varies from 0.4 to 2.0 ml per 24 hr. Sodium and chloride levels are particularly high in

Table 14   Concentrations of Some Inorganic Constituents of Estrous Rabbit Oviductal Fluid Collected Continuously via Cannulas into Chambers[a]

Constituent	Concentration (mEq/l)
Sodium	123–144
Chloride	109–119
Bicarbonate	27.4–30.6
Potassium	5.11–10.2
Calcium	1.97–3.98
Zinc	0.050
Magnesium	0.043–0.078
Phosphate	0.021

[a] From Brackett and Mastroianni, *in* Johnson and Foley (eds.), *The Oviduct and its Functions*, p. 133, Academic Press, New York (1979).

Table 15 Composition of Blood Plasma and Oviductal Fluid of Rabbit Does Obtained 1–2 Hours After Coitus[a, b]

Constituent	Plasma[c]	Oviductal Fluid[c]
Sodium	142.4 ± 1.5	149.2 ± 2.4
Chloride	104.0 ± 2.1	104.2 ± 3.2
Potassium	3.7 ± 0.2	9.2 ± 0.5
Bicarbonate	23.3 ± 0.6	43.5 ± 1.6
pCO_2 (mM)	37.5 ± 0.2	60.0 ± 3.0
pH	7.39 ± 0.01	7.47 ± 0.02
Osmolarity (mOsm)	292.8 ± 2.4	310.3 ± 3.7

[a] From Brackett and Mastroianni, *in* Johnson and Foley (eds.), *The Oviduct and its Functions*, p. 138, Academic Press, New York (1979).
[b] Values are reported as mean ± standard error.
[c] Concentrations of electrolytes are given in mEq/l.

the oviductal fluid of estrous does (Table 14). Slightly different values for ions were reported after mating (Table 15). The sodium and potassium levels of rabbit oviductal fluid are approximately the same as those found in blood, as is the pH, but the levels of potassium and bicarbonate and the pCO_2 are much higher in oviductal fluid than those of blood. The osmolarity of oviductal fluid

Table 16 Content of Total Free Amino Acids in Rabbit Oviductal Fluid[a]

Amino Acids	Mean ± S.D. (μmol/ml)
Alanine	0.469 ± 0.020
Arginine	0.064 ± 0.013
Aspartic acid	0.024 ± 0.005
Cysteine + proline	0.086 ± 0.002
Cystine	0.015 ± 0.001
Glutamic acid	0.192 ± 0.032
Glycine	2.766 ± 0.308
Histidine	0.067 ± 0.027
Isoleucine	0.069 ± 0.001
Leucine	0.129 ± 0.006
Lysine	0.165 ± 0.047
Methionine	0.022 ± 0.002
Phenylalanine	0.065 ± 0.005
Serine	0.318 ± 0.034
Taurine	0.123 ± 0.015
Threonine	0.125 ± 0.006
Tryptophan	not detected
Tyrosine	0.079 ± 0.006
Valine	0.172 ± 0.010
Total	4.936 ± 0.266

[a] From Iritani *et al.*, *J. Anim. Sci.* **33**, 829 (1971).

is slightly higher than that of blood. The sodium levels decrease and the calcium levels increase after ovulation.

The protein concentration of rabbit oviductal fluid was reported to vary from 2.1 to 2.7 mg/ml. A total of 19 amino acids has been found in rabbit oviductal fluid (Table 16). Glycine is present in highest concentrations, followed by alanine, serine, and glutamic acid. Tryptophan could not be detected. Threonine and serine are present in higher concentration in the oviductal fluid of the rabbit than in that of the sow.

Various enzymes have been found in rabbit oviductal fluid. These include amylase, alkaline phosphatase, catalase, diesterase, lysozyme, and lactate dehydrogenase. Additional biochemical constituents found include citric acid, carboxyl ester, lipid aldehyde, lipid choline, and glucose.

Inhibitors of enzymes have also been described in oviductal fluid. Recently, inhibitors of the sperm proteinase acrosin (Chapter 5), which have acidic isoelectric points, were identified in the tubal fluid of both the rhesus monkey and rabbit. Rabbit oviductal fluid was found to contain at least four inhibitors of trypsin. Total inhibitor concentration varies with the hormonal state of the animal; low levels are present at estrus and high levels occur several days after ovulation. These enzyme inhibitors may play a role in the regulation of fertilization by preventing the fusion of the sperm and the egg (requires acrosin activity) at any other than the optimum time in the life of the gamete. The inhibitors may also protect the oviductal lumen from proteolytic enzymes released from dying and degenerating spermatozoa.

REFERENCES

1   Preti, G., and Huggins, G. R., Organic Constituents of Vaginal Secretions, *in* Hafez, E. S. E., and Evans, T. N. (eds.), *The Human Vagina*, p. 151, Elsevier/North Holland, New York (1978).

2   Perl, J. I., Miller, G., and Shimonzato, Y., Vaginal fluid subsequent to panhysterectomy. *Am. J. Obstet. Gynecol.* **78**, 285 (1959).

3   Huggins, G. R., and Preti, G., Volatile constituents of human vaginal secretions. *Am. J. Obstet. Gynecol.* **126**, 129 (1976).

4   Michael, R. P., Bonsall, R. W., and Sacco, A. G., Proteins of human vaginal fluid. *Fertil. Steril.* **28**, 1345 (1977).

5   Sorensen, Jr., A. M., Animal Reproduction—Principles and Practices, McGraw-Hill, St. Louis (1979).

6   Prins, G., Zaneveld, L. J. D., and Schumacher, G. F. B., Functional Biochemistry of Cervical Mucus, *in* Hafez, E. S. E. (ed.), *Human Ovulation*, p. 313, North Holland, New York (1979).

7   Odeblad, E., Biophysical Techniques of Assessing Cervical Mucus and Microstructure of Cervical Epithelium, *in* Elstein, M., Moghissi, K. S., and Borth, R. (eds.), *Cervical Mucus in Human Reproduction*, p. 58, Scriptor, Copenhagen (1973).

8   Lee, W., Blandau, R. J., and Verdugo, P., Laser Light-Scattering Studies of Cervical Mucus, *in* Insler, V., and Bettendorf, G. (eds.), *The Uterine Cervix in Reproduction*, p. 68, Georg Thieme, Stuttgart (1977).

9   Schumacher, G.F.B., Cervical Secretions—A Product of a Target Organ for Estrogens and Gestagens, *in* Insler, V., and Bettendorf, G. (eds.), *The Uterine Cervix in Reproduction*, p. 101, Georg Thieme, Stuttgart (1977).

10 Schumacher, G. F. B., and Yang, S. L., Cyclic Changes of Immunoglobulins and Specific Antibodies in Human and Rhesus Monkey Cervical Mucus, *in* Insler, V., and Bettendorf, G. (eds.), *The Uterine Cervix in Reproduction*, p. 187, Georg Thieme, Stuttgart (1977).

11 Bhushana Rao, K. S. P., Barbier, B., Masson, P. L. *et al.*, Susceptibility of Cervical Mucus to Enzymes, *in* Hafez, E. S. E. (ed.), *Human Semen and Fertility Regulation in the Male*, p. 237, Mosby, St. Louis (1976).

12 Beyler, S. A., and Zaneveld, L. J. D., The role of acrosin in sperm penetration through human cervical mucus. *Fertil. Steril.* **32**, 671 (1979).

13 Schumacher, G. F. B., Holt, J. A., and Reale, F. Approaches to the Analysis of Human Endometrial Secretions, *in* Beller, F. K., and Schumacher, G. F. B. (eds.), *Biology of the Fluids of The Female Genital Tract*, p. 115, Elsevier/North Holland, New York (1979).

14 Tauber, P. F., *Biochemical Components of the Human Endometrium*. Frauenklinik and Poliklinik, Universitäs Klinikum der Gesamthochschule, Essen, West Germany (1979).

15 Olds, D., and VanDemark, N. L., Composition of luminal fluids in bovine female genitalia. *Fertil. Steril.* **8**, 345 (1957).

16 Fahning, M. L., Schultz, R. H., and Graham, E. F., The free amino acid content of uterine fluids and blood serum in the cow. *J. Reprod. Fert.* **13**, 229 (1967).

17 Dixon, S. N., and Gibbons, R. A., Proteins in the uterine secretions of the cow. *J. Reprod. Fert.* **56**, 119 (1979).

18 Beier, H. M., Endometrial Secretion Proteins—Biochemistry and Biological Function, *in* Beller, F. K., and Scumacher, G. F. B. (eds.), *Biology of the Fluids of The Female Genital Tract*, p. 89, Elsevier/North Holland, New York (1979).

19 Brackett, B. G., and Mastroianni, Jr, L., Composition of Oviducal Fluid, *in* Johnson, A. D., and Foley, C. W. (eds.), *The Oviduct and Its Functions*, p. 133, Academic Press, New York (1974).

20 Stanke, D. F., Sikes, J. D., DeYoung, D. W., *et al.*, Proteins and amino acids in bovine oviducal fluid. *J. Reprod. Fert.* **38**, 493 (1974).

CHAPTER EIGHT

CAPACITATION, ACROSOME REACTION, AND FERTILIZATION

B. JANE ROGERS

BARBARA J. BENTWOOD

Department of Obstetrics and Gynecology
John A. Burns School of Medicine
and Pacific Biomedical Research Center
University of Hawaii
Honolulu, Hawaii

The authors acknowledge Ms. Hana Van Campen for the preparation of the drawings and Dr. Sally Perreault for the photographs and helpful critique of the text. We also express sincere thanks to Ms. Cathy McCarville and Ms. Andrea Matsuda for typing and word processing, respectively, of the manuscript.

1   INTRODUCTION

The study of mammalian reproduction has led to an appreciation of the myriad of developmental and functional processes and changes that gametes must go through for fusion between them to occur. This chapter presents the current information specific to spermatozoal changes that take place following their introduction into the female reproductive tract. The metabolic and structural changes that have been observed *in vivo* can, in many instances, now be imitated *in vitro*. Such procedures have become key factors in enabling researchers to modulate the gamete-fusion process. Spermatozoa undergo changes that are grouped into three phases: capacitation, the acrosome reaction, and fertilization.

In 1951 two biologists, Austin and Chang, independently discovered that spermatozoa must experience changes that give them the "capacity" to fertilize. The process was called capacitation. Although 30 years have elapsed since its identification, capacitation is still not well understood. It does appear to be a universal requirement of mammalian spermatozoa that normally occurs in the uterus and fallopian tubes of the female. The time required *in vivo* seems to vary greatly between species, usually ranging from 1 to 6 hr.

Once capacitation is complete, spermatozoa undergo a morphological change called the acrosome reaction. The acrosome is a cytoplasmic vesicle in the sperm head (Fig. 1). The contents of this vesicle are released by a membrane-fusion process between the acrosomal membrane and the external plasma membrane. The acrosome contains a variety of enzymes, several of which are postulated to be directly involved in the fertilization process. The acrosome reaction has great functional significance, since only acrosome-reacted spermatozoa are capable of fusing with the egg membrane and effecting fertilization. Also, the enzymes released during the acrosome reaction play important roles in facilitating the passage of spermatozoa through the layers surrounding the egg.

The newly ovulated egg consists of an egg cytoplasm or vitellus surrounded by a plasma membrane or vitelline membrane, all covered by the zona pellucida, and a cloud of follicle cells, the cumulus oophorus. In the process of fertilization, the spermatozoon must first penetrate the cumulus and the zona and fuse with the vitelline membrane. Once sperm–egg fusion has occurred, a

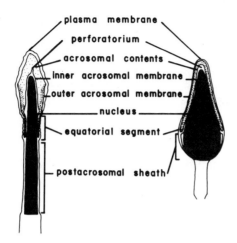

plasma membrane
perforatorium
acrosomal contents
inner acrosomal membrane
outer acrosomal membrane
nucleus
equatorial segment
postacrosomal sheath

Figure 1 Head region of hamster and human spermatozoa. The various membrane and structural components of these two spermatozoa clearly show that the acrosome is much larger and visible in the hamster spermatozoon on the left than in the human spermatozoon on the right.

block to polyspermy is established by the discharge of cortical granules from the egg cytoplasm. The fertilizing spermatozoon is incorporated into the egg, and the nucleus decondenses and transforms into the male pronucleus.

In the following, these phases are further described and discussed in light of our current knowledge. Details are presented for the biochemical and structural changes that have been documented, and speculations or hypothetical models are presented where conclusive information awaits further research.

2 CAPACITATION

Capacitation is exhibited as the lag time following ejaculation prior to the time when spermatozoa can fertilize. It is considered to be a period of physiological differentiation. The original definition of capacitation included the morphological change of the acrosome reaction. Now, however, it is regarded as the period just prior to the acrosome reaction. The physiological changes that take place during capacitation are neither well defined nor understood, as it is difficult to directly correlate changes with the actual acquisition of fertilizing competency. Several biochemical or metabolic changes can be documented concomitant with the capacitation time period, but these may or may not be directly linked to fertilizing ability. The changes seen during capacitation can be correlated to three basic parameters: (1) metabolism, (2) changes in cell surface components, and (3) membrane alterations. Metabolism is probably a pivotal factor for most of the functional modifications observed during the capacitation period. Sperm metabolism affects such factors as motility, viability, and membrane conditions. Yet the possibility that membrane reactions (through effector–receptor interactions) may trigger metabolic changes must be considered as well. Research is being directed toward a dissection of each of these parameters *in vitro* to assess their functional relationships to the capacitation process. Membrane changes probably occur as a consequence of metabolic changes, yet surface activators can also elicit metabolic activation. It is most likely a delicate interplay of both features that future research will, we hope, delineate.

2.1 Changes in Cellular Metabolism

Metabolic changes coupled to capacitation are reflected in the spermatozoal motility, respiration, and substrate utilization. The ubiquitous "second messengers," cyclic nucleotides, are also implicated in these metabolic changes. The inherent limitation to studying capacitation is that it cannot yet be effectively monitored directly. Because it is a necessary prerequisite to the acrosome reaction and, therefore, fertilization, capacitation is assessed by a quantitation of these structural and functional changes. In some species such as the hamster, the acrosome reaction can be observed by light microscopy. In species such as man, in which the acrosome reaction is difficult to visualize, capacitation is measured in terms of actual fertilization. Capacitation and acrosome reaction have been functionally segregated in only one species, guinea pig, which is discussed later. Our knowledge of the capacitation process, therefore, is based mainly on measurements of events that occur as the end products of capacitation. Other basic functional differences can, as yet, only be correlated as occurring simultaneously with the capacitation period.

2.1.1 Motility

Both the velocity and pattern of sperm motility have been shown to increase during the capacitating period. After incubation in capacitating media, hamster spermatozoa exhibit activation after 2–3 hr, that is, they move more vigorously and with greater flagellar amplitude. Although the mechanism causing this activation is unknown, it may be influenced by a change in the availability of exogenous metabolites or enzymatic activators.

2.1.2 Respiration

Direct metabolic studies done both *in vivo* and *in vitro* have demonstrated an increase in respiration associated with capacitating conditions. Rabbit spermatozoa, after 6 hr of incubation in the rabbit uterus, take up oxygen at a rate that is four times greater than that of freshly ejaculated spermatozoa. Some of the respiratory stimulation by female secretions is probably due to the presence of bicarbonate. Respiration in fowl, sheep, and pig spermatozoa is also stimulated when they are incubated within the uterus or oviduct, and in uterine or follicular fluids. Most metabolic studies to date have utilized natural fluids, consequently the metabolic effects in spermatozoa undergoing capacitation *in vitro*, using a specifically defined medium, have not been fully studied.

2.1.3   Substrate Utilization

Metabolic changes that may be associated with capacitation occur when the energy sources are modified *in vitro*. The presence and relative concentrations of specific metabolic precursors affect the efficiency and timing of capacitation in mouse, rat, hamster, and guinea pig. To accomplish *in vitro* capacitation, ejaculated or epididymal spermatozoa are washed and incubated in a balanced salt solution containing some combination of glucose, pyruvate, and lactate. In mouse, the omission of lactate from the standard medium significantly reduces the percentage of fertilized eggs (a measure of capacitation). The addition of pyruvate to the culture medium increases the proportion of motile, acrosome-reacted mouse spermatozoa and increases the proportion of eggs fertilized. The best medium for mouse sperm capacitation includes albumin, lactate and pyruvate. Glucose, however, appears to be the major energy source. Although rat fertilization was found to be supported by both glucose and mannose, but not fructose or galactose, hamster spermatozoa were found to require pyruvate specifically as an energy source and to capacitate most effectively with both glucose and pyruvate. Guinea pig spermatozoa appear unique in that they require pyruvate and lactate for capacitation, whereas glucose has actually been shown to retard their acrosome reaction. This retarding effect is paralleled by an inhibition of respiration. Reduced respiration may be linked to an inhibition of capacitation, suggesting that capacitation may be coupled to a specific type of metabolism, that is, glycolysis or oxidative respiration. The optimal energy source for capacitation may be different for different species, but it seems clear that the energy source chosen affects metabolism, and the metabolic state affects capacitation.

2.1.4   Cyclic Nucleotides

Changes in sperm metabolism, respiration, and motility are known to be affected by agents that increase levels of cellular cyclic nucleotides. Since each of these processes also undergoes changes during the capacitation period, it has been speculated that cyclic nucleotides may modulate the capacitation process. Studies in a variety of mammalian species suggest that a relationship exists specifically between the levels of cAMP in spermatozoa and capacitation. Early studies reported an increase in cAMP production by hamster spermatozoa when incubated in a capacitating medium. Subsequently, the exogenous application of cAMP or its more-membrane-permeable dibutyryl derivative reportedly augmented capacitation in rat and rabbit spermatozoa. These studies, however, utilized additional capacitation-inducing factors in the media and thus could not definitively demonstrate a direct cAMP effect.

Because of the limited permeability of cyclic nucleotides into intact cells, a widely used method of raising cellular cAMP levels is to inhibit its metabolic degradation by the use of phosphodiesterase (PDE) inhibitors. Micromolar amounts of the phosphodiesterase inhibitors papaverine, theophylline, and dibutyryl-cAMP have been shown to activate both motility and acrosome reaction in hamster spermatazoa. Increased cAMP was assumed to be responsible for this effect. Similarly, human spermatozoa have recently been shown to capacitate more rapidly in the presence of either caffeine or theophylline. The correlation between cellular cAMP levels and the metabolic state of the spermatozoa is still not well understood, but the relative cytoplasmic ratio of cAMP: ATP and the amounts of available energy sources probably modulate the operations of the glycolytic and respiratory pathways. As in the metabolic-substrate studies, guinea pig spermatozoa are unusual with respect to cAMP. In this species, incubation with PDE inhibitors actually inhibits capacitation and fertilization, whereas imidazole (a PDE stimulator) produces the opposite effect. For most species studied, however, PDE inhibitors have a stimulatory effect on metabolism and subsequently enhance motility and augment the capacitation rate.

As is discussed in the following sections, capacitation also appears to involve changes in membrane permeability, particularly to calcium ions. A calcium-binding protein found in spermatozoa, calmodulin, has been shown to govern the synthesis and degradation of cAMP in a variety of cell types by regulation of both adenylyl cyclase and PDE, depending upon calcium-ion availability. If such a system does function in spermatozoa, increased calcium flux into the spermatozoa could serve to activate calmodulin molecules, which in turn would stimulate adenylyl cyclase activity. The resulting cAMP synthesis could then effect the observed metabolic changes and capacitation.

2.2   Seminal Plasma Components and the Sperm Cell Surface

In transit from the epididymis and through the female reproductive tract, the sperm surface experiences several modifications that may be linked to capacitation. The constituents of seminal plasma, for example, appear to have significant effects on sperm capacitation. Structural and biochemical changes in the sperm plasma membrane itself may be either initiated or inhibited in association with any number of components the spermatozoa encounter en route to the fertilization rendezvous. If bound to the sperm surface, such components may function as enzyme inhibitors, and their removal may then act as stimuli for functional and structural modifications.

One such constituent, a proteinase inhibitor, that inhibits sperm acrosin (Chapter 5), is found in seminal plasma and on ejaculated spermatozoa, apparently after being adsorbed during transport through the male genital tract. Direct treatment of capacitated spermatozoa with proteinase inhibitors from male reproductive tract secretions has been shown to reduce their fertilizing ability. This reduced fertilization may be due to inhibited acrosome reaction and/or to reduced acrosomal enzyme activity necessary for sperm–egg fusion (see Section 3.4 of this chapter). Capacitation may require removal of these proteinase inhibitors, or removal may be necessary for the proteinase, acrosin, to function in sperm penetration of the zona pellucida. Epididymal rabbit spermatozoa accumulate the proteinase inhibitor when incubated in seminal plasma. Acrosomal extracts from epididymal and capacitated rabbit spermatozoa have high acrosin activity, whereas ejaculated spermatozoa extracts do not unless the inhibitor is removed by purification procedures. Chang demonstrated in 1957 a factor in seminal plasma that, when mixed with capacitated rabbit spermatozoa, reversibly inhibited their ability to fertilize ova. The spermatozoa were said to be decapacitated and the factor was called decapacitation factor (DF). It was noted that DF did not affect sperm motility and that decapacitated sperm populations could be recapacitated, that is, become fertile again, after incubation within the uterus. Since its discovery in rabbits, DF has been demonstrated in the seminal plasma of the bull, stallion, boar, monkey, dog, and man, and also in the epididymal fluid of the rabbit and bull. Attempts to isolate and characterize DF are motivated by its potential use as a contraceptive. Dukelow and associates reported that rabbit DF is resistant to digestion by pronase, α-amylase, glucose oxidase, hyaluronidase, and lysozyme, but susceptible to β-amylase digestion. DF is heat stable and can withstand long-term storage. It is now believed that the decapacitating activity of seminal plasma may be attributed to several different substances. Davis purified rabbit DF as a high-molecular-weight 86 S glycoprotein that is present in two distinct classes of membrane vesicles from seminal plasma. Rabbit seminal plasma fractionated on Sephadex G-200 had DF activity in the first two high-molecular-weight fractions of the five fractions that separated. SDS electrophoretic protein patterns showed that the only component present in the active fractions had a molecular weight of 115,000. This component stained positively with the periodic acid Schiff's (PAS) stain for carbohydrate, suggesting a glycoprotein composition compatible with the results of the enzyme studies of Dukelow. Bull seminal plasma contains a very different DF, which can be hydrolyzed into very low molecular weight peptides that retain high antifertility activity.

Because of the variable experimental results, a clear mechanism of action by decapacitating factor(s) has not been defined. It is critical to determine whether DF actually acts on the capacitation processes of the spermatozoa, or simply by inhibiting enzymes necessary for fertilization. Rabbit DF has been shown to act by a specific inhibition of the acrosomal corona-penetrating enzyme (CPE) (Chapter 5). Capacitated DF-treated rabbit spermatozoa cannot pass through the corona radiata but will fertilize the ovum if the corona is removed. Human DF prevents the fertilization of mouse eggs as long as the zona pellucida is intact (mouse eggs do not possess a corona radiata). It does not inhibit sperm acrosin or hyaluronidase, however. The decapacitation factor may also act by protecting or stabilizing the plasma membrane, thus preventing the normal acrosome reaction.

2.3 Plasma Membrane Changes

Plasma membrane changes during the capacitation period include changes in fluidity, addition and/or removal of membrane components, both protein and lipid, and changes in the overall membrane architecture. These changes have been studied with ultrastructural techniques utilizing membrane markers such as labeled antibodies, lectins, and membrane probes.

Changes in the distribution of intramembranous particles occur in rabbit spermatozoa when incubated under conditions that promote capacitation. Koehler and co-workers, using a hemocyanin-labeled antibody, reported that such particles were initially arranged uniformly on the plasma membrane but became aggregated into patches after only 30 min in a capacitating medium. Both isotonic and hypertonic media apparently removed components from the plasma membrane overlying the acrosomal region, yielding "bare" patches in the field of hemocyanin-labeled antibody. *In vivo* incubation in the uterus also has been reported to promote a redistribution of receptors for labeled anti-sperm antibody prepared against whole spermatozoa. Initially, these receptors were evenly distributed over the sperm head. Following incubation, there was a patchy distribution or complete absence of label over the acrosomal region. Similar results have been obtained using air-dried surface replicas. The change in the labeling pattern may result from a redistribution, masking, or progressive loss of surface antigenic sites.

O'Rand extended these studies by preparing a monospecific antibody to a single, intrinsic rabbit-sperm membrane glycoprotein (MGP). The localization of this MGP was monitored during prolonged *in vivo* incubation in the uterus using fluoresceinated whole antibody or purified Fab fragments. With Fab labeling, freshly ejaculated or epididymal spermatozoa showed a diffuse distribution of MGP, which could be induced to form patches when treated with whole immunoglobulin. Capacitated spermatozoa, that is, those incubated 18 hr *in utero*, exhibited the same pattern of Fab labeling, but could *not* be induced to form patches with whole immunoglobulin. These results indicate that the lateral mobility of this antigen decreased with capacitation, which probably reflects a decrease in membrane fluidity. Although O'Rand's results seem to conflict with the hemocyanin-labeled antibody studies, the use of Fab fragments for localization avoids immunoglobulin-induced aggregation. Also, the longer incubation period may be critical for visualizing a decrease in membrane fluidity. O'Rand and others have suggested that decreased membrane fluidity may result from the removal of seminal plasma components, changes in calcium influx, changes in the membrane lipids, or a combination of these and other factors. Decreased fluidity most likely reflects the membrane destabilization, which is necessary for the fusion events of the acrosome reaction (see Section 3 of this chapter).

Membrane glycoprotein or glycolipid moieties have also been monitored using labeled lectins. Concanavalin A (Con A), a specific label for α-D-glucose and α-D-mannose, has been used to follow sperm membrane changes during capacitation. Rabbit and hamster spermatozoa taken directly from the epididymis exhibit a uniform distribution of fluorescent-labeled Con A over the entire head region. Spermatozoa taken from the uteri of mated female rabbits (presumed to be capacitated) and spermatozoa incubated 3 hr in a capacitating medium show large areas devoid of Con A labeling over the acrosomal region, whereas the postacrosomal region retains a heavy distribution of label. The patchiness may be again indicative of either a removal or redistribution of lectin-binding molecules. Also, an increase in lectin-induced sperm agglutination concomitant with capacitation has been observed with guinea pig spermatozoa exposed to soybean agglutinin and Con A. This indicates that lectin-binding sites, that is, monosaccharides, are probably exposed during capacitation. Spermatozoa pretreatment with trypsin results in lectin binding and agglutination equivalent to that seen after capacitation. These data indicate that a trypsin-like enzyme may modify and expose carbohydrate-containing molecules on the spermatozoa surface during the *in vivo* course of capacitation.

More-subtle ultrastructural changes during capacitation have been demonstrated by freeze fracture and etching. Guinea pig spermatozoa show a loss of fibrillar intramembranous particles from the exterior faces of the plasma and outer acrosomal membranes and a clearing of globular particles from the interior face of the plasma membrane. In fact, three of the four membrane halves that will be involved in fusion, including both faces of the plasma membrane and the exterior face of the acrosomal membrane, acquire particle-free areas. The interior face of the outer acrosomal membrane is virtually devoid of particles from the start. As a result of these studies, it is believed that membranes become receptive to fusion at particle-deficient interfaces.

A scheme for the changes that take place in the membrane and surface coat during capacitation is suggested in Fig. 2. The spermatozoon is initially coated with DF, proteinase (acrosin) inhibitors, or other stabilizing components that come from the ejaculate while the spermatozoon is residing in the vagina. The intramembranous particles in the plasma membrane of the head area are somewhat evenly spaced (*a*). As the spermatozoon passes into the cervix on its way

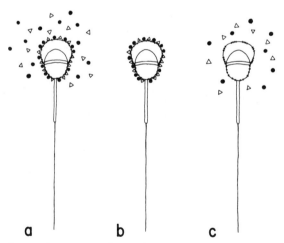

Figure 2 Membrane changes associated with capacitation. (*a*) Ejaculated spermatozoa residing in the vagina: ●, DF or other stabilizing protein, Δ, proteinase inhibitory substance, –●–●–, intramembranous particles. (*b*) Spermatozoa in the cervix. (*c*) Capacitated sperm in the uterus: –●–●–●–, clustered intramembranous particles.

to the uterus, it leaves behind the unbound seminal fluid components. It retains surface-coating factors and stabilizing proteins and maintains its original membrane architecture (b). While the spermatozoon is in the uterus undergoing capacitation, surface components are unmasked or released (c). Changes in the characteristics of the plasma membrane occur, specifically in the area directly over the acrosome. Changes occur in the distribution of calcium ions, the integral protein components may change or become redistributed, and the membrane fluidity decreases, making the membranes receptive to fusion. The sperm is primed to undergo the acrosome reaction.

3   THE ACROSOME REACTION

The changes associated with capacitation culminate in the fusion-secretion event called the acrosome reaction. The functional significance of this event is evidenced by the fact that the enzymes released or activated facilitate passage of the spermatozoon through the outer investments of the egg, and only acrosome-reacted spermatozoa are capable of fusing with the egg and effecting fertilization.

The acrosome is a cytoplasmic organelle that partially originates from the fusion of Golgi-derived vesicles (Chapters 1 and 5). As such, it is considered to be a modified lysosome. Among the enzymes reported to be acrosome-associated (Chapter 5), some have been shown to have specific functions in the fertilization process. Many, however, have undefined functions or may simply be residual lysosomal enzymes that are not specifically involved in sperm function. For an enzyme to be considered active in fertilization, three criteria must be fulfilled: (1) the enzyme should be released or available at the time of fertilization, (2) a natural substrate should be demonstrable in the sperm or egg, and (3) there should be an inhibition of fertilization coupled to a specific inhibition of the enzyme. Evidence has been presented that establishes a role in the fertilization for at least seven acrosomal enzymes, including hyaluronidase, acrosin, proacrosin, corona-penetrating enzyme (CPE), a neuraminidase-like enzyme, ATPase, and phospholipase. Examples of possible residual lysosomal enzymes are esterases, aryl sulfatases, β-N-acetylglucosaminidase, acid proteases, and collagenase. Some of these may play a role in some sperm function that is not yet understood.

In mature spermatozoa, the acrosome is a cap-like, membrane-bound structure located in the anterior portion of the sperm head, atop and around the nucleus. The plasma membrane of the spermatozoon is continuous over the entire cell but is functionally demarcated at different areas relative to the nucleus (Fig. 1). Acrosomes vary in shape and size between species, but all undergo similar structural changes when the acrosome reaction occurs. Analogous to other vesicle-secretion processes, the acrosome reaction is thought to occur by a series of fusions at discrete areas between the plasma membrane and the outer acrosomal membrane, causing a vesiculated, acrosome–plasma-membrane complex over the entire sperm head excluding the area over the equatorial segment.

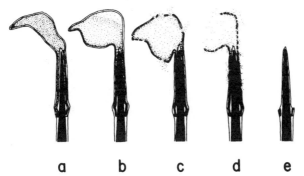

a   b   c   d   e

Figure 3 Acrosome reaction in guinea pig spermatozoa. (*a*) Intact acrosome prior to initiation of reaction, (*b*) swelling of the acrosome, (*c*) initiation of vesiculation and loss of acrosomal contents, (*d*) extensive vesiculation of plasma membrane and outer acrosomal membrane with subsequent loss of vesicles, (*e*) acrosome-reacted spermatozoa with complete loss of vesicles.

A model for the sequence of morphological events during the acrosome reaction has been reconstructed from guinea pig oviductal-insemination experiments. As illustrated in Fig. 3, the following events take place: (*a*) as the spermatozoon comes near or in contact with the egg, there is an initial swelling The punctate distribution of fusion sites results in a fragmentation of the membrane complex, which liberates the acrosomal contents and ultimately exposes the inner acrosomal membrane surface (Fig. 3 and 4).

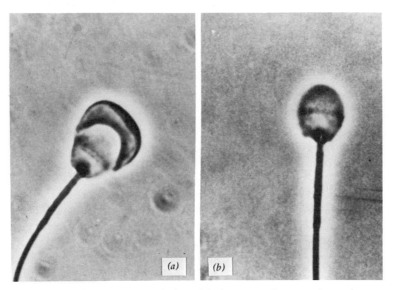

Figure 4 Guinea pig acrosome reaction. (*a*) Acrosome-intact guinea pig spermatozoon, (*b*) swelling of the acrosome, (*c*) initiation of vesiculation and loss of acrosomal contents, (*d*) extensive vesiculation of plasma membrane and outer acrosomal membrane with subsequent loss of vesicles, (*e*) acrosome-reacted spermatozoa with complete loss of vesicles.

of the acrosome, (*b*) some enzymes (e.g., hyaluronidase) are liberated prior to contact of the spermatozoon with the zona, while the sperm membranes appear structurally intact, (*c*) vesiculation by means of fusion of the plasma and outer acrosomal membranes proceeds rostrally over the sperm head while the spermatozoon is on the surface of the zona, (*d*) the fusion-complex vesicles formed from the membranes are denuded as the spermatozoon enters the zona matrix, and (*e*) secondary vesiculation at the equatorial segment (acrosome collar) occurs during passage into the more compact area of the zona.

3.1   Modulation of the Acrosome Reaction

The acrosome reaction appears akin to the process of exocytosis in secretory cells, in which cytoplasmic granules are induced to fuse with the plasma membrane for the liberation of their contents. Such degranulation reactions frequently depend upon changes in both the plasma and vesicle membranes, priming them for fusion. Such changes are usually stimulus-dependent and affected by ion gradients, membrane-enzyme activation, and architectural changes in the membrane proteins and/or the membrane lipids. The membrane-fusion process of the sperm acrosome reaction can apparently be similarly modulated by a diverse variety of natural and artificial substances. The goal of current research is to determine the mechanism(s) for acrosome-reaction initiation and inhibition for potential use in fertility control. To date a number of acrosome-reaction initiators have been identified in tissue fluids from the female reproductive system and from the outer layers of the ovum itself. Blood serum factors and calcium ions have also been shown to trigger the acrosome reaction *in vitro*. However, the true initiator(s) of the acrosome reaction *in vivo* has not yet been identified.

3.2   Types of Initiators

Secretions or tissues that have been claimed to stimulate the acrosome reaction include follicular and tubal fluids, cumulus cells, the zona pellucida, and serum factors. Most information is available for hamster spermatozoa, because the acrosome reaction in this species can be assessed directly by phase contrast microscopy.

    In the hamster, both homologous and heterologous follicular fluids, that is, from the same and different species respectively, tubal fluids, and a heat-stable factor from bovine follicular fluid will initiate the acrosome reaction. Rabbit spermatozoa will react to the components of follicular fluid from rabbits, cats, pigs, cows, and women. The active agent in this fluid is non-dialyzable but heat labile above 56°C for 20 min or more. Conflicting results have been reported regarding the nature of the active follicular fluid component. Albumin, which is found abundantly in follicular fluid, was originally shown to be capable of inducing the acrosome reaction. Oliphant, however, demonstrated that a follicular fluid protein other than albumin, possibly a component of the complement system, was responsible for the observed acrosome-reaction-initiating activity. Fertilization in the hamster and rabbit has been achieved in the absence of follicular fluid despite its *in vitro* activity. Oviduct fluid, with and without egg products, is also effective in inducing the acrosome reaction in hamster and rabbit spermatozoa, as is postovulatory oviductal fluid (partially derived from follicular fluid). This suggests that the *in vitro* incubation conditions are somehow mimicking an effect which may depend upon these reproductive fluids *in vivo*.

Because spermatozoa must penetrate the layers surrounding the oocyte, first the cumulus oophorus, and then the zona pellucida, which coats the egg itself, their components are prime suspects for acrosome-reaction initiators. Unwashed cumulus cells have, in fact, been shown to induce the acrosome reaction in human spermatozoa. However, as it can in the absence of follicular fluid, fertilization can occur in the absence of the cumulus. A direct role for the zona pellucida in initiating the acrosome reaction has not yet been clearly demonstrated. Spermatozoa necessarily must acrosome react in order to penetrate the zona, but questions remain regarding the status of the spermatozoon when it first encounters the zona. Gwatkin has suggested that the acrosome reaction of *fertilizing* spermatozoa occurs on the surface of the zona. This he concluded from observing penetration of the zona by spermatozoa preincubated with cumulus before most of the free-swimming spermatozoa were acrosome-reacted. Thus, although previously acrosome-reacted spermatozoa may be able to bind to the zona, actual penetration may require acrosome reaction "on the spot" to release the penetrating enzymes where they will be effective. If the zona does specifically induce the acrosome reaction, this inducing effectiveness should be inhibited if the zona is destroyed. However, rabbit eggs treated with either neuraminidase, trypsin, or chymotrypsin failed to inhibit the acrosome reaction of capacitated spermatozoa incubated with these eggs. Although these results oppose a concept of zona involvement, this possibility cannot be completely eliminated on the basis of these enzyme treatments alone. A specific induction stimulus could be resistant to these enzymes.

Under experimental conditions, heat-inactivated serum and some serum fractions can be utilized to induce the acrosome reaction. This inducing mechanism may simply be due to the effect of albumin. Yet, the mechanism for serum activation of the acrosome reaction may be similar to that seen with exocytotic inflammatory cells. Cross-linking of membrane components by globulin molecules or receptor activation by a heat-stable factor may sufficiently destabilize the sperm membrane to artificially stimulate the acrosome reaction.

Each natural component discussed appears to be capable of inducing the acrosome reaction under specific conditions. Although the acrosome reaction can be artificially induced *in vitro* in the absence of these natural elements, they may still be essential initiators in a natural environment. There is probably more than one mechanism by which the acrosome reaction can be triggered, or different agents may act through one common mechanism. Much current research is aimed at defining such a common mechanism.

3.3 The Role of Calcium in the Acrosome Reaction

Calcium has been well documented to play a major role in membrane-fusion processes. Membranes are known to stabilize or destabilize depending upon calcium-ion fluxes or the cytoplasmic calcium concentration. Much new information is now available regarding the initiation and/or enhancement of fusion-secretion reactions in a wide variety of cell types. Likewise, the acrosome reaction in spermatozoa is critically sensitive to calcium ions, possibly at the level of acrosomal enzyme activation or by a direct effect on the membrane proteins and/or phospholipids.

Capacitating media generally contain calcium at a concentration that is about four orders of magnitude greater than the spermatozoal cytoplasmic concentration. These conditions establish a steep electrochemical gradient for calcium across the plasma membrane. The calcium ionophore A23187 is believed to bypass calcium transport mechanisms and is widely used to test the direct effect of calcium influx on cellular functions. In sperm studies, preincubation with A23187 results in an increase in cytoplasmic free calcium immediately preceding the acrosome reaction. Guinea pig spermatozoa incubated in a capacitating medium for 10–12 hr will normally acrosome react unless calcium is excluded from the medium. Under calcium-free conditions, capacitation will ensue without the acrosome reaction, but the acrosome reaction can be induced within minutes by the addition of calcium. Further, incubation of guinea pig spermatozoa with A23187 in the presence of calcium will induce the acrosome reaction at a greater-than-normal rate. Acrosome-reaction inhibition under these conditions can be demonstrated using calcium antagonists such as lanthanum and verapamil. The immediate rise in cytoplasmic calcium probably reflects an increase in the membrane's calcium permeability. The mechanism of this change is unknown. Inducers of the acrosome reaction may act differently at their initial membrane-interaction step, but all act secondarily in the common step of permeability change. In a model proposed by Green, stimulus-receptor sites are associated with calcium channels in the plasma membrane of the acrosomal region. Capacitation proceeds by removing antagonists that occupy the receptor sites. Subsequent stimulus–ligand interaction opens the calcium-ion channels and triggers the acrosome reaction.

Exciting new evidence is accumulating regarding the presence and activity of calmodulin in spermatozoa. This protein is now known to modulate a variety of calcium-dependent systems in somatic cells, particularly secretory cells. A relatively large amount of calmodulin has been found in mammalian spermatozoa. Biochemically, a large portion of the sperm calmodulin can be isolated from the acrosomal cytosol. Direct immunofluorescence localizes the calmodulin around the acrosome and on the lower part of the spermatozoon head. The acrosomal localization is lost following membrane vesiculation and shedding, suggesting that the calmodulin is located between the plasma membrane and the acrosomal membrane. Calmodulin could be the key factor in the membrane-fusion process following calcium influx at the start of the acrosome reaction. It may also play a role in acrosin-proacrosin activation (see Section 3.4.2 of this chapter) and in the spermatozoon–egg fusion process (see Section 4 of this chapter).

3.4 Function and Localization of Specific Acrosomal Enzymes

3.4.1 Hyaluronidase

Hyaluronic acid is the most abundant of the acid mucopolysaccharides that act as flexible cement to hold the cells of the cumulus oophorus together. These cells are readily dispersed upon exposure to sperm hyaluronidase (Chapter 5). This is also the enzyme most readily released from spermatozoa, either as a consequence of aging or in the acrosome reaction. Previously, it was believed that sperm hyaluronidase when released from the acrosome, acted to denude the ova of all its cumulus cells, but fertilizing spermatozoa can be found at the zona surface before the cumulus is fully dispersed, indicating that the spermatozoa may use the enzyme simply to slide through the cumulus matrix, that is, to digest a small path in front of the spermatozoon. An exact localization of the enzyme during the penetration process would help to clarify its mode of action. The results of such studies so far are conflicting. The data suggest that hyaluronidase is located either on the surface of the plasma membrane and/or is contained within the acrosome. Hyaluronidase associated with the plasma membrane could be functional (digest the cumulus matrix) before completion of the acrosome reaction, that is, while the sperm membranes are intact. However, the acrosome reaction needs to occur before enzymes contained within the acrosomal vesicle can be released and be functional.

Spermatozoa have been examined using immunocytochemical techniques for hyaluronidase localization. Living spermatozoa were incubated with antihyaluronidase Fab fragments, then fixed and stained with a second labeled anti-Fab antibody. No surface-localized hyaluronidase could be demonstrated above background. Spermatozoa that were fixed and dried prior to staining (making the membrane permeable to antibodies) demonstrated hyaluronidase labeling over the acrosomal region. Spermatozoa are known to retain about 20% of their original hyaluronidase activity when the outer acrosomal membrane and acrosomal contents are removed. Although the bulk of the hyaluronidase is released from the acrosome during the acrosome reaction, some residual hyaluronidase may be bound to the inner acrosomal membrane. Hyaluronidase may also be released to the exterior of the cell and bind to the cell surface, where it can continue to be active.

Although the acrosome appears to house the largest portion of the hyaluronidase, whether a full acrosome reaction is required for its release is uncertain. Rogers and Yanagimachi have shown hyaluronidase release in parallel with the acrosome reaction induced in capacitated spermatozoa by the addition of calcium. Hamster spermatozoa incubated in serum, however, showed release of hyaluronidase into the media within 1 hr with no acrosomal changes visible by light or electron microscopy. This suggests that a portion of the enzyme has diffused out of the acrosome or that it is actively released from the sperm surface prior to acrosome vesiculation. More critical localization studies will depend upon high-specificity labeling of hyaluronidase under rigidly controlled conditions for the acrosome reaction.

3.4.2   Acrosin

Classically, acrosin (Chapter 5) has been implicated in zona penetration based on its proteinase activity and ability to dissolve zona components. Proteinase inhibitors will inhibit fertilization presumably by inhibiting acrosin. Additionally, acrosin inhibitors have been shown to block initiation of the acrosome reaction, but not the capacitation process, in hamsters. This inhibition process was studied further with guinea pig spermatozoa, whose ionophore-induced acrosome reaction is delayed by acrosin inhibitors. When examined by the electron microscope spermatozoa treated with ionophore A23187 in the presence of acrosin inhibitors such as benzamidine did undergo plasma membrane–acrosome membrane vesiculation, but the acrosomal matrix did not detach. It was concluded that acrosin may be necessary for solubilization of the acrosomal matrix (but not the acrosome reaction per se) and for dissolution of zonal components.

Most of the acrosin on the spermatozoon is present in its zymogen form, proacrosin (Chapter 5). In order for this proacrosin to function, it must be transformed to the active form, acrosin. Zymogens can be activated by cleavage at a specific lysine residue, removing an N-terminal peptide. Proacrosin can be converted to acrosin in this way by trypsin or by acrosin itself. The *in vivo* mechanism of proacrosin activation is uncertain, probably being modulated by the acrosomal pH, the presence of inhibitors, and/or the presence of another activating enzyme. Besides acrosin, another acrosomal enzyme, acrolysin, has also been implicated as such an activating enzyme (Chapter 5).

Knowing the location of acrosin in the intact acrosome and its distribution during or after the acrosome reaction would lend considerable insight into understanding how it functions. It was once assumed that the acrosomal contents were completely solubilized after the acrosome reaction. This view has changed, based primarily on two observations: (1) not all of the acrosin is released after inducing complete acrosome reaction in a whole spermatozoa population, and proacrosin remains bound to the sperm head, and (2) the vesiculated acrosome–plasma membrane complex is shed before zona penetration. Thus after acrosome reaction the inner acrosomal membrane is the remaining exposed surface and retains a portion of the proacrosin. The enzymes that act to allow further penetration must somehow remain in close association with this new "leading" membrane surface.

It has been proposed that acrosin (proacrosin) is associated with both the inner and outer acrosomal membranes as well as with the internal matrix. Acrosin has been localized in electron microscope specimens by the use of inhibitors. Ferritin-labeled proteinase inhibitor was used as an acrosin label and was shown to associate with the outer surface of the inner acrosomal membrane in bull spermatozoa. This binding is retained even after the acrosome reaction. Sites of proteolytic enzyme activity have been localized in rabbit spermatozoa by silver proteinate staining. These sites extend diagonally up one flat side of the spermatozoal head and diagonally down the opposite side (Fig. 5). If staining is carried out in a very hypotonic solution, these linear arrays are opened into a semirigid umbrella-like structure. These observations suggest that some proteolytic enzyme is in close association with the outer acrosomal membrane or with some structure in the parietal area of the matrix. Similar arrays of silver proteinate staining have been observed in the penetration tunnel through the zona pellucida of rabbit ova fertilized *in vivo*. It appears that acrosin and/or proacrosin, or some protease, does remain associated with the spermatozoa following the acrosome reaction. It was proposed that the nonzymogen acrosin on the spermatozoon is involved in the induction of the acrosome reaction. When the spermatozoon comes in contact with the zona pellucida, the acrosome reaction takes place, causing the activation of a portion of the proacrosin to acrosin which is released and initiates digestion of the zona. After entry of the spermatozoon into the zona, the proacrosin on the inner acrosomal membrane is activated, lysing a path in front of the spermatozoon through the zona.

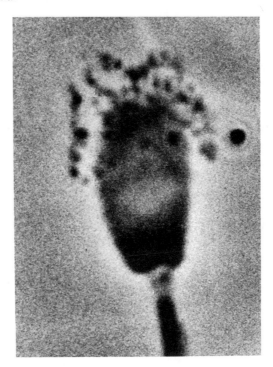

Figure 5 Proteinase activity on a rabbit epididymal spermatozoon. The silver proteinate stain for proteolytic activity shows activity localized in definite microtubule structures in the acrosomal cap of rabbit spermatozoa. From Stambaugh, *Gamete Research* **1**, 65, 1978.

3.4.3   Other Acrosomal Enzymes

A variety of other acrosomal enzymes have been identified, (Chapter 5), but their roles in the fertilization process are less clearly understood. Each is probably active at a discrete step depending upon the species being considered. Viable hypotheses will depend upon a definition of the natural substrates being affected. ed.

The corona penetrating enzyme (CPE), discovered in 1968 by Zaneveld, is considered to be the acrosomal enzyme used by spermatozoa to penetrate the matrix between the cells of the corona radiata in a manner analogous to penetration of the cumulus matrix with the aid of hyaluronidase. Originally found in the rabbit, this enzyme acts specifically on the corona with no effect on the zona. It is inhibited by both whole seminal plasma and purified DF. CPE has been shown to be functionally distinct from hyaluronidase. Hyaluronidase has no effect on the corona, and is not inhibitable by DF. This enzyme may act in conjunction with hyaluronidase in those species whose ova possess the corona cell layer (Chapter 9).

A mechanism for the active transport of ions across the sperm membrane has been suggested, based on the cytochemical localization of at least one membrane-associated ATPase. Gordon was able to demonstrate phosphatase activity at the plasma membrane and at the acrosomal membrane of both rabbit and guinea pig spermatozoa. The deposition of enzyme reaction product observed in these electron microscope studies was particularly sensitive to the processing conditions utilized. The enzyme activity was sensitive to washing of the spermatozoa, the pH of the incubation media, and the presence of cations. Gordon could delineate one ATPase, active at pH 7.0, at the external surface of the plasma membrane. This neutral phosphatase did not require divalent cations. A second ATPase was active at pH 9.0 and did require calcium in the incubation medium. This alkaline phosphatase was seen only in unwashed sperm specimens and was associated with the acrosomal membrane. These results are somewhat controversial, as Yanagimachi has reported the ATPases of these membranes in guinea pig spermatozoa to be magnesium dependent and inhibitable by calcium. If ATPase is associated with the plasma membrane, it may act as a pump to maintain an ion gradient and particular membrane potential. Spermatozoa undergoing capacitation, experience an increase in plasma membrane permeability to ions. Subsequently, the activated calcium-specific ATPase of the acrosomal membrane may transport calcium ions into the acrosome and directly effect the acrosome reaction. Critical biochemical analysis of these membranes is necessary for further consideration of this scheme.

3.5   Hypothetical Sequence of Acrosome-Reaction Events

To summarize the previous discussion, the following is a possible scheme for the events of the acrosome reaction.

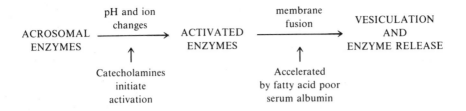

1 Calcium-ion influx is increased in the area between the plasma membrane and the outer acrosomal membrane. The β-adrenergic effect of catecholamines on calcium influx may result from stimulating a membrane adenylcyclase and subsequently increasing uptake (or permeability) of calcium.

2 The increase in calcium-ion concentration in the periacrosomal space activates the calcium-ion uptake into the acrosome itself by the activation of a calcium-dependent ATPase.

3 Calcium within the acrosome activates proacrosin to acrosin, possibly by a displacement of hydrogen ions in the acrosome proper, increasing the pH from acidic conditions that prevent proacrosin activation, to a more neutral or basic pH. Calcium may also enhance acrosin activity by a similar mechanism.

4 Acrosin, localized on the inner acrosomal membrane hydrolyzes an inactive phospholipase to an active enzyme which hydrolyzes certain lipids of the outer acrosomal membrane, producing free fatty acids or other lysophospholipid by-products. Activation of cAMP may also stimulate such phospholipase activity.

5 The phospholipase products initiate membrane perturbations, membrane fusion, and final vesiculation.

4   FERTILIZATION

Details regarding the chemical composition of the egg before and after fertilization can be found in Chapter 9. The interaction and fusion of male and female gametes in the fertilization process is a multistep phenomenon that is still not well understood at the molecular and biochemical level. The process has been studied *in vivo* and *in vitro*, and the combined information is helping to clarify the sequence of events and critical steps as they occur naturally. At this level of interaction of the mature gametes, there are several parameters that make the interaction highly sensitive and consequently susceptible to control and manipulation. First, the spermatozoa must be able to penetrate the outer egg layers and arrive at the vitelline membrane in a state that is appropriate for fusion with this membrane. Controversy still exists regarding the difference between the function and morphology of nonfertilizing spermatozoa and the one actual fertilizing spermatozoon. Second, some mechanism lends discriminative species-specificity control to sperm–egg membrane fusion. Finally, there is a critical block to polyspermic fertilization, that is, fertilization by more than one spermatozoon, by a factor, or combination of factors, that is not yet fully understood. Great insight has been gained from *in vitro* fertilization studies, but the *in vivo* situation may be quite different. Successful fertilization probably requires that each of the above steps be completed in proper sequence. It may be possible to inhibit or enhance the process if we have a clear definition of the molecular events involved.

4.1   Sperm Penetration of the Cumulus and Zona

A freshly ovulated egg presents two or three formidable barriers to spermatozoa: (1) the cluster of surrounding follicle cells, the cumulus oophorus, and (2) the noncellular coating, the zona pellucida (see Fig. 6). In some species, such as the rabbit, another cellular layer is present between the cumulus and the zona, the corona radiata. The presence of these layers lends a degree of selectivity to the egg, since only those spermatozoa that have undergone capacitation will stand any chance of penetrating these layers.

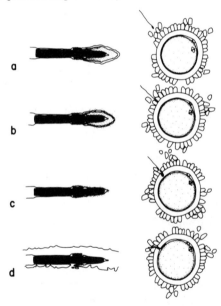

Figure 6   Sperm–egg interaction: early steps of fertilization. (*a*) Capacitated sperm has intact acrosome at time of initial contact with cumulus cells. (*b*) Acrosome-reacting spermatozoon with vesiculating membrane at the time of cumulus penetration and/or on the surface of the zona. (*c*) Acrosome-reacted spermatozoon at the time of penetration of the zona pellucida. (*d*) Acrosome-reacted spermatozoon fusing with the vitelline membrane.

It is generally believed that the acrosome reaction is triggered near or within the cumulus oophorus. Evidence for this comes from *in vivo* rabbit studies in which the majority of spermatozoa found in the oviduct 13–15 hours after mating have intact acrosomes, as seen in electron microscope preparations, whereas those spermatozoa observed among the cumulus cells have begun or completed the acrosome reaction. It is still uncertain which, if any, of the observed spermatozoa are the fertilizing ones. It seems reasonable that if the sperm enzymes are important for penetration of the spermatozoa through the layers of the egg, spermatozoa that are capacitated and ready to acrosome react will have an advantage over noncapacitated spermatozoa in their "ability to penetrate." The timing of the acrosome reaction may need to be just right with respect to the fertilizing spermatozoon. Those that acrosome react prematurely may lose factors that enable them to fertilize. The enzymes so liberated by these excess spermatozoa may function cooperatively for the penetration of a fusion-competent spermatozoon. Passage through the cumulus is most likely facilitated by enzymes such as hyaluronidase, which is activated or released as the sperm membrane vesiculates. Most spermatozoa lose the remnants of the vesiculated membrane complex during transit through the cumulus. Whether the loss of this complex allows for more or less "fusion-competent" spermatozoa remains to be shown.

Sperm penetration through the zona pellucida in the species that have been studied, requires an initial period of time during which the spermatozoa attach and bind to the zona surface. Attachment appears to be a loose interaction that, as shown in hamster spermatozoa, stimulates an alteration in the spermatozoa, affecting their subsequent binding affinity. The binding of spermatozoa to the zona is a tight, species-specific interaction. This may be the reason for the inability of spermatozoa from a different species than the egg, to pass through the zona. Removal of the zona pellucida, allows cross-species fertilization, for instance, human spermatozoa fertilizing hamster eggs. Therefore, the zona gives the egg species specificity.

Spermatozoa experimentally removed during the attachment phase will more rapidly bind to eggs newly introduced. Unfortunately, little is known about the mechanism of the binding process. It is still not even clear what anatomical part of the sperm cell actually accomplishes the binding. Spermatozoa may bind by an intact plasma membrane, a vesiculated surface, the inner acrosomal membrane, or by a combined sequential reaction involving all three.

The nature of the binding receptors is presently under intensive study. It has been speculated that one of the sperm membranes, either the plasma membrane or the inner acrosomal membrane, recognizes the zona by terminal oligosaccharide residues, similar to an antigen–antibody or receptor–ligand interaction. Capacitated spermatozoa have been shown to bind to both the inside and outside of isolated hamster zonas, suggesting that the binding ligands are located throughout the entire zona substance. These ligands are apparently altered after fertilization in a way that prohibits further binding. Recently, Wasserman has isolated three unique zona glycoproteins, one of which will specifically block sperm–egg binding.

After a period of binding at the zona surface, the spermatozoa progress through the zona matrix, aided by vigorous motion of the tail. The time required for penetration as documented for the hamster is from 4 to 22 min, averaging about 6 min. Most spermatozoa have been reported to pass through the zona in an oblique path. Recently, however, Blandau has recorded in a cinematography study mouse spermatozoa passing through the zona in a clearly perpendicular path. In either case, during zona penetration, spermatozoa are seen with the inner acrosomal membrane in direct contact with the zona. The posterior portion of the acrosome has been reported as intact, vesiculated, or absent.

In most species, the zona pellucida changes its biochemical properties as soon as the first spermatozoon penetrates this layer. This is called the "zona reaction" and the altered composition of the zona prevents passage of other spermatozoa. This block to polyspermy (fertilization by more than one spermatozoon) is important because polyspermy almost always leads to the death of the developing fetus. A proposed mechanism for the induction of the zona reaction is presented in Section 4.3 of this chapter and Chapter 9. The rabbit egg is an exception because a number of spermatozoa can penetrate the zona into the space between the zona and the vitelline membrane (perivitelline space). In this species, the vitelline membrane appears to block polyspermy.

4.2 Fusion of Spermatozoa With the Egg Vitelline Membrane

Upon penetrating the zona, the acrosome-reacted spermatozoon traverses the perivitelline space and contacts the plasma membrane of the egg. This vitelline membrane has numerous projecting microvilli, which appear to be the most-fusion-competent areas of the egg. The interaction between the male and female gametes depends upon a very special membrane-recognition system. Localized areas of the sperm membrane are discriminately recognized by the egg, either as areas for fusion or as areas to be actively engulfed through a phagocytic process. This delicate concert reflects regional differences in the sperm membrane components and the dual capability of the egg of membrane fusion and phagocytosis.

The mechanism governing this membrane-fusion process between mammalian gametes is not well understood. Spermatozoa gain fusion competency by either capacitation and/or the acrosome reaction. Initially, fusion is limited specifically to the spermatozoon's post-acrosomal region (see Fig. 6d), but subsequently proceeds over the tail. The inner acrosomal membrane does not fuse with the egg membrane but is rather the area "protectively" engulfed by the egg's microvilli. Experimental studies using ultrastructural cytochemistry have demonstrated actual mixing of membrane components following gamete fusion, indicating that part of the sperm membrane is incorporated into the egg's surface. The subsequent fate of this membrane is unknown. The inner acrosomal membrane, however, after becoming located within a phagocytic vesicle, is apparently digested by lytic processes within the ooplasm (egg cytoplasm). The remaining components of the sperm tail are, in most cases, incorporated into the ooplasm, and the midpiece, mitochondria, and axial filament are disassembled.

As shown for the hamster, the vitelline membrane retains its ability to fuse with a spermatozoon even after treatment with proteinases or lectins. This implies that protein or sugar moieties of peripheral membrane components are not directly involved with sperm–egg fusion. The vitelline membrane also apparently lacks strong species-specificity with respect to fusion. Zona-free hamster eggs can be penetrated by spermatozoa of the guinea pig, mouse, rat, and man. The zona-free eggs of the mouse, rat, and rabbit are also penetrable by spermatozoa from various other species. There is, however, some limitation to indiscriminate fusion, since human spermatozoa do not fuse with zona-free rat or mouse eggs. *In vitro* cross-species fusion has been shown to occur by the same mechanism as that of *in vivo* monospecific fusion.

4.3 Cortical Granule Breakdown and Block to Polyspermy

Sperm–egg fusion initiates a series of rapid structural and functional changes in the egg that include activation of the egg in order for embryonic development to take place and alterations that help prevent polyspermy. The most dramatic structural change is the fusion and exocytosis of the egg's cortical granules. Although the control mechanisms are poorly understood, it is generally accepted that this "cortical reaction" endows the egg with a block to further sperm penetration. Via exocytosis, the material contained within the cortical granules is released into the perivitelline space. The block to polyspermy involves changes in both the vitelline membrane and the composition of the zona. The possibility of controlling reproduction directly at the fertilization step motivates researchers to answer several basic questions. (1) What initiates the cortical granule reaction? (2) What materials are exposed or released from the cortical granules? And (3) how do these materials exert their effect upon the zona and/or vitelline membrane.

It appears that the initiating mechanism for cortical granule release is similar to that of other secretory cells. Several artificial agents have been used to induce the cortical reaction in hamster eggs, including electrical stimulation, the ionophores boromysin and A23187, neuraminidase, and certain lectins. Each of these artificial stimuli probably acts by altering membrane polarity and/or permeability to ions, which then activates the granule-fusion system. Under natural conditions, the fusion of a spermatozoon with the vitelline membrane probably elicits similar membrane changes. Granule discharge starts at the point of sperm entry and appears to propagate around the entire egg. This has a similar appearance to the membrane-depolarization phenomenon of somatic cells that results in degranulation. A popular concept is that sperm fusion causes a change in calcium permeability and/or localization that affects not only the granule-secretion mechanism but activates various metabolic pathways in the egg cytoplasm as well.

The cortical granule reaction both releases granule constituents into the perivitelline space and adds a new membrane to the egg surface because the cortical granule membranes fuse with the vitelline membrane. Both of these actions are thought to be sources of the polyspermic block. The relative importance of either factor seems to vary among species. Cortical granule contents have been shown to affect the zona reaction directly, destroying spermatozoa-binding affinity and making it significantly less permeable to spermatozoa. The active components may be trypsin-like proteases or mucoid and glycoprotein substances that are present in abundance. It has also been suggested that cortical granule proteinases alter the glycoprotein molecules on the vitelline membrane, blocking fusion receptor sites. Other blocking modifications after the cortical granule reaction may include molecular reorganization and loss of fusion competence due to the incorporation of new granule membrane or activation of an entirely independent mechanism.

Evidence has also been presented for a sperm-induced mechanism for the induction of the zona reaction. The neuraminidase-like enzyme (Chapter 5) of the sperm acrosome causes hardening of the zona, that is, a decrease in digestibility by enzymes, similar to that seen after fertilization. It remains to be established whether the cortical granule- or sperm-induced zona reaction mechanism is the primary one functionally, or if both are essential.

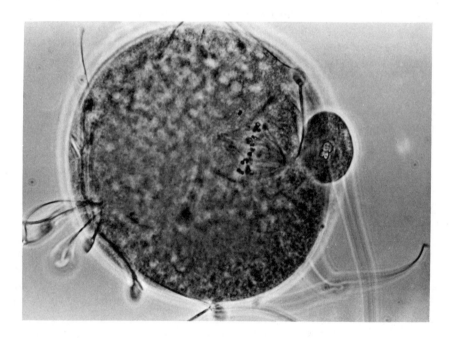

Figure 7 Unfertilized guinea pig egg. Egg matured in culture shows polar body and chromosomes in metaphase I of meiosis.

4.4   Sperm Head Swelling and Pronucleus Formation

As the spermatozoon contacts and penetrates the vitelline membrane, the relatively metabolically dormant egg activates and completes its second meiotic division, producing a second polar body and a mature ovum. The final stage of gamete interaction entails a sequence of changes in the genetic material resulting in the formation of a male pronucleus (from the sperm head chromosomes) and a female pronucleus (from the egg chromosomes). In mammals, the chromosomes from these pronuclei come together just prior to metaphase of the first mitotic division. There appears to be an interplay of physical and biochemical factors particularly within the ooplasm, that controls pronuclear formation for both the male and female gametes. Fig. 7 and 8 show the structural contrasts before and after spermatozoa penetration of an oocyte.

As the sperm nucleus submerges into the ooplasm, the nuclear envelope vesiculates and moves away from the nuclear material. The densely packaged sperm chromatin begins to swell and appears to unravel (decondensation). Fine strands of this material can be seen when prepared for electron microscope observations. This decondensed stage correlates with the swollen sperm head seen within the egg cytoplasm using the *in vitro* fertilization test (see Fig. 8). The factor(s) that initiate nuclear decondensation originate from the egg and are

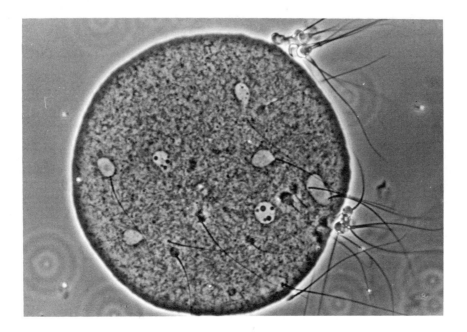

Figure 8   Polyspermic egg. A zona-free hamster egg is fertilized by at least five guinea pig spermatozoa, visualized as clear areas in the cytoplasm with attached tails.

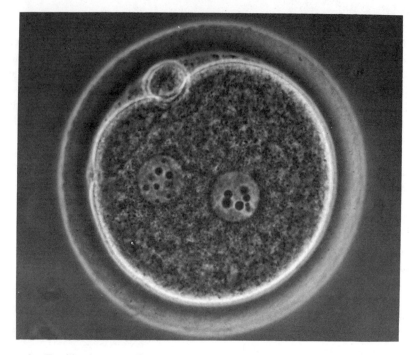

Figure 9 Fertilized mammalian egg. A hamster egg fertilized *in vivo* reveals one polar body and two pronuclei. This egg was flushed from the hamster oviduct 12–14 hr after ovulation.

available in limited amounts. The availability of these factors in an active form seems to correlate with the age of the egg. Although nuclear decondensation can be artificially achieved by disulfide reducing agents (Chapter 5), very little knowledge is available regarding the natural phenomenon.

The pronuclear status develops as membrane vesicles from the egg cytoplasm move together to surround the male genetic material and form a new nuclear envelope. Numerous dark spherical nucleoli develop within the male pronucleus, giving it the characteristic spotty appearance on phase contrast microscopy. Formation of the female pronucleus occurs by a similar process as a result of activation but does not require the decondensation step (Fig. 7).

Following the appearance of the nucleoli there is a period of DNA synthesis (chromosome replication). The male and female pronuclei both increase in size and move to the center of the egg (Fig. 9). Fusion between the pronuclei is facilitated by an interdigitation of numerous projections from the nuclear envelopes. The chromatin of both nuclei then condenses, and the chromosomes organize at a metaphase plate for the first zygotic division, thus completing the fertilization process.

REFERENCES

1   Barros, C., Capacitation of Mammalian Spermatozoa, in *Proceedings of the International Symposium on Physiologic and Genetic Aspects of Reproduction*, p. 3, Salvador, Bahia, Brazil, (1974).

2   Bedford, J. M., Sperm capacitation and fertilization in mammals. *Biol. Reprod.* Suppl. **2**, 128 (1970).

3   Bedford, J. M., Mechanisms Involved in Penetration of Spermatozoa through the Vestments of the Mammalian Egg, *in* Coutinho, E. M., and Fuchs, F. (eds.), *Physiology and Genetics of Reproduction*, Part B, p. 55, Plenum Press, New York, (1974).

4   Bedford, J. M., and Cooper, C. W., Membrane Fusion Events in the Fertilization of Vertebrate Eggs, *in* Poste, F., and Nicolson, G. L. (eds.), *Cell Surface Reviews*, Vol. 5, p. 65, North Holland, Amsterdam, (1978).

5   Cheung, W. Y., Calmodulin plays a pivotal role in cellular regulation. *Science* **207**, 19 (1980).

6   Green, D. P. L., The Mechanism of the Acrosome Reaction, *in* Johnson, M. H. (ed.), *Development in Mammals*, Vol. 3, p. 65, North Holland, Amsterdam (1978).

7   Gwatkin, R. B. L., Fertilization, *in* Poste, F., and Nicholson, G. L. (eds.), *The Cell Surface in Animal Embryogenesis and Development*, p. 1, North Holland, Amsterdam (1976).

8   Gwatkin, R. B. L., *Fertilization Mechanisms in Man and Mammals*. Plenum Press, New York (1977).

9   Hartree, E. F., Spermatozoa, eggs and proteinases. *Biochem. Soc. Trans.* **5**, 375 (1976).

10 Jones, H. P., Lenz, R. W., Palevitz, B. S., and Cormier, M. J., Calmodulin localization in mammalian spermatozoa. *Proc. Nat. Acad. Sci.* **77**, 2772, (1980).

11 Koehler, J. K., The mammalian sperm surface: Studies with specific labeling techniques. *Int. Rev. Cytol.* **54**, 73, (1978).

12 McRorie, R. A., and Williams, W. L., Biochemistry of mammalian fertilization. *Ann. Rev. Biochem.* **43**, 777 (1974).

13 Meizel, S., The Mammalian Sperm Acrosome Reaction, a Biochemical Approach, *in* Johnson, M. H. (ed.), *Development in Mammals*, Vol. 3, p. 1, North Holland, Amsterdam (1978).

14 Morton, D. B., Immunoenzymic Studies on Acrosin and Hyaluronidase in Ram Spermatozoa, *in* Edidin, M., and Johnson, M. H. (eds.), *Immunobiology of Gametes*, p. 115, Cambridge University Press, Cambridge, (1977).

15 Rogers, B. J., Mammalian sperm capacitation and fertilization *in vitro*: A critique of methodology. *Gamete Res.* **1**, 165 (1978).

16 Rogers, B. J., and Garcia, L., The effect of cAMP on acrosome reaction and fertilization. *Biol. Reprod.* **21**, 365 (1979).

17 Stambaugh, R., Enzymatic and morphological events in mammalian fertilization. *Gamete Res.* **1**, 65 (1978).

18 Williams, W. L., Biochemistry of Capacitation of Spermatozoa, *in* Moghissi, A. S., and Hafez, E. S. E. (eds.), *Biology of Mammalian Fertilization and Implantation*, p. 19, Charles C. Thomas, Springfield (1972).

19 Yanagimachi, R., Specificity of Sperm-egg Interaction, *in* Edidin, M., and Johnson, M.H. (eds.), *Immunobiology of Gametes*, p. 115, Cambridge University Press, Cambridge (1977).

20 Yanagimachi, R., Sperm-Egg Association in Mammals, *in Current Topics in Developmental Biology*, Vol. 12, p. 83, Academic Press, New York (1978).

21 Zaneveld, L. J. D., The Biochemistry of Mammalian Capacitation, *in* Thibault, C. (ed.), *La Fécondation*, p. 50, Masson et Cie, Paris (1975).

22   Zaneveld, L. J. D., The Proteolytic Enzyme Systems of Mammalian Genital Tract Secretions and Spermatozoa, *in* Reich, E., Lifkin, D. B., and Shaw, E. (eds.), *Proteases and Biological Control,* Vol. 2, p. 683, Cold Spring Harbor Laboratory, New York (1975).

CHAPTER NINE

THE OVUM BEFORE AND AFTER FERTILIZATION

DON P. WOLF

Division of Reproductive Biology
Department of Obstetrics and Gynecology
University of Pennsylvania School of Medicine
Philadelphia, Pennsylvania

1 INTRODUCTION

Insights into the biochemical events underlying fertilization, the activation of development, and embryogenesis are critical to an understanding of, and there-

The author expresses appreciation to Ms. Patricia Park for secretarial assistance. Supported by a grant from USPHS HD-06274.

fore an ability to influence, these important physiological processes. Historically, characterization of the mammalian ovum and preimplantation embryo relied on descriptive approaches, owing to the difficulties encountered in obtaining sufficient quantities of these cells for direct chemical analysis. Our heritage includes a number of elegant light and electron (both transmission and scanning) microscopic depictions of early reproductive phenomena (1). With the advent of superovulation protocols and microanalytical techniques, augmented by the development of *in vitro* culture media and insemination procedures that adequately simulate the *in vivo* condition, it is now possible to contemplate, indeed to complete, a process that has already begun: the direct chemical analysis of mammalian ova and embryos and a definition of those biochemical changes that accompany early development.

Studies conducted with the mouse and, to a lesser extent, the rabbit provide the vast majority of existing information on the biochemistry of the mammalian ovum. In this review, the species under discussion is the mouse, unless specifically stated otherwise. Finally, a review of the literature of the mammalian ovum cannot be made without reference to the innumerable pioneering studies conducted in lower vertebrates and marine invertebrates. Attention is drawn to several appropriate examples in the course of this review. So often, information generated from nonmammalian systems is invaluable in directing experimentation in mammals, where the availability and accessibility of material are limiting.

2 GENERAL FEATURES OF THE OVULATED OVUM

The ova of eutherian mammals, at approximately 80 μm diameter, are large compared to somatic cells but small when contrasted with the yolk-laden eggs of amphibians and birds. At ovulation, this cell, arrested in its second meiotic division, is surrounded by an acellular zona pellucida (Section 7.3 of this chapter) and cumulus cells embedded in a hyaluronic-acid-rich matrix (Fig. 1). In some species such as the rabbit, the innermost layer of cells, called the corona radiata, is more tightly packed and radially arranged. Cumulus cells can readily be dispersed *in vitro* with hyaluronidase, and these cells serve no known postovulatory function. Prior to ovulation, cumulus and corona cell processes extend into and through the zona pellucida, forming intimate contacts with microvillar processes from the egg surface. Such cells are normally dispersed upon limited exposure to the environment of the oviduct. Morphologically, changes that accompany fertilization include dispersion of the cumulus, extrusion of the second polar body, activation of meiosis, and the exocytotic release of the egg's cortical granules.

In the unfertilized condition, the ovum's lifetime is limited; *in vivo* ovulated ova show morphologically distinct signs of degeneration within a day, whereas their fertilizable life either *in vivo* or *in vitro* is of somewhat shorter duration (2). Following sperm-induced or parthenogenetic (factors other than spermatozoa)

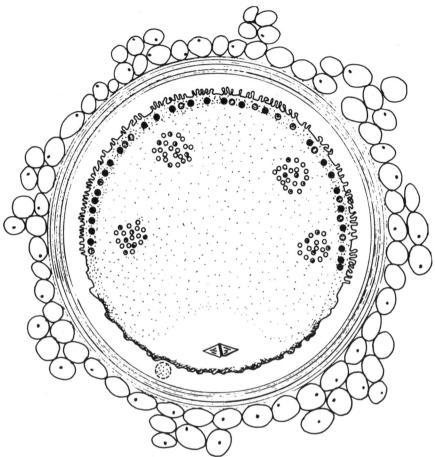

Figure 1 Schematic representation of the ovulated mammalian egg. The egg is surrounded by cumulus cells and an acellular glycoprotein-rich-layer, the zona pellucida. A first polar body is seen in the perivitelline space. The polar body is associated with the egg cortex, that contains the meiotic spindle. The egg surface that overlies the area of the cytoplasm containing cortical granules, is rich in microvilli in contrast to the smooth surface associated with the meiotic spindle area. Clusters of vacuoles in the egg represent Golgi complexes. Taken from Nicosia, Wolf, and Inoue, *Develop. Biol.* **57**, 56 (1977).

activation, the embryo or parthenote must derive its nutritional and developmental requirements from stored components accumulated in the ovum during oogenesis or from the external environment. The latter is the biochemically complex tubal fluid or, *in vitro*, a balanced bicarbonate-buffered salt solution containing energy sources and protein, usually bovine serum albumin. The preimplantation period in the mouse lasts approximately five days, corresponding at daily postfertilization intervals to one- or two-cell, eight-cell, morula-early blastocyst (30–60 cells), blastocyst, and late blastocyst stages, respectively. The rabbit ovum, although considerably larger, cleaves much more rapidly (3).

In studies with one-cell embryos, the possibility of contamination with unfertilized ova is usually ignored.

The physical characteristics of mammalian ova are summarized in Table 1. Table 2 itemizes those biochemical constituents that have been quantitated by direct chemical analysis. Although this chapter focuses on the changes that accompany or closely follow fertilization, relatively little information is avail-

Table 1   General Characteristics of the Ovulated Mammalian Ovum

Characteristic	Comments	Species	Reference
Size	Diameter of eggs vary over the range of 60–140 μm		
	60 μm	Mouse	Engel and Franke (8)[a]
	140 μm	Cow	Engel and Franke (8)
Volume	1.92×10^5 pl	Mouse	Lewis and Wright, *Contrib. Embryol.* **25**, 113 (1935)
	1.89×10^5 pl	Mouse	Abramczuk and Sawicki, *J. Exp. Zool.* **188**, 25 (1974)
	~5×10^5 pl	Rabbit	Brinster (4)
Wet weight	318 ng	Mouse	Loewenstein and Cohen, *J. Embryol. Exp. Morphol.* **12**, 113 (1964)
Dry weight	32.0 ng (26 ng for zona-free)	Mouse	Loewenstein and Cohen, *J. Embryol. Exp. Morphol.* **12**, 113 (1964)
	34.6 ng	Mouse	Abramczuk and Sawicki, *J. Exp. Zool.* **188**, 25 (1974)
Subcellular organelles			
Mitochondria	92,500 highly compacted mitochondria in unfertilized egg or early embryo	Mouse	Pikó and Matsumoto, *Develop. Biol.* **49**, 1 (1976)
Ribosomes	Only 20–25% of total ribosomes present as polysomes: inactive ribosomes may be present in stored form in morphologically distinct lattices.	Mouse	Backvarova and DeLeon, *Develop. Biol.* **58**, 248 (1977)
Cortical granules	4,500 in unfertilized mouse eggs at a concentration of 50/100 μm^2; 0.2–0.6 μm diameter	Mouse	Nicosia, Wolf, and Inoue, *Develop. Biol.* **57**, 56 (1977)
	8–15,000 in unfertilized hamster egg at a concentration of 50/100 μm^2; 0.1–0.5 μm diameter can be visualized with light microscope	Hamster	Austin, *Exp. Cell Res.* **10**, 533 (1956)

[a] In all tables, the references marked with a number only can be found in the general reference section at the end of the chapter.

Table 2 Biochemical Constituents of the Ovum or One-Cell Embryo

Constituent	Amount	Comments	Species	Reference
DNA	45.8 pg	This value is considered high by other authors	Mouse	Reamer, Ph.D. thesis, Boston University (1963)
	23.8 pg	Based on fluorometric quantitation of purified DNA	Mouse	Olds et al., J. Exp. Zool. **186**, 39 (1973)
	8 pg		Mouse	Pikó and Matsumoto, Develop. Biol. **49**, 1 (1976)
RNA	1.75 ng	Total RNA; this value is considered high by other authors	Mouse	Reamer, Ph.D. thesis, Boston University (1963)
	0.55 ng	Total RNA; based on 200 nm absorbing alkali-labile material	Mouse	Olds et al., J. Exp. Zool. **186**, 39 (1973)
	20 ng	Total RNA	Rabbit	Manes, J. Exp. Zool., **172**, 203 (1969)
	0.2 ng	rRNA and 0.16 ng (tRNA + 5S RNA); based on SDS extraction, PAGE, scan gels at 260 μm	Mouse	Backvarova, Develop. Biol. **40**, 52 (1974)
	0.4 ng	rRNA; based on phenol extraction, PAGE, UV scan of gels	Mouse	Young et al., J. Cell Biol. **59**, 372a (1973)
	1.9 pg	mRNA; putative mRNA based on poly(A) quantitation	Mouse	Levy et al., Develop. Biol. **46**, 140 (1978)
		Putative mRNA reported as 0.25% of total RNA	Rabbit	Schultz, Develop. Biol. **44**, 270 (1975)
		Translational activity *in vitro* of mRNA described	Mouse	Braude and Pelham, J. Reprod. Fert. **56**, 153 (1979)
Protein	20 ng	zona-intact	Mouse	Lowenstein and Cohen, J. Embryol. Exp. Morph. **12**, 113 (1964)
	18.4 ng	zona free	Mouse	
	27.8 ng	zona-intact	Mouse	Brinster, J. Reprod. Fert. **13**, 413 (1967)

(Table 2 continued on p. 236)

Table 2 (*continued*)

Constituent	Amount	Comments	Species	Reference
	24.5 ng	zona-intact fluorescamine assay	Mouse	Schiffner and Spielmann, *J. Reprod. Fert.* **47**, 145 (1976)
	36.6 ng	zona-intact fluorescamine assay	Rat	Schiffner and Spielmann, *J. Reprod. Fert.* **47**, 145 (1976)
Lipid	100 ng		Rabbit	Brinster, *Experientia* **27**, 371 (1971)
	3.3 ng	Based on indirect measurement; dry weight before and after organic extraction	Mouse	Loewenstein and Cohen, *J. Reprod. Fert.* **47**, 145 (1976)
Carbohydrate	8 ng	Based on indirect measurement		Loewenstein and Cohen, *J. Reprod. Fert.* **47**, 145 (1976)
Glycogen	0.11 ng		Mouse	Stern and Biggers, *J. Exp. Zool.* **168**, 61 (1968)
Amino acid pools				
Leucine	11.9×10^{-14} moles	These pools increase significantly after fertilization.	Mouse	Brinster *et al.*, *Develop. Biol.* **51**, 215 (1976)
Methionine	7.4×10^{-14} moles	These pools increase significantly after fertilization.	Mouse	Schultz *et al.* (7)
Glycolytic intermediates				
Glucose-6-phosphate	9 fmole		Mouse	Wales (5)
Fructose-1,6-diphosphate	<1 fmole		Mouse	Wales (5)

TCA cycle inter-mediates			
Citrate	459 fmole	Mouse	Wales (5)
Isocitrate	11 fmole	Mouse	Wales (5)
α-Ketoglutarate	37 fmole	Mouse	Wales (5)
Malate	75 fmole	Mouse	Wales (5)
Pyrimidine nucleotides			
NAD+	98 fmole	Mouse	Wales (5)
NADH	38 fmole	Mouse	Wales (5)
Purine nucleotides			
ATP	0.65 pmole	Mouse	Quinn and Wales, *J. Reprod. Fert.* **25**, 133 (1971)
	1.1 pmole	Mouse	Clegg and Pikó, *Develop. Biol.* **58**, 76 (1977)
UTP	0.05 pmole	Mouse	Clegg and Pikó, *Develop. Biol.* **58**, 76 (1977)
Internal ion concentrations			
Potassium	240 mmole/l	Mouse	Powers and Tupper, *Develop. Biol.* **56**, 306 (1977)
Sodium	83 mmole/l	Mouse	Powers and Tupper, *Develop. Biol.* **56**, 306 (1977)
Chloride	67 mmole/l	Mouse	Powers and Tupper, *Develop. Biol.* **56**, 306 (1977)

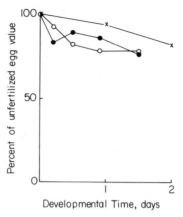

Figure 2   Decreases in the dry mass, volume, and protein content of mouse embryos during early development. Open circles: dry mass per embryo. Closed circles: volume per embryo. Data obtained from Abramczuk and Sawicki, *J. Exp. Zool.* **188**, 25 (1974). Crosses: protein content per embryo. Data obtained from Brinster (4).

able within this restricted time span; therefore, preimplantation development is often discussed. This lack of information results from the fact that nearly all studies have utilized *in vivo* fertilized ova, in which substantial asynchrony in penetration is recognized and the timing of precleavage development is not always adequately defined. Decreases in ovum volume, dry mass, and protein content have been associated with a response to fertilization or very early development in the mouse (Fig. 2), although neither their applicability to other species nor their significance is fully appreciated. Certainly, changes of this nature that occur within an hour of sperm penetration are at least partly attributable to the loss of a second polar body and to the release of the ovum's complement of cortical granules. After this initial period, further changes in protein content could reflect a preponderance of catabolic over anabolic processes.

3   METABOLIC PROFILE OF THE UNFERTILIZED OVUM AND EARLY EMBRYO

Metabolically, the mammalian ovum is depicted as an inert cell poised to fulfill a variety of biosynthetic functions in response to the activating stimulus provided by capacitated-sperm penetration or exposure to parthenogenetic agents. Thus unfertilized and newly fertilized ova of the mouse and rabbit show low levels of metabolic activity (see O_2 uptake and CO_2 production in Fig. 3). Extensive efforts have been devoted to defining the energy-substrate requirements for the early mouse embryo and determining the fate of various radiolabeled precursors (4, 5). The embryo requires an amino nitrogen and an energy source for development, and its glycolytic ability increases gradually during

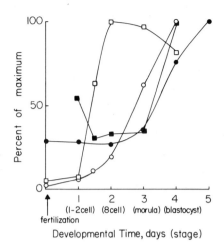

Figure 3 Changes in several biochemical properties of mouse embryos during early development. Open circles: CO_2 per embryo per hour produced from glucose. Open squares: glycogen content. Closed circles: O_2 uptake per embryo per hour. Data obtained from Brinster (4). Closed squares: citrate content. Data obtained from Wales (5).

early cleavage stages, with a marked increase preceding blastulation (blastula formation). Pentose-shunt activity is low in unfertilized ova and early preimplantation embryos, the citric acid cycle being the major supplier of energy. The levels of some glycolytic and citric acid cycle intermediates in the unfertilized ovum are listed in Table 2. Interestingly, glycogen, a potential energy source, is accumulated during early development (Fig. 3). Although substrate permeability may control metabolic processes in the unfertilized ovum (4), more-recent attention has focused on the relatively high levels of citrate found in one-cell embryos (Fig. 3). If this citrate were localized in the cytoplasm, inhibition of the activity of phosphofructokinase, a key glycolytic enzyme, would be expected, thereby accounting for a low glycolytic rate (5). Increases in metabolic activity during the preimplantation period have been correlated with increased anabolic metabolism, the expression of additional metabolic pathways, and increased protein and nucleic acid synthesis.

4 PROTEINS AND PROTEIN BIOSYNTHETIC ACTIVITY

On a dry weight basis, proteins account for approximately 70% of the mammalian ovum (Table 2) and include a complement of enzymes detectable *in vitro* (Table 3), structural components, and nuclear and membrane proteins. Additionally, the existence of yolk proteins that serve a nutritional role during early embryogenesis must be considered as well as, inevitably, a host of proteins whose function has not yet been clarified. In lower vertebrate and marine invertebrate ova, several proteins have been studied because of their obvious

Table 3 Enzymatic Activities Demonstrable in Mammalian Ova

Enzyme[a]	Species	Reference
Glycolysis		
Hexokinase	Mouse	For primary reference, see Engels and Franke (8), unless otherwise indicated
Phosphoglucose isomerase	Mouse	
Phosphofructokinase	Mouse	
Aldolase	Mouse	
Triosephosphate isomerase	Mouse	
Glyceraldehyde-3-phosphate dehydrogenase	Mouse	
Phosphoglycerate kinase	Mouse	
Phosphoglyceromutase	Mouse	
Enolase	Mouse	
Pyruvate kinase	Mouse	
Lactate dehydrogenase	Mouse, rat, Chinese and golden hamsters, rabbit, cow, ferret, rhesus and squirrel monkies, human, guinea pig	
Pentose Shunt		
Glucose-6-phosphate dehydrogenase	Mouse, rabbit	
6-Phosphogluconate dehydrogenase	Mouse, rabbit	
TCA Cycle		
NAD-Malate dehydrogenase	Mouse	
Cytoplasmic Activities		
NADP-Isocitrate dehydrogenase	Mouse	
NADP-Malate dehydrogenase	Mouse	
Nucleotide Metabolism		
Uridine kinase	Mouse	
Guanine deaminase	Mouse	
Hypoxanthine-guanine-phosphoribosyltransferase	Mouse, rabbit	

Table 3   (*continued*)

Enzyme[a]	Species	Reference
Adeninephosphoribosyltransferase	Mouse, rabbit	
NAD Metabolism[b]		
Nicotinamide mononucleotide pyrophosphorylase	Mouse	Kuwahara and Chaykin, *J. Biol. Chem.* **248**, 5095 (1973)
NAD-glycohydrolase	Mouse	
Amino Acid Metabolism		
Glutamic dehydrogenase	Mouse	
Aspartate aminotransferase	Mouse	
Glycogen Metabolism		
Glycogen synthetase	Mouse	
Miscellaneous		
RNA Adenylation	Mouse	Young and Sweeney, *Biochemistry* **17**, 1901 (1978)
ADP-Ribosylation of protein	Mouse	
Membrane-Bound Enzymes[c]		
Mg^{2+}-ATPase	Mouse	Vorbrodt *et al.*, *Develop. Biol.* **55**, 117 (1977)

[a] The enzyme was detected by biochemical assay of freeze-thawed cells, unless otherwise indicated.

[b] Inferred from *in vivo* conversion of [7-14C]nicotinamide to NAD.

[c] Ultrastructural cytochemistry.

Enzymes Reportedly Absent or Very Low in Unfertilized Mammalian Ova

Enzyme	Comments	Reference
RNA polymerase	Based on direct assays in embryo homogenates	Siracusa and Vivarelli, *J. Reprod. Fert.* **43**, 567 (1975)
Alkaline phosphatase	Ultrastructural cytochemistry—mouse	Vorbrodt *et al.*, *Develop. Biol.* **55**, 117 (1977)
5'-Nucleotidase	Ultrastructural cytochemistry—mouse	Vorbrodt *et al.*, *Develop. Biol.* **55**, 117 (1977)
Transport ATPase	Ultrastructural cytochemistry—mouse	Vorbrodt *et al.*, *Develop. Biol.* **55**, 117 (1977)
Cyclic AMP phosphodiesterase	Ultrastructural cytochemistry—mouse	Vorbrodt *et al.*, *Develop. Biol.* **55**, 117 (1977)
Adenylyl cyclase	Ultrastructural cytochemistry—mouse	Vorbrodt *et al.*, *Develop. Biol.* **55**, 117 (1977)

importance to development: histones, DNA polymerases, RNA polymerases, and microtubule proteins (3). Historically, the characterization of specific proteins in mammalian embryos began with the quantitation of enzymatic activities in freeze-thawed extracts in which activity changes were defined as a function of age. The developmental significance of these observations, however, was not always immediately obvious, in part because *in vitro* assay conditions do not necessarily reflect the *in vivo* situation and because the expression of latent or inactive enzymes is difficult to evaluate. A case in point is glucose-6-phosphate dehydrogenase, an indicator of pentose-shunt activity. By *in vitro* assay, the activity of this enzyme in unfertilized mouse ova is relatively high, yet the $[^{14}C]CO_2$ production in intact cells from C-1 of specifically labeled glucose is approximately the same as from C-6, suggesting that oxidation by way of the pentose-shunt is minimal (4).

Proteins present in the ovum are synthesized and accumulated in the oocyte during oogenesis, and interest has centered on the degree to which the embryo depends upon these stored materials for its nutritional and developmental needs. The alternative involves protein synthesis by the embryo, and, if it occurs, it can be asked: when, to what degree, and does it result from the translation of maternal mRNA or from the expression of the embryonic genome? One obvious approach to these questions involves the use of protein synthesis inhibitors such as cyclohexamide or puromycin. The conclusions corroborated by several investigators are that these inhibitors arrest development at all preimplantation embryonic stages and, therefore, that embryonic protein synthesis is essential to embryo development (6).

The direct quantitation of embryonic protein synthesis was initiated with measurements of the incorporation of radioactive amino acids into newly synthesized protein. Incorporation was detected at all stages from fertilization onward, with a marked increase observed during the third day of development; however, the absence of information on the specific radioactivity of intracellular amino acid pools rendered these results only suggestive (6). More recently, absolute rates of protein synthesis have been determined during oogenesis, fertilization, and early embryogenesis in the mouse (Table 4, Fig. 4). In general, intracellular pools of amino acids increase in size during early development, and absolute-rate determinations corroborate the early conclusions that protein synthesis occurs in the unfertilized ovum and that marked increases in synthetic activity are associated only with the transition from the eight-cell to the blastocyst stage of development. Reported changes in the absolute rate of protein synthesis at fertilization are variable, and conclusions in this regard are not yet warranted. It should also be noted that protein turnover rates have been evaluated during the early preimplantation period in the mouse (Table 4).

A definition of the nature and quantity of proteins synthesized in the unfertilized ovum and early embryo is essential to evaluating function. The need to separate and quantify is obvious, and high-resolution two-dimensional gel electrophoresis has been the method of choice. The proteins in ova or embryos are pulse labeled—exposed to radioactive amino acids for finite periods of time—extracted, subjected to electrophoresis, and then visualized by fluorography.

Table 4 Protein Synthetic Activity in Mouse Ova and Preimplantation Embryos (pg protein/hr/embryo)

| | | | Developmental Stage | | | |
| | | | Embryos | | | |
Reference	Dictyate Oocyte	Unfertilized Ovum	1-Cell	2-Cell	8-Cell	Blastocyst
Epstein (6)	—	—	—	—	56	200
Modified from Brinster et al., Develop. Biol. 51, 215 (1976), assuming a molecular weight for leucine of 110 and that leucine accounts for 10% of the amino acids in the embryonic protein[a]	—	95	81	76	—	620
Modified from Abreu and Brinster, Exp. Cell Res. 114, 135 (1978), with the same as itemized above[a]	—	60	52	45	132	413
Schultz et al. (7)[a]	43	33	45	—	51	—
Protein Degradation						
Brinster et al., Develop. Biol. 51, 215 (1976)	—	—	—	Approximately 10% of protein turns over with half-life of 18.3 hr	—	approximately 35% of protein turns over with half-life of 11.2 hr
Brinster et al., J. Biol. Chem., 254, 1920 (1979)	—	—	—	—	—	mean half-life of 12.4 hr for newly synthesized proteins; range 1–30 hr

[a] Absolute rates of protein synthesis.

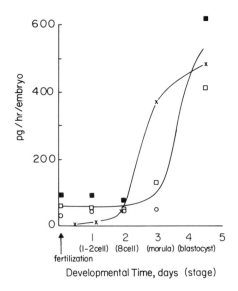

Figure 4   Absolute rates of protein and RNA synthesis in mouse embryos during early development. Protein synthetic rates: open squares, data obtained from Abreu and Brinster, *Exp. Cell Res.* **114**, 135 (1978); closed squares, data obtained from Brinster *et al.*, *Develop. Biol.* **51**, 215 (1976); open circles, data obtained from Schultz *et al.*, (7). RNA synthetic rates: crosses, data obtained from Clegg and Piko, *Develop. Biol.* **58**, 76 (1977).

The sensitivity of these techniques is high enough that only a few cells are required. In this manner, both quantitative and qualitative changes in the patterns of newly synthesized proteins have been defined in the oocytes and embryos of the mouse, rabbit, and pig. The patterns obtained, although complex, represent only the fastest-turning-over population of proteins—the vast majority, for instance, maternal protein, being inaccessible. Although these patterns change during preovulatory oocyte maturation, many common proteins are synthesized by the oocyte, ovum, and early embryo, and only limited changes in protein patterns are associated with fertilization (7). Marked qualitative changes in embryonic-protein patterns have been observed between the first and second day of development (6).

The nature of the newly synthesized protein has been examined in several cases, that for tubulin, actin, and lactate dehydrogenase-1 (LDH-1) (Table 5). Tubulin synthesis closely parallels the total protein synthetic activity during oocyte maturation and early embryogenesis. The synthesis of actin is low in the unfertilized ovum and early embryo, and relatively high at the eight-cell and blastocyst stages of development. The pattern of synthesis of LDH-1 is unique, with a high relative rate initially that drops sharply after ovulation. This drop has been dissociated from fertilization but may be related to ovulation.

Table 5   Synthesis of Actin, Tubulin, and LDH-1 in Mouse Ova and Early Embryos[a]

Protein	Dictyate Oocyte	Unfertilized	1-Cell	2-Cell	8-Cell	Morula-Early Blastocysts	Blastocysts	Reference
LDH-1	5.1	1.6	0.75	—	0.4	0.1	0.2	Mangia et al., Develop. Biol. **54**, 146 (1976)
Tubulin	1.4	1.1	1.3	—	1.3	—	—	Schultz et al. (7)
Tubulin	—	1.0	1.1	1.4	1.7	—	1.9	Abreu and Brinster, Exp. Cell Res. **114**, 135 (1978)
Actin	—	0.5	0.4	0.6	2.0	—	5.7	Abreu and Brinster, Exp. Cell Res. **114**, 135 (1978)

Developmental Stage

[a] Expressed as percentage of total newly synthesized protein during time period under study (usually 1 hr).

245

5   RNA AND RNA SYNTHESIS

Quantitative estimates of the total RNA content of the mammalian ovum, although available, are variable (Table 2). The more recent studies in which specific RNA types have been quantitated are undoubtedly more reliable. The RNA content of the unfertilized ovum, predominantly ribosomal RNA (rRNA) with significant amounts of transfer RNA (tRNA), is approximately 100 times that of an adult somatic cell. Indirect evidence for the existence of messenger RNA (mRNA) in the ovum or embryo has recently been substantiated by direct quantitation of poly(A)-containing RNA in both the rabbit and mouse ovum (Table 2). The presence of poly(A) sequences covalently linked to the 3'-terminus of RNA is characteristic of eukaryotic heterogeneous nuclear and ribosomal RNA and is equated with the existence of mRNA. Mitochondrial-associated RNA is present in embryos from the 8- to 16-cell stages of development, but its significance to embryogenesis has been discounted by inhibitor studies (6).

The presence of mRNA in the unfertilized and newly fertilized ovum again raises the question of whether the proteins necessary for early cleavage are produced upon translation of maternal mRNA or whether the embryonic genome contributes to the earliest stages of development. Parthenotes can develop to the blastocyst stage, indicating that a paternal contribution to the embryonic genome is not essential during this time. If development is absolutely dependent upon embryonic mRNA production, then inhibitors of RNA synthesis should, when added to viable embryos, arrest their development. Actinomycin D, which inhibits rRNA synthesis, does not influence development to the late blastula stage in echinoderms and amphibians, and it is commonly accepted that early development is directed by maternal mRNA in these species (3, 8). Cleavage of one-cell and later-stage mouse embryos is inhibited by $1 \mu M$ actinomycin D; and α-amanitin, a specific inhibitor of Type II RNA polymerase (responsible for heterogeneous RNA production that gives rise to mRNA), produces modest inhibition of cleavage of one-cell embryos with marked inhibitory effects at later developmental stages. Although these results suggest a dependence on RNA synthesis, some authors have raised questions of specificity, as undesirable side effects of these drugs have been described. Interpretations are further complicated by permeability considerations and the absence of direct quantitation of their intracellular effects (3). In the rabbit, the rate of protein synthesis is insensitive to α-amanitin during the one-cell to eight-cell stage of development (8). Taken together, these studies minimize the contribution of the embryonic genome to very early mammalian development.

The direct approach, that of quantitating RNA synthesis, has also received attention. The initial experimental protocol involved quantitating the uptake of radiolabeled precursors into RNA without regard to the specific radioactivity of the intracellular pool. Uridine uptake into RNA, which was high during oogenesis, was undetectable in the unfertilized ovum and did not become detectable until the eight-cell stage of development (6).

Table 6   RNA Synthetic Activity in Early Embryos of the Mouse

Measurement	Developmental Stage					
	1-Cell	2-Cell	4-Cell	8-Cell	Morula, Early Blastocyst	Expanded Blastocyst
Total perchloric acid precipitable [^{3}H] uridine, pmole/hr/embryo[a]	2.2×10^{-6c}	1.6×10^{-5}	6×10^{-5}	6×10^{-4}	7×10^{-3}	1.9×10^{-2}
Total uridine incorporated, pmole/hr/embryo (corrected for changes in specific radioactivity of UTP pool)[a]	3.0×10^{-3c}	1.8×10^{-3}	3.1×10^{-3}	1.2×10^{-2}	0.8×10^{-1}	0.9×10^{-1}
Total uridine incorporated, pmole/hr/cell[a]	3.0×10^{-3c}	0.9×10^{-3}	0.8×10^{-3}	1.5×10^{-3}	2.8×10^{-3}	1.5×10^{-3}
Estimated rate of RNA synthesis[a]						
pg/hr/embryo	—	2.5	—	20	150	—
pg/hr/cell	—	1.25	—	2.5	5	—
Estimated rate of RNA synthesis[b]						
pg/hr/embryo					50	
pg/hr/cell		—		—	1.4	—

[a] Modified from Clegg and Pikó, *Develop. Biol.* **58**, 76 (1977).

[b] Epstein (6)

[c] Overestimate, as not all perchloric acid-precipitable material was alkali labile.

The absolute rate of RNA synthesis has been determined in more recent studies (Table 6). An approximately 10,000-fold increase in [³H]uridine incorporation per hour per embryo into material precipitable by trichloroacetic acid (TCA) was seen when unfertilized ova were compared with day 4 blastocysts; however, when the increase was corrected for changes in the specific radioactivity of the uridine triphosphate (UTP) pool and allowance was made for the base composition of RNA, the increase was only 60-fold when expressed per embryo or fourfold when expressed per cell. Most of the labeled material, tentatively identified as nascent RNA in one-cell pronucleate ova, was not alkali labile, and therefore the value expressed in the table is most certainly an overestimate. The increase in RNA synthesis that occured during days 3–4 of development was largely attributed to the production of rRNA, 4 S RNA, and new ribosomes; however, putative mRNA synthesis has been reported in preimplantation embryos of both the mouse and rabbit. A microsystem for measuring the translational activity *in vitro* of either maternal or embryonic mRNA has recently been described (see references in Table 2).

If maternal mRNA is present in the unfertilized ovum, it must be preserved in some way for later use during development. In invertebrates, it has been suggested that activated tRNA is limited or that mRNA is associated with proteins in an inactive ribonucleoprotein particle (8). The existence of inactive ribosomes has been demonstrated in the unfertilized mouse ovum (Table 1). Moreover, one of the characteristic features of the cytoplasm of the rodent ovum is the presence of ordered arrays of fibrous strands that are lattice-like in appearance. These structures contain RNA and protein; they gradually disappear during preimplantation development and may represent a store of inactive maternal ribosomes (1).

6   DNA IN THE OVUM AND EARLY EMBRYO

Estimates for the DNA content of the mouse ovum range from 8 to 46 pg, the lower value probably being the most representative as it is based on the quantitation of material isolated by buoyant density centrifugation in an ethidium bromide–CsCl gradient (Table 2). Of this 8 pg, approximately half is mitochondrial DNA. The remaining, presumably nuclear DNA (4–5 pg), corresponds to the amount of nuclear DNA found in adult somatic cells of the mouse.

In the unfertilized ovum, the chromosomes are in a condensed state, and the DNA is transcriptionally inactive. DNA synthesis in fertilized ova begins during pronuclear formation several hours after sperm penetration. Unequivocal expression of the embryonic genome during preimplantation development is evidenced by the detection of paternal isozyme variants and from studies involving specific mutations affecting preimplantation development (6).

7 BIOCHEMICAL FEATURES OF SPECIFIC OVUM ORGANELLES OR STRUCTURES

7.1 Cortical Granules

The mature ovum of most invertebrate and vertebrate species including mammals contains a number of membrane-limited secretory organelles (cortical granules) located rectilinearly or in irregular rows below the plasma membrane. In mammals, cortical granules vary in size from 0.2 to 0.8 μm in diameter, are unevenly distributed in the cortex—they are excluded from the cortex overlying the meiotic spindle— and constitute different populations based on their electron density. The molecular constituents of cortical granules from divergent species summarized by Schuel (9) include sulfated mucopolysaccharides, structural proteins, and several enzymes. The isolation and characterization of intact granules or purified granule contents from mammalian ova has not yet been accomplished.

Cortical granules are assembled in the Golgi complex, from which they migrate to the cortex during preovulatory maturation. The granules are released upon sperm penetration of the ovum or in response to some, but not to all, parthenogenetic treatments. Changes in intracellular concentrations of free calcium ions have been implicated in triggering the exocytotic release of granules in invertebrate (10) and mammalian (11, 12) ova (Chapter 8). The role of these structures in preventing or blocking multiple-sperm penetration into the ovum has received considerable attention (9, 10). In invertebrate systems, a cortical-granule–derived structural protein is involved in the transformation of the vitelline layer to its sperm-refractile counterpart: the fertilization layer. Two trypsin-like protease activities, a sperm-receptor hydrolase and a vitelline delaminase of cortical granule origin, have been isolated and characterized, as has an ovoperoxidase that cross-links tyrosine residues during fertilization-layer formation.

Indirect evidence supports a comparable role for cortical granules in mammalian fertilization. Firstly, these structures are released at fertilization in all mammalian ova examined. Secondly, in the mouse and hamster, cortical-granule-exudate preparations obtained from activated zona-free ova are capable of reducing the penetration of freshly treated intact ova, and this activity is susceptible to inhibition by trypsin inhibitors (11). However, a note of caution is warranted, since species differences are apparent in the degree to which the ovum relies on a zona reaction (Section 7.3 of this chapter) versus a plasma membrane block to polyspermy (Section 7.2 of this chapter). Thus it is difficult to conceive of a mechanism by which cortical granule exudate would act exclusively at the zonal level in one species and at the plasma membrane level in another. Moreover, in the mouse and human, it appears that some cortical granules are lost even before ova are exposed to sperm. When premature loss of

Table 7 Plasma Membrane Changes at Fertilization in Mammalian Ova

Property	Rabbit	Mouse	Hamster
Block to polyspermy	Yes, based on direct observations of ova recovered from natural mating or *in vivo* inseminations, Austin (2)	Yes, Wolf, *Develop. Biol.* **64**: 1 (1978)	No, Barros and Yanagimachi, *J. Exp. Zool.* **180**, 251 (1972); evidence for a block seen only 2½–3 hr following estimated penetration time
Sperm binding	—	Yes, decreased binding to zona-free ova correlated with penetration, Wolf and Hamada, *Biol. Reprod.* **21**: 205 (1979)	Yes, cortical reaction correlated with decreased binding and penetration, Gwatkin et al., *J. Reprod. Fert.* **47**, 299 (1976)
Lectin Binding	Increased binding of Con A after fertilization, Gordon et al., *Anat. Record* **181**, 95 (1975); but Con A does not prevent fertilization, Gordon and Dandekar, *J. Exp. Zool.* **198**, 437 (1976)	No difference in Con A binding between unfertilized and fertilized, Pienkowski, *Proc. Soc. Exp. Biol. Med.* **145**, 464 (1974), and Solter, *Immunobiology of Gametes*, Cambridge (1977)	No evidence for changes in oligosaccharide structure or organization as probed by a spectrum of lectins, Yanagimachi, in Edidin, M., and Johnson, M. H. *Immunobiology of Gametes*, Cambridge, University Press, London (1977)
Lectin Agglutinability	—	Con-A-induced agglutinability low in unfertilized, high in fertilized, parthenote or pronase-treated unfertilized ova, Pienkowski, *Proc. Soc. Exp. Biol. Med.* **145**, 464 (1974) agglutinability perturbed by protein-synthesis inhibitors, Siracusa et al., *Develop. Biol.* **62**, 530 (1978)	—

Property		
Chemical properties	—	Number of negatively charged groups increases after fertilization, Cooper and Bedford, *J. Reprod. Fert.* **25**, 431 (1971)
Surface topography	—	Jackowski and Dumont observed changes that were correlated with specific stages of the cell cycle rather than with fertilization, *Biol. Reprod.* **20**, 150 (1979)
Lateral diffusion rate	—	Both protein and lipid markers diffuse at greatly reduced rates after fertilization, Johnson and Edidin, *Nature* **272**, 448 (1978)
Permeability	—	Internal K^+ and Cl^- concentrations decrease while that of Na^+ increases upon fertilization; membrane permeability to both Na^+ and K^+ increases in two-cell embryos, Powers and Tupper, *Develop. Biol.* **56**, 306 (1977); glycerol permeability increases after fertilization, Jackowski and Leibo, *Cryobiology* **14**, 67 (1977); no differences in carrier-mediated transport of amino acids noted between unfertilized and one-cell fertilized eggs, Holmberg and Johnson, *J. Reprod. Fert.* **56**, 223 (1979)

granules is triggered in zona-free mouse ova, no reduction in fertility is detected upon subsequent *in vitro* insemination (13).

7.2 Vitelline Membrane

The vitelline, or plasma, membrane that limits the egg's cytoplasm, is an active participant in many events of early development, including sperm incorporation, cortical granule release, the block-to-polyspermy response, precursor transport, and cell–cell interaction phenomena. As is appropriate to its widespread involvement, its characterization has been approached by investigators from widely differing disciplines. Moreover, since the functional properties of the membrane are known to change drastically at fertilization, comparisons between unfertilized ova and one-cell embryos are extensive. For this discussion, three species have been selected, based on their differential responses to sperm (see also Sections 7.1 and 7.3 of this chapter): the rabbit, which relies heavily on the vitelline membrane but which, unfortunately, has not been studied extensively; the mouse, which is capable of a moderately efficient zona reaction but also displays an effective vitelline-membrane-block response; and the hamster, which relies exclusively on a zona reaction and does not display a vitelline membrane block (Table 7). Characteristic properties of the vitelline membrane that might reflect a change in function at fertilization have been probed, including surface topography, sperm and lectin (plant-derived glycoproteins that bind to specific sugar residues) binding capacity, lectin agglutinability, and specific chemical constituents and membrane lateral-diffusion rates (Table 7). The fragmentary nature of the data, however, does not yet allow the establishment of cause-and-effect relationships or an evaluation of the functional importance of each membrane alteration.

A transient, electrically-mediated block-to-polyspermy response has been described in the eggs of marine invertebrates (10), and the mammalian ovum may employ a similar mechanism to control its penetrability. The electrical properties of the mouse ovum have received attention, the consensus being that the unfertilized cell has a resting membrane potential of -14 to -23 mV (Table 8). By comparison, sea urchin eggs show a resting potential in the range of -10

Table 8 Electrical Properties of the Ovulated Mouse Ovum

Resting Membrane Potential	Resistance	Specific Capacitance	Reference
+1.9 mV	—	—	Cross *et al.*, *Develop. Biol.* **33**, 412 (1973)
−14 mV	2610 ohm cm²	1.6 μF cm⁻²	Powers and Tupper, *Develop. Biol.* **38**, 320 (1974)
−23.1 mV	—	1.5 μF cm⁻²	Okamoto *et al.*, *J. Physiol.* **267**, 465 (1977)
−14 mV	—	—	Fulton and Whittingham, *Nature* **273**, 149 (1978)

Table 9 Biochemical Characteristics of the Zona Pellucida of Unfertilized Mammalian Ova

Species	Dry Mass	Protein	Carbo-hydrate	Number of Poly-peptides	Size of Polypep-tides[a]	Reference
Mouse	—	1.8 ng	—	—	—	Loewenstein and Cohen, *J. Embryol. Exp. Morphol.* **12**, 113 (1964)
	—	4.8 ng	PAS-positive material detected	3	83–200 K	Bleil and Wassarman, *Cell* **20**, 873 (1980)
	—	—	PAS-positive material detected	5	28–240 K	Repin and Akimova, *Biochemistry* (Moscow) **41**, 39 (1976)
	4.96 ng	—	—	—	—	Abramczuk and Sawicki, *J. Exp. Zool.* **188**, 25 (1974)
Rat	—	~ 5 ng	PAS-positive material detected	3	107–182 K	Repin and Akimova, *op. cit.*
Pig	—	50 ± 10 ng	10 ± 2 ng neutral hexose, 5 monosaccharides detected by quantitative analysis	15	40–60 K	Dunbar, Wardrip, and Hedrick, *Biochemistry*, **19**, 356 (1980)
Pig	—	—	—	4	17–85 K	Menino and Wright, *Proc. Soc. Exp. Biol. Med.* **160**, 449 (1979)
Rabbit	2.9 μg	0.62 μg ninhydrin-positive material	PAS-positive material detected	4	—	Gould *et al.*, *Proc. Soc. Exp. Biol. Med.* **136**, 6 (1971)

[a] Molecular weight estimates based on SDS-PAGE.

Table 10   Alterations in the Zona Pellucida at Fertilization in Mammalian Ova

Ovum Type and Example	Sperm Penetrability	Changes in Zona Characteristic after Fertilization[b]					Reference[c]
		Sperm Binding	Solubility[a]	Lectin Binding	Morphology	Biochemical Composition	
1—Rabbit	None to limited	−	+ Changes hold no obvious relationship to sperm penetration	−	−	−	Overstreet and Bedford, *Develop. Biol.* **41**, 185 (1974)
2—Rat	Moderate	−	+	−	−	+	Repin and Akimova, *Biochemistry* (Moscow) **41**, 39 (1976)
Pig	Moderate	−	−	−	−	−	Cited by Gould *et al.*, *Proc. Soc. Exp. Biol. Med.* **136**, 6 (1971)

							References
Mouse	Moderate	+	Calcium dependency established, changes in zona sperm receptor defined	+	−	+	+	Baranska et al., *J. Exp. Zool.* **192**, 193 (1975); Jackowski and Dumont, *Biol. Reprod.* **20**, 150 (1979); Inoue and Wolf, *Biol. Reprod.* **13**, 340 (1976); Saling et al., *Develop. Biol.* **65**, 515 (1978); Bleil and Wassarman *Cell* **20**, 873 (1980)
3—Hamster	Strong	+	−	Changes in lectin-binding properties not dramatic	−	−	−	Yanagimachi and Nicolson, *Exp. Cell Res.* **100**, 249 (1976); Yanagimachi et al., *Fertil. Steril.* **31**, 562 (1979)
Sheep	Strong	−	−		−	−	
Dog	Strong	−	−		−	−	

[a] Includes disulfide-bond-reducing agents, low pH, and proteases.

[b] + = Change. − = No change.

[c] See Inoue and Wolf, *Biol. Reprod.* **12**, 535 (1975), for primary references in addition to those cited.

to -80 mV, depending on the species. The vitelline membrane of the mouse ovum is electrically excitable, and the existence of calcium channels has been established. Moreover, changes in internal ion concentrations and in membrane ion permeabilities have been associated with fertilization (Table 7). However, to date, although technically possible, changes in membrane electrical properties at fertilization or activation have not been recorded for the mammalian ovum.

7.3 Zona Pellucida

The mature, ovulated ovum is surrounded by a 5–15 μm thick acellular layer, the zona pellucida, which has been implicated in several unrelated physiological roles: physical protection of the ovum or preimplantation embryo, a recognition or mechanical function in ovum pickup and transport by the oviduct, and a barrier to sperm penetration after fertilization.

Immediately prior to ovulation, in some species like the mouse and rat, cumulus cell processes that extend through the zona are retracted; in the rabbit ovum such processes persist, and the cells with such processes constitute the corona radiata. Disruption of these contacts may have physiological significance for the resumption of egg maturation. The resultant ovum still retains cumulus and corona cells and is surrounded by a relatively porous zona (1). Zonae can be isolated at this time from ovarian or ovulated oocytes upon disruption of vitelli; however, limited availability of material has restricted the direct chemical analysis of zonae with the few exceptions summarized in Table 9. These results confirm the glycoproteic nature of the zona suggested by histochemical observations, although the number, size, and structural characteristics of the glycoproteins that constitute the zona remain to be clarified. A soluble zona receptor for capacitated sperm has been isolated from hamster (11) and from mouse (12) eggs and used to titrate the *in vitro* fertilizing or egg-binding capacity of sperm. This zona sperm receptor apparently is present only in zonae surrounding unfertilized eggs (12).

A definition of those changes in the zona that accompany fertilization and the dehiscence of the ovum's cortical granules has been sought because of their functional significance to embryonic development. Multiple-sperm penetration (polyspermy) is a lethal event in mammals, and insights into the mechanism by which the sperm receptivity of this structure is controlled could be very valuable in considerations of fertility and sterility (2). Species differences in the zona response at fertilization are, however, highly variable, and a unifying hypothesis concerning its mechanism or function is not immediately obvious. Three categories of response are apparent (Table 10). Type 1, typified by the rabbit, is characterized by the occurrence of changes in solubility properties of the zona at fertilization (zona hardening) but not in sperm penetrability; multiple sperm are commonly recovered in the perivitelline space of monospermic embryos. In Type 2, typified by the mouse or rat, changes in zona solubility and sperm-binding properties are correlated with the occurrence of a zona reaction (a decreased receptivity of the zona to sperm penetration), although these reactions can be separated temporally. In Type 3, typified by the hamster, a zona reaction occurs in the absence of concomitant changes in zona solubility. The relevance of the studies summarized in Table 10 conducted *in vitro* to events occurring *in vivo*, of course, remains to be clarified. It is recognized in the mouse and hamster that the efficiency of the zona reaction observed following *in vitro* insemination does not approach that operative *in vivo*. Based in part on extrapolation of results from marine invertebrates and partly on indirect evidence available in mammals, a cortical granule trypsin-like protease and an ovoperoxidase have been implicated in the zona reaction (9–11) and in zona hardening (13) respectively.

8   MECHANISMS OF ACTIVATION

The molecular mechanisms operative in the activation of the ovum at fertilization have been studied extensively in marine invertebrates. Three primary effectors have been described: transient changes in membrane potential, increased intracellular concentrations of free calcium ions, and, somewhat later in developmental time, the activation of a Na^+-H^+ exchange mechanism, a hydrogen ion efflux, and an increased intracellular pH (10). At the present writing, calcium is the only effector implicated in mammalian ovum activation. Thus mouse and hamster ova can be activated by exposure to the calcium ionophore A23187. The spontaneous rate of activation of mouse ova is augmented markedly by the use of Ca^{2+}-free medium. Iontophoretic (drive by electromotive force) injection of calcium into the mouse ovum results in activation (14).

9   SUMMARY

Fertilization and the activation of embryonic development in mammals have been under active investigation in both *in vivo* and *in vitro* systems for some 20 years, and there are morphological descriptions of the more dramatic events: sperm incorporation, cortical granule dehiscence, polar body release, pronuclear formation, syngamy, and cleavage. Over the years, the nutritional and energy requirements of the early embryo have been defined, and a number of the enzymes present in the unfertilized ovum have been assayed. Biochemically, although metabolically relatively inert, the unfertilized ovum is well endowed with maternally derived storage products and is capable of sustained protein synthetic activity in the absence of detectable RNA synthesis. It is presumed that this protein synthesis, which includes the formation of actin, tubulin, and LDH-1, represents the translation of maternal mRNA. After fertilization, synthesis of all RNA classes becomes apparent at the eight-cell stage followed approximately 24 hr later by a marked enhancement in the protein synthetic rate (Fig. 4). Increases in the latter have been correlated with the disappearance of cytoplasmic lattice-like structures that may represent inactive complexes of ribosomes.

Fertilization or parthenogenetic activation in marine invertebrates sets into motion a cascade of changes in the ovum, including alterations in membrane potential, increased intracellular free-calcium levels, and increased pH, which, in turn, act as effectors of the ovum's block-to-polyspermy response, metabolic activation, and preparation for syngamy (the union of paternal and maternal genomes). In mammals, a number of these phenomena have been probed. As summarized here, they include studies of sperm- and lectin-binding changes in the zona pellucida and plasmalemma, intracellular changes in calcium-ion concentrations and the role of cortical granules in the block to polyspermy, and metabolic activation of the ovum; and a detailed description of protein and RNA synthetic activity. At present, we stand on the verge of what promises to be an active and exciting stage in the biochemical approach to fertilization, which, in view of the recent success with *in vitro* fertilization in man, seems most timely.

REFERENCES

1 Van Blerkom, J., and Motta, P., *The Cellular Basis of Mammalian Reproduction*. Urban & Schwarzenberg, Baltimore (1979).

2 Austin, C. R., Patterns in Metazoan Fertilization, *in* Moscona, A. A., and Monroy, A. (eds.), *Current Topics in Development*, Vol. 12, p. 1, Academic Press, New York (1978).

3 Davidson, E. H., *Gene Activity in Early Development*, 2nd ed. Academic Press, New York (1976).

4 Brinster, R. B. L., Metabolism of the Ovum Between Conception and Nidation, *in* Gibian, H., and Plotz, E. J. (eds.), *Mammalian Reproduction*. Springer-Verlag, New York (1970).

5 Wales, R. G., Maturation of the mammalian embryo: biochemical aspects. *Biol. Reprod.* **12**, 66 (1975).

6 Epstein, C. J., Gene expression and macromolecular synthesis during preimplantation embryonic development. *Biol. Reprod.* **12**, 82 (1975).

7 Schultz, R. M., Letourneau, G. E., and Wassarman, P. M., Program of early development in the mammal: Changes in patterns and absolute rates of tubulin and total protein synthesis during oogenesis in the mouse. *Develop. Biol.* **68**, 341 (1979).

8 Engel, W., and Franke, W., Maternal storage in the mammalian oocyte. *Current Topics Pathol.* **62**, 29 (1976).

9 Schuel, H., Secretory function of egg cortical granules in fertilization and development: Critical review. *Gamete Res.* **1**, 299 (1978).

10 Epel, D., Mechanisms of Activation of Sperm and Egg during Fertilization *in* Sea Urchin Gametes, *in* Moscana, A. A., and Monroy, A. (eds.), *Current Topics in Developmental Biology*, Vol. 12, p. 186, Academic Press, New York (1978).

11 Gwatkin, R. B. L., *Fertilization Mechanisms in Man and Mammals*. Plenum Press, New York (1977).

12 Bleil, J. D., and Wassarman, P. M., Mammalian sperm-egg interaction: identification of a glycoprotein in a mouse egg zona pellucida possessing receptor activity for sperm. *Cell* **20**, 873 (1980).

13 Schnell, E. D., and Gulyas, B. J., Ovoperoxidase activity in ionophore ion treated mouse eggs. I. Electron microscopic localization. *Gamete Res.* **3**, 267 (1980); II. Evidence for the enzyme role in hardening the zona pellucida. *Gamete Res.* **3**, 279 (1980).

14 Whittingham, D. G., and Siracusa, G., The involvement of calcium in the activation of mammalian oocytes. *Exp. Cell Res.* **113**, 311 (1978).

15 Wolf, D. P., Nicosia, S. V., and Hamada, M., Premature cortical granule loss does not prevent sperm penetration of mouse eggs. *Develop. Biol.* **71**, 22 (1979).

CHAPTER TEN

THE OVARY: FOLLICLE DEVELOPMENT, OVULATION, AND LUTEAL FUNCTION

GEULA GIBORI

Department of Physiology and Biophysics
University of Illinois at the Medical Center
Chicago, Illinois

JOSEPHINE MILLER

Departments of Obstetrics and Gynecology,
and Anatomy
University of Illinois at the Medical Center
Chicago, Illinois

1   INTRODUCTION

The mammalian ovaries are paired organs found within the abdominal or pelvic cavity, enclosed in either a peritoneal capsule (for instance, woman and non-human primate) or ovarian bursa (for instance, rat and mouse). The ovary consists of three distinct units: follicles, corpus luteum, and stroma, encapsulated by the surface epithelium (Fig. 1). It is a complex structure in which several compartments interact to secrete hormones and produce an ovum in response to gonadotropin stimuli.

2   FOLLICULAR DEVELOPMENT (1)

2.1   General Characteristics

The first step in follicular development is oogenesis, the formation of the oocyte. Oocytes arise from oogonia, which are extragonadal in origin. Oogonia are distinguishable from somatic cells early in fetal life by means of histochemical markers such as glycogen, alkaline phosphatase, and esterase. Upon their arrival in the developing gonads, they undergo several mitotic divisions. In most mammals, mitosis of oogonia ceases during fetal life, the cells enter the first stage of meiosis, and by definition, at that point become primary oocytes. Meiosis proceeds to, and is arrested at, the diplotene stage of the first meiotic division.

The primary oocytes are invested with an incomplete ring of squamous stromal cells, which give rise to the granulosa cells of the follicle. The oocytes remain in arrested meiotic prophase within their ring of granulosa cells for various periods of time, which, in the woman, can approach 50 years in length. The time at which follicle formation is initiated varies in different species. In some species, follicles form during embryonic life, whereas in others, folliculo-genesis begins during the neonatal or during the immature period. Beginning early in life, some oocytes begin to enlarge while their nuclei remain arrested in meiotic prophase. The granulosa cells surrounding these oocytes become cuboidal in appearance and begin to proliferate. As the oocyte enlarges and the

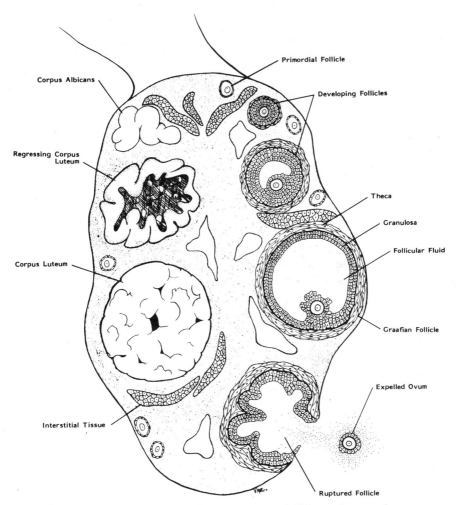

Figure 1 Schematic representation of the ovary, follicle, and corpus luteum.

encircling granulosa cells proliferate, a basement membrane external to the granulosa cells is formed; it is probably secreted by the granulosa cells. "Disk-like" cells, arising from the surrounding stroma, organize along the basement membrane and give rise to the theca layer of the follicles. Thus the growing follicle consists of an oocyte surrounded by an inner layer of granulosa cells and an outer layer of theca cells. As the follicles continue to grow, small pockets of fluid appear among the proliferating granulosa cells. These pockets eventually coalesce and form a single, large, fluid-filled cavity called the antrum. Antral fluid appears to be secreted initially by the granulosa cells; however, during later stages of growth, most of the fluid and the proteins contained therein are derived as an ultrafiltrate of the blood. The theca interna of these follicles is highly vascularized, but the capillaries in the theca do not penetrate the basement membrane. Thus the granulosa layer has no direct blood supply and remains avascular throughout development. In a follicle that has attained full growth, the oocyte is surrounded by a cluster of granulosa cells known as the cumulus oophorus, and is suspended in the follicular fluid, attached to the follicle wall by a stalk of cells. The size of the mature follicle varies among species, and seems to be correlated with body size.

2.2 Hormonal Control (1, 2)

The signal that stimulates one resting follicle to begin to grow while its neighbor remains quiescent is a mystery. Since only a few oocytes respond while others remain dormant, it is not likely that the signal is hormonal in nature. Furthermore, hypophysectomy does not stop the entry of oocytes into the growing pool. It has been suggested that the initiation of growth of oocytes may be a probabilistic event, similar to that of radioactive decay. Consistent with this hypothesis is the observation that the number of oocytes entering into the committed pool appears to be a function of the total number of oocytes in the ovary at that time. Thus the entry of oocytes into the growing pool may be spontaneous and uncontrolled. In addition, it appears that extraovarian hormones may not be necessary for several initial steps in the process of follicular development. Granulosa cell proliferation, secretion of the basement membrane, and the beginning of thecal development can proceed in the absence of the pituitary gland. However, follicular development cannot proceed beyond these early stages in the absence of the pituitary. It is now well established that two of the pituitary hormones, follicle-stimulating hormone (FSH) and luteinizing hormone (LH), and ovarian steroid hormones, are required for continued follicular development.

Granulosa cells from small follicles contain both FSH and estradiol receptors, and both FSH and estrogens are known to be involved in the early stages of follicular development. Treatment of neonatal rats with FSH will stimulate follicular growth, increase the number of granulosa cells, and stimulate antrum formation. FSH also stimulates the synthesis of its own receptor and the appearance of LH receptor in granulosa cells of rats pretreated with estradiol. Finally, FSH appears to be important in regulating the number of follicles destined to ovulate, since sustained high levels of FSH can cause an increased number of large follicles to develop. Estrogen is also involved in follicular growth and has long been known to stimulate ovarian growth and enhance ovarian responses to gonadotropins.

The intrafollicular site of estrogen action appears to be the granulosa cells, since nuclear estradiol-receptor has been shown to be preferentially localized in nuclei of isolated granulosa cells. The effect of estradiol on granulosa cell function appears to be mediated by means of the translocation of the cytosol estradiol–receptor complex to the nucleus. Thus elevated levels of nuclear estrogen receptor were associated with periods of follicular growth, whereas low levels were associated with follicular atresia or early luteinization. Estrogen acts on granulosa cells of developing follicles to increase their responsiveness to both FSH and LH. Estrogen administration to hypophysectomized rats promotes granulosa cell proliferation and increases the number of cells that contain FSH receptors. In this way, estradiol promotes the ability of FSH to increase receptor for both FSH and LH. Estrogen also appears to enhance the responsiveness of granulosa cells to FSH by increasing the ability of FSH to stimulate cAMP accumulation. Thus estradiol appears to play an important role in determining the ability of a given follicle to respond to FSH and LH.

Significant quantities of androgens are produced in the ovaries, and recent studies have implicated androgens in regulating follicular development. Testosterone receptor has been localized in the granulosa cells, and androgen can stimulate follicular growth in both the immature rat and the mature cycling rat and can induce atresia of preantral follicles. The role of progesterone in follicular development has not been clearly elucidated. Injection of progesterone during the estrous cycle of rats delays follicular maturation. Functional corpora lutea that secrete progesterone, or progesterone treatment in the absence of corpora lutea, inhibit the effect of FSH on follicular growth in hypophysectomized hamsters. Administration of progesterone to rabbits increases the number of antral follicles. In the rhesus monkey, an active corpus luteum completely supresses follicular development on both ovaries. These studies suggest that progesterone may retard the growth of follicles.

The effect of LH in early stages of ovarian follicular development has been investigated less extensively. However, it appears that LH plays, by means of its action on theca cells, an important role in early follicular cell differentiation. LH promotes the folliculogenic effect of FSH by enhancing the ability of FSH to stimulate an increase in granulosa cell content of LH receptor. This LH effect is not mediated by either estrogen or androgen.

In summary, ovarian follicular development involves both steroid and pituitary protein hormone regulation. Follicle growth is terminated either by LH stimulation of luteinization (corpus luteum formation) or by follicular atresia.

3 FOLLICULAR ATRESIA

Few follicles complete the final stages of growth just described; most degenerate by a process known as atresia. Several investigators have reported that atresia occurs at all stages of follicular development. Atresia has been reported in hamster and human preantral follicles. However, in the mouse and rat, atresia was reported to be rare in small follicles but marked in antral follicles. In all species, atresia of antral follicles appears to begin in the granulosa cells. Pyknotic granulosa cells become visible, whereas the oocyte appears to remain intact. Three progressive stages of atresia have been described in the mouse. Stage I is characterized by up to 20% pyknotic granulosa cells. In stage II, the proportion of pyknotic granulosa cells increases, leukocytes invade the granulosa cells, and the basement membrane between the granulosa and theca layers begins to disintegrate. By stage III, the follicle has shrunk, few granulosa cells are visible, and the basement membrane is hardly recognizable.

There have been few attempts to study the biochemical correlates of atresia. Atresia is associated with an increase in the activity of (1) hydrolytic enzymes, reflecting lysosomal activity and tissue degradation; (2) acid phosphatase and aminopeptidase within granulosa cells; and (3) in the rat, mouse, and guinea pig, greatly increased activity of 3β-hydroxysteroid hydrogenase. In some species, transient storage of lipids in the granulosa cells in large atretic follicles is common. When atresia is induced in rats, the granulosa cells first lose the capacity to bind LH and develop acid phosphatase activity, then show a decreased capacity to bind FSH, a reduction in mitosis, and pyknosis. Follicles undergoing atresia produce little or no estradiol. It seems likely that in these follicles, receptors and steroidogenesis either fail to be induced or cannot be maintained adequately.

The cause of atresia is a subject open to speculation. It has been suggested that a deficient theca layer of large follicles may lead to follicular atresia. A deficient theca might not be able to produce enough androgenic substrate for estrogen synthesis by the granulosa cells, and in the absence of estrogen, granulosa cells stop proliferating and degenerate. McNatty proposed the hypothesis that several intrafollicular factors influence the follicles to either ovulate or undergo atresia. The failure to develop seems to be the fate of granulosa cells exposed to follicular fluid that is devoid of FSH and estrogen and that contains a high level of androgen or prolactin. The presence of FSH in follicular fluid prevents granulosa cells from undergoing atresia, presumably by inducing the aromatization of androgen to estrogen. Follicles with very high concentrations of estrogen also contain FSH, but the level of androgen in the follicular fluid is low. Thus follicles emerging from the preantral stages of growth when FSH levels in blood are low do not accumulate high concentrations of FSH in their follicular fluid. These follicles have low aromatase activity and high androgen concentrations. The mechanism of androgen-induced atresia is unknown but is presumed to act locally by means of its receptor to stimulate the degeneration of the follicle.

4   FOLLICULAR STEROIDOGENESIS (2)

4.1   General Characteristics

The follicle synthesizes three steroids: progestagens, androgens, and estrogens. Information concerning follicular steroidogenesis has been obtained by several techniques: (1) measurement of steroid content of ovarian vein blood and of follicular fluid at various stages of the reproductive cycle; (2) *in vitro* studies of the steroidogenic capacity of the immature ovary composed mainly of small follicles; and (3) studies of the whole follicle or isolated thecal and granulosa cells *in vitro*.

Since the isolated follicle is a multicomponent system, the relative contribution of the theca cells and the granulosa cells to the overall steroid pattern has been studied by isolating cell types. The thecal cells of rabbit, rat, hamster, and man synthesize principally androgens. However, estrogen production has also been reported in some species. Thecal cells have 17α-hydroxylase activity and 17–20 lyase activity and are the main source of androgen secretion in the follicle. However, the aromatization enzyme system in thecal cells isolated from preovulatory follicles of several species is not as active as in the granulosa cells from the same follicle. Granulosa cells appear to be the principal source of estrogen and progesterone, since they have a weak 17α-hydroxylating system and little or no 17–20 lyase activity, but have the capacity to actively aromatize C_{19} steroids. Granulosa cells that appear incapable of androgen secretion probably convert the androgenic precursor synthesized by the thecal cells to estrogen. It has been shown in the horse, the human, the rabbit, the rat, and the hamster that both thecal and granulosa cells are required for the production of estrogen. Follicular estrogen biosynthesis could therefore occur through the cooperation of the thecal cells that produce androgen and the granulosa cells that convert that androgen to estrogen (Fig. 2). The granulosa cells have been shown to be the principal source of progesterone production in the sow, woman, and mare. However, granulosa cells do not become active in progestin synthesis until just before ovulation.

4.2 Hormonal Control (1, 3)

Studies *in vivo* and *in vitro* indicate that steroid secretion by granulosa cells of the developing follicle is modulated by gonadotropins and by the steroid hormones themselves. The physiological factors involved in the regulation of estrogen secretion have been extensively investigated. FSH and LH regulate ovarian estrogen secretion by acting at biochemically distinct sites. LH stimulates the synthesis of androgenic precursor by the thecal cells and FSH stimulates the conversion of that androgen to estrogen acting at the level of the granulosa cells. However, since in some species estrogen is secreted by the thecal cells alone, FSH and LH could also act on a single type of cell to stimulate estrogen synthesis. LH transitorily stimulates and then inhibits the secretion of androgen and estrogen and then stimulates progesterone secretion by the preovulatory ovarian follicles. The preovulatory surge of LH is followed by a rapid increase in estrogens, androgens, and progestins, and thereafter steroidogenesis is terminated for a period of some 36 hours.

There is considerable experimental evidence indicating that many of the effects of LH and FSH in follicles are mediated by cAMP. LH, and FSH free of LH contamination, promote an increase in intracellular cAMP concentration in mature follicles. These gonadotropin-induced elevations in follicular cAMP concentration seem to be due to activation of the adenylyl cyclase system rather than to changes in the activity of phosphodiesterases. In rats, pigs, and rabbits, the responsiveness of the cyclase system to LH increases during the latest stage

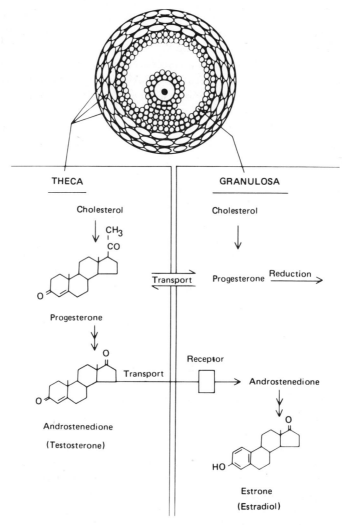

Figure 2 Follicular secretion of estradiol: interaction between thecal and granulosa cells.

of follicular maturation just prior to the ovulatory surge of gonadotropins. However, small antral follicles also contain an active adenylyl cyclase coupled to FSH receptors, and are capable of forming cAMP upon FSH stimulation. The appearance of an LH-responsive adenylyl cyclase system is thought to be due first to the generation of LH receptors followed by coupling of these receptors to the cyclase system. Exposure of preovulatory follicles to the LH surge results in a decreased number of LH receptors and a rapid loss in the capacity of the adenylyl cyclase system to by stimulated by LH. This desensitization of the adenylyl cyclase system to further gonadotropin stimulation is accompanied by a decrease in cAMP levels and in steroid secretion.

In summary, an ovulatory dose of LH initially stimulates the adenylyl cyclase system, promoting a large increase in cAMP accumulation and steroidogenesis. Thereafter, LH decreases the content of its own receptor and desensitizes the adenylyl cyclase system, which leads to a cessation of cAMP formation and steroid secretion. However, the steroidogenic capacity of the follicle is not regulated by gonadotropins only. Results of several studies have demonstrated that androgen stimulates progesterone production by porcine follicles or granulosa cells. In the rat, androgen enhances the stimulatory effect of FSH on progesterone secretion.

5 FOLLICULAR FLUID (4)

The follicular fluid is a liquid that accumulates in extracellular spaces within the follicles. During the early stages of follicular growth, fluid-filled spaces appear between the granulosa cells, and these later coalesce to form the follicular antrum as the follicle enlarges. The presence of follicular fluid in so many species is of great importance in ovarian physiology. Because the concentrations of proteins, steroids, carbohydrates, and mucopolysaccharides in follicular fluid are not constant throughout follicular growth, the granulosa cells are exposed to an environment that may be different from that found in adjacent follicles and in serum.

The follicular fluid originates from two sources, plasma and the cells of the follicles. It is a slightly viscous solution with a pH above 7.0, similar to that of serum or plasma. In women, the concentrations of sodium and potassium in follicular fluid and serum are similar, but they differ in cows and sows. However, all the other ions measured (magnesium, chloride, calcium, zinc, copper, and inorganic phosphate) are found in similar concentrations in the follicular fluid and the serum. Two mucopolysaccharides identified in follicular fluid, hyaluronic acid and chondroitin sulfate, are probably responsible for the local changes in viscosity of the follicular fluid. A heparin-like substance has also been found in the follicular fluid of women, sows, cows, bitches, and rabbits. This polysaccharide is of physiological importance, since it inhibits the clotting of the follicular fluid.

The total protein concentration in the fluid is either similar to or lower than that in serum and is similar in all follicles irrespective of size. There are at least two sources of protein in follicular fluid: one originates from the plasma and the other from the follicular cells. Many enzymes are present, and their concentrations change as the follicle develops. Some enzymes (i.e., aminotransferases, phosphatases, esterases, and dehydrogenases) are considered to be intracellular, and their presence in the follicular fluid may reflect a turnover of follicle cells. However, these enzymes are not believed to play a role in ovulation. The concentration of trypsin, cathepsin, and collagenase-like enzymes increases in follicular fluid before ovulation. The presence of plasminogen, plasminogen activator, and plasmin has also been demonstrated in bovine follicular fluid.

Plasminogen, a glycoprotein, is present in follicular fluid at concentrations similar to those in serum, and it appears to be derived from serum. Plasminogen activator, which is secreted by the granulosa cells after stimulation with FSH causes the conversion of plasminogen to plasmin. Plasmin has been shown to decrease the tensile strength of the follicle wall, to degrade the basement membrane, and to activate procollagenase (see Section 7 of this chapter).

Both protein and steroid hormones are found in follicular fluid. Protein hormones (LH, FSH, and prolactin) are present in some, but not all, antral follicles. LH and FSH levels increase in larger follicles, whereas prolactin concentrations in the follicular fluid are high in small follicles and low in large follicles. The concentration of steroids changes throughout follicular growth in all species examined. During the growth phase, androgens and estrogens are present at high levels in the follicular fluid. Since steroids are thought to diffuse from the follicle into serum, this was an enigma. However, high-affinity binding proteins from both androgens and estrogens have been identified in the follicular fluid, and this could explain this apparent contradiction. It is not known whether these high concentrations of steroids have physiological significance to the follicle. Prior to ovulation, follicles that are destined to ovulate also contain very high levels of progestins. During the luteal phase, large amounts of progestins and androgens are present in the follicles of women, ewes, and mares.

Recently it has been demonstrated that polypeptides capable of inhibiting a number of different processes are present in follicular fluid. Those inhibitors include oocyte-maturation inhibitor, luteinization inhibitor, and ovarian inhibin.

The oocyte maturation inhibitor prevents the spontaneous resumption of meiosis. This inhibitor, which is not species-specific, has been partially characterized and purified in porcine follicular fluid and seems to be produced by granulosa cells. The inhibitory effect of follicular fluid on oocyte maturation can be overcome by LH.

An inhibitor of granulosa cell luteinization has been found in the follicular fluid of small porcine follicles. This inhibitor not only prevents morphological luteinization of granulosa cells from large follicles but also depresses progesterone secretion and depresses the LH stimulation of cAMP accumulation in the follicle.

Follicular fluid possesses also inhibin-like activity. Inhibin is a substance that specifically suppresses FSH secretion in ovariectomized rats, monkeys, and hamsters. The cellular origin is not known, but the inhibin-like substance is also thought to originate from the granulosa cells.

6 INTERSTITIAL GLAND TISSUE

The interstitial gland tissue of the mammalian ovary originates from the theca interna and the surrounding stroma of degenerating follicles. This tissue is invariably present in the ovaries of both young and adult mammals and possesses ultrastructural, histochemical, and biochemical features of steroid-secreting tissue. The interstitial tissue is rich in lipids, including esterified cholesterol, fatty-acid-free cholesterol, and phospholipid. The interstitial tissue of the mature rabbit secretes progesterone and 20α-hydroxypregn-4-en-3-one, whereas androgens are the principal steroids secreted by the human ovarian interstitial cells during the follicular phase of the cycle and during pregnancy. The steroidogenic activity of rabbit and possibly rat interstitial cells is dependent upon circulating LH.

7 OVULATION

7.1 General Characteristics (5–8)

After undergoing a series of maturational changes, the "ripe" follicle can be stimulated to ovulate. In both induced and spontaneous ovulators, follicle rupture is preceded by a large, short-lived surge of LH, which is triggered either by rising blood concentrations of estradiol or by coital stimulation. This LH surge is usually accompanied by a surge in FSH (except in rabbits) and also by a surge of prolactin in the rat. LH alone is sufficient to initiate ovulation and is generally considered to be the ovulatory hormone; however, FSH in high concentrations has been shown to initiate ovulation in the rat and hamster. Although it is possible that FSH may synergize with LH in the initiation of ovulation, its major effect is thought to involve the maturation of follicles destined to ovulate at some later time.

Ovulation follows the LH surge after a time period that is characteristic for each species (see Table 1). During this interval, there is a surge of steroid secretion, the follicle and ovary become hyperemic, capillary permeability increases, and the volume of the follicle nearly doubles. The collagen ground substance, found almost exclusively in the theca interna, tunica albuginia, and surface epithelial layers of the follicle, begins to disintegrate and the cells of the granulosa layer begin to dissociate. These degenerative changes, observed with the microscope, are associated with increased distensibility and decreased tensile strength of the follicle wall. Although the follicular fluid volume is increasing during this period of time, intrafollicular fluid pressure does not increase and may even decrease slightly just prior to ovulation. Thus the major factor that allows rupture of the mature follicle is degradation of the follicle wall rather than a buildup of intrafollicular pressure.

Many biochemical changes take place in the follicle during the periovulatory period, and the possible role of each factor in the mechanism of ovulation will first be considered individually. This will be followed by an attempt to formulate a comprehensive model for the mechanism of ovulation that suggests possible interrelationships between each of these factors.

7.2 LH Binding and Adenylyl Cyclase Activation (3, 9)

LH, produced through the preovulatory surge of this hormone, is quickly bound to high-affinity receptors in the theca and granulosa layers of the ripe

Table 1 Characteristics of the Periovulatory Period of Various Mammalian Species

Species	Length of Cycle (days)			Duration of Estrous	Duration of LH Surge	Hours After Peak LH	Time of Ovulation
	Type of Ovulation	Whole Cycle	Follicular Phase				With Respect to Cycle
Cat	Induced	14–21	5–6	4 or 9–10 days[a]		25–26	24–36 hr after coitus
Cow	Spontaneous	21–22	3–5	18–19 hr	10 hr	24	25–30 hr after onset estrus
Dog (beagle)	Spontaneous	203–224	13–14	7–9 da	24–40 hr	38–44	1–3 da after onset of estrus
Ferret	Induced	None[b]		Continuous			30 hr after coitus
Fox	Spontaneous	None[b]	15	2–4 da			1–2 da after onset of estrus
Gerbil	Spontaneous	4–6		12–18 hr			6–10 hours after coitus
Goat	Spontaneous	20–21	2–3	32–40 hr			30–36 hrs after onset of estrus
Guinea pig	Spontaneous	16–17	3–4	6–11 hr			10 hr after onset estrus
Hamster (golden)	Spontaneous	4		20 hr	3 hr	10–12	8–12 hr after onset of estrus
Horse	Spontaneous	19–25	4–8	4–8 da	10 da		1–2 da before end of estrus
Human	Spontaneous	28	12–14	None	24–48 hr	16–24	14 da prior to onset of menses
Mink	Induced	8–9		2 da			40–50 hr after coitus
Monkey (rhesus)	Spontaneous	28–29	12–14	None		28	11–15 da after onset of menses
Mouse	Spontaneous	4–6	2–3	10 hr		10–12	2–3 hr after onset of estrus
Opossum	Spontaneous	29		1–2 da			12 hr after onset of estrus
Pig	Spontaneous	20–21	3–5	40–60 hr	12–16 hr	40–41	38–42 hr after onset of estrus
Rabbit	Induced	None		Continuous	2–3 hr	9–10	10–11 hr after coitus
Rat	Spontaneous	4–5	2–3	13–15 hr	2–3 hr	10–11	8–10 hr after onset of estrus
Sheep	Spontaneous	16–17	1–2	24–36 hr	10 hr	21–23	24–27 hr after onset of estrus
Squirrel	Induced	16		3 da			8–12 hr after coitus

[a] Four days with male, 9–10 days without male.
[b] One heat; ferret: March to August; fox: January to March.

follicle and to interstitial cells of the ovarian stroma. Although the duration of the LH surge may be as long as 48 hr (see Table 1), it is clear that the mature follicles are adequately stimulated by gonadotropin and are programmed to proceed toward eventual ovulation within the first hours of the LH surge. The primary effect of LH binding is a very rapid stimulation of membrane-bound adenylyl cyclase. Peak elevations in cAMP are reached within 30 min after exposure to LH and then decline, although LH levels remain elevated. After this initial stimulation, adenylyl cyclase becomes refractory to further LH stimulation and remains "desensitized" for two to three days. This "desensitization" of the LH-stimulable cyclase is not due to occupancy of the LH receptor and may involve inactivation of the adenylyl cyclase enzyme by a cAMP-independent protein kinase. A second elevation in follicle cAMP levels has been observed 4–5 hr after mating in the rabbit, and is thought to be secondary to changes in the levels of prostaglandins in the follicle, which are elevated following the initial ovulatory LH surge.

7.3 Steroidogenesis (6, 9, 10)

The LH-induced elevation in cAMP is followed at approximately 1 hr by a surge in the secretion of estradiol and, to a lesser extent, progesterone. This is also a short-lived effect, and by 2–4 hr after the LH surge, steroid secretion falls dramatically and remains at a low level until several days after ovulation. cAMP mediates this steroidogenic effect of LH, which requires both protein and RNA synthesis. This suggests that the surge represents *de novo* synthesis of steroid rather than the release of stored steroid.

It has been suggested that the preovulatory steroid release may play an important role in ovulation by, in some way, regulating the synthesis and/or release of enzymes responsible for breaking down the ground substance of the follicle wall. This conclusion is based on the observations that large doses of aminoglutethemide or cyanoketone, which inhibit steroidogenesis, can also prevent ovulation and that estradiol and progesterone can, to some extent, increase the distensibility of the follicle *in vitro* (see below). In addition, injections of antisera to progesterone, and to a lesser extent antisera to testosterone, can also prevent ovulation. However, Bullock and Kappauf (10) found that the effectiveness of aminoglutethemide and cyanoketone in inhibiting ovulation was not correlated with their inhibition of progesterone secretion. Thus the role of steroids in the mechanism of ovulation remains to be determined.

7.4 Role of Prostaglandins (11, 12)

The levels of prostaglandins $F_{2\alpha}$ ($PGF_{2\alpha}$) and E_2 (PGE_2) are significantly elevated in the follicle 4–5 hr after gonadotropin stimulation, and they remain elevated until ovulation. It is thought that prostaglandins probably mediate the secondary elevation in cAMP and the subsequent stimulation of steroid secretion observed in the preovulatory rabbit follicle. However, it is clear that prostaglandins do not mediate the effect of LH on steroidogenesis, since saturating levels of LH and prostaglandins have additive effects on the levels of cAMP. Thus cAMP levels are elevated within 5 min after the LH surge, whereas prostaglandin levels are not elevated for 4–5 hr, and LH can stimulate steroidogenesis in the presence of indomethacin, an inhibitor of prostaglandin synthetase.

It has now been demonstrated that prostaglandins play an essential role in the mechanism of ovulation. Intrafollicular administration of indomethacin or antisera to $PGF_{2\alpha}$ was found to inhibit LH-induced ovulation in rats, rabbits, and monkeys, whereas antisera to PGE_2 was only partially effective. However, these agents did not prevent luteinization or the acute stimulation of steroidogenesis by LH. In addition, intrafollicular injection of $PGF_{2\alpha}$ caused not only ovulation, but induced oocyte maturation. Thus the acute actions of LH, that is, steroidogenesis and luteinization, are not mediated by prostaglandins, whereas the long-term effects of LH, that is, ovum maturation and ovulation, are dependent upon prostaglandin synthesis.

The mechanism through which prostaglandins contribute to ovulation is not known. However, their participation is apparently not required for the early preovulatory events initiated by LH, since ovulation can be prevented by blocking the action or synthesis of prostaglandins as late as 5 hr after LH stimulation. Since an intrafollicular injection of prostaglandins is associated with oocyte extrusion through the injection site, prostaglandins may participate in follicle rupture by causing the contraction of ovarian "muscular" tissue. This is supported by the observation that injections of either LH or prostaglandins will initiate pulsatile increases in intrafollicular pressure. Although it is not clear whether the follicle wall contains true smooth muscle cells, there are several contractile elements in the ovary, including fibroblasts, collagen fibrils, and vascular smooth muscle, that may be important in ovulation. Since intrafollicular pressure does not increase prior to ovulation, it is clear that the contraction of ovarian muscular elements alone is insufficient to cause ovulation. However, the preovulatory follicle becomes very distensible, thus the ovarian contractile tissue may influence ovulation by providing the force necessary to induce rupture of the follicle once this preovulatory loss of tensile strength has taken place.

7.5   Role of Proteolytic Enzymes (6, 9, 12, 13)

The degradative changes observed in the follicle prior to ovulation support the hypothesis that ovulation results from enzymatic destruction of the collagenous ground substance of the follicle wall.

The breakdown of collagen in other tissues is regulated by a number of factors. Collagenase is synthesized *de novo* and secreted as procollagenase. The inactive procollagenase can be activated by a number of agents, including trypsin, plasmin, catecholamines, and ascorbic acid. In addition, estrogens and progesterone appear to decrease collagenolytic activity, at least in the uterus. Once activated, collagenase digests collagen fibrils, resulting in their depolymerization and breakdown. The destruction of collagen by collagenase depends not only on its activity but upon the rate of collagen repolymerization. Serum-derived proteins found in follicular fluid can inhibit collagenase, and both prostaglandins and ascorbic acid released prior to ovulation have been found to decrease the rate of collagen repolymerization. Thus many factors must be considered in evaluating the effectiveness of an enzyme with collagenolytic activity as a mediator of ovulation. Of the many enzymes found in the follicular fluid, two enzymes, collagenase and plasminogen activator, have been implicated as having important roles in the process of ovulation. Both enzymes are elevated in the follicle prior to ovulation and can degrade follicular connective tissue both *in vivo* and *in situ* if injected into follicular fluid.

Plasminogen activator is synthesized in culture primarily by granulosa cells, and its synthesis and release are increased by treatment with FSH but not LH. This action of FSH appears to be mediated by cAMP but not by cGMP or prostaglandins. It is thought that after stimulation with FSH, plasminogen activator is released into the follicular fluid, where it converts plasminogen (a serum protein transudate in follicular fluid) to plasmin, which then degrades follicular connective tissue.

It is probable that the activation of plasmin in follicular fluid is not in itself sufficient to result in ovulation since (1) FSH is not the primary ovulatory hormone and does not appear to be released in rabbits prior to ovulation; (2) the number of FSH-binding sites and the amount of FSH-stimulable adenylyl cyclase is low in granulosa cells from preovulatory follicles; (3) dissolution of the follicle wall is first evident in the tunica albuginea, theca externa, and the theca interna, whereas the granulosa cell layer is the last to deteriorate; and (4) plasmin can only attack collagen if the native helix has first been nicked with collagenase. Beers and Strickland (13) have found that *in vitro* there are two populations of granulosa cells, only one of which (25–30% of the cells) synthesizes and secretes plasminogen activator. This could explain the low FSH-binding and cyclase activity. In addition, ovarian tissue contains basal levels of collagenase, which may have sufficient activity to provide a suitable substrate for plasmin. However, the level of ovarian collagenase becomes elevated during the preovulatory period and may have sufficient activity in itself to cause breakdown of the follicle wall.

Ovarian collagenase is thought to be released either from surface epithelial cells or from fibroblasts. Surface epithelial cells contain dense granules that are distinct from lysosomes, increase in number prior to ovulation, and are thought to be the source of collagenase, since preovulatory degradation begins at the surface of the follicle. However, the surface epithelium is not always intact in all animals prior to ovulation, and follicles can ovulate after the surface epithelium has been scraped away.

Espey (9) supports a major role for fibroblasts as the source of ovarian collagenase. This is consistent with the known role of fibroblasts from other tissues in the regulation of collagen synthesis and metabolism. Large numbers of ovarian fibroblasts are layered between bundles of collagenous connective tissue in the tunica albuginea and theca externa. Collagen fibrils extend between these layers of cells to connect both fibroblasts and other collagen fibrils. Ovarian fibroblasts proliferate after the LH surge and are characterized by unusual multivesicular structures that protrude from their surface. As ovulation approaches, these vesicular structures increase in number and are frequently seen at the edge of cytoplasmic processes that extend from the fibroblasts into dense collagen. The collagenous ground substance that surrounds the fibroblasts is gradually destroyed. Since the tensile strength of the follicle wall is probably related to the interconnecting layer of fibroblasts and collagen fibrils, a localized release of collagenase from fibroblasts, causing digestion of the surrounding ground substance, would be expected to produce major changes in the tensile strength of the follicle and allow rupture to occur.

7.6   A Model for Ovulation

Prior to the ovulatory gonadotropin surge, the mature preovulatory follicle has acquired the appropriate receptors and intracellular machinery necessary for ovulation. The following model for ovulation can be proposed (Fig. 3). The quiescent fibroblasts of the theca externa synthesize and secrete procollagenase, which becomes bound to collagen fibrils in the ground substance. With the gonadotropin surge, LH becomes bound to receptors in the granulosa, theca interna, and interstitial cells of the ovary, resulting in the stimulation of adenylyl cyclase, elevations of intracellular cAMP, and increased steroid synthesis. FSH binding to receptors in granulosa cells may result in the production of plasminogen activator, and this is also thought to be a cAMP-mediated event. LH-induced hyperemia, thought to be due to histamine release from mast cells, is another early event associated with the ovulatory gonadotropin surge. At this time, the fibroblasts of the theca externa begin to proliferate, either under the direct influence of LH or through steroid mediation. These transformed fibroblasts synthesize increased amounts of procollagenase, which is activated by some unknown mechanism(s), possibly involving the action of plasmin, proteolytic enzymes derived from serum transudates, or the release of some activating enzyme from the fibroblasts themselves. Once activated, collagenase can also contribute to the conversion of procollagenase, causing a cascading effect and the rapid breakdown of the collagenous ground substance of the follicle wall.

This proteolysis is associated with an acute inflammatory response in the tissue, resulting in the release of histamine and prostaglandins. Prostaglandins contribute to the degradation of the follicle wall through an unknown mechanism by inhibiting the repolymerization of collagen fibrils and by releasing ascorbic acid. Prostaglandins also activate adenylyl cyclase, causing an elevation in the intracellular levels of cAMP and a secondary increase in the secretion of estradiol and progesterone. A third role for prostaglandins in the mechanism of ovulation is thought to involve the stimulation of contractile elements in the ovary, which include not only fibroblasts, collagen fibrils, and smooth muscle cells in the follicle wall (although there is some controversy as to whether true

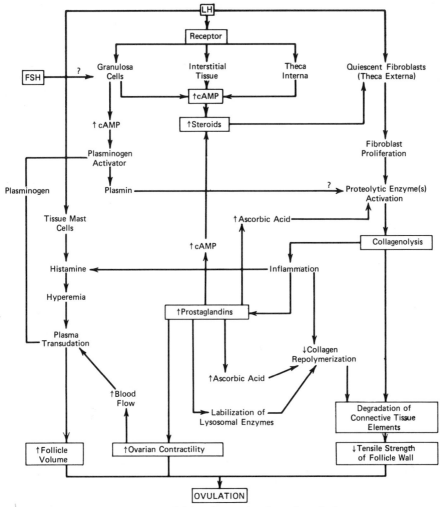

Figure 3 A model for the mechanism of ovulation.

smooth muscle cells exist in the follicle wall), but also smooth muscle cells of the ovarian vasculature. The contraction of arteriolar smooth muscle contributes to increased ovarian blood flow and follicular volume, whereas contractile activity in the follicular wall may make a significant contribution toward the development and maintenance of tension in the degrading follicle wall.

Ovulation is the culmination of these events. As the ground substance of the follicle wall is degraded, the tensile strength of the follicle wall is lost. When the disintegrating follicle wall can no longer withstand the tension created by an expanding follicular fluid volume and the activity of contractile elements in the follicle wall, the follicle ruptures and the oocyte is extruded. Although the entire follicle shows evidence of degradation, rupture is thought to occur through the

Table 2 Characteristics of the Luteal Period of Various Mammalian Species

Species	Length of Cycle or Pregnancy (days)			Estrogen Secretion by Corpus Luteum	Luteotropic Hormone[d]	Uterine Luteolysin	Day of Implantation	Number of Days that Pregnancy is Dependent on	
	Cycle	Ppreg[d]	Pregnant					CL[d]	Pituitary
Cat	14–21	30–40	58–65	?		No	13–14	50	1st half
Cow	21–22	None	278–293	No	LH, PRL, E?	Yes—local	40	200	
Dog (beagle)	203–224	59–68	59–68	No	PRL	No	11–12	All	All
Ferret	None[b]	35–40	42	Yes[c]		No	12	All	All
Fox	None[b]	50–55	51–54	?		No		All	All
Gerbil	4–6	13–18	24–26	?			6		
Goat	20–21	None	146–151	No				All	All
Guinea pig	16–17	None	67–68	No	LH, FSH, PRL	Yes—local	6–7	21–25	40
Hamster (golden)	4	8–10	16	Yes	FSH, PRL, LH, E	Yes—local	4	All	All
Horse	19–25	None	330–345	No	E	Yes—local	60	155–200	
Human	28	None	280	Yes	LH	No	6–7	40	40
Mink	8–9	Variable	40–75[a]	?			Variable[a]	All	
Monkey (rhesus)	28–29	None	163	Yes	LH	No	8–9	27–28	32–40
Mouse	4–6	10–12	19–20	Yes	LH, PRL, E	Yes	6	All	11–12
Opossum	29		12–13	Yes		No		All	
Pig	20–21	None	112–115	No	LH, PRL	Yes—local	11–20	All	All
Rabbit	None	16–18	30–32	No	E	Yes	7	All	All
Rat	4–5	12–14	20–22	Yes[c]	LH, PRL, E	Yes	6	All	12
Sheep	16–17	21	147–155	No	LH, PRL	Yes—local	18	55	50
Squirrel	16			No	LH, PRL	No		All	All

[a] Delayed implantation: 30 days between implantation and parturition.
[b] See Table 1.
[c] Can synthesize estrogens *in vitro* from an androgen precursor.
[d] LH, luteinizing hormone; FSH, follicle-stimulating hormone; PRL, prolactin; E, estradiol; Ppreg, pseudopregnant; CL, corpus luteum.

apex of the follicle, because it has the least structural support from the ovarian stroma and is, thus, the site of least resistance.

8   LUTEINIZATION (3, 14)

Luteinization is characterized by a change in the morphology, steroidogenic capacity, and function of the granulosa cell. It results in transformation of the follicle, which has been primarily estrogen secreting, into the corpus luteum, which synthesizes large quantities of progesterone. Although ovulation is always accompanied by luteinization, luteinization can occur in the absence of ovulation. The initiation of luteinization by LH is mediated by cAMP, and requires both RNA and protein synthesis. During the luteinization process, which is not completed until after ovulation, the granulosa cells assume the morphological characteristics of steroid-secreting cells. They hypertrophy, accumulate glycogen granules and lipid inclusions, and develop an abundance of smooth endoplasmic reticulum, an extensive Golgi complex, and mitochondria with tubular cristae. In some species, the cells of the theca interna also luteinize and become incorporated within the granulosa lutein cells. It is generally assumed that these theca "lutein" cells are required for the development of corpora lutea, which secrete estradiol as well as progesterone (see Table 2). Corpora lutea from species that secrete only progesterone are thought not to contain thecal cells, although this question has not been carefully examined.

For two to three days following the preovulatory gonadotropin surge, there is a generalized loss in the level of receptors for LH, FSH, and estradiol, and the luteinizing follicle is unresponsive, or "desensitized," to hormonal stimulation in terms of adenylyl cyclase activation.

It has recently been suggested that LH receptor "desensitization" involves activation of a cAMP-independent protein kinase. However, the physiological significant of this period of desensitization is unknown, but may be characteristic of the mature follicle, since preantral follicles do not become desensitized to gonadotropins following the ovulatory gonadotropin surge. During this period of luteal cell differentiation, steroid secretion by the follicle is very low or undetectable, which may reflect either receptor-adenylyl cyclase uncoupling or a generalized shutdown in the steroidogenic pathway. As the corpus luteum becomes fully formed, steroid and gonadotropin receptors reappear and this reflects the hormonal requirements for maintenance of luteal structure and continued progesterone secretion.

9   CONTROL OF LUTEAL FUNCTION (14, 15, 16)

The length of time of elevated progesterone secretion is characteristic for each species. In induced ovulators (see Table 1), coitus triggers the gonadotropin surge, which results in ovulation and the formation of corpora lutea. If these animals become pregnant, the period of elevated progesterone secretion is usually lengthened compared to that found in pseudopregnancy, that is, an infertile mating. In the rat, mouse, and hamster, which have four-to-five-day estrous cycles, the corpus luteum of the cycle synthesizes very little progesterone. However, if these animals experience a nonfertile mating or experimental cervical stimulation, they become pseudopregnant and develop corpora lutea, which secrete levels of progesterone comparable to those found during the early part of pregnancy. Other animals with estrous cycles may also become pseudopregnant following a nonfertile mating.

In the pig, the corpora lutea formed during the estrous cycle are independent of further hormonal support. However, in most species, including the pig during pregnancy, pituitary hormones are required to maintain the structure of, and progesterone secretion by, the corpus luteum, and in many animals, several hormones are required for full luteotropic support. With the exception of the hamster and possibly the guinea pig, FSH is not essential for the maintenance of luteal function. In contrast, LH is of major luteotropic significance in most species that have been studied (see Table 2). LH binds to specific receptors in the luteal cell membrane and, by a cAMP-mediated mechanism, stimulates steroidogenesis by increasing the rate at which cholesterol is converted to pregnenolone.

Although the rabbit corpus luteum also contains specific LH receptors, LH cannot directly support luteal function. Instead, LH stimulates estradiol secretion from ovarian follicles. Estradiol in turn binds to specific receptors within the corpus luteum that can be translocated to the nucleus, and estradiol is absolutely required to maintain progesterone secretion. Thus estradiol is the "ultimate" luteotropic hormone in the rabbit. The luteotropic role of estradiol has also been well documented in the rat, and estradiol is thought to be an important component of the "luteotropic complex" in other species as well (see Table 2).

In the rat, the source of estradiol is thought to be the corpus luteum itself, which is rich in aromatase activity. LH acts either on the ovarian follicles or the corpus luteum to stimulate the production of androgen, which is subsequently aromatized to estradiol within the corpus luteum. Thus estradiol appears to be an essential luteotropic hormone for the rat as well as the rabbit. Prolactin is the third component of the "luteotropic complex" in the rat. Cervical stimulation results in twice-daily surges of prolactin, the appearance of luteal LH receptors, and transformation of the "nonfunctional" corpus luteum of the cycle into a functional corpus luteum of pregnancy or pseudopregnancy. The number of estrogen receptors in the corpus luteum is also increased in response to prolactin. Thus prolactin, by regulating the availability of LH receptors, acts in concert with LH to maintain luteal estradiol synthesis and ultimately progesterone secretion.

10 LUTEAL REGRESSION (14, 15)

Corpora lutea have a finite life span, but the mechanisms regulating luteal regression (luteolysis) are largely unknown. In some animals, it has been shown that progesterone secretion can be prolonged by the administration of exogenous luteotropic hormone. However, even with the use of sensitive assay systems, it has not been possible to demonstrate significant decreases either in the physiological levels of luteotropic hormones or in the levels of receptors for these hormones, which are adequate to explain the cause of luteal regression. The uterus, however, does appear to be important in the regulation of luteal life. In many species other than primates, hysterectomy is associated with prolonged luteal progesterone secretion. It has been clearly shown in the ewe that the uterus secretes $PGF_{2\alpha}$ into the uterine vein just prior to the time of luteal regression. Through a countercurrent mechanism, $PGF_{2\alpha}$ is thought to pass from the uterine vein into the closely apposed ovarian artery to cause regression of the corpus luteum by an unknown mechanism. Because of this countercurrent exchange of $PGF_{2\alpha}$ from the uterine vein into the ovarian artery, the uterus is said to have local control over the corpus luteum. Thus the removal of one uterine horn will allow prolonged luteal function in the ipsilateral, but not the contralateral, ovary.

In other species, in which the uterine vein is not closely apposed to the ovarian artery, prostaglandins must enter the general circulation before reaching the corpus luteum. Since prostaglandins are removed from the circulation very rapidly, it is not clear whether uterine prostaglandins could function as physiological luteolysins in these species. However, corpora lutea of the cow, hamster, rat, ewe, and human contain specific membrane receptors for $PGF_{2\alpha}$, and, in all species studied, the administration of pharmacological doses of $PGF_{2\alpha}$ results in luteal regression. In the cow, it has been shown that the uterus releases large quantities of arachidonic acid prior to the time of luteal regression. This compound is very stable in the systemic circulation and is a precursor for prostaglandin synthesis. Prostaglandin synthetase activity in the bovine corpus luteum is high prior to luteal regression, and it is thought that the uterine release of arachidonic acid provides a precursor for intraluteal prostaglandin synthesis that ultimately results in luteal regression. Estradiol may play an important role in regulating the synthesis and release of prostaglandins at the time of luteal regression. In the ewe, guinea pig, and monkey, estradiol treatment results in luteal regression, and this effect of estradiol is dependent upon increased $PGF_{2\alpha}$ release from the uterus (ewe and guinea pig) or corpus luteum (monkey). It is interesting to speculate that in other species, estradiol may also increase prostaglandin synthesis and thus inhibit luteal function.

During pregnancy in most species, progesterone secretion by the corpus luteum is prolonged by mechanisms that are only partially understood. In the ewe, it is clear that the preimplantation embryo in some way prevents the release of $PGF_{2\alpha}$ from the uterus. In other species, in which hysterectomy prolongs progesterone secretion (see Table 2), the release of a uterine luteolysin is also thought to be inhibited by the conceptus. In addition, the placenta may synthesize hormones that stimulate progesterone secretion by the corpus luteum. In many species, the tissues of the syncytiotrophoblast secrete a glycoprotein hormone that has many of the properties of LH and has clearly been shown to maintain luteal progesterone secretion. Other placental luteotropins are biologically similar to pituitary prolactin or have combined LH- and prolactin-like activities. The mare is unique in that the luteotropic hormone of pregnancy is secreted by the endometrium rather than the chorion and has a biological activity similar to both FSH and LH. When present, these placental luteotropic hormones can stimulate and maintain luteal progesterone secretion in the absence of hormonal support from the pituitary. In addition, the placenta may also secrete progesterone in sufficient quantities to substitute for luteal progesterone. Thus a successful pregnancy may be independent of the pituitary or the ovary if the placenta secretes luteotropic hormones or sufficient quantities of progesterone, respectively (see Table 2).

The cycle of follicle development, ovulation, and corpus luteum formation is a continuous one, and whether the length of progesterone secretion is short or prolonged by pregnancy, diminishing progesterone levels signal the follicles to enter the final phase of follicular development, which once again leads to estrogen secretion, LH release, and ovulation.

REFERENCES

1   Richards, J. S., Uilenbroek, J. Th. J., and Jonassen, J. A., Follicular Growth in the Rat: a Reevaluation of the Roles of FSH and LH, *in* Channing, C. P., Marsh, J., and Sadler, W. A. (eds.), *Ovarian Follicular and Corpus Luteum Function*, p. 11, Plenum Press, New York (1979).

2   Dorrington, J. H., Steroidogenesis *in vitro, in* Zuckerman. L., and Weir, B. J. (eds.), *The Ovary*, p. 359, Academic Press, New York (1977).

3   Birnbaumer, L., Day, S. L., Hunzicker-Dunn, M., and Abramowitz, J., Ontogeny of the Corpus Luteum: Regulatory Aspects in Rats and Rabbits, *in* Hamilton, T. H., Clark, J. H., and Sadler, W. A. (eds.), *Ontogeny of Receptors and Reproductive Hormone Action*, p. 173, Raven Press, New York (1979).

4   McNatty, K. P., Follicular Fluid, *in* Jones, R. E. (ed.), *The Vertebrate Ovary*, p. 215, Plenum Press, New York (1978).

5   Schwartz, N. B., and Ely, C. A., Role of Gonadotropins in Ovulation, *in* Moudgal, N. R. (ed.), *Gonadotropins and Gonadal Function*, p. 237, Academic Press, New York (1974).

6   Lipner, H., Mechanism of Mammalian Ovulation, *in*, Greep, R. O. (ed.), *Handbook of Physiology, Endocrinology*, Vol. 2, Part 1, p. 409, American Physiological Society, Washington (1973).

7   Bjersing, L., and Cajander, S., Ovulation and the mechanism of follicle rupture. IV. Ultrastructure of membrana granulosa of rabbit graafian follicles prior to induced ovulation. *Cell Tiss. Res.* 153, 1 (1974).

8　Rondell, P., Biophysical aspects of ovulation. *Biol. Reprod. Suppl.* **2**, 64 (1970).

9　Espey, L. L., Ovulation, *in* Jones, R. E., *The Vertebrate Ovary*, p. 503, Plenum, New York (1978).

10　Bullock, D. W., and Kappauf, B. H., Dissociation of gonadotropin-induced ovulation and steroidogenesis in immature rats. *Endocrinology* **92**, 1625 (1973).

11　Armstrong, D. T., Moon, Y. S., and Zamecnik, J., Evidence for a Role of Ovarian Prostaglandins in Ovulation, *in*, Moudgal, N. R., *Gonadotropins and Gonadal Function*, p. 345, Academic Press, New York (1974).

12　Espey, L. L., Ovarian contractility and its relationship to ovulation: A review. *Biol. Reprod.* **19**, 540 (1978).

13　Beers, W. H., and Strickland, S., Involvement of Plasminogen Activator in Ovulation, *in* Spilman, C. H., and Wilks, J. W. (eds.), *Novel Aspects of Reproductive Physiology*, p. 13, S. P. Medical and Scientific Books, New York (1977).

14　Hilliard, J., Corpus luteum function in guinea pigs, hamsters, rats, mice and rabbits. *Biol. Reprod.* **8**, 203 (1973).

15　Fajer, A. B., Luteotropic and Luteolytic Factors, *in* Balin, H., and Glasser, S. (eds.), *Reproductive Biology*, p. 572, Excerpta Medica, Amsterdam (1972).

16　Keyes, P. L., Yuh, K.-C. M., and Miller, J. B., Estrogen Action in the Corpus Luteum, *in* Channing, C. P., Marsh, J. M., and Sadler, W. A., (eds.), *Ovarian Follicular and Corpus Luteum Function*, p. 447, Plenum Press, New York (1979).

PART TWO

REPRODUCTIVE ENDOCRINOLOGY

Edited by

ROBERT T. CHATTERTON, JR.

CHAPTER ELEVEN

HYPOPHYSIOTROPIC RELEASING AND RELEASE-INHIBITING HORMONES

CRAIG W. BEATTIE

Department of Surgery
University of Illinois at the Medical Center
Chicago, Illinois

1   INTRODUCTION

The anatomical basis underlying central nervous system (CNS) regulation of anterior pituitary function is neurovascular (Fig. 1). Hypophysiotropic release and inhibiting hormones (factors) are synthesized in hypothalamic neurons and transported by their axons to circumscribed areas of the median eminence.

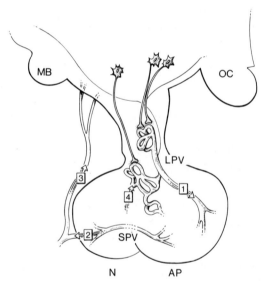

Figure 1   Hypothalamic–pituitary–neurovascular link. Two groups of hypophyseal portal veins reach the anterior pituitary (*AP*). The long portal vessels (*LPV*) originate in the capillary plexus of the median eminence and pass down the pituitary stalk, and the short portal vessels (*SPV*) connect the neurohypophysis (*N*) and the anterior pituitary. It has been proposed that releasing hormones entering the LPV reach the anterior pituitary (*1*), stimulating the release of anterior pituitary hormones, which may be released into the systemic circulation or enter the neurohypophysis (*2*) via the SPV. From the neurohypophysis they may reach hypophyseal arteries (*3*) that can serve as efferent channels, or flow from the pituitary stalk to the median eminence from connecting capillaries (*4*). *PN*, peptidergic neurons; *OC*, optic chiasm; *M*, mammillary bodies.

Here they are released into the capillary plexus of the portal vascular system and transported to the anterior pituitary gland.

The hypothalamus regulates the secretion of the gonadotropins, thyroid-stimulating hormone (TSH), and prolactin (PRL) through a complex molecular interaction between the peptides, gonadotropin-releasing hormone (GnRH), thyrotropin-releasing hormone (TRH), what appear to be prolactin-releasing (PRF) and prolactin-inhibiting (PIF) factors, the more-conventional neurotransmitters, and the gonadal steroids. GnRH, TRH, gonadal steroids, and quite possibly the putative candidates for control of PRL secretion influence anterior pituitary hormone secretion through specific, high-affinity receptors on the membrane of specific cells within the anterior pituitary. In addition, gonadal steroids and perhaps even the anterior pituitary hormones themselves exert regulatory or "feedback" effects on the secretion of releasing factors at the level of the hypothalamus and other CNS structures.

The identification and cellular location of the releasing hormones as well as the neural molecular mechanisms subserving their secretion are the subjects of this chapter.

2 GONADOTROPIN-RELEASING HORMONE

2.1 Structure

GnRH is a straight-chain decapeptide (10 amino acids; see Fig. 2) that was isolated and purified from porcine hypothalami. It is a pyroglutamic acid peptide, and although it has not been established whether pyroglutamic acid peptides are of natural occurence or are formed as an artifact by cyclization of glutamine, it may be significant that free pyroglutamic acid and pyroglutamyl-peptidase are found in mammalian tissues. Porcine GnRH is structurally identical to the porcine, ovine, bovine, rodent, and human hypothalamic peptide. The amino-terminal tripeptide and tetrapeptide fragments of GnRH do not stimulate gonadotropin release. The carboxy-terminal octapeptide and nonapeptide also have negligible gonadotropin-releasing activity. Although minimal-size active fragments apparently cannot be obtained, individual amino acid substitution drastically alters the gonadotropin-releasing activity of the native molecule. In general, the amino acids in positions 1 and 2 and from positions 4 to 10 appear to be involved only in the binding of GnRH to its target-tissue receptors and in exerting conformational effects (1, 2).

However, histidine-2 and tryptophan-3 exert an important functional effect, since substitution of these L-amino acids with other D-amino acids produces analogs of GnRH with a significant loss in gonadotropin-releasing activity and even potent antagonistic properties. Deletion of positions 2 and 3 also leads to dramatic losses in gonadotropin-releasing activity. Substitution of D-amino acids in position 6 (glycine) has produced molecules that are extremely potent agonists of the native decapeptide in terms of their ability to release luteinizing

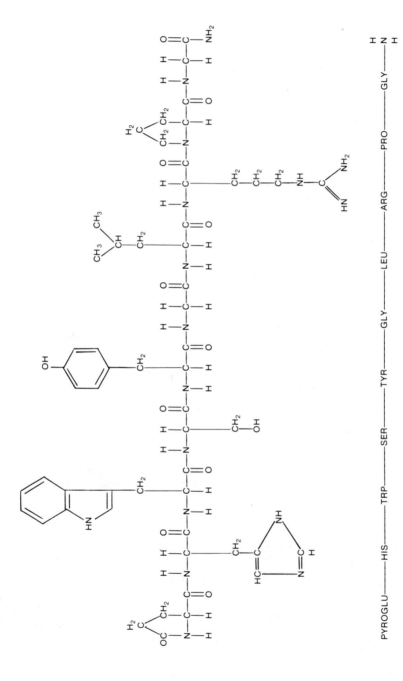

Figure 2 Amino acid sequence of gonadotropin-releasing hormone (GnRH).

hormone (LH) and follicle-stimulating hormone (FSH) from the anterior pituitary. This is apparently due to an enhanced stabilization of the β-bond configuration at the center of the molecule (1, 2), an increased affinity for the pituitary receptor compared with native GnRH, and an apparent decrease in metabolic degradation.

GnRH has been found in the central nervous system of all species of vertebrates examined. Both the natural and synthetic decapeptide release LH and FSH and induce ovulation in all mammalian species so far examined. Passive immunization of a variety of mammalian species with an antiserum to GnRH reduces blood levels of both FSH and LH and blocks ovulation. Collectively, these observations indicate that the decapeptide represents the moiety responsible for the release of LH and FSH.

2.2 Biosynthetic and Metabolic Sequences

2.2.1 CNS Site of GnRH Synthesis

GnRH as been identified immunohistochemically in a region extending from the preoptic area posteriorly through the suprachiasmatic region and ventrally through the medial basal hypothalamus (MBH) to the lateral aspects of the median eminence and proximal pituitary stalk by way of the tuberoinfundibular tract. GnRH is localized in perikarya within the arcuate nucleus, preoptic area, septum and paraolfactory region of most mammalian species examined (3, 4, 5). The concentration of GnRH within the hypothalamus of the rat is detailed in Table 1. Within the medial preoptic area, the largest concentrations of immunoreactive GnRH are found in the supraoptic crest.

Complete surgical isolation of the MBH from the rest of the brain significantly reduces the GnRH content of the MBH in rodents and monkeys, but does not decrease the amount of GnRH in the supraoptic crest in rodents or alter pituitary LH secretion in monkeys. These observations suggest that in rodents, but not primates, most if not all of the GnRH in the MBH is synthesized or under trophic influence of cells rostral to this region.

It has recently been demonstrated that hypothalamic GnRH is sequestered in two specific types of submicroscopic particles identified by density gradient centrifugation and electron microscopy. GnRH particle concentration is 13 times higher in the neuron-terminal-rich MBH than in the more anterior portion of the hypothalamus where the neuroendocrine cell somas are believed to be located. The larger (90—120 nm) GnRH particles appear to be synaptosomes (pinched-off presynaptic nerve terminals that have sealed spontaneously) as identified by their sedimentation properties, susceptibility to hypo-osmotic shock, and solubility by detergents. They are located in the axon terminals of the pallisade area of the median eminence. The smaller particles (40–80 nm) apparently arise in nerve terminals abutting on the portal plexus of vessels. Although different in size, the particles are of the same density, since they sediment as a single peak under equilibrium conditions. The smaller particles are not synaptosomes, since they do not meet the above criteria. Larger parti-

Table 1   Gonadotropin-Releasing Hormone in the Brain of the Rat[a]

Brain Region	GnRH (ng/mg protein)
Hypothalamus[b] (total content ≈ 2.7 ng)	
Median eminence	22
Arcuate nucleus	3
Ventromedial nucleus (lateral part)	0.6
Suprachiasmatic nucleus	Trace (<0.1)
Supraoptic nucleus	Trace (<0.1)
Posterior hypothalamic nucleus	Trace (<0.1)
Ventromedial nucleus (medial part)	Trace (<0.1)
Preoptic area (total content ≈ 0.2 ng)	
Supraoptic crest	14
Tissue surrounding the supraoptic crest	0.3
Medial preoptic nucleus	0.15
Circumventricular organs	
Subfornical organ	4.2
Subcommissural organ	5.9
Area postrema	10.2

[a] From Brownstein (3).
[b] GnRH could not be detected in the periventricular, anterior hypothalamic, paraventricular, dorsomedial, perifornical, dorsal premammillary, or ventral premammillary nuclei, or in the medial forebrain bundle.

cles are demonstrable in the hypothalamus of rats during the second day of neonatal life, and a mature profile of large and small particles appears during the second and third weeks of life. The ontogeny of subcellular compartmentalization correlates well with that of synapses and glial elements in the developing central nervous system (6, 7).

2.2.2   Possible Routes of Synthesis

A priori, several possible mechanisms exist for peptide bond formation for the releasing hormones. Classical ribosomal synthesis of either the active peptide or a precursor peptide (prohormone) and posttranslational modification to yield the active releasing hormone is a strong possibility for a decapeptide like GnRH. Another possibility is synthesis by an RNA-independent protein template such as a "soluble" synthetase. In the case of a large peptide such as GnRH, this route is improbable because of the critical sequencing of its 10 amino acids. Enzymatic synthesis by an RNA-dependent ribosome-independent mechanism occuring by means of an aminoacyl RNA species is also a possibility, but again in the case of GnRH, is not practical (8).

Recent *in vivo* and *in vitro* evidence is consistent with prohormone conversion to the biologically active moiety, but the methodologies employed are indirect and fraught with the problem of high levels of inactivating peptidases, which makes the isolation and purification of the "synthesizing activity" from a cell-free system essential to future studies (9).

2.2.3 Metabolism

The peptidases involved in inactivating neural peptides have varying substrate specificities and pH optima (5.5–8.0). They include neutral endopeptidases that cleave peptides internally, neutral and acid proteinases that cleave internal bonds and N- and C-terminal groups, and dipeptidases and tripeptidases. Although the peptidases that inactivate GnRH are found in greatest concentration within the hypophysiotropic area and median eminence, they, like GnRH, are also found in other brain regions. In mammals, GnRH-inactivating enzymes are found in soluble and particulate fractions of brain homogenates. Over 90% of the activity is associated with the soluble fraction, with a higher activity present in the soluble fraction of male brain homogenates, suggesting a sex difference. Castration of either sex significantly reduces peptidase activity. Administration of estrogen and testosterone to castrate females and males respectively restores enzyme activity to intact levels. This intracellular regulation of peptidase activity by gonadal steroids has obvious physiological implications in the control of hypothalamic neurosecretion.

The primary enzymatic attack on GnRH is an initial internal cleavage between glycine-6 and leucine-7 by a neutral endopeptidase, followed by the secondary action of peptide hydrolases. A slower removal of the C-terminal glycinamide by a second enzyme acting at the C-terminus has also been demonstrated (10).

2.3 Factors Affecting Secretion

2.3.1 Neurotransmitters

Since the existence of central neurohumoral inhibition in the control of anterior pituitary function was first postulated over 30 years ago, central catecholamine neurons have been repeatedly implicated in the control of gonadotropin secretion.

In the rodent, norepinephrine (NE) pathways terminating in the hypothalamus appear to exert a facilitory influence on GnRH release. NE nerve terminals are concentrated in those hypothalamic nuclei that contain GnRH. Fluorescence histochemical and immunofluorescence studies have shown that the NE nerve terminals within the lateral portion of the subependymal layer of the median eminence and medial preoptic area of the hypothalamus are activated during the critical period of the ovarian cycle in the rodent when GnRH pathways are activated (11). Other indirect evidence also suggests that the synapses controlling the release of GnRH in the preoptic and anterior hypothalamic area as well as the medial hypothalamus are not activated in ovariectomized animals in which tonic secretion of high levels of gonadotropin occurs. Collectively these observations suggest that the NE systems are mainly involved in the phasic events of GnRH secretion and not the tonic release of gonadotropin. Direct biochemical and anatomical evidence for axoaxonic contact between specific NE synapses and adjacent GnRH neurons remains to be presented.

The role of dopamine (DA), the catecholamine precursor of NE, in the control of GnRH secretion is the subject of considerable debate. Evidence for a stimulatory and an inhibitory role has been presented. DA, like NE, neurons have cell bodies in the hypothalamus but have been shown to be distinct from GnRH-containing neurons. Recently, biochemical and fluorescence histochemical estimates of DA turnover in tuberoinfundibular and lateral median eminence neurons have demonstrated that DA turnover is high during events that accompany tonic inhibition of pituitary LH release such as pregnancy, lactation, and high doses of contraceptive steroids, supporting the postulated role of DA as an inhibitor of gonadotropin release (12).

Evidence that supports a stimulatory role for DA on GnRH release is complicated by the fact that the drugs selected to stimulate DA receptors in the CNS (i.e., ergot derivatives) and anterior pituitary decrease PRL secretion. The facilitory effect of DA on GnRH secretion could therefore result from an event secondary to the inhibition of PRL secretion and subsequent interference with the ovarian steroid–gonadotropin feedback system that forms the basis of cyclic reproductive phenomena (13). The site of the inhibitory or stimulatory control appears to reside in the tuberoinfundibular dopaminergic neurons. Iontophoretic studies suggest that DA, in addition to the NE present in the mediobasal hypothalamus, modulates neural activity of the arcuate nucleus whose axons project to the median eminence. There is suggestive evidence for axoaxonic synaptic contact at the level of the lateral palisade area of the rat median eminence, an area high in GnRH content, supporting the view that DA controls GnRH release at this level (14).

Like DA, the indoleamines 5-hydroxytryptamine (5-HT or serotonin) and melatonin have also been repeatedly shown to exert an inhibitory influence on GnRH secretion in experimental animals and in man. As yet no definitive mechanism of action of these putative CNS transmitters on GnRH neurons has been demonstrated.

Table 2 lists the stimulatory or inhibitory action of monoamines on anterior pituitary function as derived from experimental evidence in animals and man.

2.3.2 Pituitary Gonadotropin (Short-Loop Feedback)

When FSH and LH are implanted into the MBH of the rat, they inhibit gonadotropin secretion from the anterior pituitary. Hypothalamic NE synthesis and metabolism (turnover) is enhanced when FSH is administered to hypophysectomized/gonadectomized animals. Further indirect evidence for a possible "short loop" feedback effect of pituitary gonadotropin on the central nervous system structures mediating their secretion is provided by anatomical studies that demonstrate retrograde blood flow to the hypothalamus by means of a reverse portal circulation and high concentrations of LH in these vessels (15, 16).

Table 2   Effect of Neurotransmitters on Hypothalamic Releasing Hormone Secretion[a]

Hypothalamic-Releasing Hormones	Subject	NE	DA	5-HT
TRH	Exp. An[c]	↑[d]	↓[d]	↓↑?[d]
	Human	→	↓[e]	↓[e]
Gn-RH	Exp. An	↑	↓↑?	↓
	Human	→	↓↑?	—
PIF	Exp. An	↑	↑	↓
	Human	→	↑	↓
PRF	Exp. An	—	—	↑
	Human	—	—	—

[a] Adapted from Muller *et al.* (17).
[b] Key to symbols: ↑ stimulation, ↓ inhibition, → no effect, — not determined.
[c] Experimental animal.
[d] Cold-induced TSH secretion.
[e] Hypothyroid subjects.

2.3.3   Peripheral Target-Organ Feedback on Neural Centers

It is now abundantly clear that gonadal steroids serve as modulators of gonadotropin release, with both neural and anterior pituitary sites of action. Estradiol and testosterone are selectively taken up by and bound to cytosol receptors in the arcuate, ventromedial, and ventropremamillary nuclei of the hypothalamus and the preoptic area. The sites of estrogen and androgen binding conform to the sites where implanted crystals of steroid modulate gonadotropin secretion and facilitate reproductive behavior. Among the possible mechanisms for these effects is an interaction between gonadal steroids and the axons of any of the classical CNS neurotransmitters found in high concentration in these areas.

Male   Several lines of indirect biochemical evidence support a role for gonadal steroids' modulating catecholaminergic and GnRH function in the hypophysiotropic area in the male. Castration of male rats produces an increase in both hypothalamic NE turnover and GnRH content. The activity of tyrosine hydroxylase, the rate-limiting step in catecholamine synthesis, is also increased in the median eminence following gonadectomy and reduced when testosterone is administered.

Female   In the female, the ovariectomy-induced increase in hypothalamic NE turnover and tyrosine hydroxylase activity is reduced by progestational steroids and increased with low doses of estrogen. During vaginal proestrus in the rat, when estrogen levels are high, anterior hypothalamic NE accumulation peaks. Estrogen also appears to decrease the passive efflux of NE from synaptosomes whereas progesterone inhibits it. At high doses, estradiol markedly decreases tonic LH secretion and increases DA turnover in all species examined. At low doses, NE turnover is increased, circulating gonadotropins rise, and DA turnover is inhibited (14). Gonadal steroids also influence monoamine metabolism

by inhibiting intraneuronal monoamine oxidase activity during the preovulatory LH surge, and by conversion of estrogen in the CNS to a catechol through
2-hydroxylation of ring A, which inhibits extraneuronal catechol-*O*-methyl
transferase (17). It appears therefore that inhibition of catecholamine metabolism during periods when estrogen secretion is high might prolong the action of
NE on GnRH neurons and increase gonadotropin release. In addition, the
hypothalamic increase in cyclic adenosine-3', 5'-monophosphate (cAMP) accumulation stimulated by estrogen and blocked by adrenergic antagonists suggests that after estrogen binding to catecholamine neurons, cAMP may be
released and affect GnRH secretion from adjacent peptidergic neurons (17).
Taken together, these observations suggest that the inhibitory feedback control
of tonic LH secretion exerted by estrogen may, at least in part, be mediated by
tuberoinfundibular dopaminergic neurons.

3   THYROTROPIN-RELEASING HORMONE

3.1   Structure

The first hypothalamic releasing hormone to be isolated, characterized, and
have its structure confirmed was TRH. It is a tripeptide containing the amino
acids histidine, proline, and glutamic acid in equimolar amounts (Fig. 3). It
induces the release of TSH and PRL from the anterior pituitary.

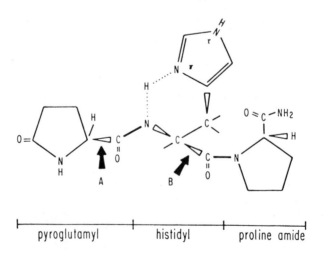

Figure 3   Primary structure of ovine and porcine TRH. The diagramed model contains
a proposed hydrogen-bonded interaction of the TRH imidazole group. Heavy lines
indicate the planar peptide bonds. *A* and *B* are the bonds that rotate without restriction.
From Vale *et al.*, *Frontiers in Neuroendocrinology*, 1973.

3.2 Biosynthetic and Metabolic Sequence

3.2.1 Sites of Synthesis

In most mammals and man, 25–30% of the TRH found in brain is located in the hypothalamus (Table 3). It is also present in the hypothalamic and extrahypothalamic brain structures including the spinal cord of a variety of species (Table 4) including, in addition, the lamprey, amphioxus, and snail, which have no

Table 3 Thyrotropin-Releasing Hormone in the Hypothalamus, Preoptic Area, Septum, and Circumventricular Organs of the Rat[a]

Brain Region	Concentration (ng/mg protein)
Hypothalamus	
Periventricular nucleus	4.3
Suprachiasmatic nucleus	1.8
Supraoptic nucleus	0.9
Anterior hypothalamic nucleus	0.8
Medial forebrain bundle (anterior)	0.7
Paraventricular nucleus	2.6
Arcuate nucleus	3.9
Ventromedial nucleus (medial)	9.2
Ventromedial nucleus (lateral)	3.0
Dorsomedial nucleus	4.0
Perifornical nucleus	2.0
Medial forebrain bundle (posterior)	1.2
Posterior hypothalamic nucleus	1.8
Dorsal premammillary nucleus	1.5
Ventral premammillary nucleus	1.3
Median eminence	38.4
Mammillary body	0.3
Preoptic area	
Medial preoptic nucleus	2.0
Supraoptic crest	1.8
Septum	
Medial nucleus	0.4
Dorsal nucleus	1.9
Dorsal nucleus (intermediate)	0.5
Fimbrial nucleus	0.6
Triangular nucleus	0.5
Lateral nucleus	30.0
Circumventricular organs	
Subfornical organ	0.7
Subcommissural organ	1.1
Area postrema	0.9

[a] From Brownstein (3).

Table 4 Distribution of TRH in Brain Areas of Several Vertebrates[a]

Species	Brain Stem	Cerebellum	Olfactory Lobe	Cerebral Cortex (Forebrain)	Hypothalamus	Pituitary Complex
Rat (Rattus norvegicus)	5[b] (4–5)[c]	2 (1–3)	6 (5–8)	2 (1–3)	280[e] (240–300)	Neurohypophysis[d] 155 (150–160)
Chicken (Gallus domesticus)	9[b] (8–10)	1 (1–2)	17 (16–17)	2 (1–3)	41 (34–49)	Neurohypophysis 168 (95–240)
Snake (Thamnophis sirtalis)	283[b] (129–359)	135 (85–186)	757 (750–764)	338 (264–381)	564 (393–731)	865 (410–1340)
Frog (Rana pipiens)	55[b] (40–85)	520 (368–724)	326 (220–444)	111 (71–150)	2.270 (1520–3620)	>5000 (>5000)
Tadpole (Rana pipiens)	303[b] (290–310)	Not examined	209 (59–461)	447[f]	947 (568–1225)	764 (508–995)
Salmon (Salmo sebago)	13[f]	3 (2–3)	165 (154–181)	37 (22–52)	235 (188–264)	150 (31–366)

[a] From Jackson and Reichlin, *Endocrinology* **95**, 854 (1974).
[b] In pg/mg tissue wet weight.
[c] The mean TRH concentrations of other mammalian hypothalami studied were 480 (hamster) and 500 (pig); human stalk median eminence (SME) contained up to 300 pg/mg tissue wet weight.
[d] Values for rat and chicken pituitary complex for neurohypophysis alone.
[e] Range of values.
[f] Only one fragment extracted.

pituitary or TSH. TRH has been visualized in axons and terminals in the spinal cord, brain stem, mensencephalon, and hypothalamus, but has been difficult to demonstrate in specific populations of neural cell bodies. In mammals, TRH, like GnRH, appears to be sequestered in two sets of particles that differ in size but not in electron density. The subcellular distribution of TRH within the hypothalamus is also similar to that of GnRH in the adult, but unlike GnRH, particles, small TRH particles exist during prenatal life and a mature profile of subcellular compartmentalization is achieved between the second and third weeks of life (6, 7).

3.2.2 Possible Precursor Molecules

TRH, like GnRH, could be synthesized *de novo* from amino acids by a nonribosomal (soluble) synthetase enzyme such as that for glutathione (18). Synthesis by this route necessitates supplementary reactions such as aminolysis of the C-terminal amide and formation of N-terminal pyroGlu, although pyroGlu formation could occur nonenzymatically by cyclization of $Glu-NH_2$. Inhibitors of cytoplasmic and mitoribosomal synthesis do not inhibit [^{3}H]proline incorporation into TRH. In cell-free systems from whole-brain homogenates however, [^{3}H]proline incorporation is inhibited by ribonuclease A. Inactivated ribonuclease A has no effect on TRH synthesis. From these results it has been suggested that enzymatic synthesis of TRH may involve aminoacyl-RNA precursors. Inhibition of inactivating peptidases with nonmetabolizable tripeptide substrates of antibiotic inhibitors such as bacitracin indicates that postmitochondrial supernatant solutions can synthesize TRH. Synthesis rates are low, on the order of 0.01—1.0 pmole/min per brain (8). To date, there is no good evidence for the formation of TRH by prohormone breakdown.

3.2.3 Metabolism

TRH, like GnRH, has an extremely short plasma half-life (~4 min), large volume of distribution, and rapid elimination (25% in urine after 60 min). The mechanisms for TRH inactivation by soluble and particulate extracts of brain or neurosecretory regions and serum remain unclear. Studies utilizing the release of amino acids as an end point of inactivation have shown that cleavage of the pyroGlu-His and $His-Pro-NH_2$ bonds occur during incubation of TRH with rat brain extracts (Fig. 4). Among enzymes potentially capable of inactivating TRH is a pyroglutamylpeptidase. This enzyme degrades TRH with the release of $His-Pro-NH_2$, indicating cleavage at the pyroGlu-His bond. In rodent hypothalamus, the enzyme is soluble and separable into two fractions, one of which displays the pyroglutamylpeptidase activity and the other a separate amidase, releasing the deamide forms of TRH and His-Pro. Enzymes inactivating TRH are inhibited by high concentrations of other peptidylamides such as GnRH, suggesting that the enzyme(s) involved in TRH degradation are highly specific. Another potential mechanism for inactivation of TRH is deamidation. The deamido form of TRH, H-Glu-His-Pro-OH, and $His-Pro-NH_2$, has been found

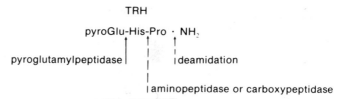

Figure 4   Sites of established (solid arrow) and postulated (dashed arrows) points of cleavage for TRH by brain enzymes. Adapted from Marks, *in Peptides in Neurobiology*, Gainer (ed.), Plenum, New York, p. 221–258 (1977).

in blood following a systemic injection of labeled TRH, as have pyroGlu-His and proline, as the major breakdown products in rat serum (8). An enzyme in the hypothalamus that is absent from the pituitary of rodents also appears to be capable of deamidating TRH. Human, like rat, serum differs from rat brain in that one of the end products is proline. As with GnRH, the inactivation of TRH in rat serum exceeds that in man; also, rates of degradation are lower for TRH than for GnRH(8). In spite of the rapid clearance of TRH from the circulation by the kidneys, the plasma enzymes inactivating it may have physiological importance, since in experimental animals their activity appears to be influenced by the level of thyroid activity and the age of the animal.

3.3   Factors Affecting Secretion

3.3.1   Neurotransmitters

The work of several groups indicates the existence of a facilatory NE mechanism(s) in the control of TRH release. Morphological analysis reveals a dense innervation of NE nerve terminals into the hypothalamic areas believed to be rich in TRH perikarya. Thyroidectomy produces a selective increase in NE turnover in the parvicellular portion of the paraventricular nucleus, where the cell bodies of TRH-containing neurons that innervate the external layer of the median eminence and discharge TRH into the portal circulation are believed to be located. Since no effect on DA turnover was noted in the medial palisade zone of the median eminence, an area also rich in TRH nerve terminals, it is difficult to determine what, if any, role dopamine has in the release of TRH. It also appears that the facilitory effects of DA on TRH release from synaptosomes *in vitro* is mediated only after dopamine is converted to NE. Although NE appears to play a facilitory role in the phasic release of TRH *in vivo*, such as during acute cold stress, tonic TRH release is apparently unaffected by central NE tone.

Serotonin consistently inhibits the release of TRH from fragments of median eminence and hypothalamus *in vitro*. An inhibitory role for serotonin on TRH release *in vivo* is based on animal and human studies in which 5-hydroxytryptophan, the amino acid precursor of serotonin, decreased plasma TSH in hypothyroid but not euthyroid subjects and produced a marked inhibition of a cold-stress-induced rise in serum TSH. More recent data, however, suggests that inhibition of indolamine synthesis prevents a cold or propylthiouracil-induced rise in TSH, and intraventricular infusion of serotonin enhances TSH release (17).

In summary, to date it would appear the central NE tone plays a stimulatory or at least a facilitory role in the phasic control of TRH release, with a role for DA yet unclear. Although no conclusion can be drawn for serotonin, the weight of evidence suggests that serotonin inhibits TRH release.

3.3.2 Pituitary Thyrotropin

Pituitary TSH increases the uptake of thyroxine (T_4) by the median eminence and decreases its efflux. The presence of a large mobile pool of free T_4 within the median eminence can serve as a potential regulatory influence on TRH–TSH interactions, since any increase in T_4 at the level of the median eminence diminishes the rate of TRH release into the portal vasculature. TSH, by inhibiting the efflux of T_4 from the median eminence, reduces the amount of free T_4 reaching the anterior pituitary and increases T_4 levels within the median eminence. This indirectly exerts a negative short-loop feedback on TRH release (19). This mechanism is supported by the observation that TSH, like the gonadotropins, is found in portal vessels supplying the hypothalamus in concentrations greatly exceeding that in the peripheral circulation (16).

3.3.3 Peripheral Target-Organ Feedback

Neural Centers Although thyroid hormones appear to exert their major feedback effects on TSH secretion at the level of the pituitary, thyroid hormones also exert direct inhibitory effects on TSH secretion at a hypothalamic level in subhuman primates. It is also apparent from preceding paragraphs that thyroid hormones may regulate not only CNS transmitter effects on TRH release, but TRH itself.

Pituitary The paramount role of the pituitary in negative-feedback control of thyroid hormones on TSH release is evidenced by the dose-related blockade of TRH-stimulated TSH release by T_4 and triiodothyronine (T_3) and the reversal of that block by increased levels of TRH. TRH and thyroid hormones apparently do not interact directly, and the inhibitory effect of T_3 on TRH-induced TSH release requires normal protein and RNA synthesis. Thyroid hormones, by acting inside the thyrotrope, could induce the formation of mRNA that controls the synthesis of a protein regulator. Such a regulator could inhibit the sequence of events stimulated by the binding of TRH to its membrane receptor on the thyrotrope. In addition, T_4 is also known to inhibit the binding of TRH to its receptor (20).

4   PROLACTIN-INHIBITING FACTOR

4.1   Putative Candidates for PIF

Unlike the secretion of gonadotropins and TSH, PRL secretion is under tonic inhibitory control by the hypothalamus; this is overcome by acute stress or cyclic events. Since the original demonstration that lesions within the median eminence and transplantation of anterior pituitaries under the kidney capsule of rats produce a significant and continuous release of PRL, two theories on the chemical nature of PIF have been advanced. One is that hypothalamic dopamine either directly or indirectly inhibits PRL release. Initially, dopamine was believed to act primarily by releasing an additional substance (PIF), which then inhibited PRL release. This hypothesis received a serious challenge when it was demonstrated that (1) high levels of dopamine were present in portal blood and were altered by phasic physiological events, (2) a specific high-affinity membrane receptor for dopamine was present in the anterior pituitary, (3) dopamine inhibited PRL release *in vitro*, and (4) PRL secretion increased *in vivo* and *in vitro* following blockade of DA receptors by drugs or estradiol (a known stimulant of PRL release). Dopamine may exert its inhibitory action through an increase in cAMP, since anterior pituitary cAMP is elevated following dopamine administration, and cAMP directly inhibits PRL release.

The second theory proposes that there is an as-yet unidentified PIF distinct from dopamine. This hypothesis is supported by the continued suppression of anterior pituitary PRL release *in vitro* by hypothalamic fragments in the presence of dopamine receptor blockade and repeated demonstration of PIF-like activity in purified hypothalamic fractions that do not contain dopamine. Candidates for a nondopaminergic PIF are γ-aminobutyric acid (GABA) and prostaglandin E_1, but it is hard to reconcile the supraphysiological amounts of GABA necessary to inhibit PRL secretion with any biochemical or physiological event. Prostaglandin E_1 is postulated to work only at a neural level, but the evidence is indirect and equivocal.

4.2   CNS Localization of PIF Activity

PIF activity appears to reside chiefly in the lateral preoptic area and large areas of the ventral hypothalamus. Although precise localization of PIF(s) is not yet clear, the tuberoinfundibular dopaminergic neurons must be considered a leading site for PIF activity.

4.3   Factors Affecting PRL Secretion

4.3.1   Neurotransmitters

Hypothalamic NE as well as DA and serotonin have been demonstrated to influence PRL secretion. High doses of NE inhibit PRL release at the pituitary

level by means of the DA receptor. A facilitory role for NE is discussed below
(section 5.1 of this chapter). Serotonin's effect on basal circulating levels of PRL
appears minimal and of doubtful physiological significance. In contrast, serotonin's role in the neurogenically mediated suckling response is well established. The suckling response resulting from mammary stimulation appears to
be activated by means of the bundle of Schutz, whereas the ascending serotonergic fibers within the median forebrain bundle sensitize the hypothalamus to
this peripheral input (21). Acetylcholine and histamine have also been implicated in PRL secretion, but no clear mechanisms have yet evolved.

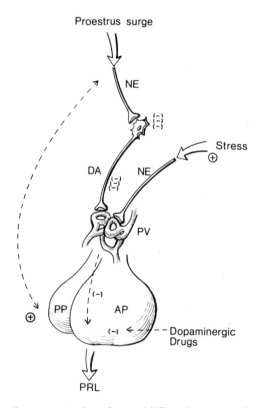

Figure 5 Schematic representation of central NE pathways exerting an episodic inhibitory influence on DA pathways controlling tonic PRL secretion by way of a PIF, or
directly at the pituitary gland. It is postulated that stress-induced PRL release or the
PRL surge at proestrus may be mediated by a functional activation of central NE
pathways impinging on DA-inhibitory pathways. Dopaminergic drugs acting directly at
the pituitary gland can block the increased PRL release. *DA*, dopamine neuron; *NE*,
norepinephrine neuron; *PV*, portal vessels; *AP*, anterior pituitary; *PP*, posterior pituitary, *PRL*, prolactin; (−), inhibition. Adapted from Muller *et al.*, (17).

4.3.2   Pituitary Prolactin (Short-Loop Feedback)

Prolactin exerts an inhibitory control over its own secretion. The mechanism of the inhibition appears to reside in the ability of PRL to inhibit its own release when implanted into the median eminence in the area of the tuberoinfundibular dopaminergic neurons. PRL, which is normally found in high concentrations in portal plasma, increases dopamine turnover following its administration to hypophysectomized and intact rats. The physiological stimulus of suckling also increases PRL and decreases dopamine release from the MBH.

4.3.3   Peripheral Target-Organ Feedback on Prolactin Secretion

Estradiol and testosterone appear intimately involved with PRL secretion. Estradiol is essential for the periovulatory surge of PRL in the rat. Estradiol and testosterone, the latter probably through conversion to estradiol, maintain PRL levels in ovariectomized animals. In addition, estrogen induces surges of PRL in ovariectomized animals. A clearer picture of the biochemical mechanism(s) involved in the steroid control of PRL release at a hypothalamic level is emerging, with evidence that estrogen, which is essential for any rise in circulating PRL, forms a portion of a complex steroid–neurotransmitter interaction. In this scheme (Fig. 5) estrogen would activate a NE pathway with ensuing inhibition of the activity of tuberoinfundibular dopaminergic neurons. Recent evidence of α-adrenergic blockade of estrogen-induced PRL release supports this hypothesis (17).

5   PROLACTIN-RELEASING FACTOR

5.1   Putative Candidates for PRF

PRL secretion in birds, unlike secretion in mammals, reptiles, and amphibians is not under tonic inhibitory control. Although no satisfactory explanation for this difference in the hypothalamic regulation of PRL is available, the scarcity of dopamine in the outer layer of the avian median eminence has been suggested to account for the lack of PIF activity in birds. Certainly, estrogen and other nonpeptidergic agents that stimulate PRL release at the pituitary level in mammals have no effect in birds (17).

In mammals, hypothalamic extracts initially inhibit then stimulate PRL release *in vitro*, suggesting the presence of a releasing factor. The leading candidates for PRF are TRH, NE, and serotonin. TRH releases PRL *in vitro* after binding to the membranes of rat pituitary tumor cells, and *in vivo* after binding to the normal pituitary in man, monkey, sheep, cattle, and birds. TRH is localized in neurons in areas similar to those that show PRF activity. It activates PRL release following binding to receptor, although the concentration necessary to increase release by 50% exceeds its binding affinity. Administration of an antiserum to TRH also appears to produce a loss in PRL secretion.

However, evidence that a separate PRF exists is found in the observations that (1) PRL and TSH are released independently under various physiological and pathophysiological states such as stress and hypothyroidism, (2) T_3 inhibits TRH-induced TSH release but not PRL, (3) partially purified hypothalamic extracts free of TRH have PRF activity, and (4) neural transmitters affect the secretion of TSH and PRL in opposite manners.

The facilitory role of NE in a neuroendocrine reflex responsible for stimulation of PRL secretion remains complex and inferential. Neuropharmacologic dissection of the increase in PRL secretion following stimulation of central NE receptors or blocking central DA neurons after NE neuron destruction in ovariectomized estrogen-treated rats suggests a NE inhibition of tuberoinfundibular dopaminergic neurons. In sum, a stimulatory NE pathway would increase PRL release by inhibiting the tonic inhibition exerted by tuberoinfundibular DA neurons (inhibiting the inhibitor). Evidence exists for a PRF localized in the anterior median eminence and medial suprachiasmatic nucleus, the latter an area rich in serotonergic nerve terminals. In contrast, PIF activity is found predominantly in the lateral preoptic area, lateral hypothalamus, and tuberoinfundibular dopaminergic neurons.

6 SUMMARY

It is apparent that most, if not all, releasing hormones (factors) synthesized, metabolized, and secreted from the hypophysiotropic area of the central nervous system are under physiological control. GnRH and TRH are essential to the release of the gonadotropins and TSH, and are certainly under the control of the neurotransmitters and peripheral target hormones that modulate their release. Gonadal steroids (particularly estrogen) and the thyroid hormones exert a negative-feedback control at the level of the hypothalamus and the anterior pituitary. GnRH and TRH are also apparently subject to regulatory control by the gonadotropins and TSH, although the physiological relevance for this negative feedback has not been established.

PRL, unlike the gonadotropins and TSH, is under tonic inhibitory control at the pituitary level. Inhibition is probably mediated by a tuberoinfundibular secretion identified as PIF. Release of PRL may come about through a decrease in PIF or the activation of an as-yet-unidentified PRL releasing factor (TRH?).

In contrast to the secretion of GnRH and TRH, PRL may rely on a dual mechanism for its control. The inhibition or disinhibition of tuberoinfundibular neurons may be achieved by classical neurotransmitters in response to neurogenic stimuli and by a long-lasting estrogen-dependent mechanism responsible for determining the baseline activity of the system (17).

Although theories concerning the activation and secretion of releasing hormones (factors) abound, conclusive evidence of a direct biochemical and anatomical link to the secretion of any releasing hormone (factor) has remained elusive and will remain an active area of investigation.

REFERENCES

1   Schally, A. V., Coy, D. H., and Arimura, A., Hypothalamic Peptide Hormones and Their Analogues, in Cox, B., Morris, I. D., and Weston, A. H. (eds.), *Pharmacology of the Hypothalamus*, p. 161, University Park Press, Baltimore (1978).

2   Coy, D. H., Sperodi, J., and Vilchez-Martinez, J. A., Structure-Function Studies and Prediction of Conformational Requirements for LH-RH, in Collo, R., Ducharme, J. R., Barbeau, A., and Rochefort, J. G. (eds.), *Central Nervous System Effects of Hypothalamic Hormones and Other Peptides*, p. 317, Raven, New York (1979).

3   Brownstein, M. J., Biologically Active Peptides in the Mammalian Central Nervous System, in Gainer, H. (ed.), *Peptides in Neurobiology*, p. 145, Plenum, New York (1977).

4   Wilber, J. F., Montoya, E., Plotnikoff, N. P., *et al.*, Gonadotropin-releasing hormone and thyrotropin releasing hormone: Distribution and effects in the central nervous system. *Rec. Prog. Horm. Res.* **32**, 117 (1976).

5   Neill, J. D., Daily, R. A., Tsou, R. C., *et al.*, Secretion of Luteinizing Hormone Releasing Hormone in Monkeys, in Porter, J. C., (ed.), *Hypothalamic Peptide Hormones and Pituitary Regulation*, p. 203, Plenum, New York (1977).

6   Barnea, A., Ben-Jonathan, N., Colston, C., *et al.*, Differential subcellular compartmentalization of thyrotropin releasing hormone (TRH) and gonadotropin releasing hormone (LRH) in hypothalamic tissue. *Proc. Nat. Acad. Sci. USA* **72**, 3153 (1975).

7   Barnea, A., Oliver, C., and Porter, J. C., Subcellular Compartmentalization of Hypothalamic Peptides: Characteristics and Ontogeny, *in* Porter, J. C. (ed.), *Hypothalamic Peptide Hormones and Pituitary Regulation,* p. 45, Plenum, New York (1977).

8   Mckelvy, J., Biosynthesis of Hypothalamic Peptides, in Porter, J. C. (ed.), *Hypothalamic Peptide Hormones and Pituitary Regulation*, p. 77, Plenum, New York (1977).

9   Miller, R., Achnelt, C., Rossier, G., *et al.*, Evidence for the existence of higher molecular weight precursor of LH-RF. *IRSC—Med. Sci.* **3**, 603 (1975).

10 Marks, N., Conversion and Inactivation of Neuropeptides, in Gainer, H. (ed.), *Peptides in Neurobiology,* p. 221, Plenum, New York (1977).

11 Fuxe, K., Lofstrom, A., Agnati, L., *et al.*, Functional Morphology of the Median Eminence. On the Involvement of Catecholamines in the Control of FSH, LH, and Prolactin Secretion, in Hubinot, P.O., Hermite, M.L., and Robyn, C. (eds.), *Progress in Reproductive Biology*, Vol. 2, Clinical Reproductive Neuroendocrinology. p. 41, Karger, Basel (1977).

12 Fuxe, K., Hokfelt, T., Jonsson, G., *et al.*, Neurosecretion—The Final Neuroendocrine Pathway, *in* Knowles, S., and Vollrath, L. (eds.), *Sixth International Symposium on Neurosecretion,* p. 269, Springer-Verlag, New York (1973).

13 Fluckiger, E., Ergot Alkaloids and Modulation of Hypothalamic Function, in Cox, B., Morris, I. D., and Weston, A. H. (eds.), *Pharmacology of the Hypothalamus*, p. 137, University Park Press, Baltimore (1978).

14 Fuxe, K., Hokfelt, T., Andersson, K., *et al.*, The Transmitters of the Hypothalamus, in Cox, B., Morris, I. D., and Weston, A. H. (eds.), *Pharmacology of the Hypothalamus*, p. 31, University Park Press, Baltimore (1978).

15 Bergland, R. M., and Page, R. B., Pituitary-brain vascular relations: A new paradigm. *Science* **204**, 18 (1979).

16 Oliver, C., Mical, R. S., and Porter, J. C., Hypothalamic-pituitary vasculature: Evidence for retrograde blood flow. *Endocrinology* **101**, 598 (1977).

17 Muller, E. E., Nistico, G., and Scapagnini, U. Brain Neurotransmitters and the Regulation of Anterior Pituitary Function, in *Neurotransmitters and Anterior Pituitary Function*. Academic, New York (1977).

18 Jackson, I. M. D., and Reichlin, S., Distribution and Biosynthesis of TRH in the Nervous System, in Collu, R., Barbeau, A., Ducharme, J. R., and Rochefort, J.-G., (eds.), *Central Nervous System Effects of Hypothalamic Hormones and Other Peptides,* p. 3, Raven, New York (1979).

19 Knigge, K. M., Joseph, S. A., Silverman, A. J., *et al.,* in Zimmerman, E., *et al.* (eds.), *Drug Effects on Neuroendocrine Regulation,* p. 7, Elsevier, Amsterdam (1973).

20 Schally, A. V., Arimura, A., Bowers, C. Y., *et al.*, Hypothalamic neurohormones regulating anterior pituitary function. *Rec. Prog. Horm. Res.* **24**, 497 (1968).

21 Enjalbert, A., Ruberg, M., and Kordan, C., Neuroendocrine Control of Prolactin Secretion, in Robyn, G., and Harter, M. (eds.), *Progress in Prolactin Physiology and Pathology,* p. 83, Elsevier, Amsterdam (1977).

CHAPTER TWELVE

STRUCTURE, CHARACTERISTICS, AND BIOSYNTHESIS OF PROLACTIN, LUTEINIZING HORMONE, AND FOLLICLE-STIMULATING HORMONE

GARY L. JACKSON

TSUEI-CHU LIU

Department of Veterinary Biosciences
University of Illinois
College of Veterinary Medicine
Urbana, Illinois

The authors' research is supported by NIH Grant HD 09659.

1   INTRODUCTION

The three anterior pituitary (AP) hormones of primary importance in reproduction are prolactin (PRL), luteinizing hormone (LH), and follicle-stimulating hormone (FSH). Although the precise role of PRL varies among species, its general function is regulation of lactation and gonadal steroidogenesis. LH and FSH regulate production of gametes and hormones by the gonads.

The basic sequence by which these AP hormones are synthesized is similar to that of most other secretory proteins. The polypeptide chain forms on the membrane-bound polysomes, quickly passes into the cisternae of the rough endoplasmic reticulum (RER), and then on to the smooth endoplasmic reticulum and the Golgi apparatus, where it matures and forms secretory granules. During this passage, the protein chains undergo many changes through enzymatic cleavage of specific peptide fragments, folding, or attachment of carbohydrate moieties (1). The product of these processes is the mature or *native* hormone—the predominant form of the hormone released from the cell.

Native hormones are formed from larger precursor chains as they pass from the polysome to the storage granule. The immediate product of translation is the *prehormone*. This transitory form contains a leader segment of 15–30 amino acids at the N-terminus of the nascent (forming) polypeptide chain. The leader segment apparently is required for recognition of the endoplasmic reticulum and for vectorial transfer of the nascent chain from the polysome into the lumen of the RER. *In vivo*, the leader segment is removed either before or soon after the entire polypeptide chain is completed. However, in cell-free heterologous systems that lack the cleaving enzymes, the completed prehormone can be isolated intact from the system. *In vivo*, the prehormone is cleaved rapidly either to the native hormone or to a relatively stable intermediate form called a *prohormone*.

The prohormone is, in most cases, the immediate precursor to the native hormone. Conversion of prohormone to native hormone is accompanied by cleavage of peptide segments from either the N- or C-terminals or the interior of the prohormone molecule. This conversion occurs within the Golgi complex.

These enzymatic conversions are coordinated with folding of the polypeptide chains and, in some cases, combination of subunits. The specific changes and sequence of changes varies among hormones. Salient features of the structure and biosynthesis of PRL, LH, and FSH are the topics of this chapter.

2 PROLACTIN

2.1 Structure of Prolactin

PRL is a single-chain protein synthesized by specific acidophilic cells called lactotrophs. Ovine PRL contains 198 amino acids, has a molecular weight of approximately 22,500, and an isoelectric point of pH 5.73 (Fig. 1). There are three disulfide bridges, one located centrally and one near each terminus. PRL from different mammals differs slightly in composition and in relative potency in bioassays (2).

There is evidence of a size heterogeneity for human PRL. A "big" PRL with significantly higher mobility on Sephadex G-100 columns than highly purified native PRL ("little" PRL) has been reported in plasma and extracts of pituitary. Big PRL constituted from 8 to 31% of the total immunoreactive PRL in the plasma; however, the relative amounts were not changed following exposure to PRL-releasing and PRL-inhibiting factors. Although the nature of big PRL remains to be established, it does not appear to be a prohormone of PRL.

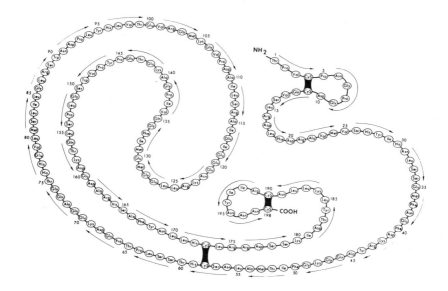

Figure 1 The amino acid sequence of ovine prolactin. From Li, *in* Knobil and Sawyer (eds.), *Handbook of Physiology*, Sect. 7, Vol. 4, Part 2, p. 103, American Physiological Society, Washington (1974).

2.2   Biosynthesis of Prolactin

2.2.1   Methodology

Two basic approaches have been followed in studying PRL biosynthesis. The first utilizes either incubations of tissues or cultures of cells from either normal APs or mammotropic tumors to quantitate PRL synthesis directly. The tissues are incubated or cultured with the desired regulatory factors and in some cases with radiolabeled amino acids. PRL synthesis is monitored either by measuring total immunoreactive PRL in the system or by measuring incorporation of radiolabeled precursor into PRL. In the latter case, the PRL is isolated from other radiolabeled proteins either by specific immunoprecipitation or by gel electrophoresis. The amount of labeled PRL then is quantitated by measuring the radioactivity of the isolated product.

The second approach, the use of heterologous cell-free systems, has been used to study both the basic sequence of PRL synthesis and the mechanisms by which various regulatory agents exert their effects on PRL synthesis. In this procedure, mRNA is extracted from pituitary fragments or cells. The mRNA is then translated in a cell-free translation system, either wheat germ or reticulocyte lysate, containing radiolabeled amino acid. The radiolabeled translation product is isolated by immunoprecipitation and gel electrophoresis and then quantitated by determining the radioactivity of the product. It is assumed that the amount of the product formed, for example, prePRL is proportional to the amount of mRNA coding for that product.

2.2.2   Sequence of Prolactin Biosynthesis

There is no evidence for a proPRL. The immediate precursor of PRL, prePRL (Fig. 2), contains 30 more amino acids at the N-terminus than native PRL. Enzymatic cleavage of this leader peptide from the nascent chain results in formation of PRL rather than an intermediate form (3). After release of the nascent chain, the PRL molecule is transported to the Golgi apparatus for packing into secretory granules.

2.3   Factors Controlling Biosynthesis and Release of Prolactin

2.3.1   General Considerations

Many factors affect PRL secretion. Among these are estrogen, photoperiod, ambient temperature, physical or psychological stress, and suckling. Estrogen (E) acts directly on the AP, possibly indirectly, by way of the hypothalamus. The other factors alter the release of hypothalamic modulators of PRL secretion.

In mammals, two of the important hypothalamic modulators are dopamine and thyrotropin-releasing hormone (TRH). The physiological role of these two substances is unclear, but much evidence suggests that dopamine is a major PRL-inhibiting factor, whereas TRH is a PRL-stimulating factor. Other specific

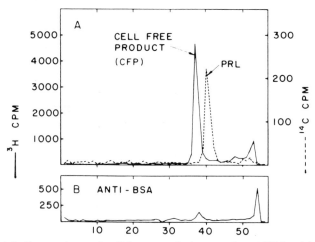

Figure 2 (*A*) Comparison of cell-free translation product (CFP) with [^{14}C]-labeled marker PRL after sodium dodecylsulfate polyacrylamide-gel electrophoresis. RNA extracted from rat pituitaries was used to direct protein synthesis in a wheat germ cell-free system. An aliquot of the reaction mixture was immunoprecipitated with anti-PRL. The immunoprecipitate was combined with [^{14}C]-PRL and subjected to electrophoresis. The gel origin is at the left of the figure. (*B*) An aliquot of the reaction mixture immunoprecipitated with anti–bovine serum albumin and then subjected to electrophoresis. This experiment demonstrates that a molecule larger than native PRL is the product of translation. From Maurer *et al.*, *J. Biol. Chem.* **251**, 2801 (1976).

hypothalamic PRL-inhibiting and PRL-stimulating factors have been proposed but not clearly demonstrated (4).

2.3.2 Actions of TRH, Dopamine, and Estrogen

TRH stimulates both synthesis and release of PRL. It is not clear whether TRH affects these two components of PRL secretion through single or multiple hormone–receptor interactions. Studies of the binding of radiolabeled TRH to AP membrane preparations indicate only a single class of receptors for TRH (5). In contrast, studies of the relationships between different doses of TRH analogs and PRL synthesis compared with PRL release suggest two TRH–receptor interactions (6). The doses of TRH analogs required for half-maximal stimulation of PRL synthesis (K_mS) by rat pituitary cells differed significantly from the doses required for half-maximal stimulation of PRL release (K_mR). With some analogs, the K_mS was much smaller than the K_mR, whereas with others the relative values were reversed ($K_mR < K_mS$). These results suggest that TRH regulates synthesis and release of PRL by means of different hormone–receptor interactions.

The biochemical sequence by which TRH, E, and dopamine alter PRL secretion is only partially known. There is indirect evidence that cyclic adenosine-3′, 5′-monophosphate (cAMP) is a mediator of the stimulatory effect of TRH on

PRL secretion. Addition of TRH to AP cell cultures led to an increase in cAMP accumulation that paralleled changes in PRL secretion, and addition of cAMP analogs to cultures of AP cells stimulated PRL synthesis. In addition, TRH appeared to increase cAMP-dependent protein kinase in cells of mammotrophic AP tumors (7). The time course of the kinase activation paralleled that of cAMP formation and PRL release.

However, other observations suggest that the correlations between cAMP and PRL secretion may be incidental rather than causative. For example, the doses of TRH required for half-maximal stimulation of PRL release and synthesis were much lower than those required for half-maximal elevation of cAMP concentrations. Furthermore, the doses of dibutyryl-cAMP required for half-maximal stimulation of PRL synthesis and release were much higher than that required for half-maximal stimulation of cAMP-dependent kinase activity (7). Overall, these observations strongly suggest, but do not establish, an obligatory role for cAMP in TRH-induced PRL secretion.

TRH, E, and pimozide, a drug that blocks dopamine receptors, increase the activity of mRNA coding for prePRL. Addition of TRH to clonal strains of rat AP tumor cells increased four-fold the activity of mRNA coding for prePRL. A small increase was observed at 12 hr after addition of TRH with maximal (four-fold) increase occuring at 48 hr. Significantly, mRNA for growth hormone was decreased (8).

Injection of either estradiol-17β or pimozide *in vivo* greatly increases incorporation of labeled amino acids into PRL by APs subsequently incubated *in vitro*. The effects of both drugs apparently are mediated by increased levels of mRNA activity for prePRL. Treatment of male rats with E for four days increased mRNA activity approximately three times, with the maximal increase observed seven days after onset of therapy. Estrogen treatment of retired female breeder rats resulted in a five-fold increase in mRNA for prePRL (9). Addition of estradiol to sheep AP cells *in vitro* also increased mRNA activity for prePRL within 24 hr (Fig. 3). Thus E acts directly on the AP, although an additional indirect effect has not been ruled out. It is not known whether E increases transcription or reduces RNA turnover. In male rats treated with E for seven days, prePRL mRNA activity had declined by only 50% at two days after the last injection of E. This suggests that the turnover of the mRNA for prePRL is relatively slow (9). A single injection of pimozide into male rats caused a two-fold increase in prePRL mRNA activity within 24 hr. Thus dopamine may inhibit prePRL mRNA activity. Whether the action of pimozide or dopamine is direct or secondary to an effect on PRL release is unknown (9).

Figure 3   Effect of estrogen on three parameters of PRL secretion by ovine pituitary cells maintained *in vitro*. Cells were cultured for six days before the addition of estradiol ($5 \times 10^{-9}\ M$). Note that increased mRNA activity paralleled increased prolactin secretion. From Vician *et al.*, *Endocrinology* **104**, 736 (1979).

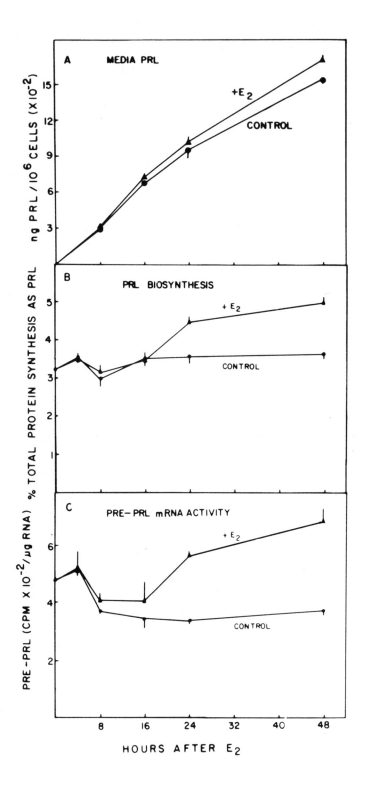

315

Estrogen increases the number of receptors for TRH on the AP cell. This is one probable mechanism by which estrogen facilitates TRH-induced PRL release. In contrast, E inhibits dopamine-induced inhibition of PRL release. This effect also may be exerted by means of changes in dopamine receptors (5).

3 LUTEINIZING HORMONE

Discussion of LH biosynthesis within a physiological perspective requires review of changes that occur in LH secretion during the estrous cycle. The equilibrium among synthesis, storage, release, and degradation of LH shifts drastically during the estrous cycle. In the rat, levels of LH in the AP (Fig. 4) increase gradually during the interval from estrus to proestrus, whereas levels in the serum remain relatively constant. Thus under these conditions, relatively more LH is synthesized and stored than is released. At late proestrus, the equilibrium shifts toward release, and there is a large increase—"surge"—of LH in the blood coincident with a fall in the AP. A similar series of changes in AP LH concentrations has been documented in the sheep and presumably occur in many other species.

Examination of the data shown in Fig. 4 raises the following questions: (1) What factors control LH biosynthesis and release? (2) How do these factors exert their effects? (3) How do these factors regulate the equilibrium between synthesis and release? Stated another way, How is the equilibrium shifted from synthesis and storage of LH during metestrus and diestrus toward synthesis and release during late proestrus? Answers to these questions require knowledge of LH biosynthesis, release, their interrelations, and their control—topics of the following discussion.

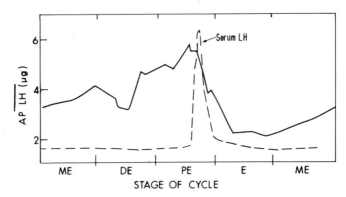

Figure 4 Schematic representation of changes in concentrations of LH in the anterior pituitary gland and serum of the rat during the normal estrous cycle.

3.1   Structure of LH

Detailed discussions of the structure of LH are presented in several reviews (10–12), thus only major points are presented here. LH, like FSH, TSH, and hCG, is a glycoprotein, that is, the core of the molecule consists of two non-identical polypeptide chains to which are attached carbohydrate side chains. LH has a molecular weight of approximately 28,000–30,000. The two noniden-tical subunits, α and β, are held together by noncovalent forces. The subunits can be dissociated by several chemical procedures, for example, exposure to mild acid solutions or to high concentrations of urea or guanidine. Although the individual subunits have immunological activity, they have little or no biological activity. Each subunit is highly cross-linked internally by disulphide bonds, of which there are five in the α-subunit and six in the β-subunit (Fig. 5). The disulphide bonds confer those structural characteristics to each subunit required for combination with the other subunit. The amino acid composition and sequences and the carbohydrate composition have been determined for the

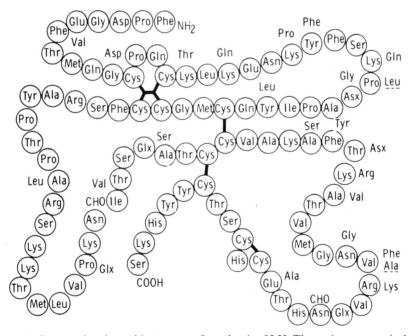

Figure 5   Proposed amino acid sequence of α-subunit of LH. The main sequence is that of ovine-bovine LH; the residues outside the circles show the substitutions reported in human and porcine α-chains. Those underlined are substitutions in porcine LH-α. The CHO indicates sites of attachment of carbohydrate moieties. From Pierce *et al.*, in *Polypeptide Hormones: Cellular and Molecular Aspects*, CIBA Foundation Symposium (New Series) Elsevier/Excerpta Medica/North-Holland, NY, p. 225 (1976).

Table 1   The Composition of Luteinizing Hormone[a, b]

Constituent	Ovine LH[c] Lamkin et al. (1970)[f]	Human LH[d] Ward et al. (1973)[g]	Murine LH[c] Ward et al. (1971)[h]	Equine LH Lande-feld and McShan (1974)[i]	Bovine LH[e] Liao and Pierce (1970)[j]
Amino Acid					
Lysine	12.7	8.9	10.5	9	13.3
Histidine	5.5	5.0	4.9	3	6.5
Arginine	10.1	12.9	11.4	9	11.9
Aspartic acid	11.1	11.9	12.7	9	12.2
Threonine	15.3	14.8	14.6	13	16.7
Serine	14.1	14.8	13.6	15	13.7
Glutamic acid	14.3	16.8	12.3	12	15.3
Proline	26.7	22.7	26.3	17	28.7
Glycine	11.9	12.9	11.3	10	11.6
Alanine	16.5	9.8	14.9	13	16.0
Half-cystine	17.8	21.6	17.1	13	23.1
Valine	13.4	20.6	13.6	9	12.7
Methionine	7.1	5.0	3.6	2	7.1
Isoleucine	6.0	5.8	7.1	8	6.3
Leucine	12.7	11.9	15.5	10	13.6
Tyrosine	6.8	5.8	6.8	3	7.4
Phenylalanine	5.8	5.8	6.4	7	8.0
Tryptophan		1.2			
Carbohydrate					
Galactose	1.1	3.3	1.4	6.0	0.2
Mannose	7.0	7.3	5.9	8.1	9.9
Fucose	1.5	1.8	1.5	1.2	1.2
Glucosamine	8.2	9.9	12.3	6.9	9.5
Galactosamine	3.3	1.7	2.8	4.2	3.2
Sialic acid	N.R.	2.0		7.8	N.R.

[a] Adapted from Liu and Ward, *Pharmac. Therap. B.* **1**, 545 (1975).

[b] Results expressed in number of residues per 30,000.

[c] Recalculated from the values given as residues per 31,000.

[d] Amino acid composition from Ward et al. (1973). Carbohydrate composition from Hartree, *in Hormone glycoproteiques hypophysaires*, Justisz (ed.), p. 71. Inserm (CNRS), Paris (1972).

[e] Recalculated from the values given as residues per 28,000.

[f] *Biochim. Biophys. Acta* **214**, 290.

[g] *Rec. Prog. Horm. Res.* **29**, 533.

[h] *Biochemistry* **10**, 1796.

[i] *Biochemistry* **13**, 1389.

[j] *J. Biol. Chem.* **245**, 3275.

LH of several species (Table 1). Approximately 70% of the residues in the α-subunits of different species are identical and an additional 21% have been replaced by only a single amino acid. There is considerable heterogeneity among species in the N-terminus of the α-subunits. In general, the α-subunits of the AP glycoprotein hormones (TSH, FSH, LH) of a single species are more similar than the α-subunits from different species. A comparison of the primary structure of LH-β from several species reveals great similarity. Indeed, ovine and bovine LH are identical, and they differ little from porcine LH.

Two carbohydrate moieties are found on the α-subunit, linked to asparagine (Fig. 5). Location of the one carbohydrate moiety on the β-subunit varies among species as does composition of the carbohydrate side chains in the whole molecule. In human LH, asparagine–CHO linkage occurs at position 30, whereas in the LH of other species it occurs at position 13. Both human and equine LH contain significant amounts of sialic acid, but LH of other species has little or none. In general, glucosamine and mannose are the predominant carbohydrates (Table 1) in LH of all species studied.

Specific structures of the carbohydrate chains are unknown, however, a tentative proposal for the structure of the asparagine-linked carbohydrate moieties of ovine and bovine LH was presented recently (Fig. 6). The function of the carbohydrate chains is unclear. They are not necessary for binding to the LH receptor or for immunological activity. They are required, however, for full biological activity. They also probably prolong the life of the molecule in the circulation.

3.2 Biosynthesis of LH

3.2.1 Methodology

Relatively little is known about LH biosynthesis. Only recently have approaches been developed to permit investigation of this topic. The basic approach has been to allow either rat pituitary quarters or cells to incorporate radiolabeled precursors into LH *in vitro*. Radiolabeled amino acids such as alanine facilitate the monitoring of synthesis of the protein moiety of the molecule. Radiolabeled glucosamine facilitates the monitoring of attachment of the carbohydrate moieties (glycosylation). After incubation, the LH in the APs and

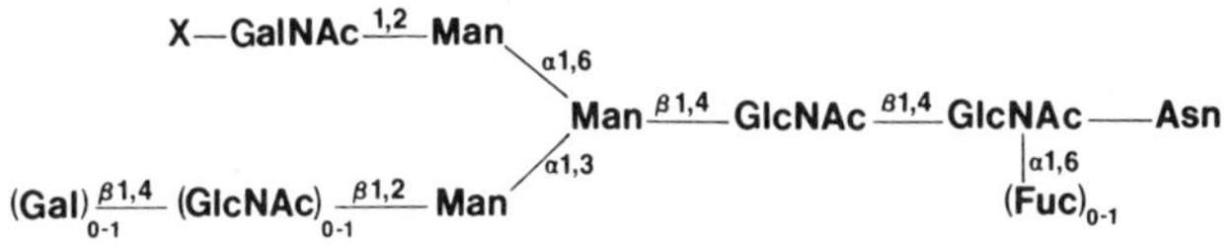

Figure 6 Proposed structure of the asparagine-linked carbohydrate units of ovine and bovine LH. *GlcNAc, N*-acetylglucosamine; *GalNAc, N*-acetylgalactosamine; *Man,* mannose; *Asn,* asparagine; *Fuc,* fucose. Adapted from: Bahl *et al., Biochim. Biophys. Res. Comm.* **96,** 1192 (1980).

media is isolated by immunoprecipitation with specific antibodies to the LH β-subunit. Quantitation is achieved by measuring the radioactivity of the precipitated LH.

3.2.2   Sequence of LH Biosynthesis

Whether FSH and LH are synthesized in the same or separate cells is debatable. Immunohistochemical studies strongly support the concept that some, but perhaps not all, gonadotrophs can secrete both hormones. However, clones of cells that secrete predominantly either FSH or LH have been produced.

The biosynthesis of LH and FSH probably follows a sequence similar to that of many other glycoproteins. However, the specific steps by which the protein precursor, or precursors, are transformed into the two-chained native glycoprotein are not known.

Two models have been proposed to describe the possible sequence by which the two-chained native LH molecule is formed: the two-chain model and the one-chain model (13). The two-chain model suggests that the α- and β-subunits are synthesized separately, in precursor form, and subsequently combined. It is not known whether combination of the chains occurs before or after conversion of the precursor to the native hormone chain.

Considerable evidence supports the two-chain model. Cell-free translation of bovine pituitary mRNA produced two separate proteins with molecular weights of 14,000 and 18,000, which were precipitated only with specific LH-α or LH-β antisera, respectively. These proteins probably represent precursor forms of the α- and β-subunits. Studies of TSH and hCG biosynthesis in cell-free systems have revealed separate mRNAs coding for proteins thought to be the pre-α-and pre-β-subunits. LH, TSH, and hCG are similar structurally, thus the sequence of biosynthesis of these molecules probably is similar.

The one-chain model was developed from observations that extracts of the AP contain both native and large-molecular-weight forms of LH and large amounts of free α-subunit. According to this model, synthesis involves formation of a large protein chain that, like proinsulin, is folded and then cleaved into the two-chain native LH and a connecting fragment, in this case free α-subunit. Direct support for this model is lacking, but several observations make it tenable.

Gel filtration of extracts of human pituitary glands produced three peaks with LH activity in radioimmunoassay: native LH, free α-subunit (the postulated connecting fragment), and a large molecule eluting near the void volume. The large molecule could not be dissociated by treatment with guanidine, a treatment that dissociates native LH into its subunits. It was postulated that this large molecule may be an LH precursor representing the two chains of native LH connected by an additional α-subunit.

In other studies, APs from rats were incubated *in vitro* with radiolabeled glucosamine and alanine. Following incubation, extracts of the APs and the medium were subjected to gel chromatography. The column eluant was monitored by standard radioimmunoassay, and radiolabeled immunoreactive LH in each fraction was precipitated with a specific anti-LH-β serum. Two peaks of radiolabeled LH were found in the eluant following chromatography of pituitary extracts (Fig. 7): peak "A" at the void volume and peak "B" at the same position as native LH. Only peak "B" substance, or native LH, was found in the medium, and no free β-subunit was found in either pituitary or medium. Peak "A" substance had a low carbohydrate content, very low biological activity and, like the large LH-like substance from the human AP, was highly resistant to dissociation. During pulse-chase studies, the synthesis of peak "A" substance preceded that of peak "B" substance. Although the nature of peak "A" substance is not established, it may be an LH precursor—either a proLH or unfinished LH β-subunit bound to a larger molecule. The relationship of this substance to the putative presubunits found in cell-free studies remains to be determined.

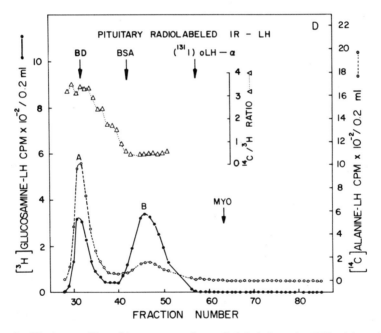

Figure 7 Elution profiles of immunoreactive radiolabeled murine LH subjected to gel filtration on Sephadex G-100. APs from spayed rats were incubated in the presence of [³H]glucosamine, [¹⁴C]alanine, and GnRH. Following the incubation, the medium and tissue were separated. The tissue was homogenized and centrifuged. The supernatant (cytosol) was applied to a column of Sephadex G-100 and eluted with phosphate buffer. Immunoreactive LH in each fraction was precipitated with specific anti-LH serum. Radioactivity in the precipitate was determined by liquid scintillation counting. Peak "B" LH eluted at the same position as native LH. Only Peak "B" LH was present in the medium. Abbreviations: *BD*, blue dextran, *BSA*, bovine serum albumin; *MYO*, myoglobin.

Attachment of the carbohydrate moieties (glycosylation) to the protein chains of LH probably is accomplished by specific glycosyltransferases in the endoplasmic reticulum, Golgi apparatus, or perhaps even the secretory granules. The observation that the peak "A" substance discussed previously contained little carbohydrate compared with native LH suggests that glycosylation is a relatively late step in LH biosynthesis. That little is known about details of glycosylation of pituitary glycoprotein hormones must be stressed. Both GnRH and estrogen stimulate glycosylation of LH, suggesting that it is an important regulatory step in biosynthesis of these hormones.

One very puzzling facet of LH biosynthesis is the relationship between free α-subunit and native LH. As noted earlier, free α-subunit is found in the pituitary and serum, particularly in postmenopausal and pregnant women (13). However, secretion of α-subunit seems to be regulated independently of native LH. Free α-subunit appears earlier than native LH during development of the human fetus. Infusion of GnRH into humans causes release of both LH and α-subunits, but the dose–response curves are nonparallel (Fig. 8). These observations are consistent with the idea of separate synthesis and release of α-subunit and native LH. However, the fact that TSH, LH, and FSH have identical α-subunits complicates interpretation of these observations. At present, there is no way to distinguish free LH-α from free α-subunits of the two other hormones.

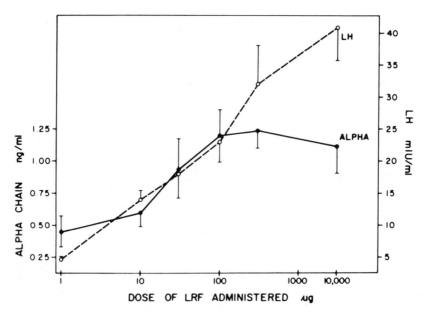

Figure 8 α-Chain and intact LH in the serum of the same individuals after administration of a bolus of GnRH. From Edmonds *et al., J. Clin. Endocrinol. Metab.* **41**, 551 (1975).

3.2.3 Relationship Between LH Synthesis and Release

LH synthesis and release appear to occur independently; that is, they are not tightly coupled. LH levels in the AP may rise in the absence of detectable changes in blood LH levels. This indicates that synthesis and storage may occur without parallel changes in release. Blockage of GnRH-induced LH synthesis with either puromycin or cycloheximide does not block GnRH-induced LH release, indicating that release can occur independently from synthesis.

Studies comparing the time course of release of newly synthesized radio-labeled LH and preexisting unlabeled LH from rat pituitaries *in vitro* showed a difference in their response to GnRH. GnRH induced release of stored LH within a few minutes. The rate of release of unlabeled LH (IRLH) increased linearly for approximately 2 hr, plateaued, then fell, although large amounts of stored, unlabeled LH remained in the tissue. In contrast, newly synthesized radiolabeled LH appeared in the medium after 1.0 hr, then the rate of release continued up to 3.5 hr and plateaued (Fig. 9). The fact that the rate of release of newly synthesized LH increased while that of preexisting stored LH fell suggests that these two pools of LH behaved independently in response to GnRH. These observations also suggest that newly synthesized LH forms a significant part of that LH released after long-term exposure to GnRH.

During either constant infusion or repeated pulsatile administration of GnRH in humans, serum LH levels often show a biphasic pattern of elevation. This pattern is characterized by an acute rise during the first minutes of infusion, followed by either a transient plateau or decrease in rate of elevation. A second sharp rise than occurs and continues for up to 3 or 4 hr. This biphasic rise in LH will occur in the presence of constant levels of circulating GnRH. The two phases of LH release also suggest the existence of two functional pools of LH in the human pituitary—one that is acutely releasable and another, or a "reserve" pool, that requires prolonged GnRH stimulation to be released. Acutely releasable LH probably represents previously synthesized LH stored in granules lying close to the cell surface. "Reserve" LH consists of both previously synthesized LH located in a storage pool and newly synthesized LH.

The cellular mechanisms governing the activity of the two-pool system of hormonal release are unknown. Undoubtedly they include a complex sequence of events involved in synthesis, transport, and release of the hormone. The complexity is clearly demonstrated by observations that gonadal steroids markedly affect the relative size of the two pools (14), and that the dose per se of GnRH affects the relative rate of release of newly synthesized and of previously synthesized stored LH (15).

3.2.4 Factors Controlling LH Biosynthesis

GnRH GnRH stimulates both synthesis and release of LH. Prolonged exposure of cultures of AP cells to GnRH increased the total amount of LH in the system. Exposure of either AP quarters or dispersed cells to GnRH for up to 4 hr increased incorporation of radiolabeled carbohydrate into LH (glycosyla-

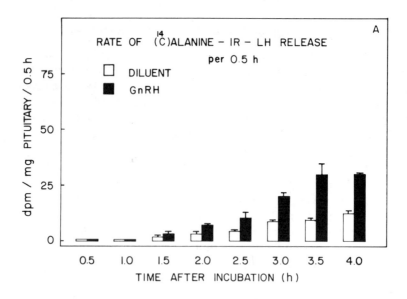

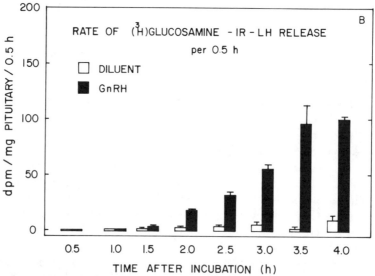

Figure 9 Difference in the time course of release of newly synthesized [¹⁴C]alanine-labeled LH (*A*) and [³H]glucosamine-labeled LH (*B*) versus total immunoreactive LH (IRLH) (*C*) in the presence or absence of GnRH during sequential 0.5 hr intervals of incubation. Quarters from either the right or left halves of 12 rat AP glands were incubated in pairs of vials in the presence of radiolabeled precursors with or without GnRH. The medium in each vial was collected and replaced by corresponding fresh medium every 0.5 hr for a period of 4 hr. Note: After 2 hr the release rate of IRLH had plateaued or started to decline while the release rate of newly synthesized radiolabeled LH was increasing. From Liu and Jackson, *Endocrinology* **103**, 1253 (1978).

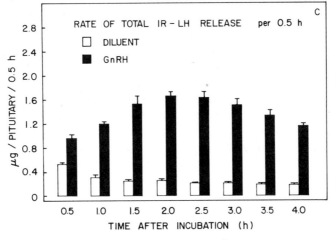

Figure 9 (*continued*)

tion) but did not affect incorporation of radiolabeled amino acids. In incorporation studies done to date, the cells were exposed to GnRH and isotopes for only 4 hr. Perhaps GnRH detectably affects synthesis of the protein moiety of LH only after several hours. Additional incorporation studies using long-term incubations or cultures may give a better understanding of this subject.

The ability of GnRH to stimulate rapidly incorporation of aminosugars into LH suggests that glycosylation is an important and readily regulatable step in LH biosynthesis. The stimulatory effect of GnRH on glycosylation appears to be specific for LH (and probably FSH)—it does not stimulate glycosylation of total proteins. How GnRH stimulates glycosylation of LH is not known. GnRH does not stimulate uptake of the radiolabeled precursors. Other possible mechanisms include stimulation of specific glycosyltransferases, of carbohydrate metabolism, and of formation of carbohydrate intermediates.

GnRH may affect synthesis and release of LH through different pathways. Evidence for this is as follows: (1) GnRH will induce release of LH even when synthesis has been blocked by cycloheximide, (2) the doses of GnRH or some GnRH agonists required for half-maximal stimulation of LH synthesis (K_mS) appear to be smaller than those required for half-maximal stimulation of LH release (K_mR), (3) the dose per se of GnRH regulates relative rates of release and of synthesis of LH. At lower doses, synthesis is preferentially stimulated, whereas at higher doses, release is preferentially stimulated (15). Explanations for this apparently dichotomous control system are largely speculative. GnRH may regulate release and synthesis through the same hormone–receptor interaction, but differences in intracellular factors such as enzymes required for synthesis and for release may respond differently to different degrees of receptor stimulation. Alternatively, there may be separate hormone-receptor interactions regulating synthesis and release.

The dichotomous effect of GnRH on synthesis and release of LH may explain partially changes that occur in LH concentrations in the AP during the estrous cycle of the rat. The fact that the K_mS is smaller than the K_mR suggests that *in vivo* relatively low concentrations of GnRH preferentially stimulate LH synthesis as opposed to LH release. This may explain why LH accumulates in the AP during diestrus. At higher concentrations of GnRH, such as may reach the AP during late proestrus, release is preferentially stimulated. This leads to the LH surge and temporary depletion of LH from the pituitary. These differential responses to GnRH are modulated by gonadal steroids—a subject discussed later in this chapter.

GnRH and Adenylyl Cyclase   The role of the adenylyl cyclase system as a mediator of GnRH action continues to be debated. There is much evidence to support the concept that cAMP mediates GnRH-stimulated LH release. Synthetic GnRH or GnRH analogs significantly increased cAMP concentrations within 4–6 min in pituitary cell cultures. Rates of LH release paralleled these changes in cAMP. Theophylline, an inhibitor of cyclic nucleotide phosphodiesterase, potentiated the effect of hypothalamic extracts on LH release. Exogenous dibutyryl-cAMP (1.5 mM) and theophylline (5–11 mM) each caused an increase in LH release after 4–5 hr of incubation with rat APs and caused a synergistic effect when added together. These and similar data suggest that cAMP plays an obligatory role in mediating the action of GnRH on LH release (5).

However, many investigators have been unable to demonstrate convincing associations between cAMP and LH release (5). The reasons for the disparate findings in different laboratories are unclear, but they may, in part, relate to subtle differences in experimental conditions and experimental animals. For example, the reproductive status and sex of the animal may significantly affect the responsiveness of the pituitary to cAMP analogues. Pituitary fragments or cells taken from either cycling or estrogen-treated ovariectomized rats appear to be more responsive to cAMP analogs than those taken from untreated ovariectomized rats.

Recent observations that cAMP analogs stimulate LH synthesis may provide a clearer understanding of the relationship between cAMP and LH secretion. After 1 hr of incubation with pituitaries from ovariectomized rats, 8-bromo-cAMP did not significantly affect LH release, whereas GnRH significantly stimulated LH release. After 4 hr of incubation, both cAMP and cAMP analogues significantly stimulated LH synthesis (glycosylation) but only slightly stimulated LH release. In contrast, GnRH stimulated both release and synthesis. Dose–response curves comparing the effects of GnRH and 8-bromo-cAMP on glycosylation after 4 hr of incubation were parallel (Fig. 10). In contrast, curves comparing the effects of these compounds on release of either total immunoreactive LH or newly synthesized [³H] glucosamine-labeled LH were nonparallel (Fig. 10). These results indicate that GnRH and cAMP have similar effects on LH glycosylation, but dissimilar effects on LH release. Thus cAMP may be the major mediator of GnRH-induced LH glycosylation, but factors in addition to cAMP may mediate GnRH-induced LH release. Calcium certainly has an important role in LH release, but its relationship to cAMP is not clear. Obviously, much remains to be learned about the mechanisms of action of GnRH.

Gonadal Steroids   The effects of gonadal steroids on LH release are well documented (16). However, relatively little is known of their specific effects on LH synthesis. Treatment of spayed rats with 0.05 μg/day of estradiol benzoate (EB) for 5 days elevated pituitary LH without suppressing plasma LH, suggesting that EB at low doses has differential effects on LH synthesis and release. Treatment of rats with E *in vivo* accelerated glycosylation and release of LH by pituitaries incubated with GnRH a few hours later. There is evidence that: (1) both E and GnRH stimulate incorporation of radioactive glucosamine into LH but have little effect on incorporation of radioactive amino acids into LH, (2) both E and GnRH stimulate release of LH labeled with either amino acid or glucosamine, and (3) E potentiates the ability of GnRH to stimulate glycosylation and LH release (17). Thus E, given *in vivo*, accelerates both release and glycosylation by APs incubated *in vitro* in the presence of GnRH.

Data available suggest that E facilitates GnRH-induced LH release more than it facilitates GnRH-induced glycosylation. This differential effect of E also may explain partially changes in LH secretion that occur during proestrus in the rat. This differential effect of E, coupled with that of high levels of GnRH, helps shift the equilibrium from synthesis and storage of LH toward synthesis and release. The result is an increase in serum LH levels concomitantly with a decrease in AP LH levels, even though synthesis of LH may be greatly stimulated at this time.

Both E and GnRH facilitate glycosylation, but the mechanism of action of the two substances differ. Estrogen markedly stimulates uptake of glucosamine, whereas GnRH has relatively little effect on uptake. Stimulation of specific glycosyltransferase is another mechanism by which E may accelerate glycosylation of LH. Estradiol significantly stimulates specific glycosyltransferase activities in the uterus, but its effect on these enzymes in the AP is unknown.

The time course of action of E on LH synthesis seems to differ from the time course of action on LH release. In time-course studies, rats were injected with EB at 5.5, 11, or 22 hr prior to sacrifice (17). Then the APs were incubated *in vitro* with labeled precursors. At all times studied, EB treatment significantly increased gylcosylation and release of labeled LH and release of total immunoreactive LH by pituitaries incubated with GnRH; however, the time course of the increased release differed from that of the increased synthesis. The amount of glucosamine-labeled LH in the total system (synthesis) plateaued at 5.5 hr after injection of EB, whereas the amount of labeled LH and IRLH released into the medium in response to GnRH increased linearly with time for up to 22 hr after EB injection. A continued E-induced increase in responsiveness, for up to 24 hr, of the release mechanism to GnRH also has been observed *in vitro* (5).

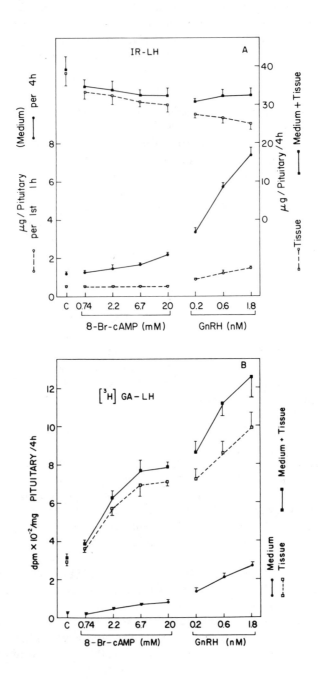

Some of the difficulties in deciphering the interactions of E and GnRH result from the biphasic effect of E on LH release and the variable responses observed in different experiments. *In vivo*, E first inhibits then facilitates GnRH-induced LH release. This facilitory effect may take 24–36 hr to plateau. However, under experimental regimens in which GnRH is infused coincidently with E, a facilitory response may not develop. On the other hand, when E was added directly to cultures of AP cells, only a facilitation of GnRH-induced release of LH was observed (5). Thus the apparent interaction of E and GnRH varies with experimental conditions. Obviously, we lack full understanding of how these two modulators of LH secretion operate under physiological conditions.

Progesterone, or its metabolites (dihydroprogesterone and 5α-pregnan-3α-ol-20-one) act on both the hypothalamus and anterior pituitary gland to modulate LH release (16). However, little is known about the role of progestins in regulating LH synthesis. The rise in LH levels in the pituitary during diestrus in the rat and during the luteal phase of the cycle in the ewe suggests that progesterone at physiological levels (either alone or in combination with estradiol) does not completely inhibit LH synthesis, whereas it effectively blocks LH release.

Whether androgens regulate LH synthesis is not known. The ability of testosterone to reduce LH levels in both the AP and serum of castrated rats suggests that it inhibits LH synthesis. However, treatment of orchiectomized rats with 125 to 500 μg of testosterone propionate for 3 days subsequently increased incorporation of [14C]leucine into LH and total protein when the APs were incubated *in vitro*.

3.3 Regulation of Gonadotropin Quality

GnRH and gonadal steroids may affect the quality as well as the quantity of LH and FSH released from the AP. The molecular size of FSH and LH in the rhesus monkey increases after ovariectomy or orchiectomy (Fig. 11). Estrogen treatment reverses this effect in the female. Ovariectomy increases the apparent size of LH in the rat AP, whereas orchiectomy decreases the size of FSH (18).

Most investigators have suggested that changes in gonadotropin size following gonadectomy or steroid therapy result from altered steroid levels per se.

Figure 10 Differential effects of 8-bromo-cAMP and GnRH on release and synthesis (glycosylation) of LH. (*A*) Changes in total immunoreactive LH (IRLH) in the medium and tissue. (*B*) Changes in [3H]glucosamine-labeled LH in the medium and tissue. Quartered rat pituitaries were incubated in the presence of [3H]glucosamine (GA) and different doses of either 8-bromo-cAMP or GnRH. IRLH was measured in the medium after 1 and 4 hr and in the tissue after 4 hr. [3H]GA-LH was measured in medium and tissue after 4 hr. The 8-Br-cAMP significantly stimulated glycosylation in a dose-related fashion but had little effect on release of either IRLH or [3H]GA-LH. In contrast, GnRH significantly stimulated all three parameters. Dose–response curves for release of IRLH and [3H]GA-LH in response to GnRH and 8-Br-cAMP were not parallel. Dose–response curves for glycosylation in response to GnRH and 8-Br-cAMP were parallel.

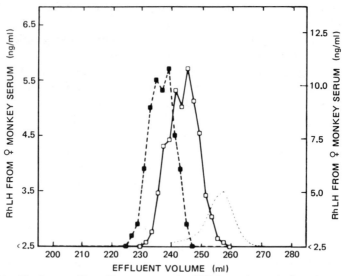

Figure 11 Elution profiles of immunoreactive LH from the exclusion chromatography of sera from an intact female monkey during the preovulatory surge (□—□) and an ovariectomized (■—■) monkey. The dashed line represents ovalbumin internal marker. From: Peckham and Krobil, *Endocrinology* **98**, 1054 (1976).

However, more-recent studies indicate that at least some of these effects may result directly from altered secretion of GnRH.

There is evidence that GnRH affects the relative carbohydrate content and biological activity of LH. In one study, rat APs were incubated with [³H]glucosamine and [¹⁴C]alanine with or without GnRH. Both the radiolabeled LH released into the medium and that remaining in the tissue were isolated by

Table 2 $^{14}C:^{3}H$ Ratio in the LH Found in Medium and Tissue
After Incubation of Rat Pituitary Glands *in Vitro* in the Presence or Absence of GnRH and in the Presence of [¹⁴C]Alanine and [³H]Glucosamine[a]

Duration of incu- bation (hr)	$^{14}C:^{3}H$ Ratio in Immunoprecipitated LH			
	Medium LH		Tissue LH	
	Diluent	GnRH	Diluent	GnRH
2.0	1.05 ± 0.10 (4)[b]	0.41 ± 0.03[c] (4)		
2.5	0.77 ± 0.08 (3)[d]	0.36 ± 0.05[c] (4)		
3.0	0.74 ± 0.09 (2)[d]	0.36 ± 0.03[c] (4)		
3.5	1.41 (1)[d]	0.31 ± 0.03[c] (4)		
4.0	1.04 ± 0.15 (4)	0.28 ± 0.02[c] (4)	4.0 ± 0.22 (3)	1.89 ± 0.10[c] (3)

[a] From Liu and Jackson, *Endocrinology* **103**, 1253 (1978).
[b] Mean ± S.E.M. Number of means given in parentheses.
[c] $P < 0.01$ (GnRH vs. diluent).
[d] Samples exhibiting very low ³H disintegrations per minute were excluded from the calculation.

Table 3   Comparison of Bioactivity: Immunoactivity (B:I) Ratios of LH Released *in Vitro* From Anterior Pituitaries of Rats in Various Hormonal States[a, b]

Source	Male	OVX	OVX + E
Medium LH (GnRH-stimulated)	3.9 ± 0.7[c]	2.1 ± 0.3	3.1 ± 0.2[d]
Medium LH (control)	2.2 ± 0.2	2.4 ± 0.3	2.1 ± 0.2
Tissue extract	2.0 ± 0.5	2.5 ± 0.5	1.7 ± 0.3

[a] Adapted from Mukhopathy *et al., Endocrinology* **104**, 925 (1979).

[b] Pituitaries from either intact male, ovariectomized (OVX), or estradiol-benzoate-treated (OVX + E) rats were incubated *in vitro* in the presence or absence of GnRH. Media and tissue extracts were assayed for bioactivity by their ability to stimulate testosterone production by dispersed mouse Leydig cells. Immunoreactivity was determined by radioimmunoassay.

[c] $P < 0.01$ vs. other male B:I ratios.

[d] $P < 0.01$ vs. other (OVX + E) B:I ratios.

immunoprecipitation. The relative radioactivity of ^{14}C and ^{3}H in the isolated LH was then determined by liquid scintillation counting. The ^{14}C:^{3}H ratio (an index of relative carbohydrate content) was significantly lower in LH synthesized in the presence of GnRH than in LH synthesized in the absence of GnRH (Table 2): GnRH appeared to increase the relative carbohydrate content of the newly synthesized LH.

Another study found that the biological to immunological activity (B:I) ratio of LH released into the medium was significantly higher in the presence than in the absence of GnRH (Table 3). The results of these studies are consistent, since there is strong evidence that the half-life and the biological activity of the glycoprotein hormones are related to the relative carbohydrate content. These results raise the intriguing possibility that hypothalamic-gonadal alteration of glycoprotein hormone "quality" is an important physiological regulatory mechanism.

4   FSH

4.1   Structure of FSH

Compared with LH, FSH has been difficult to isolate and purify. However, proposed amino acid sequences have been published for human, porcine, and equine LH (19). FSH, like LH, is a glycoprotein composed of nonidentical α-and β-subunits. The α-subunits of LH and FSH are nearly identical. The dissimilar β-subunits confer hormone-specific activities to the molecules. The estimated molecular weight of FSH ranges from 25,000 to 41,000, depending on species and investigator. Unlike other pituitary glycoproteins, FSH is acidic, containing 2–7%, by weight, of sialic acid (Table 4). Labile sialic acid residues are thought to be one factor that hinders purification of FSH. The sialic acid and other carbohydrate moieties are necessary for biological, though not im-

Table 4   The Composition of Follicle-Stimulating Hormone

Constituent	Ovine FSH Sherwood (1970)[c]	Equine FSH Nuti et al. (1972)[d]	Human FSH Reichert and Ward (1973)[e]	Bovine FSH Grimek et al. (1979)[f]
Amino Acid[a]				
Lysine	8.29	6.50	5.93	7.18
Histidine	2.91	2.83	2.55	2.80
Arginine	3.83	4.11	3.81	3.95
Aspartic acid	8.06	7.34	7.94	7.61
Threonine	9.94	9.17	8.30	9.93
Serine	6.06	5.83	7.27	6.12
Glutamic acid	8.63	8.00	9.54	9.13
Proline	5.77	7.06	7.34	5.79
Glycine	4.51	5.22	5.49	4.78
Alanine	6.57	5.56	5.64	6.43
Half-cystine	9.94	9.45	8.04	9.20
Valine	6.00	7.50	6.32	5.72
Methionine	1.82	1.22	1.57	1.87
Isoleucine	3.94	4.39	4.78	3.65
Leucine	3.88	5.06	5.76	4.47
Tyrosine	4.74	4.50	5.91	5.04
Phenylalanine	2.57	4.78	3.79	3.71
Tryptophan	2.45	1.39	N.R.	2.61
Carbohydrate[b]				
Hexoses	8.9	9.3	11.6	
Galactose				3.57
Mannose				5.62
Fucose	1.2	1.3		1.86
Hexosamines		9.1	9.1	
Glucosamine	7.1	6.5		8.21
Galactosamine	1.1	1.5		2.80
Sialic acid	6.0	4.7	5.2	3.13

[a] Amino acid is expressed in residues per 100 residues of amino acids.
[b] Carbohydrate is expressed in % dry weight.
[c] *Biochim. Biophys. Acta* **221**, 87.
[d] *Endocrinology* **91**, 1418.
[e] *Endocrinology* **94**, 655.
[f] *Endocrinology* **104**, 140.

munological, activity. The common proteolytic enzymes such as trypsin and chymotrypsin have little effect on either the biological or immunological activity of FSH (11).

4.2 Biosynthesis

The absence of highly purified FSH or FSH β-subunits in amounts required for development of specific anti-FSH antibodies has precluded significant study of FSH biosynthesis. Significant progress in this area cannot be expected until this hiatus has been bridged.

A priori it seems likely that the basic sequence of FSH biosynthesis will prove similar to that of LH and TSH. However, the hormonal regulation of FSH and LH biosynthesis probably will prove significantly different. Reasons for this prediction are (1) in the absence of GnRH, cultured AP cells produce FSH for prolonged intervals while LH synthesis and release cease (20); (2) infusion of GnRH *in vivo* at a relatively low rate may stimulate FSH release while having no discernable effect on LH release (21), and (3) the peptides "inhibin" and "gonadostatin" appear to inhibit selectively FSH release. Therefore, they likely have selective effects on FSH biosynthesis.

REFERENCES

1 Palade, G., Intracellular aspects of the process of protein synthesis. *Science* **189**, 347 (1975).

2 Li, C. H., Chemistry of Ovine Prolactin, *in* Knobil, E., and Sawyer, W. H. (eds.), *Handbook of Physiology*, Sect. 7, Vol. 4, Part 2, p. 103, American Physiological Society, Washington, D.C. (1974).

3 Lingappa, V. R., Devillers-Thiery, A., and Bloebel, G., Nascent prohormones are intermediates in the biosynthesis of authentic bovine pituitary growth hormone and prolactin. *Proc. Nat. Acad. Sci.* **74**, 2432 (1977).

4 MacLeod, R. M., *in* Martini, L. and Ganong, W. F. Regulation of prolactin secretion, *Frontiers Neuroendocrin.* **4**, 169, Raven, New York (1976).

5 Labrie, F., Drouin, J., Ferland, L., *et al.*, Mechanism of hypothalamic hormones in the anterior pituitary gland and specific modulation of their activity by sex steroids and thyroid hormones. *Rec. Prog. Horm. Res.* **34**, 25 (1978).

6 Dannies, P. S., and Tashjian, Jr., A. H., Release and synthesis of prolactin by rat pituitary cell strains are regulated independently by thyrotropin releasing hormone. *Nature (London)* **261**, 707 (1976).

7 Gautvik, K. M., Walaas, E., and Walaas, O., Effect of thyroliberin on the concentration of adenosine 3'5'-phosphate and on the activity of adenosine 3'5'-phosphate-dependent protein kinase in prolactin-producing cells in culture. *Biochem. J.* **162**, 379 (1977).

8 Evans, G. A., David, D. N., and Rosenfeld, M. G., Regulation of prolactin and somatotropin mRNAs by thyroliberin. *Proc. Nat. Acad. Sci.* **75**, 1294 (1978).

9 Stone, R. T., Maurer, R. A., and Gorski, J., Effect of estradiol-17 beta on preprolactin messenger ribonucleic acid activity in the rat pituitary gland. *Biochemistry* **16**, 4915 (1977).

10 Papkoff, H., The Chemistry of the Interstitial Cell-Stimulating Hormone of Ovine Pituitary-Origin, *in* Li, C. H. (ed.), *Hormonal Proteins and Peptides*, Vol. 1, p. 59, Academic, New York (1973).

334

11 Liu, W. K., and Ward, D. N., The purification and chemistry of pituitary glycoprotein hormones. *Pharmac. Therap. B* 1, 545 (1975).

12 Pierce, J. G., Faith, M. R., Giudice, L. C., *et al.*, Structure and Structure-Function Relationships in Glycoprotein Hormones, *in Polypeptide Hormones: Molecular and Cellular Aspects*, p. 225, CIBA Symp. 41 (new series) (1976).

13 Hagen, C., Studies on the subunits of the human glycoprotein hormones in relation to reproduction. *Scand. J. Clin. Lab Invest. 38* Suppl. **148**, (1978).

14 Hoff, J. D., Lasley, B. L., Wang, C. F., *et al.*, The two pools of pituitary gonadotropin: regulation during the menstrual cycle. *J. Clin. Endocrinol. Metab.* **44**, 302 (1977).

15 Liu, T. C., and Jackson, G. L., Modifications of luteinizing hormone biosynthesis and release by gonadotropin releasing hormone, cycloheximide, and actinomycin D. *Endocrinology* **104**, 962 (1979).

16 Fink, G., Feedback actions of target hormones on hypothalamus and pituitary with special reference to gonadal steroids. *Ann. Rev. Physiol.* **41**, 571 (1979).

17 Liu, T. C., and Jackson, G. L., Effect of *in vivo* treatment with estrogen on luteinizing hormone synthesis and release by rat pituitaries *in vitro*. *Endocrinology* **100**, 1294 (1977).

18 Bogdanove, E. M., Nolin, E. M., and Campbell, G. T., Qualitative and quantitative gonad-pituitary feedback. *Rec. Prog. Horm. Res.* **31**, 567 (1975).

19 Saxena, B. B., and Rathnam, P., The Structure and Function of Follicle-Stimulating Hormone, *in* McKerns, K. W. (eds.), *Structure and Function of Gonadotropins*, pp. 183–212, Plenum, New York (1978).

20 Sheridan, R., Loras, B., Surardt, L., *et al.*, Autonomous secretion of follicle-stimulating hormone by long term organ cultures of rat pituitaries. *Endocrinology* **104**, 198 (1979).

21 Barraclough, C. A., Wise, P. M., Turgeon, J., *et al.*, Recent studies on the regulation of pituitary LH and FSH secretion. *Biol. Reprod.* **20**, 86 (1979).

CHAPTER THIRTEEN

STRUCTURE, BIOSYNTHESIS, AND PROPERTIES OF CHORIONIC GONADOTROPIN (hCG) AND SOMATOMAMMOTROPIN (hCS, hPL)

JOHN B. JOSIMOVICH

KANTILAL H. THANKI

Department of Obstetrics & Gynecology
College of Medicine & Dentistry of New Jersey
Newark, New Jersey

1 hCG STRUCTURE

Chorionic gonadotropin is a glycoprotein hormone of about 38,000 daltons (1). It consists of two chains, α and β, linked by 11 or 12 disulfide bridges between cysteine residues. At several regions in both chains, galactose and glycoside links occur primarily to serine and to asparagine. Because isolation and purification techniques readily disrupt glycosides, it is uncertain to what extent the heterogeneity in carbohydrate chains found in purified hCG preparations is artifactual. It is clear however, that there is a natural heterogeneity in the N-terminal portion of the α-subunit (2). The carbohydrate α-chain has a structure similar to, and perhaps in some cases identical to, the α-chains of the pituitary glycoproteins, FSH, LH (see Chapter 12) and TSH of a variety of species. Its molecular weight is about 15,000 and it consists of 92 amino acids and about 4% carbohydrate (galactose, mannose, fucose, N-acetylglucosamine, N-acetylgalactosamine, and sialic acid).

The β-chain contains approximately 145–149 amino acids. The carboxylic acid end contains a "tail portion" (3) of 30 amino acids that extends beyond the usual approximately 115 amino acids found in β-chains of the other glycoprotein hormones. The β-chain is about 10% carbohydrate by weight (2), a high value for glycoprotein hormones.

Chorionic gonadotropin is produced by the syncytial trophoblast of the human placenta (1). Under certain pathological circumstances, usually involving anoxia to the placenta, there may be hyperplasia of the persisting cytotrophoblast with increased synthesis and secretion of hCG.

Originally, it was thought that hCG was only produced by certain primates such as the baboon, chimpanzee, and rhesus monkey (4).

Pregnant mare serum (PMS) hormone, a gonadotropin, appears to be of maternal decidual origin (4). A recent finding of an hCG-like peptide in rabbit blastocyst fluid raises the possibility that nonprimates may also produce trophoblastic gonadotropins (5). As reviewed by Gusseck (6), poly(A)-rich RNA from term placenta can only direct about one fifth the hCG production by ascites tumor cells of that produced by polysomes from first trimester placentas. The difference, from these and other studies, appears to be attributable to the mRNA content of the polyribosomes.

The question of whether or not hCG is produced by other normal human tissues and by bacteria as claimed by several groups is still not resolved (3). Others have been unable to confirm hCG in malignant cells in the majority of strains of bacteria obtained from patients with cancer. Other groups continue to describe the presence of hCG, albeit with lower carbohydrate content, in normal tissues and in bacteria (7, 8). Adejuwon and co-workers (9) have shown that the protease content of normal tissue may falsely give the impression of the presence of hCG, calling into question some studies in which hCG was thought to have been found (10).

In addition to production of free α-subunits, a larger-molecular-weight precursor of hCG may be produced by the placenta (11). There is agreement, however, that malignant human cells of a variety of tissue origins produce hCG (3) with a normal complement of carbohydrate (12). There may be more cleavage by-products of an α-chain precursor, of lower molecular weight than the normal α-chain (13). Differences seem to exist in the control of hCG and α-chain production in malignancies such as the choriocarcinoma in comparison with normal placentas or those transformed by viruses. Although dibutyryl-cAMP is a potent stimulus of protein-hormone synthesis in choriocarcinoma cells (14), sodium butyrate and 5-bromo-2'-deoxyuridine are less potent than in normal cells (15).

1.1   Structure and Function

It is clear that desialylation of hCG causes loss of biologic activity, presumably due to more-rapid clearance by the liver and other tissues (3); but that activity can be restored by increasing clearance through galactose oxidase action after desialylation (16). The large number of lysine and arginine residues is also important for action, since chemical masking of these two amino acids greatly reduces activity (17), whereas reduction of some disulfide bonds causes reversible decreases in activity (18).

What, then, are the biological effects of hCG, and how are they initiated biochemically? The main function of hCG is to prolong the production of progesterone and estradiol by the ovarian corpus luteum if fertilization takes place, thus assuring the development of the endometrial decidua for implantation (3). The concentration of hCG in serum lessens two months after conception, after the placenta has taken over the majority of estrogen and progesterone production needed for further growth and development of the pregnant uterus. Nevertheless, hCG remains thereafter quite elevated (at about 10-fold higher levels than the tonic levels of LH found in the circulation of nonpregnant women). Thus there is clearly a possibility that hCG, besides maintaining the corpus luteum of pregnancy, plays other roles such as helping regulate steroid production within the placenta itself (3). Furthermore, the large carbohydrate content of hCG directly on the cell surface of the trophoblast lining the intervillous circulation may function as a negatively charged barrier to prevent destruction of the conceptus by circulating maternal antibodies and cells against the antigenically foreign embryo.

Although the exact mechanisms by which hCG acts on the corpus luteum have not been delineated, the available evidence suggests a pattern of events similar to that observed with other steroidogenic hormones (19). The hCG binds to specific cell membrane receptors and in doing so activates the receptor-coupled adenylyl cyclase (20, 21). The generated cAMP attaches to binding sites on a subunit of holoenzyme protein kinase (22). Upon binding cAMP the regulatory subunit dissociates from the catalytic subunit, which now is activated to phosphorylate a number of proteins in the cytoplasm and nucleus. When phosphorylated, the enzyme cholesterol esterase converts cholesterol esters into free cholesterol (23). The entry of this increased amount of cytoplasmic cholesterol into mitochondria is a rate-limiting step (24, 25). A carrier protein, perhaps synthesized in response to phosphorylation of ribosomal proteins, facilitates the entry of cholesterol into mitochondria (26). A group of enzymes in the mitochondria cleaves the side chain of the cholesterol molecule and converts the steroid to Δ^5-pregnenolone. This metabolite is further acted upon by a number of enzymes to produce progesterone, a major product of the corpus luteum (see also Chapter 18).

Recently, it has been discovered that removal of the carbohydrate moieties from the hCG molecule does not affect its binding to specific receptors, but the ability of such carbohydrate-depleted hCG to stimulate cAMP is drastically reduced (27). Evidently, hCG also regulates its own effect on target cells. Once hCG has bound to the receptor and stimulated cAMP production, the hCG–receptor complex may be internalized and degraded by the lysosomal enzymes (28). This results in a decreased number of hCG receptors on the cell surface and, consequently, desensitization to further biologic effects of additional hCG. Whether the mechanism of internalization of hormone–receptor complex has been developed by the cells merely for getting rid of the complex or whether the complex performs other functions prior to degradation remains to be clarified by future research (29). An additional locus for desensitization by hCG may be the adenylyl cyclase system (30), though here the nature of the mechanism remains obscure. And hCG has also been reported to stimulate enzymes such as ornithine decarboxylase and 5-adenosylmethionine decarboxylase (31); but whether or not this action is mediated by cAMP is not clear.

In summary, hCG is a complex polypeptide produced by the normally developing trophoblast of the mammalian embryo. However, the hCG genes may be present in the genomes of all living matter and may be expressed under abnormal conditions such as malignant transformation. The main function of hCG appears to stimulate steroidogenesis and maintain continuation of pregnancy.

2   hCS

Chorionic somatomammotropin (hCS), commonly known as placental lactogen (hPL) is a polypeptide hormone of about 22,500 molecular weight. It consists of a simple polypeptide chain of about 191 amino acids in which a partially helical structure is linked through two disulfide bridges at adjacent cysteines (between amino acid number 53 and 165, and between 182 and 189) (1). Studies of circular dichroism have shown about 40% α-helical structure. It is this globular tertiary structure that results in a tendency for polymerization of the hormone, particularly at alkaline pHs (3). With dimerization, there is loss of the weak growth hormone-like activity, but much less reduction in prolactin-like activity.

The hormone hCS has greater similarity in amino acid sequence to human growth hormone than to human prolactin: hence its more correct name, chorionic somatomammotropin (see Chapter 23 for further discussion of its biologic effects).

C. H. Li and his colleagues have suggested that the correspondence of gene-expressed amino acids between the structure of growth hormone and hCS is close to 96–99%, and that the two hormones probably evolved from a common ancestral polypeptide (1). In support of this hypothesis, an hCS-like protein has been described in growth-hormone-producing cells of the rat pituitary (32). At any rate, the synthesis of this protein appears to be present in perhaps all mammalian species so far studied (3). Again, as in the case of hCG, human malignant cells, regardless of tissue origin, appear to have the propensity for production of hCS, although with perhaps less universality and in lower concentration than hCG. Production through transformation has not yet been described in bacteria as it has been for hCG. However, production and secretion of this hormone have been achieved in experiments utilizing mRNA-rich placental fractions.

Gusseck (6) points out the importance of distinguishing between placental content of hCS and releasability. Thus although placentas associated with dysmature (malnourished) newborns have double the placental content of the hormone found in normal placentas, only one third as much could be released by incubation in a medium. Gusseck finds that most of the intracellular hCS belongs to a pool that does not exchange with immediately releasable hCS. Furthermore, advancing placental age results in changes in protein synthetic mechanisms. Review of experiments carried out in several laboratories employing mRNA-rich, poly(A)-rich RNA or polysomal-RNA-directed synthesis in heterologous cell-free ascites tumor cells, rabbit reticulocytes, and wheat germ, or in intact *Xenopus laevis* oocytes, showed that term polysomes, especially consistent with seven to nine ribosomes per message, were responsible for a greater percentage of hCS synthesis at term than in the first trimester. The studies on fine regulation of DNA transcription in the synthesis of hCS awaits the use of molecular hybridization probe techniques.

The biologic effects of hCS reflect the chemical structure, in that it is equipotent to sheep pituitary prolactin when measured in the rabbit mammary gland, somewhat less potent in the pigeon crop sac assay, and fully potent in the estrogen- and progesterone-primed rhesus monkey mammary gland. It is also equipotent to prolactin when tested in the mouse mammary gland, *in vitro* (1, 3). Its luteotropic effects are greater than ovine prolactin in the rodent *in vivo* (3). However, the growth hormone-like effects are only 2–5% that of human pituitary growth hormone when tested in the only hormone-specific bioassay: promotion of tibial epiphysial cartilage growth in the hypophysectomized, immature female rat (7, 3). This effect can be demonstrated only when injections are made in an acid solution, when the monomer exists; it is not seen with the dimer at dilute alkaline pHs (3). Nevertheless the hCS level is at least 100-fold greater in the human maternal circulation by mid-pregnancy than that which growth hormone normally reaches. Thus the weak somatotropic potency of hCS clearly contributes to the growth hormone-like metabolic changes of late pregnancy, such as the resistance to exogenous and endogenous insulin, and the greater mobilization of free fatty acids from fat-depot lipids seen during fasting in late gestation (13). Sherwood's group has shown the interesting biologic properties of recombined fragments of hCS and growth hormone produced *in vitro* (33).

Although the biochemical mechanisms of action of hCS are thought to be similar to those of both growth hormone and prolactin, there is little direct knowledge of the details of its action (see Chapter 23).

The regulation of hCS synthesis is poorly understood. It is apparent, however, that prolonged fasting does increase placental secretion of the hormone (3, 7). Hypoglucosemia may well be the stimulus. The result, an anti-insulin effect, may cause increased ketosis and hepatic gluconeogenesis (34).

In summary, hCS is a simple protein produced by the syncytial trophoblast that exerts growth hormone-like, anti-insulin effects and has an effect on placental steroidogenesis (see Chapter 23). Although structurally related to both pituitary growth hormone and prolactin, it is weak in the former activity, which depends on circulation of a monomeric form. Malignant trophoblast (choriocarcinoma) and other malignant tissue cells often appear to produce the hormone, just as these tissues secrete hCG.

REFERENCES

1 Josimovich, J. B., Placental protein hormones in pregnancy. *Clin. Obstet. Gynecol.* **16**, 46 (1973).

2 Villee, C. A., Synthesis of protein in the placenta. *Gynecol. Invest.* **8**, 145 (1977).

3 Josimovich, J. B., Hormonal Physiology of Pregnancy: Steroid Hormones of the Placenta, and Polypeptide Hormones of the Placenta and Pituitary, *in* Gold, J. J., and Josimovich, J. B. (eds.), *Gynecologic Endocrinology*, 3rd ed., p. 147, Harper & Row, Hagerstown (1980).

4 Josimovich, J. B., Protein Hormones and Gestation, in Benirschke, K. (ed.), *Comparative Aspects of Reproductive Failure,* p. 171, Springer-Verlag, New York (1967).

5 Saxena, B. B., Ratnam, P., and Morris, P., Purification of a Luteotropic Material from Preimplanted Rabbit blastocysts, abstract 42, in *Program and Abstracts of the 60th Annual Meeting of the Endocrine Society, June 1978.*

6 Gusseck, D. J., Role of nucleic acid in the regulation of human placental lactogen synthesis. *Gynecol. Invest.* 8, 162 (1977).

7 Braunstein, G. D., Kamdar, V., Rosor, J., *et al.*, Widespread Distribution of a Chorionic Gonadotropin-like Substance in Normal Human Tissues, *J. Clin. Endocrinal. Metab.* 49, 917 (1979).

8 Maruo T., Cohen, H., Segal, S. J., and Koide, S. S. Production of choriogonadotropin-like factor by a microorganism. *Proc. Natl. Acad. Sci. U.S.A.* 76, 6622 (1979).

9 Adejuwon, C. A., Segal, S. J., and Mitsudo, S., Protease Inhibitor Abolishes the Apparent hCG-like Immunoreactivity in Human Tissues, abstract 954, *Program and Abstracts of the 61st Annual Meeting of the Endocrine Society*, June 1979.

10 Canfield, R. E., Morgan, F. J., Human Chorionic Gonadotropin (hCG), in Berson, S. A., and Yalow, R. A. (eds.) *Methods in Investigative and Diagnostic Endocrinology,* Part 3, p. 727, North Holland, Amsterdam (1973).

11 Maruo, T., Segal, S. J., and Koide, S. S., Large molecular species of chorionic gonadotropin from human placental tissues: biosynthesis and physico-chemical properties. *Acta Endocrinol. (Copenhagen)* **94**, 259 (1980).

12 Schlegel-Haueter, S. E., Robinson, J. C., and Chou, J. Y., Characterization of Human Chorionic Gonadotropin Synthesized by Human Cell Lines, abstract 43, *Program and Abstracts of the 60th Annual Meeting of the Endocrine Society*, June, 1978.

13 Benveniste, R., Lindner, J., Puett, D., and Rabin, D., Human chorionic gonadotropin alphasubunit from cultured choriocarcinoma (JEG) cells: comparison of the subunit secreted free with that prepared from secreted human chorionic gonadotropin. *Endocrinology* **105**, 581 (1979).

14 Benveniste, R., Speeg, K. V., Jr., Long, A., *et al.*, Concanavalin-A stimulates human chorionic gonadotropin (hCG) and hCG-alpha secretion by human choriocarcinoma cells. *Biochem. Biophys. Res. Comm.* **84**, 1082 (1978).

15 Chou, J. Y. Effects of sodium butyrate and 5-bromo-2′-deoxyuridine on the synthesis of human chorionic gonadotropin in human placental cells transformed by tsA mutants of simian virus 40. *Endocrinology* **107**, 1327 (1980).

16 Kalyan, N. K., Madnick, H. N., Segal, H. L., *et al.*, Effect of Modification of Galactose Residues on the Biological Properties of Asialo-Chorio Gonadotropin, abstract 946, *Program and Abstracts of the 61st Annual Meeting of the Endocrine Society*, June, 1979.

17 Swaminathan, N., Braunstein, G. D., Location of major antigenic sites of the beta subunit of human chorionic gonadotropin. *Biochemistry*, **17**, 5832 (1978).

18 Mori, K. F., Hum, V. G., Disulfide Bonds of hCG Alpha Subunit, abstract 45, *Program and Abstracts of the 60th Annual Meeting of the Endocrine Society*, June, 1978.

19 Channing, C. P., Tsafriri, A., Mechanism of action of luteinizing hormone and follicle stimulating hormone on the ovary *in vitro*. *Metabolism* **26**, 413 (1977).

20 Robison, G. A., Butcher, R. W., and Sutherland, E. W., *Cyclic AMP*. Academic, New York (1971).

21 Channing, C. P., Thanki, K. H., Lindsey, A. M., *et al.*, Development and Hormonal Regulation of Gonadotropin Responsiveness in Granulosa Cells of the Mammalian Ovary, *in* Birnbaumer, L., O'Malley, B. W., (eds.), *Receptors and Hormone Action*, 3rd ed, p. 435. Academic, New York (1978).

22 Podesta, E. J., Dufau, N. L., Solano, A. R., *et al.*, Hormonal activities of protein kinase isolated from Leydig cells. Electrophoretic analysis of cyclic AMP receptors. *J. Biol. Chem.* **253**, 8994 (1978).

23 Behrman, M. R., and Armstrong, D. T., Cholesterol esterase stimulation by luteinizing hormone in luteinized rat ovaries. *Endocrinology* **85**, 474 (1969).

24 Mahafee, D., Reitz, R. C., New, R. L., The mechanism of action of adrenocorticotropic hormone. The role of mitochondrial cholesterol accumulation in the regulation of steroidogenesis. *J. Biol. Chem.* **249**, 227 (1974).

25 Robinson, J., Stevenson, P. N., Boyd, G. S., *et al.*, Acute *in vivo* effects of HCG and LH on ovarian mitochondrial cholesterol utilization. *Molec. Cell Endocrinol.* **2**, 149 (1975).

26 Gill, G. N., Mechanism of ACTH action. *Metabolism* **21**, 571 (1972).

27 Bahl, O. P., and Moyle, W. R., Role of Carbohydrate in the Action of Gonadotropins, *in* Birnbaumer, L., and O'Malley, B. W., (eds.), *Receptors and Hormone Action*, 2nd ed., p. 261, Academic, New York (1978).

28 Ascoli, M., and Puett, D., Inhibition of the degradation of receptor-bound human choriono-gonadotropin by lysosomotropic agents, protease inhibition and metabolic inhibitors. *J. Biol. Chem.* **253**, 7832 (1978).

29 Catt, K. J., Harwood, J. R., Aguilera, G., *et al.*, Hormonal regulation of peptide receptors and target cell responses. *Nature* **280**, 109 (1979).

30 Hunzicker-Dunn, M., Bockaert, J., and Birnbaumer, L., Physiological Aspects of Appearance and Desensitization of Gonadotropin-Sensitive Adenylyl Cyclase in Ovarian Tissues and Membranes of Rabbits, Rat, and Pigs, *in* Birnbaumer, L., and O'Malley, B. W. (eds.), *Receptors and Hormone Action*, 3rd ed., p. 394, Academic, New York (1978).

31 Villee, C. A., and Loring, J. N., Effect of FSH and LH on RNA Synthesis in the testis: Role of Ornithine Decarboxylase, *in* McKerns, K. W. (ed.), *Structure and Function of the Gonadotropins*, p. 295, Plenum, New York (1978).

32 Witorsch, R. J., Evidence for human placental lactogen immunoreactivity in rat pars distalis. *J. Histochem. Cytochem.* **28**, 1 (1980).

33 Russell, J., Sherwood, L. M., Kowalski, K., *et al.,* Recombinant hormones from fragments of human growth hormone and human placental lactogen. *J. Biol. Chem.* **256**, 296 (1981).

34 Grumbach, M. M., Kaplan, S. L., and Vinik, A., Human Chorionic Somatomammotropin (hCS). Regulation of Secretion, *in* Berson, S. A., Yalow, R. A. (eds), *Methods in Investigative and Diagnostic Endocrinology,* p. 807, North Holland, Amsterdam (1973).

CHAPTER FOURTEEN

IMMUNOREACTIVE FORMS OF GLYCOPROTEIN HORMONES

STEPHEN J. ZIMNISKI

Department of Biochemistry
Vanderbilt University
Nashville, Tennessee

JUDITH L. VAITUKAITIS

Section of Endocrinology and Metabolism
Thorndike Memorial Laboratory
Boston City Hospital
Departments of Medicine and Physiology
Boston University School of Medicine
Boston, Massachusetts

1 INTRODUCTION

Each of the human glycoprotein hormones, luteinizing hormone (LH), follicle-stimulating hormone (FSH), thyroid-stimulating hormone (TSH), and chorionic gonadotropin (hCG), is composed of two different α- and β-subunits bound principally by hydrophobic binding of high affinity (1, 2). All these hormones except hCG are secreted by the normal pituitary gland. Under appropriate conditions, the glycoprotein-hormone subunits can be dissociated; highly purified subunit preparations are essentially devoid of significant intrinsic biologic activity (2). That lack of bioactivity does not reflect denaturation, since complementary subunits of the same or different hormones may be reassociated with restoration of biologic activity characteristic of the β-subunits used for reassociation (2, 3). The human glycoprotein hormones share many similarities with subhuman primate and nonprimate glycoprotein hormones. Some of these similarities are discussed below.

2 IMMUNOLOGY OF GLYCOPROTEIN-HORMONE SUBUNITS

Initial attempts to isolate and purify the glycoprotein hormones in the late 1960s and 1970s greatly accelerated the understanding of the physiological roles of these hormones. New techniques including radioimmunoassay, radioreceptor assays, and sensitive *in vitro* bioassays greatly advanced our understanding of hormone action.

Initially, antisera generated against the intact glycoprotein hormones rarely could distinguish among the variety of native glycoprotein hormones. That cross-reactivity was solely attributed to cross-contamination of the hormonal preparations (4). Even with purification of hormone preparations to apparent "homogeneity," significant cross-reactivity persisted. When the amino acid structures for the α- and β-subunits became known and after initial insights provided by homologous- and heterologous-subunit radioassays were available, it became apparent that a major part of the immunologic cross-reactivity reflected the structural homology among the α-subunits; in addition, areas of homology were noted among some portions of the β-subunits (5). Fig. 1 depicts the dose–response lines for human glycoprotein hormones in a homologous hCG-α radioimmunoassay. Note that all the dose–response lines are parallel, and that reflects complete immunological cross-reactivity among the α-subunits. On the other hand, when those same native hormones were examined in a homologous hCG-β assay (Fig. 2), the dose–response lines had different slopes, reflecting the fact that the β-subunit conferred the immunological specificity to the glycoprotein hormones (5).

Interestingly, even though the α-subunits of the glycoproteins have extensive structural homology between species, rare immunological cross-reactivity is observed (6). On the other hand, there is considerably more amino acid heterogeneity among the β-subunits, but greater immunological cross-reactivity among species (6). Those relationships are depicted in Figs. 3 and 4.

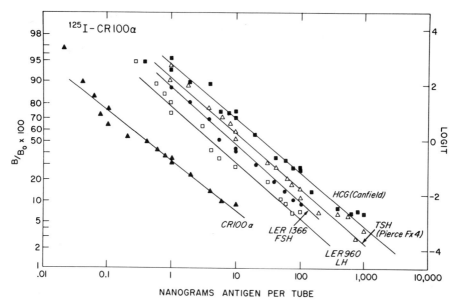

Figure 1 Dose–response curves for highly purified hCG-α (CR100-α), highly purified human pituitary LH (LER 960), FSH (LER 1366), and TSH (Pierce Fx 4) and hCG. B_0, Maximum counts with labeled ligand alone; B, counts bound in the presence of labeled and unlabeled ligand. Dose–response lines were constructed in a homologous hCG-α radioimmunoassay (RIA). From Vaitukaitis and Ross (4) with permission. Nonspecific binding was subtracted throughout this and subsequent studies.

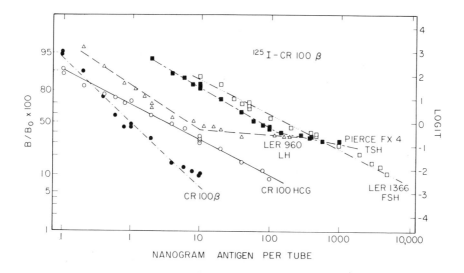

Figure 2 Dose–response lines for purified hCG-β (CR100-β), intact hCG (CR100-hCG), and highly purified human pituitary LH (LER 960), FSH (LER 1366), and TSH (Pierce Fx 4) determined in an homologous hCG-β RIA. Results are depicted as described in Fig. 1. From Vaitukaitis and Ross (4).

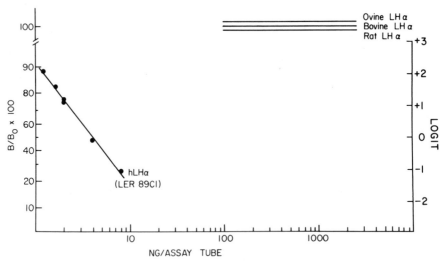

Figure 3 Dose–response lines for ovine, bovine, rat, and human LH-α in a homologous hLH-α RIA. Results are depicted as shown in Fig. 1. From Vaitukaitis *et al.* (6).

3   STRUCTURE-FUNCTION STUDIES

Comparison of immunological versus biologic activity of glycoprotein hormones revealed several interesting observations. Formation of hybrid hormones suggested that the binding sites between α- and β-subunits did not significantly contribute to the immunoreactivity of either the free subunit or the recombined hormone (7). Neutralization studies of primate chorionic gonado-

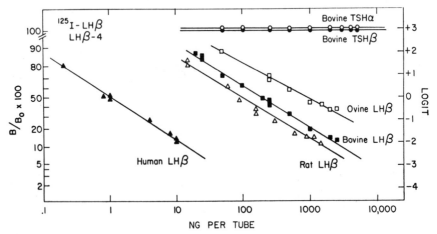

Figure 4 Dose–response lines for ovine, bovine, rat, and human LH-β. Bovine TSH-α and TSH-β failed to cross-react in the homologous hLH-β RIA. Results are depicted as described in Fig. 1. From Vaitukaitis *et al.*

tropins demonstrated that biologic activity was inhibited by the addition of either α- or β-subunit antisera generated to hCG subunits (8). That suggested considerable antigenic and possibly structural similarity among primate chorionic gonadotropins. The amino acid sequences of subhuman primate gonadotropins remain unknown.

The biologic and immunological activity of intact hormones can be altered by variable removal of sialic acid, a nine-carbon sugar that is usually found at the terminus of oligosaccharides found on the glycoprotein hormones. In general, one or two oligosaccharides are present on each hormone subunit (9). However, hCG-β may contain as many as four or five oligosaccharide side chains (9). Desialylated hCG (sialic acid removed) was as biologically effective as fully desialylated preparations when tested in an *in vitro* bioassay system (10). However, desialylated hormone has markedly less biologic activity *in vivo*. This marked difference found *in vivo* apparently is due to a reduction in plasma half-life attributed to removal of sialic acid (11). With removal of the sialic acid, the underlying galactose moieties are exposed and are recognized by the hepatic asialoglycoprotein receptors (12), and this in turn, probably results in more rapid hepatic removal of previously sialylated hormones (2).

Not only is the role of the oligosaccharides poorly understood, but nature's rationale for the subunit structure of the glycoprotein hormones is also an enigma (13). Moreover, the plasma half-lives of the hormone subunits are but a fraction of those of the native hormones (5).

4 SUBUNITS IN BIOLOGIC FLUIDS AND TISSUES

In some physiological and pathophysiological situations, free subunits may be found in tissue and biologic fluids, including blood and urine (5). In general, the higher the concentration of native glycoprotein hormone, the higher the concentration of free subunit of that hormone. The free subunit, even when present in high concentrations, circulates in relatively low concentrations when compared with its native hormone (5). Moreover, it is usually the α-subunit that circulates in excess; rarely does excess β-subunit circulate except when tumor is present and in other selected circumstances (5). The presence of free subunit does not reflect spontaneous dissociation of subunits, since one would expect to observe equimolar concentration of the complementary subunit if that situation obtained. Moreover, since complementary hormone subunits bind with high affinity, one would not expect them to dissociate spontaneously in the circulation. Several studies have confirmed that speculation. Whereas the plasma half-life of fully sialylated hCG is of the order of 24–36 hr in humans, the subunits have considerably shorter half-lives of 13 min for α- and 41 min for β-subunits in the human (14).

Finally, the concentration of free subunit can be increased under selected circumstances. For instance, administration of synthetic thyrotropin-releasing hormone (TRH) to hypothyroid subjects markedly increases circulating concentrations of TSH, TSH-α and TSH-β (15). In other circumstances, one may administer synthetic doses of gonadotropin-releasing hormone (GnRH) to individuals with primary gonadal failure and observe a significant increase in circulating free α-subunits (16). There may be little or no increase in circulating levels of β-subunit gonadotropin levels (16). In contrast, when hypogonadal or hypothyroid patients are placed on appropriate doses of replacement homone, basal levels of free α-subunit are lowered significantly.

5 DETECTION OF NATIVE HORMONE AND HORMONE SUBUNITS

In order to detect free hormone, one must usually combine immunological and chromatographic techniques, except in those rare settings in which nonchromatographed biologic fluids may be assayed for TSH-β directly (17). The reason for excess secretion of free α-subunit by both normal pituitary and placental tissues is poorly understood (5). Moreover, separate mRNAs control the translation of each hCG subunit and probably all the other pituitary glycoprotein hormone subunits (18). Not only is free hCG-α and/or hCG-β found in extracts of malignant tumors, but other altered forms of hCG may be observed in biologic fluids and tissues (5).

6 CARBOXY-TERMINAL ANTISERA

Recently, several specific antisera have been developed to the unique carboxy-terminal fragment of hCG-β as well as to synthetic hCG-β subunit carboxy-terminal peptide analogues. Matsuura et al. (19) synthesized a series of 32 carboxy-terminal analogue peptides and evaluated each for antigenic recognition by the H-93 antiserum generated against the native carboxy-terminal fragment. The degree of immunoreactivity increased with the carboxy-terminal chain length starting from the carboxy-terminal end, and immunoreactivity plateaued at 15 amino acids. The dose–response curves of the longer synthetic peptides were also parallel with that of native hCG. They concluded that the last 15 amino acid moieties of the carboxy-terminal fragment were responsible for the major population of antibodies in the antigenic recognition site (19). However, their conclusions relate only to the carboxy-terminal antibody and do not relate to other specific "β-subunit" antibodies which may be conformationally dependent, as is the first antibody described (SB6) (9). The specific hCG assay (or "β-subunit assay"), using antiserum to hCG-β, has been invaluable for monitoring patients with ectopic pregnancies and selected patients with hCG-secreting tumors (5). Antisera produced to the carboxyl fragments react poorly with native hCG and therefore have not been clinically useful for routine application.

7 "PIECES" OF GLYCOPROTEIN HORMONES

We have recently described a small urinary 7000-dalton immunoreactive substance that cross-reacts with native hCG, its β-subunit, and LH (20). From our observations, that small fragment appears to be a normal secretory fragment of normal placental tissue and some hCG-secreting tumors (20). The physiological significance of the small immunoreactive substance is not understood at this time. Other small fragments of glycoproteins have been found in the urine but have not been characterized extensively, and probably represent metabolic breakdown products of the glycoprotein hormones. A small LH-like substance has been isolated from blood (21). That small substance was isolated from a large pool of blood and contained biologic activity indistinguishable from LH in *in vitro* testing. That substance also displayed LH immunoreactivity.

8 FORMS OF GLYCOPROTEIN HORMONES IN TISSUES AND FLUIDS

Analysis of extracts of normal pituitary or placental tissues for glycoprotein-hormone subunits and intact hormones has helped to provide initial insights into the synthesis of the glycoprotein hormones. Extracts of hCG-secreting tissues normally contain an excess of α-subunits and insignificant quantities of β-subunits. During pregnancy, the concentrations of immunoreactive hCG and α-subunit decrease in placental tissues from the mid first trimester to the end of gestation. However, the relative concentration of α-subunit is considerably greater than that for hCG. Those and other observations suggest that the concentration or the translation of mRNA for the β-subunit may be one of the limiting factors in native glycoprotein-hormone synthesis. Although it is known that α- and β-subunits are translated from separate messenger RNAs, very little is known about the regulatory mechanisms involved. At least part of the differential concentration may be due to a changing ratio of subunit synthesis from early to late term (22).

Normally, free subunits are not detectable in peripheral circulation in concentrations above a few nanograms per milliliter. However, several clinical conditions, including primary hypothyroidism and gonadal failure, exist in which not only native glycoprotein hormone levels but also the levels of their respective subunits are elevated. Fig. 5 depicts normally circulating levels of native hCG and hCG-α in a newborn infant. Both native pituitary glycoprotein hormone and subunit concentrations increase concomitantly in response to appropriate releasing-hormone stimulation. However, even though a common pathway for subcellular release may exist for both hormone and subunits, the mechanisms involved are poorly understood.

In addition to secretion of intact hormone and subunits, a variety of tumors, both trophoblastic and nontrophoblastic, synthesize and secrete hCG and its subunits. Several tumors have been found to secrete altered forms of hCG and

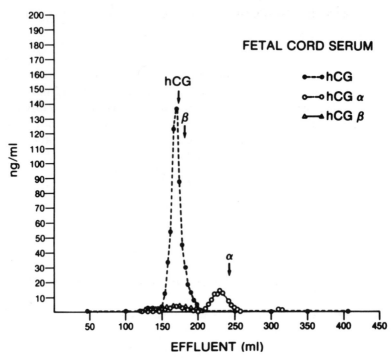

Figure 5 Elution profile of fetal cord serum chromatographed through a standardized Sephadex G-100 column. Vertical arrows depict points of peak elution for purified hCG, hCG-β and hCG-α. Each fraction was assayed in homologous hCG, hCG-α, and hCG-β RIAs. From Vaitukaitis (5).

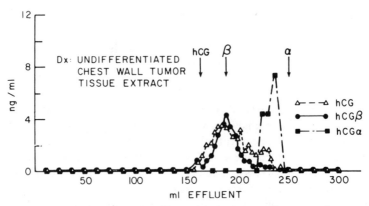

Figure 6 Tissue extract of an undifferentiated chest wall tumor chromatographed through a Sephadex G-100. Each fraction of the eluate was assayed in a homologous hCG, hCGα, and hCG-β RIAs. Arrows indicate points of peak elution of highly purified hCG and its subunits. From Vaitukaitis (5).

its subunits. For reasons not entirely clear at this time, however, tumors rarely secrete the *pituitary* glycoproteins ectopically. Fig. 6 depicts the variety of forms of native hCG and its subunits found in a tissue extract obtained from a patient with an hCG-secreting tumor. Other smaller forms of hCG and the pituitary glycoproteins may circulate in peripheral blood, but assays are either too insensitive to detect them or do not immunologically cross-react with them.

9 NORMAL INDIVIDUALS AND BACTERIA

There is now evidence to suggest that an hCG-like substance may be present in normal tissue not usually associated with its synthesis. Using hCG-β antisera, small but significant amounts of an immunoreactive hCG-like substance have been found in tissue extracts of human testes, pituitary, gastrointestinal tract, and urine. The clinical significance of that hCG-like substance is unknown. In addition, an hCG-like substance has been isolated and partially characterized from some forms of bacteria (23). The significance of the latter observation is unclear and may represent gene transfer from human cells to some bacterial species.

REFERENCES

1 Garnier, J., Molecular Aspects of the Subunit Assembly of Glycoprotein Hormones, *in* McKerns, K. W., (ed.), *Structure and Function of the Gonadotropins*, p. 381, Plenum, New York (1978).

2 Vaitukaitis, J. L., Ross, G. T., Braunstein, G. D., *et al.,* Gonadotropins and their subunits: Basic and clinical studies. *Rec. Prog. Horm. Res.* **32**, 289 (1976).

3 Hagen, C., Studies on the subunits of the human glycoprotein hormones in relation to reproduction. *Scand. J. Clin. Lab. Invest.* **38** (Suppl. 148), 1 (1978).

4 Vaitukaitis, J. L., and Ross, G. T., Antigenic Similarities Among the Human Glycoprotein Hormones and Their Subunits, *in* Saxena, B. B., Beling, C. G., and Gandy, H. M. (eds.), *Gonadotropins*, p. 435, Wiley, New York (1972).

5 Vaitukaitis, J. L., Glycoprotein Hormones and Their Subunits—Immunological and Biological Characterization, *in* McKerns, K. W. (ed.), *Structure and Function of the Gonadotropins,* p. 339, Plenum, New York (1978).

6 Vaitukaitis, J. L., Ross, G. T., Reichert, L. E., Jr., *et al.*, Immunologic basis for within and between species cross-reactivity of luteinizing hormone. *Endocrinology* **91**, 1337 (1972).

7 Vaitukaitis, J. L., Ross, G. T. and Reichert, Jr., L. E., Immunologic and biologic behavior of hCG and bovine LH subunit hybrids. *Endocrinology* **92**, 411 (1973).

8 Hodgen, G. D., Nixon, W. E., Vaitukaitis, J. L. *et al.,* Neutralization of primate chorionic gonadotropin activities by antisera against the subunits of human chorionic gonadotropin in radioimmunoassay and bioassay. *Endocrinology* **92**, 705 (1973).

9 Birken, S., Canfield, R. E., Structural and Immunochemical Properties of Human Choriogonadotropin, *in* McKerns, K. W., (ed.), *Structure and Function of the Gonadotropins,* p. 47, New York, Plenum, 1979.

10 Catt, K. J., Dufau, M.L. Tsuruhara, T., Absence of intrinsic biological activity in LH and CG subunits. *J. Clin. Endocrinol. Metab.* **36**, 73 (1973).

11   van Hall, E. V., Vaitukaitis, J. L., Ross, G. T., *et al.,* Effects of progressive desialylation on the rate of disappearance of immunoreactive hCG from plasma in rats. *Endocrinology* **89**, 11 (1971).

12   Hudgin, R. L., Pricer, W. E., Jr., Ashwell, G., *et al.*, The isolation and properties of a rabbit liver binding protein specific for asialoglycoproteins. *J. Biol. Chem.* **249**, 5536 (1974).

13   Fontaine, Y.-A., Burzawa-Gerard, E., Esquisse de l'évolution des hormones gonadotropes et thyreotropes des vertebres. *Gen. Comp. Endocrinol.* **32**, 341 (1977).

14   Wehmann, R. E., and Nisula, B. C., Metabolic clearance rates of the subunits of hCG in man. *J. Clin. Endocrinol. Metab.* **48**, 753 (1979).

15   Kourides, I. A., Weintraub, B. D., Re, R. N., *et al.,* Thyroid hormone, oestrogen and gluco-corticoid effects on two different pituitary glycoprotein hormone alpha subunit pools. *Clin. Endocrinol.* **9**, 535 (1978).

16   Edmonds, M., Molitch, M., Pierce, J. G., *et al.,* Secretion of alpha subunits of LH by the anterior pituitary. *J. Clin. Endocrinol. Metab.* **41,** 551 (1975).

17   Kourides, I. A., Re, R. N., Weintraub, B. D., *et al.,* Metabolic clearance and secretion rates of subunits of hTSH. *J. Clin. Invest.* **59**, 508 (1977).

18   Daniels-McQueen, S., McWilliams, D., Birken, S., *et al.,* Identification of mRNA's encoding the alpha and beta subunits of human choriogonadotropin. *J. Biol. Chem.* **253**, 7109 (1978).

19   Matsuura, S., Chen, H.-C., and Hodgen, G. D., Antibodies to the carboxyl terminal fragment of hCG beta-subunit: Characterization of antibody recognition sites using synthetic peptide analogues. *Biochemistry* **17**, 575 (1978).

20   Masure, H. R., Jaffee, W. L., Sickel, M. A., *et al.,* Characterization of a small molecular sized urinary immunoreactive hCG-like substance produced by normal placenta and by hCG-secreting neoplasms. *J. Clin. Endocrinol. Metab.* **53**, 1014 (1981).

21   Leidenberger, F. A., Graesslin, D., Scheel, H. J., *et al.*, A low molecular weight substance obtained from serum which has luteinizing hormone activity (mini LH). *J. Clin. Endocrinol. Metab.* **43**, 1410 (1976).

22   Chatterjee, M., and Munro, HN., Changing ratio of hCG subunits synthesized by early and full-term placental polyribosomes. *Biochem. Biophys. Res. Comm.* **71**, 426 (1977).

23   Maruo, T., Cohen, H., Segal, S. J., *et al.,* Production of choriogonadotropin-like factor by a microorganism. *Proc. Nat. Acad. Sci. USA* **76**, 6622 (1979).

CHAPTER FIFTEEN

MODULATION OF OVARIAN GONADOTROPIN RECEPTORS

STEPHEN J. ZIMNISKI

Department of Biochemistry
Vanderbilt University
Nashville, Tennessee

JUDITH L. VAITUKAITIS

Section of Endocrinology and Metabolism
Thorndike Memorial Laboratory
Boston City Hospital
Departments of Medicine and Physiology
Boston University School of Medicine
Boston, Massachusetts

1 INTRODUCTION

As is the case with other protein hormones, after initial binding to a specific receptor, the gonadotropins mediate their action through a "second messenger." The gonadotropins first bind with specific cell surface receptors, and that interaction activates a cascade of intracellular biochemical reactions, including activation of adenylyl cylcase, an enzyme that converts ATP to cAMP. Consequently, intracellular cAMP concentrations increase with hormonal stimulation. Cyclic AMP binds to the regulatory subunit of cAMP-dependent protein kinase (CDPK), a holoenzyme constituted of two regulatory and two catalytic subunits. As a result of cAMP's binding to the regulatory subunit, the holoenzyme dissociates and the catalytic subunits may now phosphorylate specific substrates, including membrane proteins, ribosomes, and subcellular enzymes. Since the "second messenger" model was first proposed 20 years ago by Haynes, Sutherland, and Rall (1), additional regulatory steps have been described, but the general mechanism is essentially unchanged. That cascade is schematically depicted in Fig. 1.

2 RECEPTOR–HORMONE INTERACTIONS

Certain requirements must be met by a membrane acceptor before it can be considered a receptor. First, the receptor must be specific. The putative receptor must bind with high affinity to only one hormone, and the addition of other hormones should not markedly affect that hormone–receptor interaction. Since the number of cell surface receptors is finite, those binding sites should be

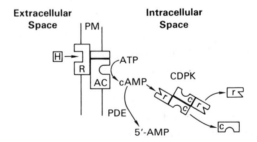

Figure 1   General mechanisms of protein hormone action. Interaction of the hormone (*H*) with its specific receptor (*R*) leads to activation of adenylyl cyclase (*AC*) in the plasma membrane (*PM*). The activation of AC increases the conversion of ATP to 3'5'-cyclic AMP (*cAMP*). In turn, cAMP interacts with the regulatory subunit (*r*) of the cAMP-dependent protein kinase (*CDPK*). This interaction induces dissociation of the regulatory subunit from the catalytic subunit (*c*), thereby activating the enzyme. The level of intracellular cAMP and CDPK activation is determined by the relative rate of synthesis of cAMP by AC and degradation by phosphodiesterase (*PDE*).

saturable. The final requirement is that the hormone–receptor interaction be of high affinity. A mathematical method commonly used to describe hormone–receptor interaction is Scatchard analysis (2). A typical Scatchard plot is displayed in Fig. 2.

When a hormone H interacts with its specific receptor R, a simple reversible reaction may be characterized at equilibrium by the following relationship:

$$[H] + [R] \underset{k_d}{\overset{k_a}{\longleftrightarrow}} [HR] \tag{1}$$

where $[HR]$ represents the concentration of the hormone–receptor complex, k_a is the association rate constant and k_d is the dissociation rate constant. For these purposes, it is assumed that both hormone and receptor are homogeneous and univalent.

At equilibrium, the association constant K_a may be represented as follows:

$$K_a = \frac{[HR]}{[H][R]} = \frac{1}{K_d} \tag{2}$$

where K_a is the equilibrium association constant in liters per mole or M^{-1} and K_d is the dissociation constant in moles per liter (M).

The relationship between total receptor concentration and those sites occupied may be depicted as follows:

$$[R_0] = [HR] + [R]$$

or

$$[R] = [R_0] - [HR]$$

where $[R_0]$ equals total receptor concentration.

By rearrangement and substitution of equation 2, the following equations are derived:

$$\frac{[HR]}{[H]} = -K_a[HR] + K_a[R_0]$$

and

$$\frac{[HR]}{[H]} = -\frac{1}{K_d}[HR] + \frac{[R_0]}{K_d}$$

These equations are the mathematical representation of a Scatchard plot. A plot of $[HR]/[H]$, the ratio of the concentrations of bound to free hormone, against

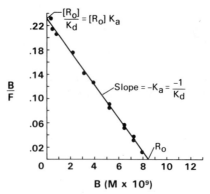

Figure 2   Representative Scatchard plot of data obtained from a radioreceptor assay.

$[HR]$, that for bound hormone, as seen in Fig. 2, yields a straight line of slope $-1/K_d$. By extrapolating the line to the x axis, a total receptor concentration $[R_o]$, can be ascertained.

Scatchard analysis can be performed *in vitro* by incubating a fixed concentration of receptor and labeled hormone with increasing concentrations of either labeled or unlabeled hormone and allowing that reaction to continue until equilibrium is attained. It is assumed that the labeled and unlabeled hormones interact with receptors identically. Occasionally, however, Scatchard analysis may result in a nonlinear plot. When that occurs, several factors may be responsible. These include nonequilibrium conditions in the assay, errors in separation of bound and free hormone, or an inaccurate estimate of nonspecific binding. In addition, there is always the possibility of multiple receptors with different affinities, or allosteric interactions between receptors. Allosteric interactions may result in either increased or decreased hormone-binding affinity of receptors adjacent to those occupied by hormone. Those interactions are commonly referred to as "negative" or "positive" cooperativity. Further discussion of the interpretation of Scatchard plots can be obtained from Dahlquist (3).

3   MODULATION OF GONADOTROPIN RECEPTORS

The concentration of receptor in plasma membrane need not be fixed. Moreover, receptors may move across the surface of the plasma membrane as suggested by the fluid mosaic membrane model of Singer and Nicolson (4). Experiments have suggested that receptors may be concentrated in localized cell membrane regions, primarily the pericapillary membrane (5). Furthermore, with high concentrations of hormone, gonadotropin–receptor complexes may coalesce and form "caps" on the cell surface, analogous to that described for other hormones, drugs, and ligands. Whether capping is a requisite for subcellular hormonal action or is an early degradative step for hormone, receptor, or both is speculative at this time. Results from studies with insulin and epidermal growth factor strongly suggest that capping aids in the internalization of hormone–receptor complexes (6).

In addition to the postulated plasma membrane lateral movement, ovarian gonadotropin receptor concentration is modulated by several factors, including circulating homologous hormone concentration, level of maturation of the follicle, and sex steroids. Those factors will be systematically examined.

3.1 Factors Affecting Induction of FSH and LH Receptors

Estrogen stimulation of immature, hypophysectomized rats results in marked proliferation of preantral follicles with markedly increased numbers of granulosa cells per ovary. Specific ovarian FSH binding is significantly increased and simply results from an increased number of granulosa cells containing the same number of specific FSH receptors with the same binding affinity for the hormone (7). Granulosa cell receptors for LH increase with spontaneous follicular maturation or may be increased experimentally with exposure of granulosa cells to FSH and estrogen, to dibutyl-cAMP, and to LH or hCG (8). Channing and Kammerman observed that porcine granulosa cells obtained from large preovulatory follicles had 10–500-fold more LH/hCG receptors than granulosa cells obtained from small follicles (9). Moreover, when granulosa cells were harvested from small, medium, and large follicles, the concentration of FSH receptors was greatest among cells harvested from small follicles and those for LH from granulosa cells of large follicles (8). There was no significant difference in the size of granulosa cells harvested from the various follicles. This transition in follicular responsiveness from FSH to LH is believed to be modulated by FSH and estrogen. Low concentrations of LH may also significantly contribute to that maturational process. Both LH and FSH are essential for maximum induction of LH receptor. When one examines FSH-induced activation of adenylyl cyclase, granulosa cells harvested from small follicles are most responsive. On the other hand, when one examines LH- or hCG-induced activation of adenylyl cyclase, granulosa cells harvested from large follicles yield the greatest concentrations of cAMP when compared with the response of granulosa cells harvested from small and medium follicles. Consequently, adenylyl cyclase responsiveness reflects the relative concentrations of FSH or LH receptors on the granulosa cell. Lee (10) suggested that the change in gonadotropin sensitivity reflects the change in surface receptors and is not due to the induction of a new cyclase system.

Increased ovarian responsiveness to LH persists throughout luteinization. Luteinization of the rat ovary can be experimentally induced by the sequential administration of pregnant mare serum gonadotropin (PMSG) and hCG. PMSG contains both LH and FSH biologic activities, but is administered predominantly for its intrinsic FSH bioactivity. The administration of these hormones produces pseudopregnant ovaries. The rat ovaries are superluteinized and contain 50–100-fold higher concentrations of LH/hCG receptors. Because of that, this model has greatly aided in the characterization of the

LH/hCG receptor and a better understanding of the biochemical events induced by those hormones.

4   MODULATION OF OVARIAN FUNCTION

Recently, several new substances have been found in ovarian tissue that affect
ovarian function. Among these is an inhibitor of LH binding to either granulosa or luteal cells (1). Moreover, that substance inhibits not only LH binding to
ovarian tissue but also blocks LH-induced progesterone synthesis. In *in vitro*
studies, approximately 90% of labeled LH could be prevented from binding to
receptor in the presence of this potent inhibitor. A similar binding inhibitor for
FSH has been isolated from follicular fluid of follicles of different levels of
maturation (12). The concentration did not correlate with the level of follicular
maturation, but merely reflected the volume of fluid present.

Other follicular fluid regulators have been described that affect oocyte maturation and luteinization of granulosa cells (8). In both cases, smaller follicles
appear to possess greater inhibitor activity (8). Another follicular fluid substance affecting gonadotropin action is "folliculostatin" or inhibin (13). Rather
than exerting a local effect on the ovary, folliculostatin decreases serum FSH
levels in selected animal models. The chemical nature and physiological importance of that substance is not yet fully defined. Its concentration is greatest in
the fluid of small follicles. Moreover, it is not yet clear whether that substance
enters the systemic circulation physiologically. Studies by DePaolo *et al.* (14)
suggest that folliculostatin may be present in ovarian venous plasma. Antral
fluid folliculostatin activity varies inversely with plasma FSH throughout the
rat estrous cycle. DePaolo *et al.* (15) suggest a pituitary site of action for
folliculostatin. However, folliculostatin, as well as the other inhibitors mentioned above, has yet to be purified to homogeneity. Therefore, whether these
compounds play any significant role in the normal physiological modulation of
ovarian function is unknown.

4.1   Gonadotropin–Receptor Interactions and Subcellular Events

The appearance of a specific receptor and its subsequent interaction with gonadotropin is the initial step for stimulation of a cell. To transduce its action
across the cell, the hormone–receptor complex must be coupled to the enzyme
adenylyl cyclase. This enzyme, an intrinsic membrane protein, is responsible for
the conversion of ATP to cAMP, an intracellular messenger. It is still unknown
whether the coupling occurs after the hormone–receptor complex is formed or
if a population of receptor-adenylyl cyclase complexes exists prior to hormonal
stimulation. Gonadotropin receptors from both solubilized and particulate
plasma membranes are associated with adenylyl cyclase. It is likely that the
receptor and the enzyme are at least in close proximity, if not in direct contact,
prior to hormonal stimulation. Clarification of that point is complicated by the

observation that only 1–2% of receptors need be occupied to maximally in- duce subcellular biochemical reactions. The remaining receptors are consid- ered to be excess or "spare" receptors (16). Teleologically, the relationship between receptor and adenylyl cyclase in that 1–2% of gonadotropin receptors probably reflects that of the total population of receptors, since hormone–re- ceptor interaction probably reflects "chance" or random binding of circulating gonadotropin to that small population of receptors.

Ryan *et al.* (17) characterized gonadotropin-sensitive adenylyl cyclase of ovarian tissue. Adenylyl cyclase may contain several regulatory sites in addition to a catalytic site. As for the other cyclase systems, catalytic activity is mod- ulated by several mechanisms. Foremost among them is the action of the nucleotides. The primary modulator of cyclase activity appears to be guanosine triphosphate (GTP). Adenosine triphosphate (ATP) as well as GTP may stimulate ovarian adenylyl cyclase. Since evidence suggests that those nucleo- tides act independently, ovarian cyclase may possess two purine regulatory sites. Ovarian cyclase may be inhibited by the sulfated mucopolysaccharides—hep- arin and chondroitin and dermatan sulfates (18). The physiological role of those mucopolysaccharides in modulating gonadotropin action is poorly un- derstood at this time.

4.2 Cyclic-AMP-Dependent Protein Kinases

The ultimate result of cyclase activation is the increased production of cAMP. That increase leads to an increase in cAMP-dependent protein kinase activity, which in turn results in an increase in the phosphorylation of a variety of specific subcellular proteins. With administration of physiological doses of gonadotropins, one may not detect increased intracellular cAMP concentra- tions. Because of that, it was suggested that gonadotropins need not act solely through the cAMP second-messenger system. However, other data strongly suggest that the failure to observe increased intracellular cAMP concentrations reflected limitations of the assays used for those studies. Although no signifi- cant change in intracellular cAMP levels were observed with low-dose stimula- tion of ovarian tissue by gonadotropin, a significant increase in cAMP binding to the regulatory subunit of protein kinase was observed (19). That observation confirmed the limitation of the cAMP assays used for other studies. As seen in Fig. 3, administration of graded doses of hCG *in vivo* results in a dose-de- pendent increase in intracellular cAMP concentration and activation of CDPK. Significant activation of CDPK occurs before one observes a significant in- crease in intracellular cAMP. That relationship clearly shows that monitoring CDPK activity is a significantly more sensitive end point of ovarian gonado- tropin stimulation than measuring intracellular cAMP concentrations.

Most mammalian tissues contain at least two cytoplasmic CDPKs desig- nated type I and type II (20). Type I may be rapidly activated by physiological levels of hormone. Type I activation occurs quite rapidly and correlates with increasing cAMP concentrations. Type II PK, on the other hand, is activated more slowly, and its role in cellular activation is uncertain.

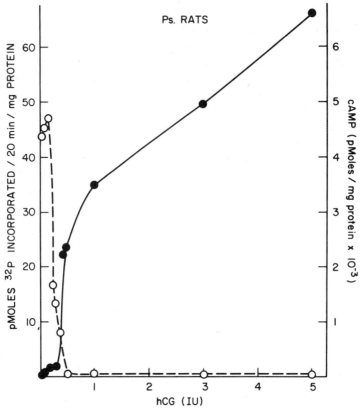

Figure 3   Effect of graded doses of hCG on the intracellular accumulation of cAMP (closed circles) and cAMP-dependent protein kinase (open circles).

In gonadal tissue, activation of the second-messenger cascade by gonadotropin ultimately leads to increased steroid synthesis. In granulosa cells, LH/hCG stimulation results primarily in estrogen synthesis, whereas in luteal cells, progesterone is the primary steroid produced. Maximal steroid production is observed prior to any detectable increase in intracellular cAMP. However, if cAMP binding to the regulatory subunit is measured, a direct correlation between the amount of cAMP bound and the degree of steroid production is observed (19).

4.3   Gonadotropin-Induced Ovarian Refractoriness

Gonadotropins regulate the concentration of receptors in their target cells. During maturation of ovarian follicles, LH receptors are induced as a result of granulosa cell stimulation with FSH and LH. However, after induction of those receptors, the cell may be "turned off" with subsequent LH or hCG stimulation. With high physiological and pharmacologic doses of hCG or LH, the concentration of LH receptors decreases in a dose- and time-dependent manner (21). After initial stimulation, "loss" of receptors requires hours. Down regulation of receptors appears to be a relatively long-term regulatory process. However, tissue refractoriness occurs prior to down regulation of receptor number (22). That rapid hormone-induced ovarian refractoriness may be due to an uncoupling of the gonadotropin-receptor complexes from adenylyl cyclase. The adenylyl cyclase of desensitized ovaries is unresponsive to gonadotropin stimulation but may be activated with other stimuli such as guanine analogs and fluoride ions that directly stimulate the catalytic sites of adenylyl cyclase (17). Hormone-induced desensitization occurs within 1 hr of gonadotropin stimulation and may persist long enough to overlap with down regulation of receptors. Therefore, it may play a more important role in the regulation of ovarian function, certainly for the short term and possibly in conjunction with down regulation for the long term. The initial phase of hormone-induced desensitization probably occurs more commonly in physiological settings than does down regulation, since the latter is observed with high physiological or pharmacologic levels of gonadotropin. The midcycle preovulatory surge of gonadotropin is probably one physiological setting in which both ovarian desensitization and down regulation occur.

Investigators have observed that at lower physiological doses of gonadotropins, tissue refractoriness, although present, is not absolute and may not be attributed to decreased receptor number. Albertson and Vaitukaitis (22) found that ovarian refractoriness can be overcome with higher doses of gonadotropin.

An additional level of ovarian refractoriness was detected by Conti *et al.* (23). If refractoriness was at the receptor or cyclase level, then the addition of cAMP should overcome the block. However, neither dibutyryl-cAMP nor cholera toxin restored steroid production in desensitized cells. These data imply that an additional lesion beyond the level of cAMP production may exist in luteal cells.

A paradox clearly exists. Granulosa cells of the developing follicle require LH and FSH to induce LH receptors. Once LH receptors are present, binding of LH or hCG to those receptors induces desensitization and possibly down regulation of LH receptors, especially if high physiological or pharmacologic doses of LH are encountered by granulosa cells.

That same relationship undoubtedly holds for FSH receptors as well. Granulosa cells harvested from small, immature follicles have the greatest concentration of FSH receptor. However, as LH receptors are induced with gonadotropin stimulation, granulosa cell FSH-receptor content is down regulated. In addition, stimulation of those cells with LH results in cellular refractoriness not only to LH, but also to FSH stimulation (24). It should be obvious that very little is understood about the mechanisms responsible for receptor biogenesis and down regulation. Moreover, a host of other subcellular events occur, including RNA and DNA synthesis. Over the next few decades some, but not all, of the foregoing will be better understood.

REFERENCES

1 Haynes, R. C., Jr., Sutherland, E. W., and Rall, T. W., The role of cyclic adenylic acid in hormone action. *Rec. Prog. Horm. Res.* **16**, 121 (1960).

2 Scatchard, G., The attractions of proteins for small molecules and ions. *Ann. N.Y. Acad. Sci.* **51**, 660 (1949).

3 Dahlquist, F. W., The meaning of Scatchard and Hill plots. *Meth. Enzymol.* **48**, 270 (1978).

4 Singer, S. J., and Nicolson, G. L., The fluid mosaic model of the structure of cell membranes. *Science* **175**, 720 (1972).

5 Abel, J. H., Jr., Chen, T. T., Endres, D. B., *et al.*, Sites of Binding and Metabolism of Gonadotropic Hormones in the Mammalian Ovary, *in* Straub, R. W., and Bolis, L. (eds.), *Cell Membrane Receptors for Drugs and Hormones*, p. 183, Raven, New York (1978).

6 Maxfield, F. R., Schlessinger, J., Shechter, Y., *et al.*, Collection of insulin, EGF and alpha$_2$-macroglobulin in the same patches on the surface of cultured fibroblasts and common internalization. *Cell* **14**, 805 (1978).

7 Louvet, J.-P., and Vaitukaitis, J. L., Induction of FSH receptors in rat ovaries by estrogen priming. *Endocrinology* **99**, 758 (1976).

8 Channing, C. P., Anderson, L. D., and Batta, S. K., Follicular growth and development. *Clin. Obstet. Gynecol.* **5**, 375 (1978).

9 Channing, C. P., and Kammerman, S., Characteristics of gonadotropin receptors of porcine granulosa cells during follicle maturation. *Endocrinology* **92**, 531 (1973).

10 Lee, C. Y., Modulation of ovarian adenylate cyclase responses to gonadotropins following luteinization of the rat ovary. *Endocrinology* **101**, 876 (1977).

11 Yang, K.-P., Samaan, N. A., and Ward, D. N., Effects of LH R-BI on the *in vitro* steroidogenesis by rat ovary and testis. *Endocrinology* **104**, 552 (1979).

12 Darga, N. C., and Reichert, L. E., Jr., Some properties of the interaction of FSH with bovine granulosa cells and its inhibition by follicular fluid. *Biol. Reprod.* **19**, 235 (1978).

13 Marder, M. L., Channing, C. P., and Schwartz, N. B., Suppression of serum FSH in intact and acutely ovariectomized rats by porcine follicular fluid. *Endocrinology* **101**, 1639 (1977).

14 DePaolo, L. V., Shander, D., Wise, P. M., *et al.*, Identification of inhibin-like activity in ovarian venous plasma of rats during the estrous cycle. *Endocrinology* **105**, 647 (1979).

15 DePaolo, L. V., Wise, P. M., Anderson, L. D., *et al.*, Suppression of the pituitary FSH secretion during proestrus and estrus in rats by porcine follicular fluid: Possible site of action. *Endocrinology* **104**, 402 (1979).

16 Catt, K. J., and Dufau, M. L., Spare gonadotropin receptors in rat testis. *Nature New Biol.* **244**, 219 (1973).

17 Ryan, R. J., Birnbaumer, L., Lee, C. Y., *et al.*, Gonadotropin interactions with the gonad as assessed by receptor binding and adenylyl cyclase activity. *Int. Rev. Physiol.* **13**, 85 (1977).

18 Salomon, Y., and Amsterdam, A., Heparin: A potent inhibitor of ovarian LH-sensitive adenylate cyclase. *FEBS Lett.* **83**, 263 (1977).

19 Sala, G. B., Dufau, M. L., and Catt, K. J. Gonadotropin action in isolated ovarian luteal cells: The intermediate role of cAMP in hormonal stimulation of progesterone synthesis. *J. Biol. Chem.* **254**, 2077 (1979).

20 Corbin, J. D., and Lincoln, T. N., Comparison of cAMP and cGMP-dependent protein kinases. *Adv. Cycl. Nucl. Res.* **9**, 159 (1978).

21 Conti, M., Harwood, J. P., Hsueh, A. J. W., *et al.*, Gonadotropin-induced loss of hormone receptor and desensitization of adenylate cyclase in the ovary. *J. Biol. Chem.* **25**, 7729 (1976).

22 Albertson, B. D., Vaitukaitis, J. L., Ovarian refractoriness not related to the "down receptor phenomenon." *Endocr. Res. Comm.* **4**, 259 (1977).

23 Conti, M., Harwood, J. P., Dufau, M. L., *et al.*, Effect of gonadotropin-induced receptor regulation on biological responses of isolated rat luteal cells. *J. Biol. Chem.* **252**, 8869 (1977).

24 Richards, J. S., Hormonal control of ovarian follicular development: A 1978 perspective. *Rec. Prog. Horm. Res.* **35**, 343 (1979).

CHAPTER SIXTEEN

THE RECEPTORS OF GONADOTROPIC HORMONES IN THE TESTIS, AND RADIORECEPTOR ASSAYS

H. G. MADHWA RAJ

Departments of Obstetrics/Gynecology and Pharmacology
University of North Carolina
School of Medicine
Chapel Hill, North Carolina

The author expresses his appreciation to Dr. G. S. R. C. Murty for helping with the manuscript and to Mrs. Liza Catino for excellent secretarial assistance. This work was supported by a grant from NIH (1-R01-HD12254-01).

1 INTRODUCTION

The two gonadotropic hormones in the male, namely follicle-stimulating hormone (FSH) and luteinizing hormone [LH; also called interstitial cell-stimulating hormone (ICSH)], secreted by the pituitary gland, are responsible for normal development and functioning of the testes. Thus ablation of the pituitary gland or neutralization of the biologic activities of these two hormones using antibodies or other pharmacologic agents leads to degeneration of the testes and loss of testicular function. The two main functions of the testis, namely spermatogenesis and production of androgens, require an optimal interaction of FSH and LH. LH acts primarily on the interstitial cells of Leydig to promote production of androgens. The target tissue for the action of FSH is the Sertoli cells of the seminiferous tubules. The presence of specific receptors capable of

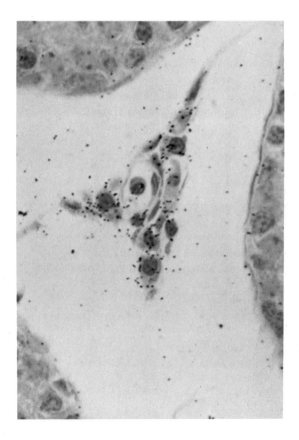

Figure 1 Autoradiographic localization of ^{125}I-labeled LH in rat testis. Note the presence of grains in the Leydig cells; parts of three adjacent tubules are seen at periphery of photograph (Courtesy of Drs. Sar, Stumpf, and Petruz).

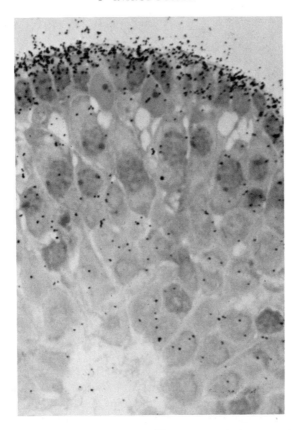

Figure 2 Autoradiographic localization of ^{125}I-labeled FSH in the testis. Localization is primarily along the outer border of the tubule (Courtesy of Drs. Sar, Stumpf, and Petruz).

binding these gonadotropins with high affinity has been clearly demonstrated in their respective target cells. Figs. 1 and 2 depict specific localization of radio-actively labeled LH and FSH in the Leydig and Sertoli cells, respectively. Fluorescent labeling, labeling with peroxidase coupled to specific antibodies, and autoradiographic techniques using radioactively labeled FSH and LH have been utilized to demonstrate the presence of such receptors. Further, the receptors have been isolated, solubilized, and well characterized by several groups of workers (1–3). The binding of LH and FSH to the receptors on the cell membrane leads to the activation of adenylyl cyclase and formation of cyclic adenosine monophosphate (cAMP) (4), culminating in steroidogenesis or spermatogenesis by processes described in Chapter 15. For example, the actions of FSH are summarized in Table 1. This chapter deals with an account of testicular gonadotropic receptors for these two hormones.

Table 1   Temporal Sequence of Events Initiated by FSH in the Testis[a]

Event	Time
Binding to plasma membrane	<5 min
Activation of adenylyl cyclase, accumulation of cyclic AMP, and stimulation of protein kinase	<15 min
Protein phosphorylation leading to transcription	45–180 min
New protein synthesis	1–2 hr
Increase in wet and dry weights	8–16 hr
Increase in mitosis	6–9 hr
Decrease in spermatogonial degeneration	9–12 hr

[a] Adapted from Ref. 4.

2   PREPARATION OF RADIOACTIVELY LABELED FSH AND LH SUITABLE FOR RECEPTOR STUDIES

It is essential to obtain a gonadotropic hormone preparation labeled to a reasonable specific activity (at least 10 μCi/μg) in order to develop a suitable receptor-binding system. However, the hormone should not be damaged during the process of labeling, as alterations in structure of the homone lead to reduction or even total loss of binding to the receptor. In general, radioactive labeling of gonadotropins has been achieved by using isotopes ^{131}I or ^{125}I. In principle, iodination involves generation of nascent radioactaive iodine in a medium containing the hormone. The amount of iodine and the duration of exposure of the hormone to iodine determine the degree of iodination. Iodination to the degree that not more than one atom of the iodine is incorporated per molecule of the hormone will yield a preparation that retains biological activity with a suitable specific activity.

A number of methods have been utilized to liberate nascent iodine from inorganic iodide for protein iodination (6–9). These are listed in Table 2.

Chloramine-T has been widely used as the oxidizing agent to prepare radio-iodinated hormones. However, preparation of biologically active hormone by this method depends on careful controlling of the various parameters during iodination. In a typical iodination procedure, 5–10μg of LH or FSH or hCG

Table 2   Iodination of Gonadotropins

Oxidizing Agent Used	Reference
Iodine monochloride	Glazer and Singer (1964) (6)
Chloramine-T	Greenwood et al. (1963) (7)
Lactoperoxidase	Catt et al. 1976 (2); Miyachi et al. (1972) (8);
Sodium perchlorate	Redshaw and Lynch (1974) (9)

are introduced into a vial containing 1 mCi of carrier-free sodium [^{125}I]iodide in 50 μl of 0.5 M phosphate buffer pH 7.2. Nascent iodine is liberated by oxidizing the sodium iodide by the addition of 10–25 μg of freshly prepared chloramine-T in 25 μl. The reaction is carried out at 4°C for 30 sec. At the end of this time, the excess iodine is reduced by the addition of a reducing agent such as sodium metabisulphite (about 100 μg). This solution now contains both the labeled protein and excess unused sodium iodide.

2.1   Radioiodination Using Lactoperoxidase as the Oxidizing Agent

In this method, hydrogen peroxide is used as the oxidizing agent. Five to 10 μg of the hormone are mixed with 1 mCi of carrier-free sodium iodide and 100–250 ng of enzyme in a small volume. The reaction is initiated by the addition of 10 μl of 30% hydrogen peroxide. The reaction is carried out from 2 to 5 min, depending on the specific activity required. At the end of the time period, the reaction is stopped by the addition of a reducing agent such as 5 mM mercaptoethanol or 1 mM dithiothreitol. The addition of the reducing agents should be omitted if a highly biologically active product is desired. As in the chloramine-T method, this mixture now contains labeled gonadotropin as well as the excess iodine that has not entered into the reaction.

2.2   Separation of Iodinated Hormone from Residual Iodide

A variety of methods have been utilized to purify the labeled hormone from the unreacted iodine. These utilize different principles. Gel filtration on Sephadex columns utilizes the difference between the molecular weights of the labeled hormone and the unreacted iodine. Ion-exchange chromatography and affinity chromatography using concanavalin A coupled to a variety of supports have been used.

None of these methods differentiates very well between the undamaged labeled hormone and the damaged labeled hormone. Several types of damage could occur to the hormone during labeling. For example, the use of oxidizing agents could also lead to oxidation of disulfide bonds in the glycoprotein hormone. Similarly, the use of reducing agents later could also reduce the hormone, leading to loss of biological activity to different degrees. To overcome these possibilities, Catt and co-workers (2) have utilized the procedure of binding and eluting the biologically active hormones from their receptors. In this procedure the labeled hormone is reacted with the receptor particles prepared from the testes or the ovaries. The biologically active hormone binds to the receptors whereas the denatured hormone does not. Later, by modifying the conditions slightly, the bound hormone is released from the receptor. This hormone now exhibits maximal biological activity. Using such a method, they could obtain an hCG preparation that would retain more than 60% binding and an FSH preparation that had more than 25% binding after iodination.

2.3 Determination of Specific Activity

Once the hormone is separated, it is essential to determine its specific activity. Different methods have been utilized for this purpose. In general, the method used for determination of specific activity is the following (3). First, the fraction of total radioactivity taken up by the hormone is estimated from the chromatographically purified material. The specific activity (SA) is then calculated as follows:

$$SA \text{ (in } \mu Ci/\mu g) = \frac{\mu Ci \ ^{125}I \text{ added} \times \text{fraction } ^{125}I \text{ incorporated}}{\mu g \text{ hormone added}}$$

where

$$\text{fraction of } ^{125}I \text{ incorporated} = \frac{\text{total counts} - \text{counts under } ^{125}I \text{ peak}}{\text{total counts}} \times 100$$

Another convenient method of determining specific activity has been described by Catt and co-workers (2). In this method, termed self-displacement, the labeled hormone is used both as tracer and sample in the radioreceptor assay, using the original unlabeled hormone as a standard. Figs. 3 and 4 represent such self-displacement curves for labeled hCG and labeled FSH, respectively. It can be seen that the labeled hormone gives a dose–response curve parallel to the curve of the unlabeled hormone. From this, the specific activity of the labeled hormone can be calculated.

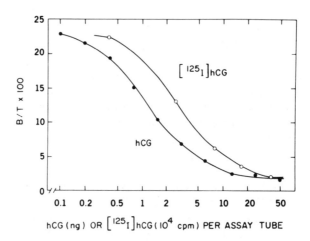

Figure 3 Determination of specific activity of ^{125}I-labeled hCG by "self-displacement" assay of increasing amounts of the tracer in the testis radioligand-receptor assay for LH and hCG. The mass of the labeled hormone is estimated by comparison with the binding-inhibition curve obtained with the unlabeled hCG employed for tracer preparation (from Ref. 2).

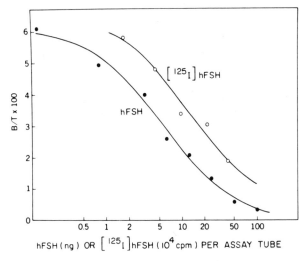

Figure 4 Determination of specific activity of ^{125}I-labeled hFSH by "self-displacement" assay of tracer mass in the radioligand-receptor assay for FSH employing rat testicular homogenate (from Ref. 2, p. 187).

A direct method of calculating specific activity of labeled hormone preparation uses sensitive and accurate *in vitro* assays for LH and FSH that have recently been developed. The assay for LH depends on the production of testosterone from Leydig cells *in vitro* in a dose-dependent manner. This assay is sensitive and can measure 1 ng of LH. Similarly, the ability of FSH to specifically stimulate cAMP production from rat testes tubules has also been utilized for *in vitro* bioassay of radioiodinated FSH.

3 TESTICULAR FSH RECEPTORS

Reichert and co-workers have extensively investigated the FSH receptors obtained from immature bovine testes (1, 10). They found that beef testes have high concentrations of FSH receptors in contrast to immature or mature rat testes. Solubilization of the FSH receptors was conveniently achieved by the use of the detergent Triton X-100. A membrane fraction obtained from the testicular tubule preparation was treated with 1% Triton X-100 in Tris-HCl buffer at pH 7.5 for 90 min at 4°C. This was followed by a 10-fold dilution with buffer and high-speed centrifugation at 300,000*g* for 1 hr. They demonstrated that the supernatant contained specific FSH receptors. Labeled human FSH specifically bound to the solubilized receptors and was not displaced by a large excess of other peptide hormones. In these studies a final concentration of 12.5% polyethylene glycol was found to precipitate receptor-bound FSH from free FSH, thus achieving separation.

The binding characteristics of receptors obtained from different sources seem to be different. The bull testis FSH receptors bind maximally at 24°C and pH 7.5, reaching equilibrium in about 4 hr. High concentrations of nucleotides like ATP, CTP, UTP, and CTP seem to significantly inhibit the specific binding of the radioligand. The affinity constant was found to be $1.2 \times 10^9 \, M^{-1}$ for the particulate receptor and about $2.1 \times 10^9 \, M^{-1}$ for the solubilized receptor. The molecular weight of the free receptor was found to be 146,000 and that of the hormone receptor complex, 183,000. The difference between these two corresponds to the molecular weight of FSH, indicating a binding ratio of 1:1.

A variety of proteolytic enzymes like trypsin, pronase, chymotrypsin, and collagenase significantly reduce the binding of FSH to the receptor (1, 5). This indicates that FSH receptor is probably protein in nature. Similarly, removal of membrane-bound phospholipids using phospholipases A and C also reduced FSH binding to the receptors, indicating the participation of membrane lipids in this process. However, removal of membrane-bound sialic acid by neuraminidase did not reduce any binding of FSH to its receptor, indicating that carbohydrates are not involved in this interaction. A variety of agents that modify the amino acids tyrosine, histidine, and tryptophan in the receptor preparations also lead to loss of binding of FSH to the receptor. This suggests that these amino acids may be involved in the receptor–FSH interaction. Similarly, membrane disulfide bonds are also involved in hormone–receptor interaction, as their reduction leads to loss of hormone–receptor binding.

3.1 Testicular Receptors for LH

These have been investigated by several investigators (see Ref. 2 for details). In general, the interstitial cells of the testis form the starting material for preparation of solubilized receptors. The interstitial cell preparation can be obtained by several methods. Physical dissection of the interstitial elements from the tubules and collagenase or trypsin digestion of decapsulated testicular material are the most commonly employed methods. After such treatment, the interstitial cells are separated from the tubules by filtration through nylon or cheesecloth. Solubilization of receptors from interstitial cells is achieved in a manner similar to that described for FSH receptors, by treatment with Triton X-100 followed by precipitation with 12% polyethylene glycol. The binding activity of the solubilized receptor preparation can be checked by reacting with ^{125}I-labeled hCG, and again separating the bound and free hCG using 12% polyethylene glycol. The solubilized LH/hCG receptor is quite labile. Further treatment with proteolytic enzymes, reducing agents, or phospholipase A all lead to significant loss of binding activity. However, the particulate receptor or the hormone–receptor complex seems to be much more stable than the free solubilized receptor.

Specific binding of ^{125}I-labeled hCG can be readily demonstrated using either particulate or soluble receptor preparations from testis. Some of the binding characteristics of the soluble receptors have been summarized in Table 3.

Table 3   Characteristics of hCG-Receptor[a]

Property	Values
Binding capacity	10^{13} mole/mg of soluble protein
Optimum pH	7.4
Association rate constant	$6.1 \times 10^5\ M^{-1}\ min^{-1}$ at $4°C$
Dissociation rate constant	$1.2 \times 10^4\ min^{-1}$ at $4°C$
Equilibrium constant at $4°C$	$0.5 \times 10^{10}\ M^{-1}$
Molecular weight of free receptor	194,300
Molecular weight of hormone–receptor complex	224,200

[a] Adapted from Ref. 2.

The receptors can be saturated with either labeled or unlabeled hCG. Increasing concentrations of unlabeled LH or hCG compete with labeled hCG for receptor-binding sites and inhibit the binding of the latter in a dose-dependent fashion. Using such systems, inhibition curves can be obtained with either the particulate or soluble receptor preparations. From such inhibition curves, Scatchard analysis can be performed to obtain the association constant. This is illustrated in Figs. 5 and 6. The receptor binding is specific, and whereas LH or hCG can displace the bound hormone, FSH, prolactin, TSH, or growth hormone are not effective (2).

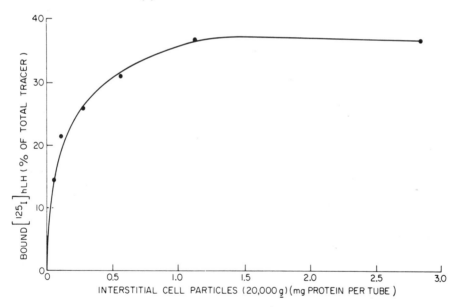

Figure 5   Binding of ^{125}I-labeled hLH to increasing concentrations of gonadotropin receptors during incubation with interstitial cell particles from the rat testis. The maximum receptor-binding activity of the labeled hormone is about 37% of the total radioactivity of the tracer preparation (from Ref. 2, p. 184, by courtesy of Marcel Dekker, Inc.)

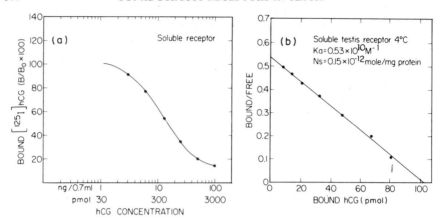

Figure 6 (*a*) ^{125}I-labeled hCG binding-inhibition curve obtained with soluble receptor. (*b*) Scatchard plot of receptor–hCG binding data (from Ref. 2, p. 230; by courtesy of Marcel Dekker, Inc.).

The molecular weight of the free LH receptor is about 194,300, and that of the receptor-hormone complex is 224,200. The difference between these two is 29,900, which corresponds approximately to the hCG molecular weight of 37,000. This indicates that 1 molecule of receptor binds 1 molecule of hormone.

4 RADIORECEPTOR ASSAYS FOR FSH AND LH AND THEIR APPLICATIONS

4.1 Assay of FSH

Radioligand receptor assay for FSH has been performed using either testis-tubule homogenates or whole-testis homogenates from adult rats (2, 11). Varying assay protocols have been used by different investigators. The protocol for a typical assay is as follows:

12 × 25 mm glass tubes are used and the following ingredients are added sequentially:

    200 μl testis homogenate containing 20 mg wet tissue

    varying amounts (0.1–100 ng FSH) of gonodogropin standards or samples to be assayed in 200 μl volume

    100 μl ^{125}I-labeled FSH (40,000–50,000 cpm) in phosphate buffer (0.01 M) containing 0.9% NaCl, 0.1% egg albumin, and 0.05% neomycin sulfate

Nonspecific binding is determined by adding to an identical set of tubes an excess amount of FSH (1–2 IU of Pergonal or other suitable FSH preparation). After mixing, the tubes are incubated at 24° C for 18 hr with shaking. At the end

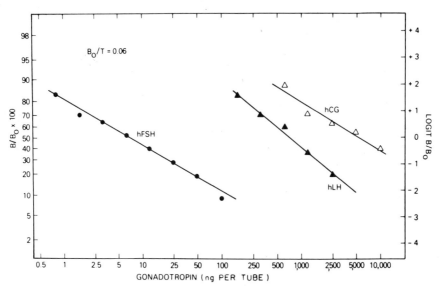

Figure 7 Radioligand-receptor assay for hFSH employing immature rat testis homogenate. The assay is performed at 24°C. The hLH and hCG reacted in the systems only at doses respectively 100 and 750 times greater than the hFSH dose (from Ref. 2, p. 207; by courtesy of Marcel Dekker, Inc.).

of incubation, the bound and the free hormone are separated by centrifugation at 2000*g* for 10 min. The supernatant is aspirated, and the tissue pellet is washed once with 1–2 ml cold phosphate buffer. The pellet is counted in a gamma spectrometer. A typical standard curve is shown in Fig. 7. It can be seen that approximately 1–200 ng of the hormone forms a linear displacement curve. A suitable reference preparation is used as a standard, and this is usually the Second International Reference Preparation (IRP) of human menopausal gonadotropin (hMG).

4.2 Binding Assays for LH and hCG

Interstitial cell particles from the adult rat testis are prepared by dissecting the decapsulated testis apart in phosphate-buffered saline, followed by filtration through cotton, wool, or cheesecloth. Membrane fractions from these cells can be prepared for radioreceptor assay. Even homogenates of the whole decapsulated testis are suitable for general binding studies and radioreceptor assays of LH and hCG, since tubules or germ cells do not interfere in the binding of labeled hCG or LH. A 1500 × g sediment from such homogenates is diluted suitably and used. The various agents are added in the following sequence:

1 100 μl testis homogenate
2 100 μl of the LH standards or samples to be assayed in phosphate-buffered saline containing 0.1% bovine serum albumin (PBS–BSA)

3   50 μl ^{125}I-labeled hCG tracer (about 50,000 cpm) in PBS-BSA containing 0.05% neomycin sulfate.

As in the FSH assay, nonspecific binding is determined by the addition of excess hCG (about 10 μg) to a duplicate set of tubes. After mixing, the tubes are incubated at room temperature for 16 hr and the reaction mixture is diluted by the addition of 3 ml of ice-cold PBS. The mixture is centrifuged at 2000g for 20 min to isolate the receptor-bound hormone. The pellet can be washed once and the radioactivity in the pellet counted. A typical inhibition curve is shown in Fig. 8. Highly purified LH or hCG preparations of known biologic activity are used as standard reference preparations.

4.2.1   Measurement of Hormone Activity as a Clinical Test

Radioreceptor assays are increasingly being employed to measure hormone activity. There are two advantages in this assay over the radioimmunoassay.

1   The radioreceptor assay probably measures biologically active, intact hormones, because not only the primary, but also the secondary and tertiary structures of the hormones must be retained to preserve the receptor-binding site.

2   It is easy to prepare stable particulate receptor preparations suitable for assays by freeze-drying the material. Further, the receptor assay can be completed in a short time of a few hours.

Measurement of gonadotropin in plasma by radioreceptor assay is a useful method of evaluating the high levels of hCG in sera of pregnant women, for

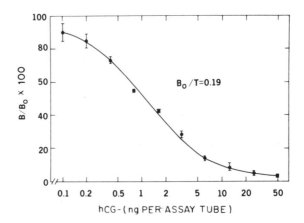

Figure 8   Radioligand-receptor assay for LH/hCG employing 1/100 of a rat testis per assay tube, performed in an incubation volume of 0.25 ml. Each point is the mean $\pm$ SD of four determinations (from Ref. 2, p. 199; by courtesy of Marcel Dekker, Inc.).

Table 4 Some Uses of Radioreceptor Assay of hCG—LH[a]

Clinical Application	No. of Cases
Routine pregnancy test	407
Positive: 114	
Negative: 293	
Suspected early ectopic pregnancies	33
Monitoring early threatened abortions	32
Detection of early pregnancy for miniabortions	55
Evaluation of therapy for infertility cases	180
Patients with intrauterine devices	44
Premenopausal women	16
Hydatiform mole, choriocarcinoma, and other tumors	27

[a] From Ref. 3.

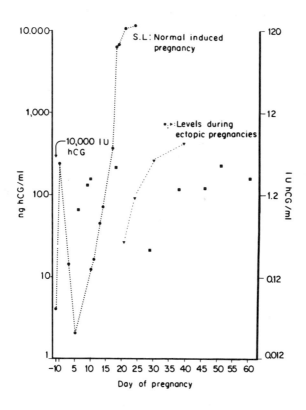

Figure 9 Comparison of the plasma levels of hCG during hCG-induced normal intrauterine pregnancy (*S.L.*) with hCG levels in 10 patients with ectopic pregnancy. The plasma levels of hCG during early induced or natural pregnancies were similar. The day of pregnancy was based on the last menstrual period reported by the patient. In one of the 10 patients (. . .), the radioreceptor assay was performed on days 21, 24, 30, and 42 of pregnancy. The initial radioreceptor assay on day 21 was associated with a negative Pregnosticon test (from Ref. 3, by courtesy of Marcel Dekker, Inc.).

377

example. Attempts to measure gonadotropins in nonpregnant women are difficult because of nonspecific interference (12) by plasma and serum proteins at
low hormone levels. However, such interference can be reduced by addition of
serum from nonpregnant women to the tubes containing standards in the assay.
Radioreceptor assays for hCG have been used to diagnose pregnancy, ectopic
pregnancy, and a variety of pathological conditions. These have been summarized in Table 4. Both ectopic pregnancies and threatened abortions are characterized by low levels of hCG output compared with normal pregnancies at the
same age. These levels are detected by radioreceptor assay and may support
clinical findings. Fig. 9 shows the levels of hCG in ectopic pregnancy compared
with the levels in a normal pregnancy (3). In Fig. 10 the low levels of hCG
output in threatened abortions is demonstrated. In addition, the hCG levels are
detected very early for the purposes of terminating unwanted pregnancies by
menstrual extractions as well as to detect induced pregnancies rapidly in subfer-
tile patients.

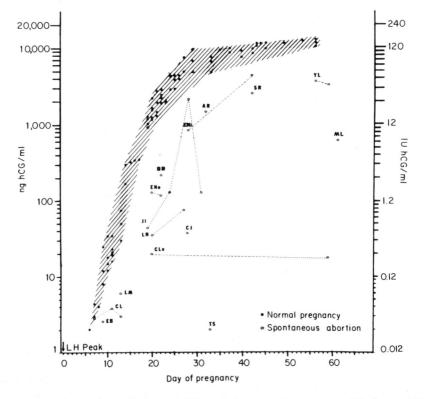

Figure 10 Comparison of plasma hCG levels in normal pregnancy with plasma hCG
levels in 15 patients who spontaneously aborted. The day of pregnancy is based on the
last menstrual period and length of cycle reported by the patient. International units of
second International Reference Preparation of hCG equivalent to nanograms of hCG
are indicated on the ordinates (from Ref. 3, by courtesy of Marcel Dekker, Inc.).

4.2.2   Structure–Activity Relationship

As the receptor binding requires a hormone that is intact in its primary, secondary, and tertiary structures, the radioreceptor assays have been used to study the effects of modifications of the hormone structure. Desialylation of hCG is known to reduce its biologic activity *in vivo*. However, it was found that *in vitro*, in the receptor-binding assay, it did not lose its activity. This led to the demonstration that the loss of activity *in vivo* was due to a more rapid clearance. Modification of LH by oxidation with performic acid, succinylation, nitration, or maleylation results in loss of receptor binding to various degrees (13, 14). Similarly, removal of other carbohydrates such as galactosamine, glucosamine, and mannosamine also resulted in loss of receptor binding and biologic activity to different degrees (15). Neither the α- nor the β-subunit of LH, hCG, and FSH has significant receptor-binding activity in the dissociated form (16). Recombination of the α- and β-subunits restores the receptor binding (2). Fig. 11 summarizes receptor-binding reactions of some of the modified LH preparations.

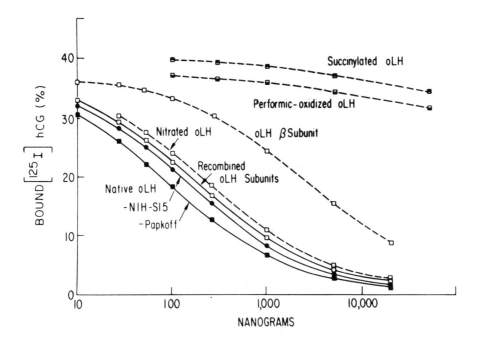

Figure 11 Radioligand-receptor assay of derivatives of ovine LH, prepared by Dr. H. Papkoff. The activity of each modified-hormone preparation measured by radioligand assay was similar to the potency estimate determined by bioassay (from Ref. 2, p. 000; by courtesy of Marcel Dekker, Inc.).

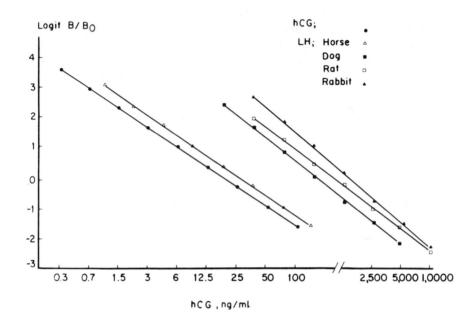

Figure 12. Competitive inhibition of the binding of ^{125}I = labelled hCG by hCG and LH from various species in the radioreceptor assay. The biological potencies of human, horse, dog, and rabbit LH were 8.9, 5.5, 0.023, and 0.9 units, respectively, in terms of NIH-LH-SI by ovarian ascorbic acid-depletion assay (from Ref. 3).

4.2.3   Relative Potencies of Gonadotropins from Different Species

LH from different species differ in structures and hence are not immunologically identical, though they retain partial similarities. For this reason, their potencies cannot be estimated using radioimmunoassay. However, they all possess common biologic activity and bind to the hormone receptor. Thus parallel inhibition curves can be generated using LH from different species, and their potencies can be compared. Such an inhibition curve is shown in Fig. 12.

These applications of radioreceptor assays illustrate the usefulness of this tool in clinical diagnosis as well as to better understand the mode of interaction of gonadotropins with their receptors in bringing about specific effects.

REFERENCES

1   Reichert, L. E., Jr., and Abou-Issa, H., The interaction of Follitropin (FSH) With Membrane-Bound and Solubilized Gonadal Receptors and Adenylate Cyclase, *in* McKerns, K. W. (ed.), *Structure and Function of Gonadotropins*, pp. 259–274, Plenum, New York (1978).

2 Catt, K. J., Ketelslegers, J. M., and Dufau, M. L., Receptors for Gonadotropic Hormones, *in* Blecher, M. (ed.), *Methods in Receptor Research,* Part 1, pp. 175–250, Marcel Dekker, New York (1976).

3 Saxena, B. B., Gonadotropin Receptors, *in* Blecher, M. (ed.), *Methods in Receptor Research,* Part 1, pp. 251–299, Marcel Dekker, New York (1976).

4 Means, A. R., Biochemical Effects of Follicle-Stimulating Hormone on the Testis, *in* Hamilton, D. W., and Greep, R. O. (ed.), *Handbook of Endocrinology,* Sect. 7, Vol. 5, pp. 203–218, American Physiological Society, Washington DC (1975).

5 Means, A. R., and Vaitukaitis, J. L., Peptide Hormone "Receptors": Specific Binding of ^{3}H-FSH to Testis. *Endocrinology* **90**, 39 (1972).

6 Glazer, A. N., and Sanger, F., The iodination of chymotrypsinogen, *Biochem. J.* **90**, 92 (1964).

7 Greenwood, F. C., Hunter, W. M., and Glover, J. S., The preparation of ^{131}I-labelled human growth hormone of high specific radioactivity, *Biochem. J.* **89**, 114 (1963).

8 Miyachi, V., Vaitukaitis, J. L., Nieschlag, E., and Lipsett, M. B., Enzymatic radioiodination of gonadotropins, *J. Clin. Endocrinol. Metab.* **34**, 23 (1972).

9 Redshaw, M. R., and Lynch, S. S., An improved method for the preparation of iodinated antigens for radio-immunoassay, *J. Endocrinol.* **60**, 527 (1974).

10 Reichert, Jr., L. E. and Abou-Issa, H., Follitropin Receptors in Rat Testis Tubule Membranes: Characterization, Solubilization, and Study of Factors Affecting Interaction With FSH, *in* Birnbaumer, L., and O'Mally, B. W. (eds.), *Receptors and Hormone Action,* Vol. 3, pp. 341–361, Academic, New York, (1978).

11 Reichert, Jr., L. E. Follicle-Stimulating Hormone: Measurement by a Rat Testes Tubule Tissue Receptor Assay, *in* Blecher, M. (ed.), *Methods in Receptor Research,* Part 1, pp. 99–118, Marcel Dekker, New York (1976).

12 Catt, K. J., Tsuruhara, T., Mendelson, C., Ketelslegers, J. M., and Dufau, M. L., Gonadotropin Binding and Activation of the Interstitial Cells of the Testis, *in* Dufau, M. L., and Means, A. R. (ed.), *Hormone Binding and Target Cell Activation in the Testis,* pp. 1–30, Plenum, New York (1974).

13 Liu, W.-K., Yang, K.-P., Burleigh, B. D., and Ward, D. N., Functional Groups in Ovine Luteinizing Hormone and Their Receptor Sites Interactions—A Chemical Study, *in* Dufau, M. L., and Means, A. R. (eds.), *Hormone Binding and Target Cell Activation,* pp. 89–108, Plenum, New York (1974).

14 Sairam, M. R., Papkoff, H., and Li, C. H. Reaction of ovine interstitial cell stimulating hormone with tetranitromethane, *Biochim. Biophys. Acta* **278**, 421 (1972).

15 Bahl, O. P., Marz, L., and Moyle, W. R., The Role of Carbohydrates in the Biological Function of Human Chorionic Gonadotropin, *in* Dufau, M. L., and Means, A. R. (eds.), *Hormone Binding and Target Cell Activation in the Testis*, pp. 125–144, Plenum, New York (1974).

16 Catt, K. J., Dufau, M. L., and Tsuruhara, T., Absence of intrinsic biological activity in LH and hCG subunits, *J. Clin. Endocrinol. Metab.* **36**, 73 (1973).

CHAPTER SEVENTEEN

INACTIVATION AND EXCRETION OF FSH, LH, AND hCG

STEPHEN J. ZIMNISKI

Department of Biochemistry
Vanderbilt University
Nashville, Tennessee

JUDITH L. VAITUKAITIS

Section of Endocrinology and Metabolism
Thorndike Memorial Laboratory
Boston City Hospital
Departments of Medicine and Physiology
Boston University School of Medicine
Boston, Massachusetts

1 INTRODUCTION

The plasma concentration of a hormone is determined by its rate of secretion and metabolic clearance. Although hormones are usually secreted from a single gland, several organs may contribute to their metabolism. Several routes of inactivation and excretion have recently been suggested for the glycoprotein hormones—follicle-stimulating hormones (FSH), luteinizing hormone (LH), thyroid-stimulating hormone (TSH), and human chorionic gonadotropin (hCG). The glycoprotein hormones are composed of two noncovalently linked subunits, designated α and β. Each subunit contains two to five oligosaccharide side chains that usually terminate with sialic acid. The glycoprotein hormones have a common α-subunit. The amino acid sequence of the β-subunit differs among the glycoprotein hormones. Consequently, biologic and immunological specificity is conferred by the β-subunit of those hormones (1).

Both FSH and LH are synthesized in cells of the anterior pituitary, whereas hCG, as the name implies, is of placental origin. In spite of the different sites of synthesis, LH and hCG share extensive structural homology and consequently have indistinguishable biologic activities (1). The carboxyl terminus of the β-subunit of hCG contains an additional 28–30 amino acids not found in LH-β nor any other human glycoprotein hormone. The synthesis of hCG subunits is regulated by separate mRNAs for each subunit (2). Several observations suggest that the rate-limiting step for glycoprotein-hormone synthesis is the rate of synthesis of its β-subunit.

2 CLEARANCE

One of the earliest methods of monitoring plasma clearance of hormones was the injection of radioactively labeled hormone. Not only could the rate of plasma clearance be calculated by this method, but also the pathways of inactivation and excretion could be examined by following the uptake of radioactive hormone by certain tissues. Tritiated biologically active hCG or FSH has been injected into female rats, and tissue distribution and plasma membrane concentration of labeled gonadotropin monitored (1). Both [^{3}H]hCG and [^{3}H]FSH were concentrated primarily in the ovary, kidney, and liver. Comparison of plasma clearance of labeled and unlabeled hormone revealed no significant difference in plasma half-life. However, when ^{131}I-labeled asialo-hCG was injected into rats, 86% of the injected hormone concentrated in the liver 30 min after injection, 10% was found in the kidneys, whereas only 0.07% of the labeled hormone concentrated in the ovary (3). In separate studies, progressive removal of sialic acid (desialylation) from hCG led to markedly increased plasma clearance of the hormone and significantly decreased biologic activity (1). When more than 62% of the sialic acid was removed, the plasma half-life decreased to less than 1 min. Moreover, desialylated hCG, LH, and FSH have markedly decreased to absent biologic activities *in vivo* (1). The latter undoubtedly reflects their markedly shortened plasma half-lives, which obviate significant concentrations of those hormones from interacting with their target tissues. The latter suggestion is supported by the observation that desialylated hCG is at least as active as its fully sialylated molecule *in vitro* (1).

A direct correlation between the sialic acid content of a glycoprotein hormone and its plasma half-life has been observed. For gonadotropins, hCG, with 12% sialic acid, has the longest plasma half-life, whereas FSH and LH, with 5% and 1% sialic acid, respectively, have proportionately shorter plasma half-lives. The plasma half-lives of highly purified hCG subunits are similarly ordered. The β-subunit, with 10% sialic acid, has a plasma half-life of 11 min (rapid component) in immature female rats, whereas its α-subunit, with 4% sialic acid, as a plasma half-life of 6 min (rapid component) in that same animal model (1).

The α-subunits of the glycoprotein hormones are very similar in structure and carbohydrate content. Metabolic clearance studies of TSH-α (4) and hCG-α (5) revealed similar clearance rates of approximately 60 ml/min · m² in human volunteers. In both rat and man, rapid hCG-α-subunit clearance accompanied by a slower hCG-β-subunit clearance was observed. This may in part be due to the higher carbohydrate content on the β-subunit. However, Wehmann and Nisula (5) observed that the metabolic clearance rates of hCG-β, intact hLH, and hTSH were similar, even though their carbohydrate contents varied considerably. They suggested that molecular weight and carbohydrate content may be important determinants of metabolic clearance.

Even though the α-subunit is cleared more rapidly, plasma levels of α-subunit generally exceed those of the β-subunit. That observation reflects the increased secretion of α-subunits into the peripheral circulation and does not reflect spontaneous dissociation of intact hormone in blood. If the latter obtained, then equimolar concentrations of α and β would result; since the β-subunit has a longer plasma half-life, one would expect higher circulating concentrations of that subunit. Plasma concentrations result not only from metabolic clearance but also from secretion.

3 TISSUE DISTRIBUTION

Tissue distributions of α- and β-subunits of hCG are similar to those of native hCG (1). The primary difference is that the subunits do not concentrate in gonadal tissue, reflecting the lack of significant intrinsic biologic activity of the glycoprotein-hormone subunits (1). However, renal and hepatic uptake of hCG subunits is similar to that of native hCG. Apparently the pathways of inactivation and excretion are the same for subunits and intact hormone.

3.1 Liver Metabolism

Several groups have observed hepatic uptake and subsequent degradation of gonadotropins (6, 7). Markkanen et al. (6) characterized the degradatory products of labeled hCG. Their results suggest that the hormone may not only be dissociated into its subunits, but also catabolized into single amino acids. Hudgin et al. (8) isolated a hepatic membrane receptor responsible for the uptake of serum glycoproteins, primarily as asialoglycoproteins. This receptor has subsequently been isolated from subcellular organelles, including the Golgi apparatus and lysosomes. The proposed mechanism included (1) binding of glycoproteins to receptor, (2) subsequent internalization by endocytosis, and (3) eventual catalysis by lysosomal enzymes (9). Whether that mechanism plays a major part in metabolism of glycoprotein hormones is unknown, since no definitive data is available on the extent to which the glycoprotein hormones are desialylated under physiological conditions.

3.2 Kidney Metabolism

The kidney also inactivates and excretes the gonadotropins. In fact, Markkanen et al. (10) found a greater concentration of injected hCG in kidney than in the liver tissue; but since the liver is a larger organ, it probably plays a greater role. The injected hormone was primarily concentrated in lysosomes and large endocytotic vesicles of proximal tubule cells (10). The degradation products were the same as in the liver. Analysis of urine from these animals revealed not only degradation products, but also intact hormone (6). After injection of hCG into humans, only 5–25% of that exogenous hormone is excreted in the urine in biologically active form.

Apparently the other gonadotropins are cleared in a similar manner. Bilateral nephrectomy dramatically increases the plasma half-life of both FSH and LH (11) and lysosomal catabolism has been suggested for LH (7). The mechanisms for uptake of these hormones into renal cells have not been determined. However, Rajaniemi et al. (12) found hCG to be concentrated primarily in renal cytosol and did not find any evidence suggesting binding to a renal plasma membrane receptor.

3.3 Gonadal Inactivation

The final tissue involved with gonadotropin inactivation is its target tissue—the gonad. Gonadotropins first interact at the cell surface with a specific plasma membrane receptor. There is increasing evidence that once gonadotropin is bound to its specific receptor, at least some of it is internalized and subsequently metabolized through pathways described for hormonal metabolism in liver and renal tissues (13). Both subunits and amino acid degradation products have been isolated from gonadal cells (14, 15). Moreover, evidence suggests that hormonal degradation is not essential for biologic activity (16).

Although the biochemical pathways of inactivation and excretion of glycoprotein hormones are still unknown, several tissues in this process have been identified, and progress is continuing in elucidating the variety of intracellular mechanisms contributing to hormonal catabolism.

REFERENCES

1 Vaitukaitis, J. L., Ross, G. T., Braunstein, G. D., *et al.,* Gonadotropins and their subunits: Basic and clinical studies. *Rec. Prog. Horm. Res* **32**, 289 (1976).

2 Daniels-McQueen, S., McWilliams, D., Birken, S., *et al.,* Identification of mRNAs encoding the alpha and beta subunits of human choriogonadotropin. *J. Biol. Chem.* **253**, 7109 (1978).

3 Birken, S., and Canfield, R. E., Labeled asialo-human chorionic gonadotropin as a liver-scanning agent. *J. Nucl. Med.* **15**, 1176 (1974).

4 Kourides, I. A., Re, R. N., Weintraub, B. D., *et al.,* Metabolic clearance rates of subunits of hTSH. *J. Clin. Invest.* **59**, 508 (1977).

5 Wehmann, R. E., and Nisula, B. C., Metabolic clearance rates of the subunits of hCG in man. *J. Clin. Endocrinol. Metab.* **48**, 753 (1979).

6 Markkanen, S., Tollikko, K., Vanha-Pertulla, T., *et al.,* Disappearance of human (^{125}I) iodo-chorionic gonadotropin from the circulation in the rat: Tissue uptake and degradation. *Endocrinology* **104**, 1540 (1979).

7 Ascoli, M., Liddle, R. A., and Puett, D., Renal and hepatic lysosomal catabolism of luteinizing hormone. *Molec. Cell Endocrinol.* **4**, 297 (1976).

8 Hudgin, R. L., Pricer, W. E., Jr., Ashwell, G., *et al.,* The isolation and properties of a rabbit liver binding protein specific for asialoglycoproteins. *J. Biol. Chem.* **249**, 5536 (1974).

9 Tanabe, T., Pricer, W. E., Jr., and Ashwell, G., Subcellular membrane topology and turnover of a rat hepatic binding protein specific for asialoglycoproteins. *J. Biol. Chem.* **254**, 1038 (1979).

10 Markkanen, S. O., and Rajaniemi, H. J., Uptake and subcellular catabolism of human choriogonadotropin in the proximal tubule cells of rat kidney. *Molec. Cell Endocrinol.* **13**, 181 (1979).

11 Gay, V. L. Decreased metabolism and increased serum concentration of LH and FSH following nephrectomy of the rat: Absence of short-loop regulatory mechanisms. *Endocrinology* **95**, 1582 (1974).

12 Rajaniemi, H. J., Hirshfield, A. N., and Midgley, A. R., Jr., Gonadotropin receptors in rat ovarian tissue. I. Localization of LH binding sites by fractionation of subcellular organelles. *Endocrinology* **95**, 579 (1974).

13 Catt, K. J., Harwood, J. P., Aquilera, G., *et al.,* Hormonal regulation of peptide receptors and target cell responses. *Nature* **280**, 109 (1979).

14 Huhtaniemi, I., Rajaniemi, H., Martikainen, H., *et al.,* Autoregulation of LH/hCG receptors and catabolism of hCG in rat testis. *Int. J. Androl.* Suppl. **2**, 276 (1978).

15 Amsterdam, A., Nimrod, A., Lamprecht, S. A., *et al.,* Internalization and degradation of receptor-bound hCG in granulosa cell cultures. *Am. J. Physiol.* **236**, E129 (1979).

16 Ascoli, M., Demonstration of a direct effect of inhibitors of the degradation of receptor-bound hCG on the steriodogenic pathway. *J. Biol. Chem.* **253**, 7839 (1978).

CHAPTER EIGHTEEN

MECHANISMS FOR STIMULATION OF STEROIDOGENESIS

ROBERT T. CHATTERTON, JR.

Department of Obstetrics and Gynecology
Northwestern University Medical School
Chicago, Illinois

1 INTRODUCTION

Preformed cholesterol is the primary substrate for steroid biosynthesis. Cholesterol is synthesized to some extent within the endocrine glands, but the formation of steroids occurs largely from cholesterol produced at other sites and taken up by the endocrine glands. Cholesterol circulates in the blood in complexes of lipoproteins. The uptake of low-density lipoproteins by cells that utilize cholesterol, including cells of the endocrine glands, is a regulated process involving specific membrane receptors. Biosynthesis of cholesterol by the steroid-producing cell is suppressed by the uptake of cholesterol and increased by depletion of cholesterol from the cell.

Although cholesterol is synthesized by many if not all cells of the body, the conversion of cholesterol to pregnenolone, that is, cleavage of a C_6 unit from the side chain of cholesterol, occurs to a significant extent only in endocrine glands. It is this conversion of cholesterol to a steroid that is stimulated by the tropic hormones in the adrenal, ovary, and testis.

Stimulation of side-chain cleavage involves hydrolysis of stored cholesterol esters to free cholesterol, transport of cholesterol to a mitochondrial binding site with the aid of a specific binding protein, and an oxidative cleavage process involving cytochrome P_{450}. The first two steps are apparently mediated by cAMP-dependent protein kinase. The actual mechanism of side-chain cleavage is still in some doubt. At least two mechanisms are probably involved, and a separate cholesterol-cleavage system apparently exists for conversion of cholesterol sulfate to pregnenolone sulfate.

The particular steroid hormones that are produced by an endocrine steroid-producing gland depend on the enzymes that exist in the secretory cells. Different steroid products may emerge from the ovary, testis, or adrenal at different times in development of the gland or in the congenital absence of some enzymes in the biosynthetic pathways. Such differences are not results of changes in the actions of the hormones that stimulate steroidogenesis. Pregnenolone formed in the process of steroidogenesis passively diffuses from the mitochondria and is converted to progesterone or 17-hydroxypregnenolone in the microsomes. Additional transformations occur to the extent that preformed enzyme systems are present in the cell.

2   CHOLESTEROL AS A PRECURSOR OF STEROID HORMONES

In experimental studies in rats, cholesterol for adrenal steroid production has been shown to be almost exclusively from extraadrenal sources. The specific activity (radioactivity per unit mass) of adrenal cholesterol was found to be equal to that of plasma cholesterol when [^{14}C]cholesterol was infused at a constant rate intravenously (1). If significant *de novo* synthesis of cholesterol had occurred within the gland, the specific activity of adrenal cholesterol would have been lower than that of plasma cholesterol even after steady state had been reached. Also, when labeled cholesterol was infused intravenously in pregnant women, the specific activities of steroids formed by the placenta were similar to that of plasma cholesterol once equilibration had been achieved (2). *In vitro* studies of placental tissue have confirmed the very low capacity of the placental trophoblast for synthesis of cholesterol from acetate. The ovary probably also obtains most of the cholesterol that is utilized for steroid biosynthesis from the plasma, but significant synthesis of cholesterol from acetate can be demonstrated (3).

Considerable progress has been made in recent years in the biochemistry of cholesterol utilization by steroid-producing tissues. This work was reviewed by Brown *et al.* (2). A specific uptake process for cholesterol that is carried in plasma lipoproteins has been described. Uptake of cholesterol from plasma lipoprotein fractions that are separated centrifugally appears to be limited almost exclusively to the low-density lipoprotein (LDL) fraction. The LDL is composed of a core of cholesterol esters and a polar coat that consists of a protein (apoprotein B), phospholipids, and a small amount of free cholesterol. This complex contains approximately 1600 molecules of cholesteryl esters and 7–10 monomers of apoprotein B. The LDL-receptor pathway was first described in human fibroblasts, and because of the ease of culturing these cells, the process of uptake is probably best delineated in fibroblasts. As shown in Fig. 1, specificity of the process is dependent upon a cell surface receptor that binds the apoprotein component of LDL. The receptor will not bind the other major lipoprotein fractions of serum. LDL appears to transport cholesterol from absorptive and synthetic sites to parenchymal cells that utilize the sterol for structural and metabolic purposes. Binding of LDL to the receptor site of the fibroblast plasma membrane results in invagination of the membrane to form a vesicle. Internalized LDL interacts with lysosomes, where the protein component is hydrolyzed, and cholesterol esters are cleaved by an acid lipase.

Regulation of uptake of LDL by fibroblasts and other cells is dependent upon the concentration of free cholesterol in the cell and is exerted by inhibition of receptor synthesis. Increasing concentrations of free cholesterol in the cell also cause a suppression of cholesterol biosynthesis within the cell. A similar effect has been demonstrated in human placental trophoblast cells (4), even though the rate of cholesterol synthesis in these cells is always very low com-

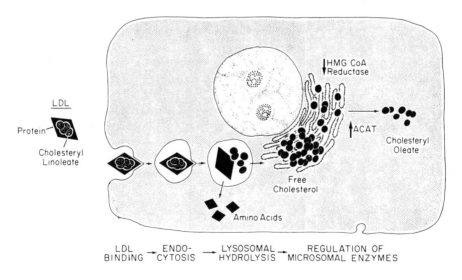

Figure 1 Sequential steps in the pathway of low-density lipoprotein (LDL) metabolism in cultured fibroblasts, from Brown and Goldstein, *Science* **191**, 150, 1976, with permission.

pared with the production of progesterone from circulating cholesteryl esters in pregnancy. Intracellular cholesterol suppresses cholesterol biosynthesis by suppressing the activity of 3-hydroxy-3-methylglutaryl-coenzyme A (HMG CoA) reductase as well as the HMG CoA synthase. The presence of free cholesterol in the cell also activates a microsomal acyl-coenzyme A cholesterylacyltransferase that esterifies the excess free cholesterol generated by the action of the lysosomal enzymes on LDL.

The uptake of cholesterol in the form of LDL has been shown in steroid-secreting cells in culture to permit maximal biosynthesis of steroid products. The steroidogenic response to adrenocorticotropic hormone (ACTH) of the Y-1 clone of adrenal tumor cells in culture was increased in proportion to the LDL concentration in the medium (5). In the presence of ACTH, more LDL entered the cell at each concentration of LDL because ACTH increased the number of LDL receptors by twofold to threefold. Whether this is a direct action of ACTH on receptor production or secondary to decreased free cholesterol in the cell is not known. LDL similarly increased production of progesterone by the cultured placental trophoblast cells (4), although no tropic hormone effect was sought. Studies on the binding of ^{125}I-labeled LDL to steroid-producing cells show that the binding is saturable, reversible (particularly in the presence of polyanions), has high affinity, and is relatively specific for the protein component of LDL. The receptor is destroyed by pronase and requires a divalent cation, Ca^{2+} or MN^{2+}, for LDL binding. The highest receptor concentrations among tissues analyzed were in adrenal, corpus luteum, and testis. In the presence of chloroquine, a nonspecific inhibitor of lysosomal enzymes, the hydrolysis of lipoprotein was blocked, and intact ^{125}I-labeled LDL accumulated within the lysosomes. Cultured Y-1 cells accumulate increasingly greater concentrations of cholesteryl esters when incubated in the presence of LDL. Free cholesterol increases initially, and reaches a steady state in the cultured cells within 24 hr.

The action of ACTH, and presumably LH and HCG on their target organs as well, results in a depletion of cholesteryl esters, probably utilizing free cholesterol first. This depletion is followed by increases in activity of HMG CoA reductase and HMG CoA synthase and in measurable increases in [14C]acetate incorporation into cholesterol. Cholesterol exists in the steroid producing cell in several forms. There is a fixed pool of free cholesterol in cell membranes that does not contribute to net synthesis of steroid; there is a small pool of metabolically active free cholesterol; and there is the larger pool of cholesteryl ester droplets in the cytoplasm that can be mobilized for steroid synthesis.

3 ACTIONS OF TROPIC HORMONES ON STEROID-PRODUCING ENDOCRINE GLANDS

The tropic hormones in this narrow definition are those hormones such as ACTH that act on their target endocrine gland to stimulate steroid biosynthesis. Angiotensin II would also be included, as a hormone acting on the adrenal glomerulosa, in this definition. LH is the primary tropic hormone that stimulates steroidogenesis in the ovary and testis. In pregnancy, structurally related hormones such as hCG produced by the human placenta would also be included in this group of hormones. Prolactin, which has been called luteotrophic hormone (LtH), is no longer considered a primary luteotropin, and therefore the name LtH is currently inappropriate, but prolactin does serve a collateral role in ovarian function, and prolactin and growth hormone may serve similar functions in adrenal function.

Each of the tropic hormones acts by binding with high affinity to membrane receptors on its target cells. These receptors are associated with adenylylcyclase (see Chapter 15 for a review of this process), and binding of the tropic hormone to its receptor results in activation of the enzyme. The enzyme in the presence of Mg^{2+} converts ATP to cyclic 3', 5'-AMP and inorganic phosphate. The activation of protein kinase by cAMP and the subsequent cascade by which amplification is achieved for conversion of inactive phosphorylase to the active form is by now classic biochemistry. The idea (6) that glycogen could serve as a source of reducing equivalents for $NADP^+$ for steroidogenesis after activation of phosphorylase by cAMP, however, now seems untenable, since only rather small amounts of glycogen are stored in steroidogenic tissues. Further, although LH stimulates glucose uptake by ovarian tissue *in vitro*, the steroidogenic response to LH can be elicited without affecting glucose uptake. LH therefore probably does not primarily stimulate steroidogenesis by any process that results in greater availability of glucose for oxidation. This is rather enigmatic, because addition of NADPH to tissue slices of corpora lutea results in an increase in the rate of progesterone formation (7). Certainly glucose, when oxidized through the pentose shunt, could yield the required cofactor. It seems, however, that under most conditions NADPH is not rate limiting in intact cells.

Although many investigators began to question the relationship between cAMP production and the action of tropic hormones on steroidogenesis, largely from observations that concentrations of LH or ACTH capable of stimulating steroidogenesis were substantially less than those required to produce a detectable increase in the concentration of cAMP in the target tissue, several lines of evidence now substantiate a physiological intermediary role of cAMP. The most likely explanation for the inability to detect increases in cAMP is that it and the protein kinase it activates are compartmentalized in the cell such that local high concentrations may not be detectable after the cells are homogenized.

Furthermore, some of the primary mechanisms by which tropic hormones have now been shown to stimulate steroidogenesis are activated by cAMP. These are (a) the activation of cholesteryl ester hydrolase and (b) the synthesis of a protein that is involved in the translocation of cholesterol to the side-chain cleavage site in the mitochondrion of steroid-producing cells (see Fig. 2).

3.1 Cholesteryl Ester Hydrolase Activation by Tropic Hormones

Herbst (8) showed that in superovulated rat ovaries, administration of LH resulted in a marked depletion of cholesteryl esters from the tissue within 3–5

MODES OF ACTION OF ACTH

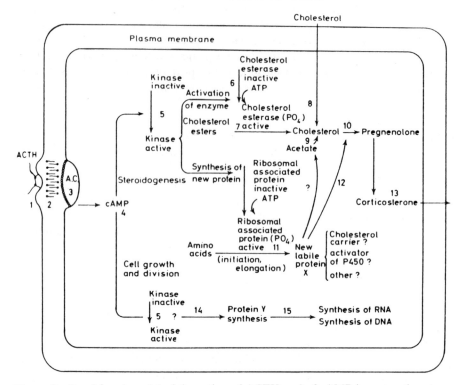

Figure 2 Provisional model of the action of ACTH and of cAMP in promoting steroidgenesis, cell growth, and division in cells of the adrenal cortex. From Sayers, Beall, and Seelig, *MTB International Review of Science*, Biochemistry Series One, Vol. 8, MTB Press, Ltd., Lancaster, England (1974), p. 27, with permission.

hr. This was followed by a reaccumulation of the esterified cholesterol within 18 hr to even higher levels than were present before LH treatment. Behrman and Armstrong (9) found that the ester hydrolase was measurably increased within an hour after LH administration. Subsequently, the activation of this enzyme has been more fully characterized (10). Cholesteryl ester hydrolase is activated by a cAMP-dependent activation of a protein kinase that phosphorylates the hydrolase in the presence of ATP. Activation of the hydrolase is not affected by blockers of protein synthesis such as cycloheximide.

Although tropic hormones, particularly LH, have been suggested to increase the synthesis of cholesterol from low-molecular-weight precursors, this occurs to a much greater extent after stores of cholesteryl esters have been depleted and suppression of HMG CoA reductase has presumably been decreased. The stimulation of [^{14}C]acetate into cholesterol is therefore considered to be a secondary response to tropic hormones.

Maintenance of the stores of cholesteryl esters seems to be a complex process. As indicated earlier, the presence of free cholesterol in the cell after uptake and hydrolysis of LDL stimulates the esterification of cholesterol itself by primarily long-chain fatty acids. Cholesterol esterified with fatty acids having the 18:1, 20:4, 22:4, and 22:5 structures appears to be preferentially hydrolyzed by the actions of gonadotropins in luteinized rat ovaries (11). Stimulation of steroidogenesis by LH invariably results in a greater than initial concentration of cholesteryl esters; the mechanism for this response is not known, but may be related to the increase in receptor-mediated uptake of LDL following gonadotropin stimulation. Other hormones are also involved in maintenance of the intracellular stores of esterified cholesterol; hypophysectomy of the rat leads to depletion of cholesterol from the ovary. Cholesteryl ester concentration in hypophysectomized rats is so low that a further lowering by administration of LH cannot be detected, and no compensatory increase in cholesteryl ester content 24 hr after LH treatment, characteristic of intact rats, is observed (3). The cholesterol depletion and other responses to LH can be partially restored by prior administration of prolactin to the hypophysectomized rats. This seems to be the primary function of prolactin in the ovary, at least as far as steroidogenic responses are concerned. The importance of prolactin for ovarian function in other species is less certain, but concomitant administration of prolactin has been shown to increase the effectiveness of hCG in extending luteal secretion of progesterone during the menstrual cycle in women.

3.2 Stimulation of the Synthesis of a Cholesterol-Binding Carrier Protein by Tropic Hormones

Since the cholesterol side-chain cleavage (SCC) system has been found to be localized in the mitochondria of adrenal, ovary, testis, and placental steroidogenic cells, a mechanism for the transfer of cholesterol to the enzymatic site from its sites of storage in the cytoplasm is required. Less work has been done on the transport system in gonadal than in adrenal tissue, but the mechanisms appear to be similar. The possibility of a transport system was proposed by Hechter (12) and was developed further by Garren and associates (13). Cycloheximide, an inhibitor of protein synthesis, was shown to block the acceleration of steroidogenesis by ACTH, and it caused an increase in the accumulation of extramitochondrial lipid droplets; inhibition is exerted before the cholesterol SCC reaction. ACTH stimulates, through activation of adenylyl cyclase, the translation of a heat-stable cholesterol-binding protein. New synthesis of mRNA is not required, since activation of the process can be achieved in the presence of an inhibitor of RNA transcription, actinomycin D. The protein has a specific binding affinity for cholesterol and stimulates the activity of the cholesterol SCC system of adrenal mitochondria. Ungar *et al.* (14) have proposed that this protein is a cholesterol-carrier protein that is involved in transport of cholesterol into the mitochondrion. Unfortunately, ACTH does not appear to increase synthesis of the protein described by Ungar *et al.* (14). Some possibility exists, however, that ACTH, by means of cAMP, activates the carrier protein (15), increasing its affinity for cholesterol for the intramitochondrial binding site.

The transport (carrier) protein may be identical to the sterol-carrier protein of liver required for cholesterol synthesis by microsomal enzymes, since the protein isolated from liver is capable of promoting SCC of cholesterol by adrenal mitochondria.

In any case, ACTH or dibutyryl-cAMP treatment of rats increases the level of free cholesterol in adrenal mitochondria and the conversion of cholesterol to pregnenolone; the effect on cholesterol uptake by mitochondria is not diminished when conversion of cholesterol to pregnenolone is inhibited by aminoglutethimide (16). It is the availability of free cholesterol within the mitochondrion that controls the rate of SCC (17). Since cycloheximide does not acutely inhibit the activity of cholesteryl ester hydrolase, it must inhibit steroidogenesis by blocking the mechanisms for transfer of free cholesterol to the SCC site within the mitochondrion.

3.3   Cholesterol Side-Chain Cleavage System

Cholesterol is cleaved to pregnenolone and isocaproaldehyde by concerted actions of a mixed function oxidase and a desmolase (lyase) system within the mitochondria of steroidogenic cells. Electrons are transported from the citric acid cycle intermediates to NAD^+ and then by an ATP-dependent pyridine nucleotide transhydrogenase to $NADP^+$. A flavoprotein (adrenodoxin reductase) oxidizes NADPH, and electrons are passed on through the nonheme-iron protein (adrenodoxin) to cytochrome P_{450}. The oxidase then effects concomitant hydroxylation of the steroid molecule and the reduction of oxygen to water. This cytochrome P_{450} is similar to that found in the liver and kidney microsomes, where it functions to hydroxylate a large number of substances before their elimination from the body. The shift in light absorption from 430 to 450 nm when combined with carbon monoxide is characteristic of the cytochrome from all of these sources; the purified protein has a molecular weight of 48,000, and no differences in biophysical characteristics between P_{450scc} of corpus luteum and adrenal sources have been observed (18). The electron transport system is described in all recent basic biochemistry texts.

The steroid-substrate binding to cytochrome P_{450} is characterized by a low- to high-spin transition in the spin state of the hemoprotein. The spin transition is accompanied by a change in the absorption maximum of the cytochrome from 416 to about 390 nm. Jefcoate et al. (19) observed two high-spin forms of cytochrome P_{450scc} from EPR spectra; the two forms apparently were not in equilibrium, since they were separately suppressible by different substrates. Other investigators also have characterized at least two cholesterol SCCs in adrenal mitochondria; one for free cholesterol and another for cholesterol sulfate and its acyl ester (20). Each substrate inhibited the others competitively, but with a much higher K_I than would be expected if they were competing for a single site.

Studies on the specificity of the SCC system reveal that changes in side-chain length or structure is less critical than modifications in ring A (21). The presence of a 25-hydroxyl group permits binding on the same site as the parent compound. Even C-20 *t*-butyl or *p*-tolyl analogs can be cleaved (22).

Several mechanisms for SCC have been proposed:

1 Sequential hydroxylation of cholesterol to (22*R*) 22-hydroxycholesterol to (20*R*, 22*R*) 20, 22-dihydroxycholesterol

2 Hydroperoxide mechanism with (20*S*) 20-perhydroxycholesterol and (20*R*, 22*R*) 20, 22-dihydroxycholesterol as intermediates

3 A reactive-intermediate mechanism

4 An epoxy-diol mechanism

The sequential hydroxylation mechanism was described by Burstein and Gut (23). The (22*R*) 22-hydroxycholesterol was the best substrate for pregnenolone formation, even better than the dihydroxy compound. The (20*S*) 20-hydroxycholesterol was the least rapidly converted. Kinetic studies confirmed the reaction sequence. However, a direct conversion of cholesterol to pregnenolone had to be assumed as an additional pathway to account for the rate of pregnenolone formation. The direct pathways were postulated to occur as "enzymatic concerted attacks by an oxygen-carrying moiety leading to products in one step, without any intermediates appearing in solution." Hoyte and Hochberg (24) have carried this idea further (Fig. 3) showing that analogs with C-22 fully substituted were still cleaved to form pregnenolone. Hydroxylation at C-22 evidently need not precede cleavage. The active complex apparently has the characteristics of a radical, since the product of a cyclopropylcarbinyl substituent at C-22 retained the cyclopropane ring after cleavage and no cyclobutanol or cyclobutanone were formed as would be expected from an electrophilic attack. A pathway that includes the hydroxylated intermediates probably also exists, since these compounds have been isolated as endogenous steroids from

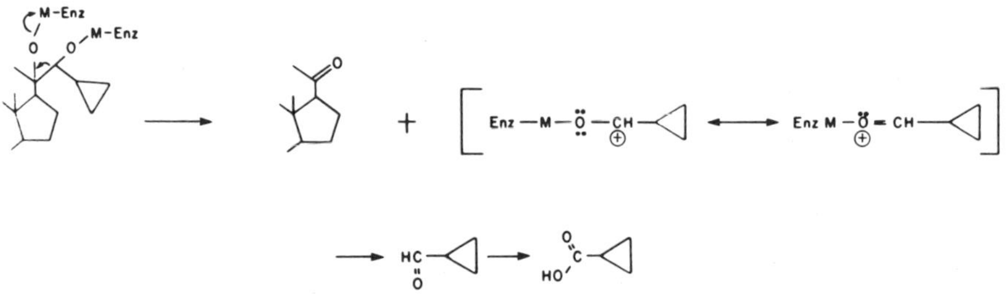

Figure 3 A proposed mechanism for C-20,22 cleavage of the cyclopropylcarbinyl analog. *M–Enz* designates a metalloenzyme. The partial structure represents a Δ^5-3β-ol steroid. From Hoyte and Hochberg, *J. Biol. Chem.* **254**, 2278, 1979, with permission.

adrenal tissue. Alternatively, these products could represent compounds that dissociated away from the enzyme complexes before the SCC was complete.

The hydroperoxide mechanism is not inconsistent with the processes mentioned above (25), and supports the concept of a ferroatomic oxygen complex as the intermediate. The possibility of an epoxide intermediate has not been supported by studies in which chemically synthesized 20, 22-epoxycholesterols were synthesized and incubated with adrenal mitochondrial preparations; none was converted to pregnenolone, and they only slightly inhibited the conversion of cholesterol itself to pregnenolone (26).

Pregnenolone in general is readily converted by existing enzymes of the ovary, testis, placental trophoblast, or adrenal to progesterone or 17-hydroxy-pregnenolone and other metabolic products. This process is presented in the following chapter.

REFERENCES

1 Schulster, D., Burstein, S., and Cooke, B. A., Biosynthesis of Steroid Hormones, *in* Schulster, D. (ed.), *Molecular Endocrinology of the Steroid Hormones*, p. 44, Wiley, New York (1976).

2 Brown, M. S., Kovanen, P. T., and Goldstein, J. L., Receptor mediated uptake of lipoprotein-cholesterol and its utilization for steroid synthesis in the adrenal cortex. *Rec. Prog. Horm. Res.* **35**, 215 (1979).

3 Armstrong, D. T., Gonadotropins, ovarian metabolism and steroid biosynthesis. *Rec. Prog. Horm. Res.* **24**, 255 (1968).

4 Winkel, C. A., Snyder, J. M., McDonald, P. C., *et al.*, Regulation of cholesterol and progesterone synthesis in human placental cells in culture by serum lipoproteins. *Endocrinology* **106**, 1054 (1980).

5 Faust, J. R., Goldstein, J. L., and Brown, M. S., Receptor mediated uptake of low density lipoprotein and utilization of its cholesterol for steroid synthesis in cultured mouse adrenal cells. *J. Biol. Chem.* **252**, 4861 (1977).

6 Haynes, R. C., and Berthet, L., Studies on the mechanism of action of the adrenocorticotropic hormone. *J. Biol. Chem.* **225**, 115 (1957).

7 Savard, K., LeMaire, W., and Kumari, L., Progesterone Synthesis from Labeled Precursors in the Corpus Luteum, *in* McKerns, K. W. (ed.), *The Gonads*, p. 119, Appleton-Century-Crofts, New York (1969).

8 Herbst, A. L., Response of rat ovarian cholesterol to gonadotropins and anterior pituitary hormones. *Endocrinology* **81**, 54 (1967).

9 Behrman, H. R., and Armstrong, D. T., Cholesterol esterase stimulation by luteinizing hormone in luteinized rat ovaries. *Endocrinology*, **85**, 474 (1969).

10 Boyd, G. S., Arthur, J. R., Beckett, G. J., *et al.*, The role of cholesterol and cytochrome P-450 in the cholesterol side chain cleavage reaction in adrenal cortex and corpora lutea. *J. Steroid Biochem.* **6**, 427 (1975).

11 Tuckey, R. C., and Stevenson, P. M., Free and esterified cholesterol concentration and cholesteryl ester composition in the ovaries of maturing and superovulated immature rats. *Biochim. Biophys. Acta* **575**, 46 (1979).

12 Hechter, O., Concerning possible mechanisms of hormone action. *Vit. Horm.* **13**, 293 (1955).

13 Garren, L. D., Gill, G. N., Masui, H., *et al.*, On the mechanism of action of ACTH. *Rec. Prog. Horm. Res.* **27**, 433 (1971).

14 Ungar, F., Kan, K. W., and McCoy, K. E., Activator and inhibitor factors in cholesterol side-chain cleavage. *Ann. N.Y. Acad. Sci.* **212**, 276 (1973).

15 Marsh, J. M., The role of cyclic AMP in gonadal steroidogenesis. *Biol. Reprod.* **14**, 30 (1976).

16 Mahaffee, D., Reitz, R. C., and Nez, R. L., The mechanism of action of adrenocorticotropic hormones. The role of mitochondrial cholesterol accumulation in the regulation of steroidogenesis. *J. Biol. Chem.* **249**, 227 (1974).

17 Mason, J. I., Arthur, J. R., and Boyd, G. S., Regulation of cholesterol metabolism in rat adrenal mitochondria. *Mol. Cell. Endocr.* **10**, 209 (1978).

18 Kashiwagi, K., Dafeldecker, W. P., and Salhanick, H. A., Purification and characterization of mitochondrial cytochrome P-450 associated with cholesterol side chain cleavage from bovine corpus luteum. *J. Biol. Chem.* **255**, 2606 (1980).

19 Jefcoate, C. R., Orme-Johnson, W. H., and Beinert, H., Cytochrome P_{450} of bovine adrenal mitochondria: Ligand binding to two forms resolved by EPR spectroscopy. *J. Biol. Chem.* **251**, 3706 (1976).

20 Wolfson, A. J., and Lieberman, S., Evidence suggesting that more than one sterol side chain cleavage enzyme system exists in mitochondria from bovine adrenal cortex. *J. Biol. Chem.* **254**, 4096 (1979).

21 Aringer, L., Eneroth, P., and Nordstrom, L., Side-chain cleavage of 4-cholesten-3-one, 5-cholesten-3α-ol, β-sitosterol, and related steroids in endocrine tissues from rat and man. *J. Steroid Biochem.* **11**, 1271 (1979).

22 Hochberg, R. B., McDonald, P. D., Feldman, M., *et al.*, Studies on the biosynthetic conversion of cholesterol into pregnenolone. *J. Biol. Chem.* **249**, 1277 (1974).

23 Burstein, S., and Gut, M., Biosynthesis of pregnenolone. *Rec. Prog. Horm. Res.* **27**, 303 (1971).

24 Hoyte, R. M., and Hochberg, R. B. Enzymatic side chain cleavage of C-20 aklyl and aryl analogs of (20S)20-hydroxycholesterol. *J. Biol. Chem.* **254**, 2278 (1979).

25 Morisaki, M., Bannai, K., Ikekawa, N., and Shikita, M., Cholesterol 20,22-epoxides: No conversion to pregnenolone by adrenal cytochrome P_{450scc}. *Biochem. Biophys. Res. Commun.* **69**, 481 (1976).

26 Van Lier, J. E., Rosseau, J., Langlois, R., and Fisher, G. J., Mechanism of cholesterol side-chain cleavage. II. The enzymic hydroxyperoxide-glycol rearrangement of the epimeric 20-hydroperoxycholesterols in ^{18}O-enriched water. *Biochim. Biophys. Acta* **487**, 395 (1977).

CHAPTER NINETEEN

STEROID HORMONES: BIOSYNTHESIS, SECRETION, AND TRANSPORT

KERRY L. CHEESMAN

Department of Obstetrics and Gynecology
Northwestern University Medical School
Chicago, Illinois

1 STEROID CHEMISTRY, NOMENCLATURE, AND FUNCTION

1.1 Introduction

Steroids are a large class of organic compounds biologically derived from six isopentenyl pyrophosphate units, and are found throughout the whole animal and plant kingdoms. They include a wide range of naturally occuring compounds, among which are the sterols proper, the bile acids, the sex and adrenocortical hormones, the cardiac glycosides, and some alkaloids and other minor groups.

The name *sterol* (Greek *stereos* = "solid") was originally given to the solid alcohols obtained from the nonsaponifiable portions of lipid extracts of tissues. The general name *steroid* was introduced in 1936 to cover all compounds with a sterol-like skeleton. Although much of steroid chemistry is based on knowledge of "stereochemistry," the similarity of these two words is purely coincidental. The term *steroid* as it is used in this chapter refers only to those steroids that occur as adrenocortical or sex hormones in the human, with minor reference to important synthetic hormones with similar properties.

1.2 General Structures

The basic steroid structure is the four-ring system of cyclopentanoperhydrophenanthrene (Fig. 1). The carbon atoms of the basic ring structure are numbered 1–17; additional carbon atoms in the sex hormones are numbered as follows: 18 for the carbon atom attached to C-13; 19 for the carbon atom attached to C-10; and 20–21 for the carbon atoms attached to C-17.

The carbon atoms of the cyclohexane rings can exist in either the boat or chair conformations (Fig. 2a). The chair form is less strained and is therefore

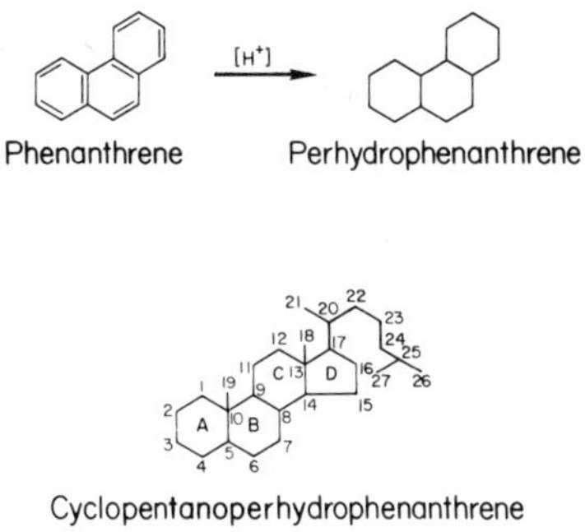

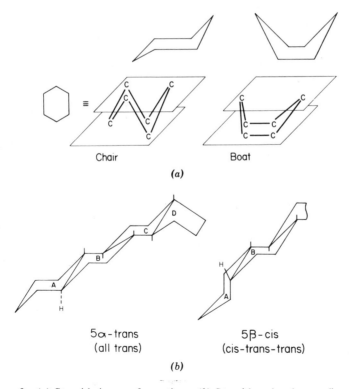

Phenanthrene   Perhydrophenanthrene

Cyclopentanoperhydrophenanthrene

Figure 1   The basic steroid molecule. Numbers refer to the individual carbon atoms; rings are lettered *A–D*.

preferred. In all naturally occuring steroids, the bond common to rings C and B is *trans*, and they are both locked in the chair conformation. The two hydrogen atoms attached to each carbon of the ring can be positioned either in the general plane of the ring (equatorial) or perpendicular to the plane of the ring (axial). By convention, the hydrogen atoms are depicted as a solid line if they project above the plane of the ring (β configuration) and with a dotted line if

Chair   Boat

(a)

5α-trans   5β-cis
(all trans)   (cis-trans-trans)

(b)

Figure 2   (*a*) Steroid ring conformations. (*b*) Steroid molecular configurations.

they project below the plane of the ring (α configuration). The bond between rings A and B can be either *trans*, giving rise to the 5α, or "allo" series, or *cis*, giving 5β, or normal, series compounds. Some of the biologically active steroids, such as dihydrotestosterone, have a *trans* junction between rings A and B. This makes the four rings almost completely planar (Fig. 2*b*).

The carbon–carbon double bond is a common feature of natural steroids. Double bonds in the hormonally important steroids generally occur between C-4 and C-5 (Δ^4) or between C-5 and C-6 (Δ^5). Conjugated double bonds are common, for instance in the A ring of the estrogens, where an aromatic ring is found. Such conjugation leads to a half-chair conformation, which is slightly less stable, and in which four carbon atoms lie in a single plane with one above and one below the plane. However, such loss of stability because of conformational strain is compensated for by the resonance energy produced by the aromatic structure.

Most steroid molecules contain one or more hydroxyl groups. Primary, secondary, and tertiary hydroxyls are found, with secondary hydroxyls the most prevalent. In the A ring, hydroxyls are common at the C-3 position, giving rise to a phenolic structure in the estrogens and an allylic alcohol in metabolites of the other major classes of steroids. The allylic hydroxyls are readily oxidized to ketones or dehydrated to a diene system *in vivo*.

Ketone groups are as prevalent among the steroids as are hydroxyl groups and are often required for biological activity. Ketone conjugation with a double bond frequently occurs, particularly as the Δ^4-3-ketone, but ketones are found at other positions in the molecule as well.

Although the aldehyde group is not common among the steroids, aldosterone, the hormone that controls the body's salt balance, contains an aldehyde group at the C-18 position. Since there are no hydrogens on the carbon atom adjacent to the carbonyl function of aldosterone, it cannot enolize, but it does form a hemiacetal with the hydroxyl group at C-11.

The carboxylic acid group occurs predominantly in the bile acids such as cholic and deoxycholic acids. These are important as surface-active agents of the digestive tract, but are not discussed further here.

1.3 Nomenclature

The complete systematic nomenclature of the steroids may be found in the IUPAC–IUB Revised Tentative Rules for Steroid Nomenclature (1) and only those portions directly relevant to the sex hormones are noted here.

Parent compounds, usually hydrocarbons, are used as a base for steroid nomenclature. Prefixes and suffixes are added to these in a systematic way in order to describe the functionality and in part the stereochemistry of the molecule. Insofar as the carbon skeleton is concerned, most or even all of its stereochemistry is fixed by the parent name. All natural steroids are named after the following saturated parent hydrocarbons: gonane (C_{17}), estrane (C_{18}), androstane (C_{19}), pregnane (C_{21}), cholane (C_{24}), and cholestane (C_{27}) (Fig. 3). Whereas

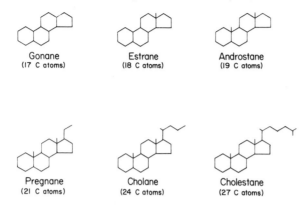

Gonane	Estrane	Androstane
(17 C atoms)	(18 C atoms)	(19 C atoms)

Pregnane	Cholane	Cholestane
(21 C atoms)	(24 C atoms)	(27 C atoms)

Figure 3 Saturated parent hydrocarbons. All steroids are derived from one of these basic units; naming of steroid compounds reflects the parental unit.

androstane and estrane refer to the 5α-isomers, pregnane refers to the 5β-isomers (see below).

Names for the partially unsaturated or aromatic steroids are derived from the saturated parent compounds by means of the systematic termination *-ene*, as recommended by IUPAC. Before a vowel, the form used is *en*. The use of "Δ" to indicate the position of unsaturation is still found in the literature, but is rapidly becoming obsolete. Locations of unsaturated bonds are shown by sequential numbering, except where that may be ambiguous, in which case the second carbon atom of the double bond is shown by a bracketed number. Unsaturation may be expressed as a number in the middle of the name of the parent hydrocarbon (i.e., pregn-4-ene) or at the beginning of the name (4-pregnene or Δ^4-pregnene).

Substituent groups are attached to the name of the parent hydrocarbon, either as prefixes or as suffixes. Only one suffix may generally be applied to a name; all other substituents are noted as prefixes. An order of priority has been attached by IUPAC to substituent groups for use as steroid suffixes: *onium salt, acid, lactone, ester, aldehyde, ketone, alcohol, amine,* and *ether*. However, since many steroids are multisubstituent molecules, this leads to a complex systematic name. For this reason, occasionally all substituent groups are listed as prefixes or suffixes to the parent hydrocarbon. Examples of steroid naming are illustrated in Fig. 4.

The position and configuration of the ring substituents are indicated by the carbon number and either α or β. Hydroxyl groups are shown by the suffix *-ol, -diol,* and so forth, or by the prefix *hydroxy-*. Thus the two alcohol substituents of cortisone may be noted as

$$17\alpha,21\text{-}dihydroxy\text{-}4\text{-pregnene-}3,11,20\text{-trione}$$

or as

$$4\text{-pregnene-}17\alpha,21\text{-}diol\text{-}3,11,20\text{-trione}$$

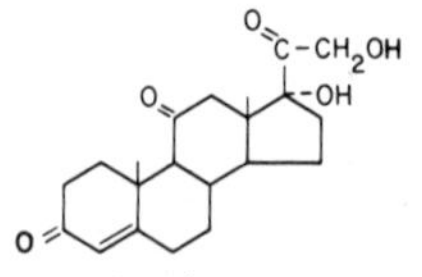

Cortisone
17α,21-dihydroxy-4-pregnene-3,11,20-trione
4-pregnene-17α,21-diol-3,11,20-trione

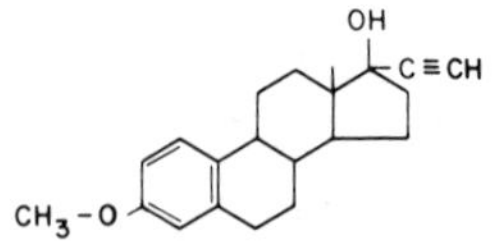

Mestranol
Ethinylestradiol 3-methyl ether
17α-ethinyl-3-methoxyestradiol
17α-ethinyl-1,3,5(10)-estratriene-3,17-diol-3-methyl ether

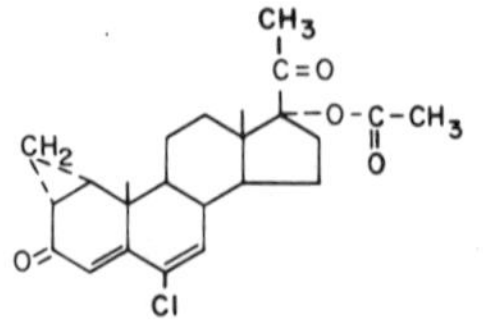

Cyproterone Acetate
17α-acetoxy-6-chloro-1α,2α-methylene-4,6-pregnadiene-3,20-dione
6-chloro-1α,2α-methylene-4,6-pregnadien-17α-ol-3,20-dione 17-acetate

Figure 4   Examples of steroid nomenclature. More than one chemical name may be correct for any given compound. For a further list, see Table 1.

Ketone groups are similarly denoted as the suffix -one, -dione, and so forth, or by the prefix oxo-. The 11-keto form of cholanoic acid may thus be written as

11-oxo-5α-cholan-24-oic acid

or as

5α-cholan-11-one-24-oic acid

For aldehyde groups, the suffix is -al, and the prefix is also oxo-, except when the change is from –COOH to –CHO, in which case the suffix is -aldehyde and the name is derived from that of the acid. For instance,

5α-cholan-24-oic-acid

becomes

5α-cholan-24-aldehyde

upon conversion of the carboxyl to the aldehyde form.

Hydrocarbon and halogen additions to the ring structure are always added as prefixes, and have no suffix forms. Commonly employed names for hydrocarbon groups are methyl-, ethyl-, vinyl-, ethynyl-, and methylene-. Chloro- is the most common halogen prefix found among steroid analogs, followed by bromo- and iodo-. Various other prefixes and suffixes are used for less-common ring substituent groups and may be found in the IUPAC nomenclature rules.

The nomenclature of the side chain at C-17 presents a special problem, as its carbon atoms are not in the same plane as the ring structure, and thus the limits of the α and β symbols for below and above the plane are exceeded.

Two systems of nomenclature are in use for these substituents—that of Fieser and Fieser (2) and that of Cahn–Ingold–Prelog (3). The first of these views the side chain from the C-17 position and looks along the plane of the side chain (Fig. 5a). From this angle, those groups to the left are arbitrarily labeled β, whereas those to the right are labeled α. This can more easily be seen if the chain is first drawn as a Fischer projection. The advantage of this system is that it maintains uniformity by retaining the α and β nomenclature, although with a different meaning for the side-chain substituents than for ring substituents.

The Cahn–Ingold–Prelog system (or sequence rule) (3) is more closely related to the three-dimensional position of the substituent group on the carbon atom (Fig. 5b). The symbols R for right and S for left indicate rotation in order of molecular weights of substituents around the carbon atom. The symbols R and S are used only in the side chain, although it is also correct (and may be found in the literature) to use them for ring substituents as well. In this system the parent hydrocarbons cholane, cholestane, ergostane (24-methylcholestane), and stigmastane (24-ethylcholestane) imply the configuration 20R, and need not be noted as such in naming the modifying compounds. Similarly, at the C-24 position, ergostane implies an S configuration and stigmastane, an R configuration.

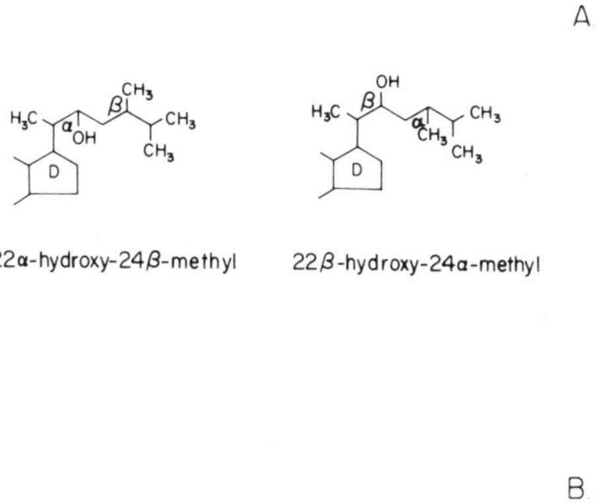

A

22α-hydroxy-24β-methyl   22β-hydroxy-24α-methyl

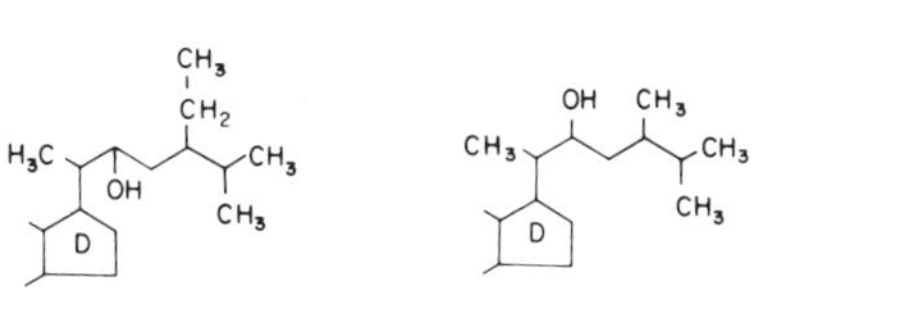

B

(22S, 24R) 22-hydroxy-24-ethyl   (22R, 24R) 22-hydroxy-24-methyl

Figure 5 Side-chain nomenclature (see text for explanation). (*A*) The system of Fieser and Fieser. (*B*) The Cahn–Ingold–Prelog system.

The prefix *allo-* always refers to 5α compounds of the pregnane series, that is, compounds in which the C-4 double bond of the parent compound has been reduced so that rings A and B assume the *trans* configuration as defined above. When there is no prefix, the 5β, or normal, series of rings A and B (*cis* configuration) is assumed. The prefix *epi-* refers to the inversion of any one of the substituents from below the ring to above or vice versa. The prefix *nor-* indicates the elimination of a carbon atom, as in nortestosterone, which lacks a methyl group at C-19, thus becoming an estrane derivative. *Homo-* indicates the enlargement of a ring. Many synthetic steroids contain six carbons in ring D (i.e., D-homotestosterone).

Most important biological and synthetic steroids have been given trivial names. The prefixes *hydro-, dihydro-, tetrahydro-, dehydro-, anhydro-, hydroxy-, dihydroxy-, tetrahydroxy-, deoxy-, dideoxy-,* and so forth refer to the addition or deletion of hydrogen and oxygen atoms from the compound referred to by its trivial name. When no carbon number indicating the exact position of this change is given, a specific position is always understood. For instance, hydrocortisone always refers to the β isomer produced by the addition of two hydrogen atoms to the ketone group at C-11. Table 1 gives the chemical names equivalent to many of the commonly employed trivial names.

1.4   Progestogens

The term progestogen encompasses a small group of naturally occurring C_{21} steroids as well as a large and continually increasing group of synthetic analogs that have the property of preventing involution of the uterine endometrium and maintaining pregnancy.

Progesterone is the primary biologically active progestogen, and is secreted in large quantities during pregnancy (4–6). In general, progestational activity is defined as a property that will cause a secretory transformation in the uterine endometrium of a spayed animal that has been primed with estrogens. Progesterone plays an essential role in establishing the decidua, a tissue that develops from the uterine endometrium at the site of attachment of the blastocyst to the uterus. During pregnancy, progesterone is believed to reduce the muscle tone of the uterus; it also helps to bring about vascularization and proliferation of the mammary gland alveolar system. The initiation of milk secretion after parturition may also be delayed until released from the inhibitory influence of progesterone.

The main extragenital effect of progesterone is thermogenesis; a metabolite of progesterone causes an increase in body temperature during the last half of the menstrual cycle. Other effects include promotion of water excretion and vascular contraction.

Progestogens were first shown to exist in 1929, when Corner and Allen extracted the corpus luteum of sows with hot alcohol to yield a material that caused a secretory change in the endometrium of spayed rabbits (7). Within six years after this discovery, four independent groups—in Germany and the

Table 1 Trivial and Systematic Names of Common Steroids and Their Derivatives

Trivial Name	Systematic Name (IUPAC–IUB)	Derivative of[a]
Aetiocholanolone	(see Eticholanolone)	
Aldosterone	11β,21-dihydroxy-3,20-dioxo-4-pregnen-18-al (11→18)lactol	P
Allocholesterol	4-cholesten-3β-ol	H
Allopregnanolone	3β-hydroxy-5α-pregnan-20-one	P
Allopregnanediol	5α-pregnane-3α,20α-diol	P
Androstanediol	5α-androstane-3β,17β-diol	A
Androstanedione	5α-androstane-3,17-dione	A
Androstenediol	5-androstene-3β,17β-diol	A
Androstenedione	4-androstene-3,17-dione	A
Androstenol	5α-androsten-3α-ol	A
Androsterone	3α-hydroxy-5α-androstan-17-one	A
Betamethasone	9α-fluoro-11β,17,21-trihydroxy-16β-methyl-1,4-pregnadiene-3,20-dione	P
Calciferol	9,10-secocholesta-5,7,10(19)-trien-3β-ol	H
Chlormadinone acetate	17-acetoxy-6-chloro-4,6-pregnadiene-3,20-dione	P
Cholaic acid	3α,7α,12α-trihydroxy-5β-cholan-24-oic acid N-(2-sulphoethyl)-amide	C
Cholanic acid	5β-cholan-24-oic acid	C
Cholecalciferol	9,10-secocholesta-5,7,10(19)-trien-3β-ol	H
Cholestanol (β-cholestanol)	5α-cholestan-3β-ol	H
Cholestanone	5α-cholestan-3-one	H
Cholesterol	5-cholesten-3β-ol	H
Cholic acid	3α,7α,12α-trihydroxy-5β-cholan-24-oic acid	C
Coprostane	5β-cholestane	H
Coprostanol (coprosterol)	5β-cholestan-3β-ol	H
Coprostenol	4-cholesten-3β-ol	H
Cortexolone (11-deoxycortisol) (S)	17,21-dihydroxy-4-pregnene-3,20-dione	P
Cortexone	(see Deoxycorticosterone)	

Table 1 (*continued*)

Trivial Name	Systematic Name (IUPAC–IUB)	Derivative of[a]
Corticosterone (B)	11β,21-dihydroxy-4-pregnene-3,20-dione	P
Cortisol (F)	11β,17,21-trihydroxy-4-pregnene-3,20-dione	P
Cortisone (E)	17α,21-dihydroxy-4-pregnene-3,11,20-trione	P
Cortol (α-cortol)	5α-pregnane-3α,11β,17,20α,21-pentol	P
Cortolone (α-cortolone)	3α,17,20α,21-tetrahydroxy-5β-pregnan-11-one	P
Cyproterone acetate	17α-acetoxy-6-chloro-1α,2α-methylene-4,6-pregnadiene-3,20-dione	P
Dehydrocortisol	11β,17,21-trihydroxy-1,4-pregnadiene-3,20-dione	P
Dehydrocortisone	17,21-dihydroxy-1,4-pregnadiene-3,11,20-trione	P
Dehydroepiandrosterone (DHA)	3β-hydroxy-5-androsten-17-one	A
Deoxycorticosterone (DOC)	21-hydroxy-4-pregnene-3,20-dione	P
21-Deoxycortisone	17α-hydroxy-4-pregnene-3,11,20-trione	P
Dexamethasone	9α-fluoro-11β,17,21-trihydroxy-16α-methyl-1,4-pregnadiene-3,20-dione	P
Diethylstilbestrol (DES)	3,4-bis (p-hydroxyphenyl)-3-hexane	—
Dihydroandrosterone	5α-androstane-3α,17β-diol	A
Dihydrocholesterol	5α-cholestan-3β-ol	H
Dihydrocorticosterone	11β,21-dihydroxy-5β-pregnane-3,20-dione	P
Dihydrocortisol	11β,17,21-trihydroxy-5β-pregnane-3,20-dione	P
Dihydrocortisone	17,21-dihydroxy-5β-pregnane-3,11,20-trione	P
Dihydrotestosterone (DHT) (5α-DHT)	17β-hydroxy-5α-androstan-3-one	A
Dydrogesterone	9β,10α-pregn-4,6-diene-3,20-dione	P
Epiandrosterone	3β-hydroxy-5α-androstan-17-one	A
16,17-Epiestriol	1,3,5,(10)-estratriene-3,16β,17α-triol	E
Epitestosterone	17α-hydroxy-4-androsten-3-one	A
Ergostanol	(24S)-24-methyl-5α-cholestan-3β-ol	H
Ergosterol	(24S)-24-methyl-5,7,22-cholestatrien-3β-ol	H
Estradiol (17β-Estradiol) (E₂)	1,3,5,(10)-estratriene-3,17β-diol	E

410

17α-Estradiol	1,3,5,(10)-estratriene-3,17α-diol	E
16α-Estradiol	1,3,5,(10)-estratriene-3,16α-diol	E
Estriol (E₃)	1,3,5(10)-estratriene-3,16α,17β-triol	E
Estrone (E₁)	3-hydroxy-1,3,5,(10)-estratrien-17-one	E
Ethinylestradiol	17α-ethinyl-1,3,5,(10)-estratriene-3,17β-diol	E
Ethisterone	17α-ethinyl-17-hydroxy-4-androsten-3-one	A
Eticholanolone (Etiocholanolone)	3α-hydroxy-5β-androstan-17-one	A
Glycocholic acid	3α,12α-dihydroxy-5β-cholan-24-oic acid N-(carboxymethyl)-amide	C
Hydrocortisone	11β,17,21-trihydroxy-4-pregnene-3,20-dione	P
Isocholesterol	3α,5-cyclo-5α-cholestan-6β-ol	H
Lanosterol	5α-lanosta-8,24-dien-3β-ol	X
Lathosterol	5α-cholest-7-en-3β-ol	H
Medrogestone	6,17-dimethyl-4,6-pregnadiene-3,20-dione	P
Medroxyprogesterone acetate (MPA)	17-acetoxy-6α-methyl-4-pregnene-3,20-dione	P
Megestrol acetate	17-acetoxy-6-methyl-4,6-pregnadiene-3,20-dione	P
Mestranol	17α-ethinyl-3-methoxy-1,3,5,(10)-estratrien-17β-ol	E
Methalone	17β-hydroxy-2β,17-dimethyl-5-androstan-3-one	A
Methadienone	17β-hydroxy-17-methyl-1,4-androstadien-3-one	A
Methenolone	17β-hydroxy-1-methyl-5α-androst-1-en-3-one	A
Methyldihydrotestosterone (methyl-DHT)	17β-hydroxy-17-methyl-5α-androstan-3-one	A
Methyltestosterone	17β-hydroxy-17-methyl-1,4-androstadien-3-one	A
Norcholanic acid	23-nor-5β-cholan-24-oic acid	C
Norcholic acid	3α,7α,12α-trihydroxy-23-nor-5-cholan-24-oic acid	C
Norethindrone	17α-ethinyl-17-hydroxy-4-estren-3-one	E
Norethynodrel	17α-ethinyl-17-hydroxy-5(10)-estren-3-one	E
D-Norgestrel	D-17α-ethinyl-17-hydroxy-18-methyl-4-estren-3-one	E
19-Nortestosterone	17β-hydroxy-4-estren-3-one	E

Table 1 (*continued*)

Trivial Name	Systematic Name (IUPAC–IUB)	Derivative of[a]
Oestradiol	(see Estradiol)	
Oestriol	(see Estriol)	
Oestrone	(see Estrone)	
Prednisolone	11β,17,21-trihydroxy-1,4-pregnadiene-3,20-dione	P
Prednisone	17,21-dihydroxy-1,4-pregnadiene-3,11,20-trione	P
Pregnanediol	5β-pregnane-3α,20α-diol	P
Pregnanedione	5β-pregnane-3,20-dione	P
Pregnanetriol	5α-pregnane-3α,17,20α-triol	P
Pregnanolone	3β-hydroxy-5β-pregnan-20-one	P
Pregnenolone	3β-hydroxy-5-pregnen-20-one	P
Progesterone	4-pregnene-3,20-dione	P
Provera	17-acetoxy-6α-methyl-4-pregnene-3,20-dione	P
Quinestradiol	3-(cyclopentyloxy)-1,3,5,(10)-estratriene-16α,17β-diol	E
Quinestrol	3-(cyclopentyloxy)-17α-ethinyl-1,3,5,(10)-estratriene-3,17-diol	E
Quingestrone	3-(cyclopentyloxy)-3,5-pregnadien-20-one	P
Retroprogesterone	9β,10α-pregna-4,6-diene-3,20-dione	P
Spironolactone	17β-hydroxy-7-mercapto-3-oxo-17α-pregn-4-ene-21-carboxylic acid γ-lactone,7-acetate or 3-(3-oxo-7α-acetylthio-17β-hydroxy-4-androsten-17α-yl) propionic acid γ-lactone	A
Squalene	2,6,10,15,19,23-hexamethyl-2,6,10,14,18,22-tetracosahexaene	X

Taurocholanic acid	5β-cholan-24-oic acid N-(2-sulphoethyl)-amide	C
Taurocholic acid	3α,7α,12α-trihydroxy-5β-cholan-24-oic acid N-(2-sulphoethyl)-amide	C
Testane	5β-androstane	A
Testosterone	17β-hydroxy-4-androsten-3-one	A
Tetrahydrocortisol	3α,11β,17,21-tetrahydroxy-5β-pregnan-20-one	P
Tetrahydrocortisone	3α,17,21-trihydroxy-5β-pregnan-11,20-dione	P
Triamcinolone	9-fluoro-11β,16α,17α,21-tetrahydroxy-1,4-pregnadiene-3,20-dione	P
Zymosterol	5α-cholesta-8,24-dien-3β-ol	H

a G, gonane; E, estrane; A, androstane; P, pregnane; C, cholane; H, cholestane; X, other carbon base. See text for structures.

United States—had produced pure crystalline preparations of this material, and the chemical structure was elucidated by Butenandt (2). This compound received the name progesterone and was shown to contain 21 carbon atoms and two oxygen atoms present as ketone groups. Several years later two other progestogens, both dihydroprogesterones, were isolated from urine of pregnant females and shown to have some progestational activity. The structures of these three natural progestogens are shown in Fig. 6.

The structure of all three is characterized by a ketone function at C-3 and a double bond between C-4 and C-5. The dihydroprogesterones differ from progesterone in having a hydroxyl group in place of the C-20 ketone, and from one another only in the spatial organization of that group. Although the C-3 ketone is found in all natural progestogens, it is not absolutely essential for biological activity, as is demonstrated by some potent synthetic progestogens (Fig. 23). Likewise, the side chain is not essential; many synthetic progestogens have been derived directly from nortestosterone, with only 18 carbon atoms, or testosterone, a C_{19} steroid (Fig. 7).

1.5   Androgens

The androgens (Greek *andros* = "man") are a diverse group of C_{19} steroids that induce and maintain characteristic secondary aspects of maleness or their vestigial remains in the female (8–10). These are growth of the penis or clitoris, growth and pigmentation of the scrotum or labia, development of prostate and seminal vesicles, deepening of the voice, and the appearance of pubic and axillary hair. Androgens also show characteristic anabolic effects such as increased protein formation, particularly of muscle and bone, causing an increased rate of linear growth in the pubescent individual. Although androgens are primarily produced in males by the Leydig cells of the testes, they are also produced in the human female by both the ovaries and the adrenal cortex.

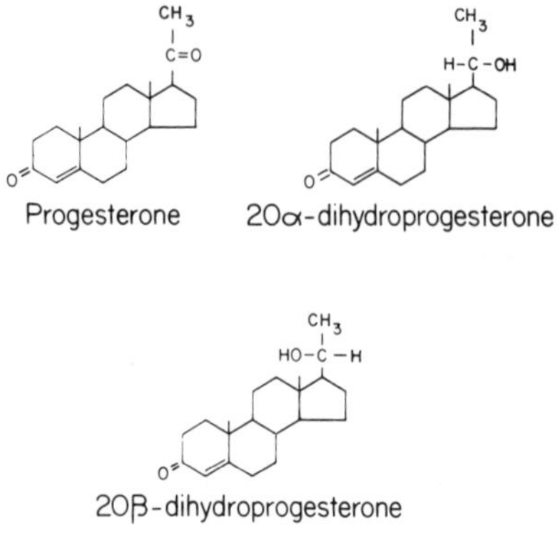

Progesterone   20α-dihydroprogesterone

20β-dihydroprogesterone

Figure 6   Structures of the natural progestogens.

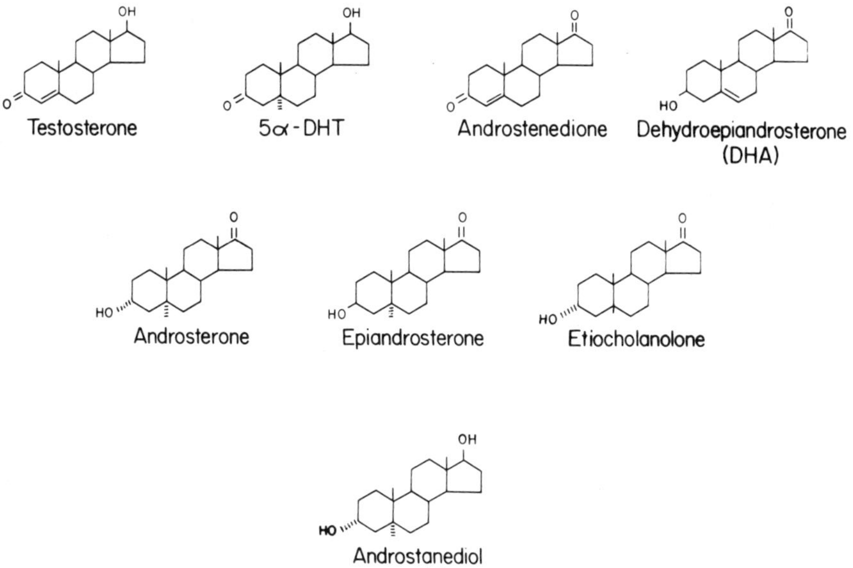

Figure 7   Structure of the common androgens in man.

Testosterone is the primary naturally occuring androgen, being found in the highest concentrations and possessing the greatest biological activity. As a whole, the androgens are among the simplest natural steroids, containing no side chain at C-17; it is replaced by a simple oxygen function (Fig. 7).

Crude extracts of testes have long been known to induce growth of secondary sexual characteristics (comb) in spayed capons (9, 10). The definition of androgenic activity in a steroid, however, is still fairly vague, since the secondary sexual characteristics are expressed by many different tissues and vary considerably between species. The extent to which these tissues are restored after androgen administration differs widely, and thus the term *androgenic activity* has arbitrarily been restricted to the effect of the compound on organs directly involved in the production and transport of semen. Androgenic activity in the female is still largely an undefined term.

The first androgen to be characterized was androsterone; its structure was suggested by Butenandt in 1932 (9, 10). Dehydroepiandrosterone (DHA) and testosterone were shown to exist in 1934 and 1935, respectively; testosterone had been isolated from bull testes, and androsterone and DHA, together with a large number of related substances, had been obtained from urine. The most potent of these natural androgens are testosterone and 5α-dihydrotestosterone. Both are physiologically active even in small tissue concentrations.

All of these C_{19} steroids contain oxygen substituents at both the C-3 and C-17 positions, and most possess double bonds between C-4 and C-5 or C-5 and C-6. More-potent androgenic activity can be induced by esterification of the hydroxyl group at C-17β, or by alkylation at the C-17α position. Both structural modifications prevent the rapid degradation of the compound and thus permit more adequate amounts to reach the target organs. However, alkylation at C-17α produces compounds with high liver toxicity, as opposed to alkylation at other positions, which does not increase toxicity, but significantly lowers the anabolic properties (10).

1.6   Estrogens

Estrogens are compounds responsible for the development and maintenance of female sex organs and secondary sexual characteristics, as well as for maintenance of cyclicity and pregnancy.

The term *estrogen* is derived from the idea that a hormone induces *estrus* (Greek *oistros* = "frenzy" + *gennein* = "to beget"), the phenomenon associated with female's biological ability for and behavioral acceptance of successful mating with a male (5, 11, 12). Many species show rapid and complete loss of mating interest following ovariectomy, but this loss can be almost completely overcome by injections of ovarian estrogens. If castration occurs before sexual maturity is reached in women, the genital tissues, mammary glands, and pubic and axillary hair fail to develop but, unlike most infraprimates, libido is not much affected, since breeding is not cyclic. The term *menstrual cycle* is used for women and higher primates that undergo a period of *menses* (Latin *mensis* = month) or bleeding at the end of each cycle from sloughing of the uterine endometrium, whereas *estrous cycle* applies to species in which behavioral cyclicity and resorption of the uterine lining are seen.

The primary role of estrogens is to control the growth and function of the uterus (5, 11). Other genital growth effects are seen on the ovaries, cervix, fallopian tubes, vagina, external genitalia, and breasts. Estrogens are necessary for development of both the duct and secretory systems of the breast and the increase in size, pigmentation, and mobility of the nipple in pregnancy. They are also mildly anabolic, promoting calcium deposition and nitrogen retention. Vasodilation, vocal cord changes, and fat deposition in the breasts and hip area are also attributable to estrogens.

Estrogens are synergistic with androgens in all the extragenital effects of both hormones, and are antagonistic to androgens in all genital effects in the female. Estrogenic effects are modified by progesterone, whereas the presence of small amounts of estrogens ensures the full effectiveness of progesterone (5).

Estrogens are C_{18} steroids that are characterized by an aromatic ring A, with its resultant phenolic C-3 hydroxyl, and the absence of a side chain at C-17. An oxygen function appears at C-17, whereas the methyl group at C-10 of progestogens and androgens is lacking. The most potent of these phenolic steroids is estradiol-17β, with estriol and estrone also showing considerable activity *in vivo* (Fig. 8).

The first estrogen isolated from human pregnancy urine was estrone isolated in 1929 by Butenandt and Doisy (12). Estriol was found in the same source the following year by Marrian, and estradiol was isolated from sow ovaries in 1935 by Doisy. The actual structure for estrone was elucidated in 1932 by Butenandt

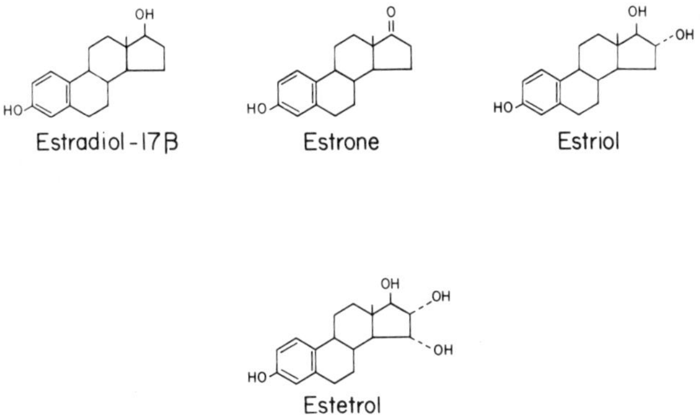

Figure 8 Structures of the naturally occuring estrogens.

and Marrian independently, and other estrogens were determined in subsequent years.

1.7 Corticosteroids

The major steroids produced by the adrenal cortex (corticosteroids) are essential for life and must be replaced following adrenalectomy. They are divided into two groups, depending upon their biologic activities. The mineralocorticoids affect the excretion of fluids and electrolytes and are principally characterized by their effect on sodium retention. The glucocorticoids affect intermediary metabolism and suppress inflammatory responses. A few steroids exist that possess both glucocorticoid and mineralocorticoid activity.

Deoxycorticosterone (DOC), synthesized in 1937 and isolated a year later by Reichstein, was the first adrenocortical steroid to show substantial *in vivo* activity (13). Cortisol and its 11-keto analog cortisone were isolated in 1937 and 1936, respectively, and in 1941 were shown to have a strong effect on carbohydrate metabolism, having anti-insulin character and producing glycosuria in large doses. However, they did not have the effect on sodium retention and survival that DOC possessed. It was on the basis of these early data that the physiological behavior of the adrenal cortex was divided into two separate phenomena into which the adrenocortical hormones were placed as they were subsequently identified (Fig. 9).

Glucocorticoid activity is closely associated with the presence of an 11β-hydroxyl group, which is essential for such activity and is dependent on the stereochemistry of the A and B rings; little if any oxidation of the 11β group occurs after reduction of the Δ^4-3-ketone to 5β-steroids (those with a *cis* A/B junction). The 11-hydroxyl group of 5α-epimers with *trans* A/B junctions is extensively oxidized, however. The reaction must occur by association of the α side of the steroid with the enzyme, and steroids with the *cis* A/B junction are not likely to bind tightly to such enzymes.

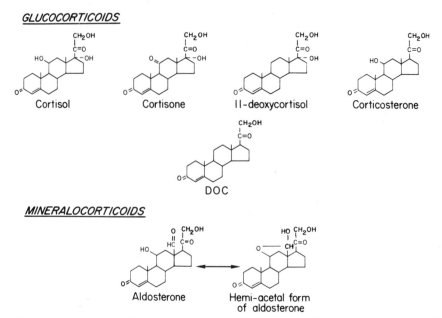

Figure 9   Structures of the major natural corticosteroids. Note the two structures for aldosterone.

A 17-hydroxyl group enhances, but is not necessary for, such activity (13). Cortisone, which is reductively metabolized to cortisol, has little or no activity of its own. Cortisol, differing only in an 11β-hydroxyl group, is the most active natural glucocorticoid. The importance of the 11β-hydroxyl group is related to its electrostatic, rather than its steric, qualities. This is demonstrated by strong potentiation of the ability to induce glycogen deposition when a halogen is placed either at C-9 or C-12, and the effect increases as the halogen becomes smaller. For instance, 9α-fluorocortisol is 11 times as active as cortisol; the *chloro-* analog is 4 times as active. The double bond between C-4 and C-5 is essential for glucocorticoid activity, but removal of carbon-21 or the 17-hydroxyl group results in only a minor loss that can be overcome by halogenization.

Glucocorticoids induce deposition of glycogen in the liver and to a smaller extent in the muscle (13). No effect on kidney glycogen results. These hormones may also lead to increased blood glucose. Glucocorticoids also lead to protein depletion from most body tissues, with the exception of the liver, and an increase in urinary excretion of sodium and urea. The overall effect is to promote the conversion of body protein to amino acids, which in the liver are converted to urea and glucose. In addition, glucocorticoids dramatically suppress the inflammatory response, as well as decreasing resistance to infection by inhibiting antibody production. Such activity seems to be dependent on the configuration around C-18 and C-20, the nonpolar 18-methyl group and the oxygen configuration at C-20 being important. Presence of a hydrogen atom at C-11 or

C-17 renders the molecule more active by comparison, whereas esters at C-21 do not appreciably alter this effect.

Mineralocorticoid behavior is induced by the presence of a 21-hydroxyl group on a 3-keto-Δ^4-steroid and is greatly enhanced by an oxygen atom on C-18 in the hydroxyl or aldehyde form (14, 15). Such activity is measured as an increase in retention of sodium ion and excretion of potassium ion. Aldosterone is the most active of these natural hormones. The presence of its 11β-hydroxyl group also produces some glucocorticoid behavior, although not as strongly as in cortisol. The principal target of aldosterone is the distal convoluted tubule of the kidney, where it causes a change in the water and ion balance. It is thought to act by stimulation of RNA synthesis and protein formation. The most serious effect of adrenalectomy is alteration of this electrolyte balance, as is also seen in adrenal insufficiencies such as Addison's disease.

2 BIOSYNTHETIC PATHWAYS FOR STEROIDOGENESIS

The pattern of steroids secreted by the respective endocrine glands is determined by the relative proportions of cell types, the anatomical organization of the gland, the blood supply, the concentration of cofactors and precursors present in the gland, and the presence of appropriate tropic stimuli.

2.1 Synthesis of Cholesterol

The biochemical reaction that initiates steroid biosynthesis is the condensation of three molecules of acetate with the aid of coenzyme A and ATP, followed by ligase, transferase, and synthetase reactions to form R-mevalonic acid. Mevalonic acid is not known to be used for anything except the biosynthesis of steroids, other terpenoids, and some alkaloids. Only the R form gives rise to these compounds; the S form is metabolically inert.

Mevalonate becomes phosphorylated in two kinase steps, ATP being the phosphate donor (16, 17). The resulting pyrophosphate is degraded by means of ATP, an isomerase, and two stereospecific proton shifts to obtain an electrophilic allyl pyrophosphate intermediate. Condensation of this intermediate, with resulting loss of pyrophosphates, produces the sesquiterpene farnesyl pryophosphate. The synthetase condensation of two molecules of farnesyl pyrophosphate yields the triterpene squalene. Cyclization of squalene uses molecular oxygen and a cyclase reaction to yield lanosterol. The stereochemical changes seen during conversion of squalene to lanosterol are attributed to the folding imposed by the enzyme on the long hydrocarbon chain, as well as the concerted transmigration of methyl groups at C-18 and C-14 and hydrogen atoms at C-13 and C-17. Lanosterol has been shown to be the triterpene that is converted to cholesterol.

The essential steps in the degradation of lanosterol to cholesterol are the removal of three ring methyl groups, together with the shift of the ring double bond from C-8 to C-5 (16, 17). The main pathway is by way of zymosterol, 5α-cholesta-7,24-dien-3β-ol, and desmosterol, although alternative pathways are known, as shown in Fig. 10. The three methyl groups are removed first, the ring double bond shifts, and then two hydrogen atoms are added at C-24. Cholesterol is the principal starting molecule for steroid hormone biosynthesis in animals, but desmosterol and lanosterol may also be used in significant quantities.

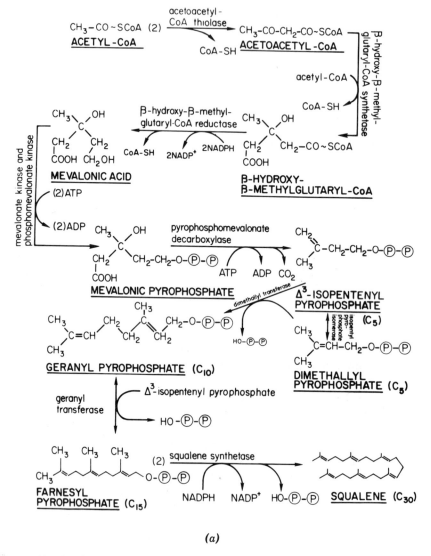

(a)

Figure 10   Outline of the biosynthetic pathways for the production of cholesterol in the human, including the major enzymes and cofactors involved in the biosynthesis.

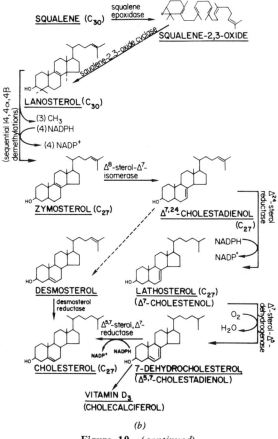

(b)

Figure 10 (*continued*)

2.2 Formation of Pregnenolone

Cholesterol is converted to pregnenolone by enzyme systems present in adreno-cortical mitochondria, with 20α-hydroxycholesterol and 20α,22β-dihydroxy-cholesterol as intermediates (Fig. 11). The supply of pregnenolone for steroid biosynthesis appears to be rate limiting, and the conversion of cholesterol to pregnenolone is regarded as the control point for the entire steroid biosynthetic process (4, 18). The splitting of the side chain of the dihydroxycholesterol to form pregnenolone and isocaproic acid is mediated by pregnenolone synthetase and probably consists of cleavage of the 20–22 carbon–carbon bond by proton expulsion and removal of the hydride ion from the 20α-hydroxyl group.

The mechanism of the cleavage of other hydroxylated intermediates, such as 20,22 R-dihydroxycholesterol, appears to involve various mixed-function oxi-dases or oxygenases (18, 19). Hydroxylation of any intermediate in the biosyn-thetic pathway requires a flavoprotein, a nonheme-iron-containing protein, cy-

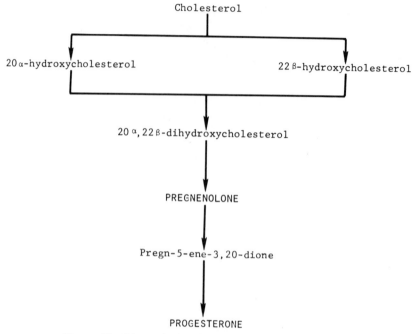

Cholesterol

20 α-hydroxycholesterol   22 β-hydroxycholesterol

20 α, 22 β-dihydroxycholesterol

PREGNENOLONE

Pregn-5-ene-3,20-dione

PROGESTERONE

Figure 11   Biosynthesis of progesterone in the human.

tochrome P_{450}, NADPH (or NADH), molecular oxygen, and the specific enzyme(s). Hydroxylation occurs when an electron, passed along the electron transport chain, reduces cytochrome P_{450}. The molecular oxygen is reduced to water, and a second activated atom of oxygen is introduced into the steroid molecule (19–21). See Chapter 18 for more details.

2.3   Synthesis of Progestogens

The further metabolism of pregnenolone to progesterone proceeds by means of microsomal pyridine nucleotide-dependent oxidation of the hydroxyl group at C-3 to a ketone group, giving rise to pregn-5-ene-3,20-dione (Δ^5-pregnenedione) (4, 6). Isomerization yields the Δ^4-derivative progesterone. The isomerization reaction consists of an internal 1,3-migration from C-6 to C-4. Most probably the same enzyme combination also operates to convert 17-hydroxypregnenolone to 17-hydroxyprogesterone, and on DHA in its conversion to androstenedione. The ovary has been extensively investigated for the detection of a Δ^5-3β-hydroxysteroid dehydrogenase. In the human, the corpus luteum of the menstrual cycle is strongly active until about the twenty-second day, and the corpus luteum of pregnancy is strongly active up to about four months, retaining at least some activity until term and becoming almost completely inactive postpartum (22).

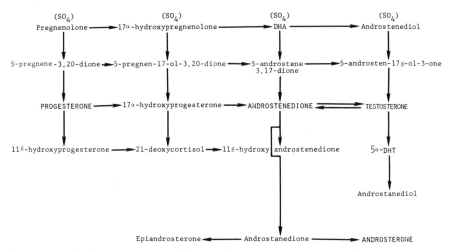

Figure 12 Outline of androgen biosynthesis in the human. A parallel pathway exists for sulfate derivatives from pregnenolone to androstenediol.

2.4 Synthesis of Androgens

Conversion of pregnenolone to testosterone and other androgens may occur by either of two pathways, as outlined in Fig. 12 (10, 23, 24). The primary pathway for testosterone biosynthesis in the testis has not been elucidated. However, unlike the adrenal cortex, where 11β-, 17-, 18-, and 21-hydroxylations all occur, testicular production of androgens proceeds exclusively through 17-hydroxylation of pregnenolone or progesterone. This hydroxylation and side-chain cleavage is accomplished by 17,20-desmolase, with 17-hydroxypregnenolone becoming DHA and 17-hydroxyprogesterone becoming androstenedione. The desmolase operates in close association with 17-hydroxylase; it is this hydroxylase and not the desmolase that is rate limiting with respect to side-chain cleavage.

The reduction of androstenedione to testosterone or 19-hydroxyandrostenedione to 19-hydroxytestosterone is the only readily reversible reaction in the androgen synthetic pathway (13, 23). The oxidoreductase, though, is apparently absent or inactive in the adrenals, and thus this conversion occurs only in the gonads (and to a small extent in the liver). On the whole, equilibration is in the direction of the more potent steroids, testosterone and estradiol, which have the 17β-hydroxyl, rather than the less potent 17-ketosteroids, androstenedione and estrone. The NADH:NAD ratio determines which steroids predominate.

As much as 60% of the testosterone produced by the testes is eventually converted to the A-ring reduction product 5α-dihydrotestosterone (5α-DHT) (10, 23–25). The highly active 5α-reductase responsible for this conversion is found in microsomal fractions of testicular and prostatic tissues and appears to be NADPH-dependent.

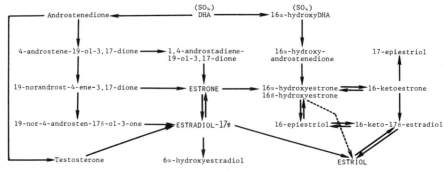

Figure 13   Outline of estrogen biosynthesis in the human.

2.5   Synthesis of Estrogens

The primary biosynthetic source of estrogens is the ovary (5, 11), which also produces androstenedione and the 17β-hydroxy intermediates, although in men and postmenopausal women the adrenals contribute the largest amount (13). Estrogens are derived by an extension of the androgen pathway (Fig. 13) in which the Δ^4-3-keto system in ring A undergoes additional α,β-desaturation and C-19 is lost, with resultant establishment of the benzoid system by enolization (Fig. 14). Testosterone is converted to estradiol in the aromatization process.

The microsomal fraction of human placental tissue also readily aromatizes steroids, according to the scheme shown in Fig. 15 (26–28). The C-19 methyl group is eliminated after hydroxylation and oxidation to produce a 19-nor-Δ^4-3-ketone. Aromatization in the presence of NADPH and O_2 results in formation of the phenolic structure in ring A. 19-Norsteroids will become aromatized in both placental and ovarian tissues.

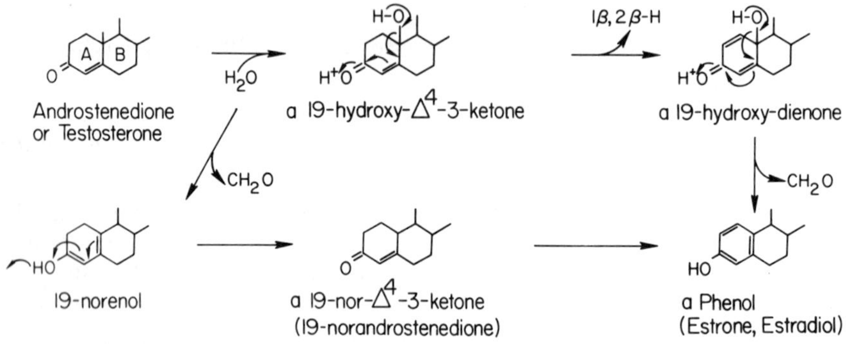

Figure 14   Probable mechanism of androgen aromatization in the human. Both pathways may coexist in the same tissue (i.e., ovary), although either one may predominate under given conditions.

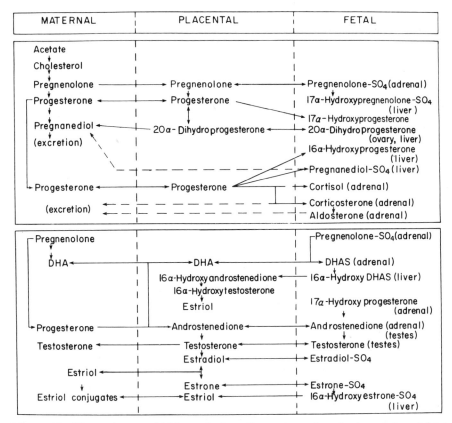

Figure 15 The major steroid biosynthetic pathways in maternal, placental, and fetal compartments during gestation in the human. Fetal glands of origin or transformation are noted in parentheses. Solid lines indicate transfer and metabolism of the steroid between tissues, whereas dashed lines indicate passive excretion from fetal to maternal compartments.

An alternative pathway involving desaturation of the 19-hydroxysteroid followed by eliminative aromatization has also been theorized and may occur *in vivo*. Estriol apparently is formed by 16α-hydroxylation of DHA prior to aromatization.

2.6 Synthesis of Corticosteroids

There appear to be at least four pathways by which adrenal corticosteroids are formed from pregnenolone (Fig. 16) (13, 14). The major pathway leads to formation of cortisol by way of 17-hydroxyprogesterone; cortexolone is then formed by action of a steroid 21-hydroxylase with subsequent 11β-hydroxylation to form cortisol. The principal mineralocorticoid pathway involves 21-hydroxylation of progesterone followed by 11β-hydroxylase action without the

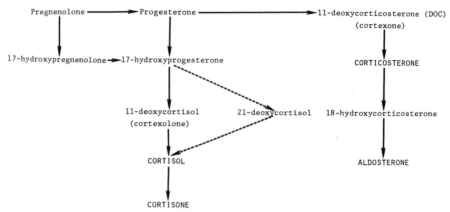

Figure 16 Outline of corticosteroid biosynthesis in the human adrenal. The dashed line indicates a minor pathway.

intermediate 17-hydroxylation. Aldosterone is formed after hydroxylation and subsequent oxidation of the angular C-18 methyl group to an aldehyde. Intermediates in this pathway, deoxycorticosterone and corticosterone, may be important secretory products in some species and under some conditions in humans.

The steroid hydroxylases all require NADPH and molecular oxygen (13, 20, 21). The 17- and 21-hydroxylases are associated with the adrenal microsomal fractions, and the 18- and 11β-hydroxylases with the adrenal mitochondrial fraction. Most steroids may be hydroxylated in a variety of positions, but a definite sequence appears to be obligatory; 17-hydroxylation may precede, but cannot follow, 21-hydroxylation, and 21-hydroxylation always precedes 11β-hydroxylation. The 11β-dehydrogenation must follow all hydroxylation steps. This oxidation enzyme is also found in the microsomal fraction and requires NADP.

2.7   Sites of Steroid Synthesis

(See Table 2)

2.7.1   Adrenal Glands

The human adrenal glands consist of two components, the medulla and the cortex, of which the cortex is the steroid-synthesizing fraction (13, 14). The adrenal cortex consists of three zones of cells. The outer zone immediately beneath the capsule is the zona glomerulosa, composed of short columnar cells. The zona fasciculata and zona reticularis are next; these cells contain large amounts of phospholipids and cholesterol with few mitochondria. These two zones seem to function as a single unit, their outermost cells containing the largest store of cholesterol available for steroid synthesis. Under the influence of ACTH, these cells produce the glucocorticoids, androgens, and estrogens of

Table 2   Steroid Production and Secretion

Site	Steroids
Adrenal cortex	Corticosteroids
	Glucocorticoids
	Mineralocorticoids
	Some androgens (DHA)
	Small amounts of estrogens and progestogens
Ovaries	Estrogens
	Progestogens
	Small amounts of androgens
Testes	Androgens
Placenta	Estrogens
	Progestogens
Liver	Cholesterol
	Steroid metabolites
Kidney	Steroid metabolites
Skin	Cholesterol
	Vitamin D
	Steroid metabolites

the adrenal cortex. The outer layer of cells in the zona fasciculata contains the 3β- and 17-hydroxysteroid dehydrogenases and is responsible for production of cortisol (or cortisone) and corticosterone. The adrenal cortex can store up to 7% of its weight in cholesterol and appears to synthesize corticosteroids on demand, as the amount of steroid stored is only about the amount that can be secreted in 3 min.

Mineralocorticoids, most notably aldosterone, are produced by the zona glomerulosa, or outer layer of the cortex. The activity of cells in this zone is not exclusively under the control of ACTH, but is controlled from the kidney by the renin-angiotensin system.

Androgen biosynthesis has been shown to occur by way of both the Δ^4 and Δ^5 pathways shown in Fig. 12 (8, 10, 13). The primary secretory product is DHA; much of it is secreted as DHA-sulfate (DHAS). Once DHA is formed, it can be converted to DHAS by endogenous sulfokinase activity, or to androstenedione and then to the 11β-hydroxy derivative. Testosterone may also be formed from androstenedione, although to a smaller extent than in other tissues. Estrone is the only estrogen produced by the human adrenal cortex. The amounts secreted are small and do not contribute significantly to the body pool in fertile women, but represent a major source of estrogen activity in men and postmenopausal women.

2.7.2 Ovaries

Steroid biosynthesis by the ovaries is complicated by the constantly changing population of cells in this tissue as the processes of follicular development, ovulation, and corpus luteum formation and regression recur with each menstrual cycle (11, 22). An outline of the overall secretion of steroids throughout the menstrual cycle is discussed later.

The production of estrogens occurs primarily during the follicular phase, and biosynthesis may occur in both the granulosa and theca interna cells (11), but *in situ* the granulosa apparently contributes little to the total estrogen production. This steroidal development is under the influence of FSH and LH. The thecal cells contain a much more active aromatizing system than the granulosa cells, and at least during the follicular phase, estrogen synthesis here occurs primarily by the Δ^5 pathway, whereas in the granulosa cells steroid synthesis occurs by the Δ^4 pathway (Fig. 13). Although DHAS can act as an ovarian precursor for estrogen biosynthesis, ovarian tissue is known to contain an active sulfatase and it is probable that the DHAS is hydrolyzed prior to conversion. The transfer of steroids between the theca interna and granulosa has been adequately demonstrated, and it is possible that modifications of the steroids produced in the theca occur in the granulosa prior to release of the products.

The corpus luteum produces significant quantities of C_{21}, C_{19}, and C_{18} steroids (22). In the absence of an active aromatizing system, the granulosa cells stay relatively inactive in steroid synthesis until after ovulation, when they become vascularized. The major compound produced then becomes progesterone, with smaller amounts of pregnenolone, 17-hydroxyprogesterone, 20α-dihydroprogesterone, androstenedione, estradiol, and estrone also being formed. The production of large amounts of progesterone, 17-hydroxyprogesterone, and androstenedione in the absence of 17-hydroxypregnenolone and DHA is further evidence that the major route of steroid synthesis in the granulosa is the Δ^4 pathway and that the Δ^5 pathway is unimportant. The biosynthetic potential of the human corpus luteum appears to be altered by the inclusion of a significant quantity of thecal cells (22). Such inclusion permits the incorporation of an aromatizing system, which has been isolated from the microsomal fraction of this tissue. The production rate of estrogens formed by the corpus luteum during the luteal phase is thus not significantly less than that formed during the follicular phase. However, species differences do occur, and the potential of the human corpus luteum to produce estrogens is somewhat unique. For instance, bovine corpus luteum appears to lack both the 17-hydroxylase and the desmolase system, so that no significant amounts of androgens or estrogens are produced. The sow and sheep appear to produce 17-hydroxyprogesterone and androstenedione in addition to progesterone, but apparently cannot aromatize androstenedione to estrone.

2.7.3   Testes

The large polyhedral Leydig cells, or interstitial cells of the testes, are the primary endocrine component of the organ (23–25). They lie in the connective tissue between the seminiferous tubules, and are interspersed with elongated, supportive, or Sertoli, cells. Leydig cells contain large quantities of fats, phospholipids, and cholesterol, and are very rich in ascorbic acid. These interstitial cells secrete the testicular androgens and are stimulated primarily by LH (ICSH) to promote steroidogenesis. The Sertoli cells are believed by some workers to secrete estrogens and to be under the control of FSH, whereas others believe that the testicular estrogens are secreted by Leydig cells and that the Sertoli cells only provide nutrition for the spermatids. The functions of FSH in the testes are not clearly understood.

The seminiferous tubules have also been shown to produce and secrete androgens (10, 23, 25). Beginning with pregnenolone, conversion to both progesterone and DHA has been shown to occur. Using 17-hydroxyprogesterone as the substrate, these tubules can synthesize testosterone and androstenedione; 5-androstenediol has likewise been shown to be an excellent precursor for testosterone in the tubule preparation. Thus both the Δ^4 and Δ^5 pathways are present. Nonflagellate germinal cells (spermatocytes and spermatids) have been shown to actively form androgens from progesterone. However, there is good evidence for reductive processes in the tubules whereby progesterone is inactivated by reduction to 20α-dihydroprogesterone; reduction of the double bond to form pregnanedione and further reduction to pregnanolone and pregnanediol also occurs.

2.7.4 Placenta

The human placenta is a multifunctional endocrine organ that secretes hCG, hCS, and numerous steroid hormones (26–29). The syncytiotrophoblast, from which the steroids are secreted, may be regarded as the endocrinologically and morphologically mature form of the trophoblast and contains a well-developed endoplasmic reticulum and Golgi bodies, together with free ribosomes. The syncytiotrophoblast is in direct contact with maternal blood, and thus both steroid precursors and products flow freely across its barrier.

The placenta utilizes maternal cholesterol and fetal DHAS as primary precursors for steroid synthesis. Large amounts of progesterone and estrogens are secreted from the placenta into the circulations of the fetus and mother; the absolute ratios of these products change with the time of gestation. An understanding of the interactions between the placenta and other organs is essential to an understanding of the role of the placenta as a steroid-producing and steroid-metabolizing organ.

2.8 Fetoplacental Biosynthesis

The human placenta carries out a variety of functions and as an endocrine organ is capable of synthesizing hormones necessary for the maintenance of pregnancy. In addition to the polypeptide hormones (hCG, hCS, and thyrotropin), the placenta is capable of synthesizing progestogens and estrogens (26–29).

It has been known since 1932 that the placenta contained progestational activity, but progesterone was not isolated from it until 20 years later. Several metabolites of progesterone have also been identified from the placenta, some of which have weak progestational activity. In the early stages of pregnancy, the corpus luteum secretes progesterone; however, as the trophoblast develops, progesterone of placental origin begins to be elaborated, increasing from about the second month of gestation until parturition (26, 28). The increase in total progesterone content appears to be related to growth and increase in weight of the placenta. Proof that progesterone is elaborated by this endocrine organ was established primarily through studies of the excretion of its principal metabolite, 5β-pregnane-3α,20α-diol (pregnanediol), which decreases to trace amounts following expulsion or surgical removal of the conceptus.

Although most other steroidal endocrine organs are capable of converting acetate to progesterone, the human placenta does not appear to have this capacity (26, 27). In 1951, it was shown that nonradioactive pregnenolone was readily converted to progesterone by the placenta, and these findings have been amply confirmed using labeled precursors. Although the fetus is capable of contributing a small portion of the precursor, most comes from the maternal circulation in the form of cholesterol.

Numerous perfusion studies carried on placentas *in situ* after disconnection of the fetus have also demonstrated the conversion of 17-hydroxypregnenolone to 17-hydroxyprogesterone. However, no 17-hydroxylase activity has been found, and pregnenolone cannot be converted to 17-hydroxyprogesterone in the placenta. It must first be passed to the fetal compartment where hydroxylation occurs. It is readily apparent from these studies, as well as from placental conversion of fetal or maternal DHA to androstenedione, that the placenta contains a large supply of 3β-hydroxysteroid dehydrogenase and isomerase.

Estrone, estradiol, and estriol, as well as other estrogens, have been isolated from human placental tissues (26, 29, 30). As do progestogens, these appear to increase in concentration according to the size and weight of the growing fetal compartment. It has not been possible to demonstrate conversion of labeled acetate, cholesterol, pregnenolone, or progesterone to estrogens in the placenta, and it has been amply demonstrated that the organ uses C_{19} precursors present in maternal and fetal circulations for synthesis of estrogens. It has been known since 1959 that the placenta has the capacity to convert androgens to estrogens; the aromatization of these neutral C_{19} steroids to phenolic compounds occurs in the microsomal fraction. Testosterone can be converted to estradiol and androstenedione to estrone in high yield. In spite of the fact that the placental tissue produces large amounts of estriol, there is no evidence that the placenta can convert estradiol or estrone to estriol; estriol is produced by aromatization of a 16α-hydroxylated androgen. Although estrogens are present in the developing fetus, they most likely come from maternal or placental sources, and the fetus is not a site of synthesis of these hormones.

The 16-hydroxylated steroids, such as 16-hydroxytestosterone, 16-hydroxyandrostenedione, 3β,16α-dihydroxy-5-androsten-17-one and Δ^5-androstene-3β, 16α,17β-triol are readily converted to estriol by placental tissues (26, 31). The fetal liver has the capacity to hydroxylate steroids at C-16, and several of these steroids have been shown to be present in the fetal circulation in high concentration, although primarily as their sulfate derivatives. This is true of most fetal C_{18} and C_{19} steroids, and many fetal tissues have been shown to have extensive sulfokinase activity (C-3, C-21, C-16, etc.). The placenta, on the other hand, contains sulfatases, and rapidly hydrolyze the sulfates coming from the fetal circulation.

Although cortisol, cortisone, aldosterone, and other corticosteroids have been shown to be present in the placenta, the placenta does not appear to be capable of synthesizing these hormones from either pregnenolone or progesterone (26). Rather, it has been suggested that these corticosteroids are trapped by the placenta; for what purpose has yet to be determined. The adrenal of the fetus close to term is capable of synthesizing corticosteroids from placental progesterone by the same pathways as in the maternal adrenals, and this may contribute to the high concentrations seen in the placental tissues. There is evidence that production of corticosteroids by the maturing fetus plays a role in the termination of pregnancy in sheep, but this process is apparently not operative in the human.

2.9   Other Sites of Synthesis

Between the dermal and epidermal layers of human skin lie the sebaceous glands. They are present everywhere on the skin except the palms and soles, and are especially concentrated on the forehead, face, chest, back, and scalp. These sebaceous glands secrete a sebum that contains fatty acids, squalene, cholesterol, and its esters and derivatives (32). They are not endocrine glands like the adrenals, ovaries, and testes, but are holocrine glands, breaking down completely to release their contents. The only sterol to be released by these glands is cholesterol, coming primarily from desquamation of keratinized cells (32). In some instances it is accompanied by trace amounts of dihydroxycholesterol and 7-dehydroxycholesterol. *In vitro* studies of human skin, however, have revealed its ability to synthesize androgens as well. Incubation of both hair-bearing and hairless skin of males and females shows that DHA is readily converted to androstenedione and androstanediol, the former serving as a precursor for testosterone. The presence of an endogenous sulfokinase has also been demonstrated, and DHAS may be formed from DHA. In addition there is evidence for the formation of less well known DHA metabolites including 7α- and 7β-hydroxy and 7-oxo derivatives, as well as for reduction of testosterone to 5α-DHT, androstenedione, and androsterone.

Although little is known about the ability of salivary glands to participate in steroid synthesis, there is good evidence that the submaxillary glands can bring about metabolism of corticosteroids and cause reduction of testosterone to 5α-DHT. Formation of androstenedione and 5α-androstanediol from these glands has also been shown to occur *in vitro*.

2.10   Control of Steroidogenesis by Product Inhibition

Much of the initial regulation of steroid biosynthesis, including effects of ACTH and the gonadotropins, has been discussed in the previous chapter, and will not be repeated here. Emphasis will instead be placed on local regulation by enzyme supply and distribution, and feedback by endogenous steroids. A summary of the knowledge in this area is found in Fig. 17.

The initial step in steroidogenesis, 20α-hydroxylation of cholesterol, is inhibited by the product as well as by pregnenolone, progesterone, and testosterone (13, 16, 18). The side-chain cleavage of 17-hydroxyprogesterone to androstenedione is also affected by the presence of endogenous steroids, although no product inhibition has been found. The C-17,20 lyase is competitively inhibited by progesterone, pregnenolone, and 17-hydroxypregnenolone, while other steroids such as 5α-DHT, androsterone, estrone, estradiol, DOC, cortisol, and corticosterone have no effect. This lyase has also been found to be competitively inhibited by the 20α-reduction product of 17-hydroxyprogesterone (17,20α-dihydroxy-4-pregnen-3-one) *in vitro*. From this evidence it has been suggested that inhibition of C-17, 20-lyase activity could be a means of regulating androgen biosynthesis. The 20α-hydroxysteroid dehydrogenase required for formation of the inhibitor is found in the cytoplasm, and thus the 17-hydroxyprogesterone would have to be transferred from the smooth endoplasmic reticulum (ER), where it is formed from progesterone, to the cytoplasm and then back to the smooth ER before the lyase could act. This sort of shuttling between intracellular compartments, however, is typical of steroid pathways.

More recent work has confirmed the competitive inhibition of the lyase by 17,20α-dihydroxy-4-pregnen-3-one in testicular microsomal preparations, as

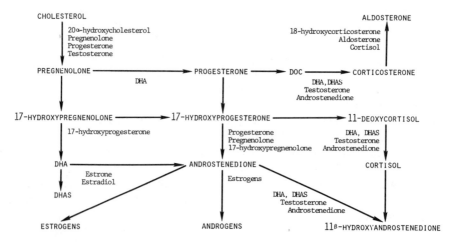

Figure 17   Known regulatory effects of endogenous steroids on human adrenal steroidogenesis. All steroids produce inhibitory effects at the levels indicated.

well as by 20α-dihydroprogesterone to a lesser extent (23, 25). Although analysis of testes from several species has clearly shown that such inhibition does not occur in the adult testes *in vivo*, it is also known that the level and activity of 20α-hydroxysteroid dehydrogenase is low in adult, compared with immature, testes. Presumably, therefore, this inhibition could be involved in decreased androgen biosynthesis in the immature animal. Furthermore, LH increases 20α-hydroxysteroid dehydrogenase activity in ovarian tissue, and this could also lead to a reduction in the synthesis of androgens.

The ovary contains a high proportion of androstenedione relative to testosterone, probably because of the direct inhibitory effect of estrogens on 17β-hydroxysteroid dehydrogenase (11, 33). Both estrone and estradiol (but not estriol) have also been shown to be noncompetitive inhibitors of the 3β-hydroxysteroid dehydrogenase–Δ^{4-5}-isomerase enzyme system, which converts DHA to androstenedione. Relatively small amounts of these endogenous estrogens can markedly inhibit the synthesis of androstenedione in the human adrenal.

The 11β-hydroxylase of adrenal mitochondria is markedly inhibited by endogenous steroids, particularly DHA, DHAS, testosterone, and androstenedione (competitive) (13). Thus the synthesis of glucocorticoids may be influenced by the level of adrenal androgens.

Aldosterone biosynthesis is particularly sensitive to regulation by endogenous steroids (14, 15). There is no doubt that aldosterone inhibits its own production and secretion, as well as the intermediary 18-hydroxycorticosteroids. Cortisol has similarly been shown to inhibit both aldosterone and 18-hydroxycorticosterone, indicating an effect on both the 18-hydroxylase and 18-hydroxysteroid dehydrogenase systems. Both DOC and progesterone have *in vivo* inhibitory effects on aldosterone production, but these are due to changes induced in the renal glomerulosa and in sodium excretion rather than directly on the biosynthetic pathway.

3 STEROID PRODUCTION AND SECRETION

3.1 The Concept of Production and Secretion Rates

Until recent years, most information about the secretion of steroid hormones was derived from studies on the excretion of their urinary metabolites (34, 35). However, there are many limitations to this approach. Particular urinary metabolites usually represent only a small and often variable fraction of their secreted precursors. Most gonadal and adrenal steroids may be produced by alternative pathways and by common precursors, making analysis of the secretory activity extremely complex. Collection of urine is not difficult, but analysis of such a pool gives only an integrated estimate, without considering episodic (pulsatile) steroid release or influence of other factors that affect short-term production. Radioimmunoassays have now made it possible to measure blood concentrations of most steroids directly. However, unless frequent samples can be obtained, the problems of episodic secretion still cannot be overcome.

The best way to determine the secretion of a hormone from a particular gland or organ might be by catheterization of its vein(s) if this could be done without disturbance of the normal rate of blood flow. The existence of a concentration gradient between the blood entering and leaving the gland constitutes the only unequivocal proof of secretion. But the calculation of secretory rates also requires knowledge of the blood flow through the organ in terms of volume per time. It should also be remembered that catheterization is not without hazard; it is technically difficult and cannot be carried out repeatedly in the same individual.

Most satisfactory solutions to this problem have arisen through the use of the isotopic dilution methods of estimation (34, 35). These methods offer more quantitatively precise estimates and have been responsible for the development of new models concerning the mechanisms involved in the production and secretion of steroids. They have also led to the formulation of several concepts, including production rate, secretion rate, and metabolic clearance rate.

The secretion rate (SR) of steroid A is the amount of A released by the endocrine organ into the circulation per unit of time. This hormone can, however, also be derived from the peripheral conversion of other hormones (B or C), secreted either by the same or another endocrine gland.

The production rate (PR) of steroid A is the total rate at which A enters *de novo* into the circulation. In the steady state it can also be defined as the rate by which the hormone is irreversibly removed from the circulation. When hormone A is derived exclusively from secretion by an endocrine organ, SR and PR for that hormone are the same; but when it is also derived from peripheral conversion of another hormone, the PR will be greater than the SR.

The metabolic clearance rate (MCR) of steroid A is a concept related to the PR, and can be defined as the ratio of the rate of irreversible removal of the hormone from the blood (V_A) to its concentration in peripheral blood (C_A). Its units are in volume per unit of time. Thus

$$MCR_A = \frac{V_A}{C_A}$$

and

$$PR_A = MCR_A \times C_A$$

in the steady state.

Since MCR for many hormones has been shown to be approximately constant in a wide range of concentrations, it is a useful index of hormone metabolism, especially as related to disorders of the metabolic pathways or endocrine glands. The measurement of both PR and MCR has been valuable in assessing the metabolic alterations associated with a wide variety of conditions. For instance, menopausal women excrete significant amounts of urinary estrogens, yet recently it has been shown that the SR of estrogens in these women is almost zero. In actuality, most of the estrogen originates by peripheral conversion of androstenedione (secreted by the adrenals) to estrone (33). The ovaries are not involved, as was once assumed from urinary studies. A similar situation exists in women with some forms of polycystic ovaries in which these ovaries secrete large amounts of androstenedione, which is then converted peripherally to estrone for excretion.

3.2 Steroid Production by the Immature Organism

The criteria or signs generally taken to denote the onset of puberty or the maturing of the individual to the adult state are those dependent upon the actions of the sex hormones (36–38). Although it was once believed that secretion of these hormones was abruptly initiated at puberty, it is now abundantly clear that this system is not quiescent or functionless prior to this time. Although it is true that an increase in the level of gonadal function produces the external manifestations of puberty, some activity on the part of the ovaries, testes, and adrenals can be detected much earlier.

The earliest rises in steroid hormone levels observed in the prepubertal individual are those that result from adrenal gland activity (36, 39). During intrauterine life, the adrenals are quite functional and participate in steroidogenesis. In the immediate neonatal period, the adrenal gland becomes somewhat refractory in its production of C_{19} steroids, although it maintains at least minimal androgen secretion throughout childhood. This decrease in function is associated with the degradation of the fetal layer of the adrenal cortex.

The next increase in secretory activity of the adrenal cortex comes several years prior to the clinical onset of puberty, at a time when the adrenal gland shows a marked increase in size and once again begins to secrete an appreciable amount of androgens, primarily DHA and DHAS (37). DHA may become detectable in urine of girls by the age of 6 or 7, and of boys by age 7 or 8. Testosterone of adrenal origin may also be found as early as age 6 in girls. These secretions occur prior to physiological maturation of the hypothalamic–pituitary–gonadal axis, and well before significant increases in serum FSH or LH are seen. No increases in adrenal cortisol production are seen during pubertal development.

The testes similarly secrete androgens from an early prenatal stage and continue to do so during infancy (36). Removal of the male gonads at an age when the reproductive tract is not fully differentiated (usually before birth) causes maldevelopment of the genital organs. The major steroid present in the testes is androstenedione, but as sexual maturity proceeds, testosterone increases so that it eventually becomes the predominant androgen.

There is no evidence for any hormonal action normally exerted by the ovary of the female fetus upon the genital tract (26, 28), although estrogens derived from the placenta are present during intrauterine life. The development of the primordia of the genital tract in the female structures seems to be independent of fetal gonadal hormones. Ovaries of the neonate and early childhood produce only minimal amounts of estrogens, and this is assumed to be due to the influence of low levels of gonadotropins present.

The major increases seen in all steroid hormones occur just prior to and during the onset of clinical puberty, as would be expected by their external manifestations (36–39). In boys, increases in serum testosterone correlate well with increases in levels of LH, beginning to rise just prior to puberty and reaching a plateau only after complete sexual development has occurred. Androstenedione increases most significantly in the later stages of puberty, whereas 5α-DHT parallels the rise in testosterone throughout the course of development.

Small increases in serum estrone, estradiol, progesterone, and 17-hydroxy-progesterone may be seen throughout puberty in the male, and represent combined adrenal and gonadal activity (36–39). DHA and DHAS show parallel increases throughout puberty, with a slightly more rapid rise in the levels of the sulfate in early puberty and in DHA levels later. These rises represent increments in adrenal androgen activity, and account for at least some of the anabolic and virilizing changes of puberty.

The physiological changes of puberty in the female are primarily the manifestation of increased production of estrogens by the ovaries (38, 39). At least two years before menarch appears, and just prior to thelarche (the beginning of breast development), estrogen secretion becomes cyclic in girls, and the intensity of these cycles gradually increases. At the same time, changes in estrogen metabolism occur, with estradiol becoming the primary product in place of estriol, as in the infantile system.

Pubarche (first growth of pubic hair) occurs in females as a direct result of interactions between estrogens and androgens, which are being secreted in response to enlargement of the adrenal cortex (37, 38). Significant increases in serum testosterone and DHT are noted, compared with prepubertal levels. These increases are closely followed by those of estradiol, and represent both adrenal and ovarian steroidal activity. Unlike males, females show a rise in DHA earlier than a rise in DHAS, with a very rapid rise occuring at thelarche. By menarche, DHAS becomes the primary androgen produced by the adrenals. Although increments in serum progesterone occur early in puberty, these are masked once ovulation has become a more regular occurence—toward the end of puberty.

Menarche (Greek *men* = "month" and *arche* = "beginning"; the onset of menses) is the climax of puberty in the human female, and occurs when estradiol levels are high enough to cause sufficient proliferation of the endometrium (37–39). This comes several months after increased FSH release, but prior to increases in LH, which is not released until ovarian follicles are capable of secreting fairly substantial amounts of estradiol. Therefore, menstrual bleeding is often irregular in frequency, duration, and quantity, and early menstrual periods are generally anovulatory. Endocrine maturation of the female may not be considered to be complete until estradiol has reached a level high enough to cause a positive feedback on the pituitary and thus the ovulatory surge of gonadotropins.

3.3 Steroid Production During the Menstrual Cycle

Since the mechanisms that control or regulate the normal menstrual cycle have been described in earlier chapters, discussion here centers only on the steroid hormone changes that occur. The role of these steroids in preparation of the genital tract epithelium is likewise excluded.

Fig. 18 illustrates the changes in plasma steroid hormones throughout the normal menstrual cycle (40–43). It can be noted that during the early period of follicular development, estradiol levels remain quite low. Approximately one week before the midcycle LH peak, there is a slow, then rapid, rise of estradiol, reaching a peak generally the day before the LH peak. A sharp drop in estradiol then occurs, with the LH peak ordinarily occuring while the estradiol concentration is falling. These events precede the actual rupture of the follicle. Following ovulation there is a brief decrease in estradiol secretion, followed by a second rise that corresponds to the function of the corpus luteum.

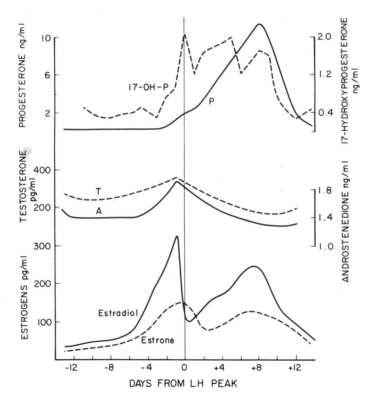

Figure 18 Average plasma steroid levels throughout the human menstrual cycle, in relation to the LH surge.

Although much less pronounced, the changes in estrone are much the same as those of estradiol (11). Estrone is secreted directly by the ovary, but most of the estrone in circulation is derived from conversion of peripheral androstenedione. This is in contrast to the production of estradiol; peripheral conversion of this hormone from testosterone is very small. The significance of the rises in estrone is still uncertain. Although estrone has much less estrogenic activity than estradiol (when measured by standard bioassay), it may still be important for activities other than uterine growth promotion.

The changes in progesterone show quite a different pattern. Throughout the follicular phase, progesterone concentration remains low. There is a small initial increase in the concentration of progesterone at the very beginning of the LH surge, followed by a second major increase that parallels the activity of the corpus luteum (40–43). The timing of the early rise suggests that this is a direct result of the high levels of LH on the unruptured follicles. The mechanism by which progesterone secretion is stimulated and estrogen secretion is decreased is not clear. However, the granulosa cells contain little 17-hydroxylase or desmolase, and thus as they hypertrophy to become the corpus luteum and receive an increased blood supply, progesterone becomes the major steroid product. Although the theca interna remains the major source of estrogen, a shift in available precursors for aromatization may result in a temporary diminution in estrogen formation after ovulation.

The secretion by the human ovary of 20α-dihydroprogesterone has been established, but the contribution relative to conversion by peripheral precursors is not clear (11, 43). It is a major product of the corpus luteum of the rodent ovary, however. There is evidence that 20β-dihydroprogesterone is also secreted by the human ovary as well as by that of some domestic animals, and although significant amounts of this steroid are present in the circulation during the luteal phase of the cycle, the majority is assumed to be derived peripherally from circulating progesterone. Changes in serum concentrations of pregnenolone and 17-hydroxypregnenolone throughout the menstrual cycle have been described, and both compounds were shown to be present at higher concentrations in the luteal phase than in the follicular phase.

Androstenedione is clearly the major androgen secreted by the ovary, although small amounts of testosterone are also secreted throughout the menstrual cycle (11, 41). Changes in serum levels of androstenedione appear to parallel the changes in estradiol, and most of the androstenedione in the circulation comes from direct ovarian or adrenal secretion. Both testosterone and androstenedione show an elevation in plasma levels during the middle of the menstrual cycle, and these are almost assuredly of ovarian origin. As the follicles undergo the normal process of atresia, thecal cells degenerate into stromal cells, thus causing steroidogenic changes such that instead of secreting estradiol, they now produce primarily androstenedione and testosterone.

Although there is evidence that the ovaries secrete small amounts of DHA and even DHAS during the cycle, virtually all of the circulating DHA comes from the adrenals (11, 13). Circulating DHA does not contribute significantly to the levels of androstenedione or testosterone, and shows approximately the same pattern as the other androgens.

There are several hormonal changes during the normal menstrual cycle that are secondary to effects of gonadal secretions on other glands. During the luteal phase, there is an increase in aldosterone which probably is a compensatory mechanism to overcome the inhibitory effect of progesterone upon the sodium-retaining activity of aldosterone (15). Renin shows both a midcycle and a midluteal peak in response to diuretic effects of ovarian secretions. The secretion of many other nonsteroidal hormones such as growth hormone, prolactin, and relaxin is also influenced by the presence of ovarian steroids.

3.4 Steroid Production and Secretion During Pregnancy

Marked changes in steroid production, secretion, and metabolism take place during human pregnancy (26–29, 30, 31, 42, 44). These effects result from the appearance of endocrine activities within the fetus and placenta. Such effects may be attributed to (1) changes in SR of various steroids by the maternal adrenals and ovaries, (2) the influence of the fetal adrenal on steroid production and metabolism, (3) the influence of the placenta on steroid metabolism, and (4) the effect of altered steroid hormone levels on hepatic and nonhepatic metabolism of steroids. Steroid production and metabolism in pregnancy are characterized by the complementary function of the fetal, placental, and maternal compartments, and this cooperational biosynthesis is the basis for consideration of the "fetoplacental unit." As is implied by this definition, the elimination of any one part of this physiological unit will significantly alter the function of the others.

Although the immediate precursors for the fetal steroids are not fully known, the major interactions of these compartments are outlined in Fig. 15. Since much of the steroidal synthesis in the fetus and placenta has already been described in Section 2.8, this section centers on changes in the maternal system throughout the course of pregnancy.

From approximately the second month of pregnancy until term, all of the plasma estrogens rise steadily (Fig. 19) (30, 31). Estradiol, estrone, and estriol all increase severalfold in plasma, reaching a peak at term, then rapidly falling off after parturition. The PR of each is significantly increased with advancing gestation. Estetrol (15α-hydroxyestriol) is formed by placental aromatization of a C_{19} precursor that is produced almost entirely within the fetal adrenal and liver. Its production is very small by comparison to that of estriol, but it also increases with gestation.

Urinary levels of estrogens show similar patterns (30, 31). Estriol excretion shows a slow increase during the first and second trimesters, then increasing sharply toward term. A correlation with the weight of the fetus has been suggested, and this seems probable since the fetoplacental unit accounts for the vast majority of estriol being secreted. Urinary levels of estrone are higher than those of estradiol, but both remain lower than estriol throughout. Many other

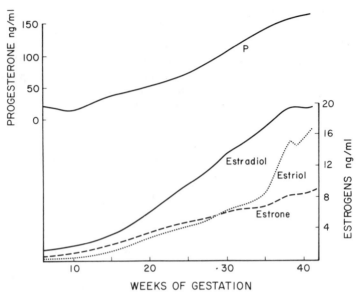

Figure 19   Average plasma concentrations of estrogens and progesterone from the fifth week of human gestation.

phenolic steroids have been identified in urine, but little else is known about them.

Serum progesterone levels begin to rise during the luteal phase of the menstrual cycle and continue to increase through week 6 (after the last menstrual period) (41, 42). During weeks 7–9 a slight decrease is seen, then a steady increase follows to term (Fig. 19). This pattern has been divided into three phases: (1) weeks 0–6, all progesterone is of ovarian origin, (2) weeks 7–9, the ovary and trophoblast contribute equally, and (3) after the ninth week, the trophoblast finally becomes the predominant source of progesterone. At delivery there is a rapid decrease in serum progesterone concentrations that corresponds to the detachment of the placenta from the uterus.

A rise in plasma 17-hydroxyprogesterone occurs only until week 5 or 6, then it decreases to undetectable concentrations within a few weeks and remains low for the remainder of pregnancy. This is as expected in light of the inability of the placenta to synthesize 17-hydroxyprogesterone and the declining capacity of the corpus luteum.

Although the SR of pregnenolone is much less than that of progesterone, its MCR increases greatly with advancing gestation, exceeding that for progesterone (45). This may be due to its high rate of extraction by the fetoplacental unit for steroid biogenesis.

Testosterone levels in plasma increase in two separate surges to a peak at term, then sharply decline to a prepregnancy level immediately postpartum (42, 45). At least 60–70% of this testosterone is from extraglandular conversion of androstenedione and DHA. The PR of testosterone remains constant, however, with a decrease in MCR alone being responsible for the rise seen. 5α-DHT follows a similar pattern in serum concentrations.

Plasma androstenedione shows an increase through pregnancy and labor, rapidly decreasing following parturition (45). By term, the ratio of androstenedione to testosterone (5:4) is more than twice the level in the nonpregnant female, but the significance of these high values, as well as the extra surge during delivery, is not yet clear.

DHA and DHAS are the only C_{19} steroids to show decreases in plasma values (33). This is accounted for by two factors: (1) the secretion rate of each is decreased, and (2) the MCR for DHAS is increased at least 10-fold. Both of these androgens are used as primary substrates for steroidogenesis in the fetoplacental compartment, with at least 40% of circulating DHAS in late pregnancy being irreversibly metabolized to estradiol by the placenta, and another 35–40% proceeding from maternal serum estriol by way of 16α-hydroxyDHAS in the fetus. A small increase is seen during labor and delivery.

Changes in the glucocorticoid system during a normal pregnancy include (1) an increase in both SR and PR of cortisol, (2) a continuous increase in both bound and free levels of plasma cortisol (the levels decline postpartum), (3) a decreased MCR for cortisol, and (4) an exaggerated plasma cortisol response to ACTH (16, 10). Many of these alterations may be attributed to the progressive rise in estrogen production throughout pregnancy and the corresponding high levels of plasma estrogens.

Aldosterone also shows an increased SR throughout pregnancy, with an increase in circulating levels (39). This may be a result of (1) hepatic and renal effects of increased circulating estrogen levels that stimulate renin production and (2) increased progesterone levels in the plasma, since progesterone competitively inhibits aldosterone-stimulated sodium retention. This increase in circulating aldosterone becomes more marked with advancing gestation. Because plasma levels of corticosteroids are high throughout pregnancy, some vascular and metabolic changes characteristic of Cushing's disease are seen in pregnant women.

3.5 Disorders of Ovarian Function

Basic disorders of ovarian function may be divided into (1) failure of sexual maturation (primary amenorrhea) and (2) failure of established sexual function (secondary amenorrhea). Both of these may be due to dysfunction of the ovary itself or indirectly due to failure of the hypothalamic-pituitary axis (46–48). In addition, diseases of the ovary may be associated with other endocrine or metabolic disturbances (congenital adrenal hyperplasia, Cushing's syndrome, obesity, hyperthyroidism, and hypothyroidism).

3.6 Disorders of Testicular Function

Disorders of testicular function can be divided into three groups: (1) primary and secondary hypogonadism, (2) male sexual precocity, and (3) testicular

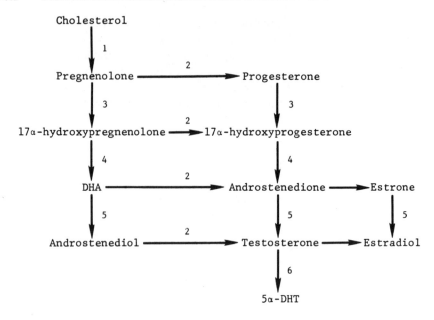

ENZYME	INTRACELLULAR LOCUS	BIOCHEMICAL MARKERS	
		Increased	Decreased
1 = 20,22-desmolase	M	(Data incomplete. Fatal; does not live to adulthood)	
2 = 3β-hydroxysteroid dehydrogenase	ER	DHA	Testosterone, Cortisol
3 = 17α-hydroxylase	ER	Corticosterone	Testosterone, Cortisol
4 = 17,20-desmolase (lyase)		Pregnanetriolone	Testosterone, Androstenedione
5 = 17-ketosteroid reductase		Androstenedione, Pregnanetriol	Testosterone
6 = 5α-reductase		Testosterone, Estradiol	5α-DHT

Figure 20 Location of known enzyme defects in testosterone synthesis and action. *M*, Mitochondria; *ER*, endoplasmic reticulum.

feminization (49). The first two types of disorders are related to hypothalamic-pituitary dysfunction, tumors, or to enzymatic defects in testosterone synthesis, (Fig. 20) (49–51) and the latter to insensitivity of target organs to testosterone and 5α-DHT because of a congenital deficiency of androgen receptors.

3.7 Disorders of Adrenal Function

Both hypofunction and hyperfunction of the adrenal cortex are known to exist in man, and such abnormal production of corticosteroids may be of primary or secondary origin.

Addison's disease (primary adrenal insufficiency) results from partial or complete destruction of all layers of the adrenal cortex, usually due to autoimmunity or tuberculosis (51–53). Decreased production of cortisol results, thus impairing glyconeogenesis and stimulating large increases in ACTH secretion because of lack of feedback inhibition. Decreased aldosterone produces a loss of renal and plasma sodium and chloride ions with consequent loss of plasma volume, pressor responses, and renal blood flow.

Three major syndromes of adrenal hyperfunction are known (13, 51–53). One of these, Cushing's syndrome, is essentially caused by excessive cortisol production by the adrenal cortex. This condition derives either from adrenal hyperplasia or tumors secreting ACTH. Cortisol SR is increased, and exaggerated responses of plasma corticosteroids are seen in response to ACTH stimula-

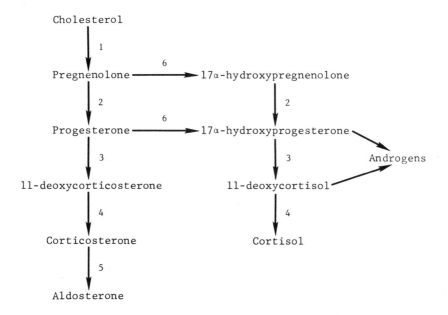

ENZYME	INTRACELLULAR LOCUS	BIOCHEMICAL MARKERS Increased	Decreased
1 = 20,22–desmolase	M	(Data incomplete. Fatal; does not live to adulthood)	
2 = 3β-hydroxysteroid dehydrogenase	ER	DHA	Testosterone, Cortisol
3 = 21-hydroxylase	ER	DHA, Testosterone	Aldosterone, Cortisol
4 = 11β-hydroxylase	M	Deoxycorticosterone	Cortisol, Aldosterone
5 = 18-hydroxylase	M	18-hydroxycorticosterone	Aldosterone
6 = 17α-hydroxylase	ER	Corticosterone	Testosterone, Cortisol

Figure 21 Deficiencies in adrenal enzymes found in human congenital adrenal hyperplasias. *M*, Mitochondria; *ER*, endoplasmic reticulum.

tion. In addition, both 11-oxysteroids and 17-ketosteroids show significant increases in plasma levels. Amenorrhea or menstrual irregularity and hirsutism often result.

Congenital adrenal hyperplasia (CAH) is a condition that includes numerous syndromes, all of which are characterized by a deficiency of specific enzymes involved in steroid biosynthesis (52–54). In each case, inadequate cortisol production gives rise to excessive ACTH output, adrenal hyperplasia, and increased production of steroids in collateral pathways. The most common form of the syndrome involves a 21-hydroxylase deficiency (Fig. 21). Here the steroid nucleus can be hydroxylated at C-11 and C-17, but not at C-21. As a result progesterone, 17-hydroxyprogesterone, and 11β,17-dihydroxyprogesterone accumulate in blood, most being converted to 17-ketosteroids.

4   STEROID BINDING AND TRANSPORT

4.1   Steroid Binding in Blood

Steroid hormones in blood are found primarily in three different physicochemical forms; they may be bound to plasma proteins, conjugated with sulfuric or glucuronic acid residues, or present as free steroids (55, 56). The protein binding of steroids in serum has important effects on their distribution, transport, excretion, and biotransformation. The bound steroid is not available to intracellular receptors except as it dissociates and diffuses through the plasma membranes. Some evidence has been presented, however, for uptake of the corticosteroid-binding globulin (CBG)–cortisol complex by the liver, but no such uptake has been found to occur in other organs.

Three characteristics of a steroid are important in determining its interaction with a protein: (1) electrostatic charge, (2) polarity (i.e., number of polar substituents), and (3) stereochemical configuration (55). Most of the natural steroids and other products of steroid biosynthesis are uncharged or electroneutral molecules at the neutral pH of blood. However, glucuronides, sulfates, and ester conjugates of the steroids are present as their salts.

The differences in the number and nature of oxygen functions (hydroxyls or ketones) and possibly other electron-attracting or -repelling "polar" groups provide another measure of potential binding of steroids by serum globulins (55, 56). According to the number of polar groups present in a steroid molecule, it may be labeled nonpolar, monopolar, dipolar, and so forth (Fig. 22). The parent hydrocarbons are thus nonpolar compounds, and we refer to an increase in polarity when the number of polar groups in the molecule increases. Consideration of steroid-protein interactions relative to the number of polar groups has led to the concept of the *polarity rule*. According to this rule, the binding affinity of a steroid for a protein decreases with an increasing number of polar groups on the molecule. This affinity-decreasing effect of the electron-attracting group is strongest for α-hydroxyl derivatives. This behavior is especially characteristic of binding to serum albumin.

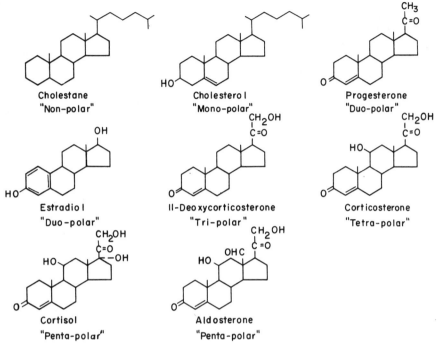

Figure 22   Structures of common estrogenic and progestational steroids used as active compounds in oral contraceptives.

Steroid-protein complexes are held together by noncovalent bonds. Three types of bonds are generally considered to prevail: (1) hydrogen bonds, (2) ionic bonds, and (3) hydrophobic bonds (55). Numerous possibilities exist for the formation of hydrogen or hydrophobic bonds between steroids and proteins. Aliphatic, alicyclic, and aromatic hydroxyl groups of the steroids as well as hydroxyl, amino, and peptide structures of the proteins can donate hydrogens; ketone, carbonyl, and other such groups can function as hydrogen acceptors. Hydrophobic bonds are formed between large areas of the steroid molecule, which is basically a nonpolar structure, and aliphatic and aromatic side chains of the proteins. The interaction of protein molecules with steroids is regulated by the law of mass action, and any mathematical formulations developed for quantification of binding must be based on this law.

The most abundant of the transport proteins in human serum is albumin (HSA) (55–57). The presence of large concentrations of this protein accounts for its enormous binding capacity in comparison with the specific binding proteins listed below. HSA enters into reversible binding equilibria with all dipolar and polypolar steroid compounds, although binding is weak and non-specific (Table 3). Interaction occurs with substances of varied chemical nature, following closely the polarity rule as described above. Two binding sites per molecule have been demonstrated for testosterone, progesterone, and corticos-terone; a single site exists for estrone and estradiol.

Table 3 Characteristics of Plasma Binding Proteins[a]

Protein	M.W.	Relative Plasma Conc. (mg/l)	Steroids Bound	Human $K_d(M)$(n = 1) 4°C	Human $K_d(M)$(n = 1) 37°C	Comments
ALBUMIN human (HSA)	69,000	38,000	Estrogens	1.6×10^{-5}		Low affinity, high capacity
			DHAS	1.3×10^{-5}		
			Progesterone	1.0×10^{-5}	0.6×10^{-5}	
			Testosterone	3.0×10^{-4}	2.1×10^{-4}	
			Corticosteroids	3.0×10^{-3}	2.5×10^{-3}	
α_1-ACID GLYCOPROTEIN human (AAG; orosomucoid)	41,000	750	Progesterone	1.1×10^{-6}	0.4×10^{-6}	Medium affinity, medium capacity
			Testosterone	5.7×10^{-5}	2.4×10^{-5}	
			Corticosterone	0.8×10^{-5}	0.4×10^{-5}	
			Estradiol	0.7×10^{-5}	0.2×10^{-5}	
			Cortisol	0.3×10^{-5}	1.6×10^{-4}	
CORTICOSTEROID-BINDING GLOBULIN human, monkey, rat, guinea pig (CBG; transcortin)	52,000	35 norm. 75 preg.	Cortisol	6.0×10^{-8}	0.4×10^{-8}	High affinity, low capacity, increases in pregnancy
			Corticosterone	1.0×10^{-9}	0.3×10^{-8}	
			Progesterone	7.0×10^{-8}	0.9×10^{-8}	
			Aldosterone	0.3×10^{-8}	0.7×10^{-7}	
			Testosterone		1.4×10^{-6}	
			Estradiol		2.0×10^{-4}	
TESTOSTERONE-ESTRADIOL BINDING GLOBULIN, human (TEBG; sex-hormone binding globulin; SHBG; SBG)	52,000–100,000	2–6 norm. 20–30 preg.	Testosterone	1.5×10^{-9}	5.0×10^{-8}	High affinity, low capacity, increases in pregnancy
			Estradiol	0.5×10^{-9}		
PROGESTERONE-BINDING GLOBULIN guinea pig (PBG; PBP)	100,000		Progesterone	$\sim 2 \times 10^{-10}$		High affinity, increases in pregnancy

[a] See references 55–57.

446

Most endogenous steroids also bind to some degree to α_1-acid-glycoprotein (AAG; orosomucoid) (55, 57). It is the most soluble of the plasma glycoproteins, and is similar in its interactions to the other binding globulins. Although its affinity for steroids is greater than that of albumin, its low concentration in blood does not permit it to bind a significant proportion of steroids in the circulation. As with HSA, binding of the Δ^4-3-ketosteroids is in accordance with the polarity rule. The affinity of AAG is highest for progesterone and testosterone, and somewhat lower for the corticosteroids. There is one high-affinity binding site per AAG molecule for each of the measured steroids except estradiol, which has five. Large positive entropy changes associated with steroid–AAG binding have been interpreted as being due to displacement or reordering of water molecules.

Binding of glucocorticoids to CBG (transcortin), an α_1-globulin, is strong and specific although its binding capacity is low (Table 3). It is by far the most important glucocorticoid binding protein in plasma, but also binds progesterone with markedly high affinity. In fact, the binding of all steroids to CBG is at least three orders of magnitude greater than the binding found for albumin or AAG. The binding of steroids to CBG is also accompanied by a much greater change in free energy, suggesting that the bonds formed are much stronger than for the less specific binding proteins. Currently available evidence suggests that all the steroids that bind to CBG compete for a single binding site, although this feature is not universal among species. The introduction of hydroxyl groups into the molecule in some cases enhances the ability of a steroid to bind to CBG. This type of binding, which is contrary to the polarity rule, is characterized by an affinity for proteins with a comparatively high number of hydroxyl groups, for example, CBG and testosterone-estradiol-binding globulin (TeBG). It is believed that hydrogen bonds form between the hydroxyl groups of the steroid and those of the protein.

Testosterone and estradiol also bind with high affinity to TeBG, also called sex-hormone-binding globulin (SHBG), which is a β-globulin present in both male and female plasma (55, 56). The presence of a 17β-hydroxyl group seems mandatory for binding of steroids to TeBG, although low-affinity binding has been reported for DHA, 17α-hydroxyprogesterone, and 17α-methyltestosterone. In general, the presence of methyl groups at C-17 (particularly in the α orientation), the presence of 17α-ethynyl groups, and unsaturation at C-1 or C-6 dramatically reduce binding to TeBG, whereas the presence of an oxygen function at C-3 is required for high-affinity binding. The 19-nor-steroids do not bind tightly to this protein, but 5α-DHT and some of the 5α-androstanediols have a higher binding affinity than even testosterone or estradiol. Although some evidence exists to show that 5α-DHT binds so tightly to TeBG that release into peripheral tissues is insignificant, all other binding to TeBG appears to be a freely reversible process.

Although not found in human serum, progesterone-binding protein (PgBP) is an important binding component in guinea pig serum. Its affinity for progesterone is greater than that of CBG in this species and accounts for most of the increased serum binding of progesterone seen in pregnancy (100-fold). Similarly, rat plasma shows a specific binding component, estrogen-binding protein (EBP), which can be distinguished from TeBG by its physical characteristics. In contrast to TeBG, EBP binds estrone as well as estradiol but not testosterone or 5α-DHT.

Numerous steroids have also been shown to bind to the cellular components of blood. Low-capacity, high-affinity binding to erythrocytes has been shown for most androgens, but only when studies were conducted *in vitro*. Whether or not such binding actually occurs *in vivo* is still under investigation. *In vivo* binding of progesterone and 20α-dihydroprogesterone to erythrocytes has been demonstrated, but association is very weak (K_d 2–4 $\times$ $10^{-4}M$) and probably represents nonspecific interactions. Further evidence suggests that erythrocytes contain a substance that binds cortisol, but with a much lower capacity than CBG. Even so, a significant uptake of cortisol (loosely bound) by erythrocytes can be demonstrated. It has been suggested that the corticosteroid molecules may diffuse into the phospholipid layer of the erythrocyte cell membranes such that the lipophilic ring structure intermixes with the aliphatic chains of the phospholipids, whereas the hydrophilic side chains of the steroid molecule remain outside. Evidence for and against association of steroids with lymphocytes and other cellular components has also been reported.

4.2   Kinetics of Steroid Binding

The binding of steroid hormones to high-affinity plasma proteins provides a means of increasing the solubility of the hormones in aqueous media (i.e., blood), as well as providing a reservoir of steroids that dissociate slowly from their complexes as needed. In the event of cessation of steroid secretion, for instance, the time required for the plasma concentration of a steroid bound to CBG to decrease is much longer than that of albumin-bound or unbound steroid. Thus binding of steroids to high-affinity serum proteins acts as a natural buffer, preventing dramatic fluctuations in serum steroid levels even when steroid secretion is irregular. Another effect of high-affinity binding proteins is to protect the steroids against metabolism; the hepatic extraction of a bound steroid is greatly reduced and the MCR decreases conspicuously.

A steroid that is bound to a serum protein is biologically inactive, and only the unbound (free) fraction of a steroid in plasma (usually less than 10% of the total plasma steroid concentration) is freely exchangeable with the extravascular compartments. Measurement of the degree of binding of these steroids, therefore, as well as the concentration of bound and free steroids in the plasma, is important in diagnosis and treatment of endocrine disorders.

The affinity of a steroid for binding to a protein in serum may be expressed in terms of the binding constant

$$K = \frac{[PS]}{[P][S]}$$

where [PS] is the concentration of protein-bound steroid, [S] is the concentration of unbound steroid, and [P] is the concentration of binding protein present in moles per liter (55). K has units of liters per mole, an expression of dilution, and describes the specific affinity of a binding protein for a particular steroid. By similar methods the association constant (K_a) may be derived and expressed as

$$K_a = \frac{[PS]}{[nP][S]}$$

where n is the number of binding sites per mole of P. The binding constant K may be expressed as

$$K = nK_a$$

When the protein has only one binding site for the ligand ($n = 1$), $K = K_a$, and K_a describes completely the strength of the interaction. The value of K_a is dependent on temperature and to some extent on pH and other solutes. If there is more than one binding site, each site has its own intrinsic association constant. If the binding sites are equivalent and independent of each other, then a single binding constant K is sufficient to characterize all the n binding sites (55). This case is rarely encountered with proteins that bind ligands at more than one site, however.

In blood there is frequently competition among steroids for the same binding site on a binding protein. In this case, the amount of S_1 bound to P is dependent on the concentrations of S_1 and S_2. The ratio of their binding will be dependent on the concentrations of S_1 and S_2 and their relative binding affinities. An example of this type of competitive binding is the interaction of corticosteroids and progesterone with CBG, which has only one binding site for these steroids. Similar competition is seen for the progesterone-binding site of AAG by both corticosteroids and testosterone (Table 3). Inhibition of binding by a second steroid does not necessarily indicate that S_1 and S_2 are attached to the same binding site (55), but allosteric effects have not been demonstrated with steroid-binding proteins in serum.

In normal human plasma, the unbound fraction of cortisol is approximately 8%; that bound to CBG at least 75%; and that bound to albumin about 15%. Progesterone is distributed as about 30% bound to CBG; 60% bound to albumin; and 5% bound to AAG. Serum of women has 80% testosterone bound to TeBG; 4% bound to albumin, and 7% bound to AAG. By contrast, serum of men has 6% bound to CBG, 69% bound to TeBG, and 14% bound nonspecifically. Estradiol-17β is bound by 50% or less to TeBG in women and is present in a greater percentage in the unbound form.

4.3 Alterations of Binding in Pregnancy and Disease

The changing endocrine conditions of pregnancy bring about a steady rise in CBG and TeBG, as well as a moderate increase in albumin concentration with advancing human gestation (55–57). CBG concentration increases approximately two-fold, as does its capacity for steroid binding. The increase in CBG follows by several days the increase in estrogen concentration, which is believed to stimulate the hepatic synthesis of this protein; a decrease to prepregnancy levels occurs within two weeks following parturition. Although elevated, plasma concentrations of CBG are still low by comparison with less-specific binding proteins, being only 10% of the concentration of AAG and less than 0.3% of the concentration of HSA.

The binding affinity of CBG also appears to be decreased in pregnancy (57), which probably results in a more rapid rate of dissociation of cortisol and progesterone from CBG. In addition, the presence of the sex steroids decreases the feedback effectiveness of cortisol on ACTH production, and a higher set-

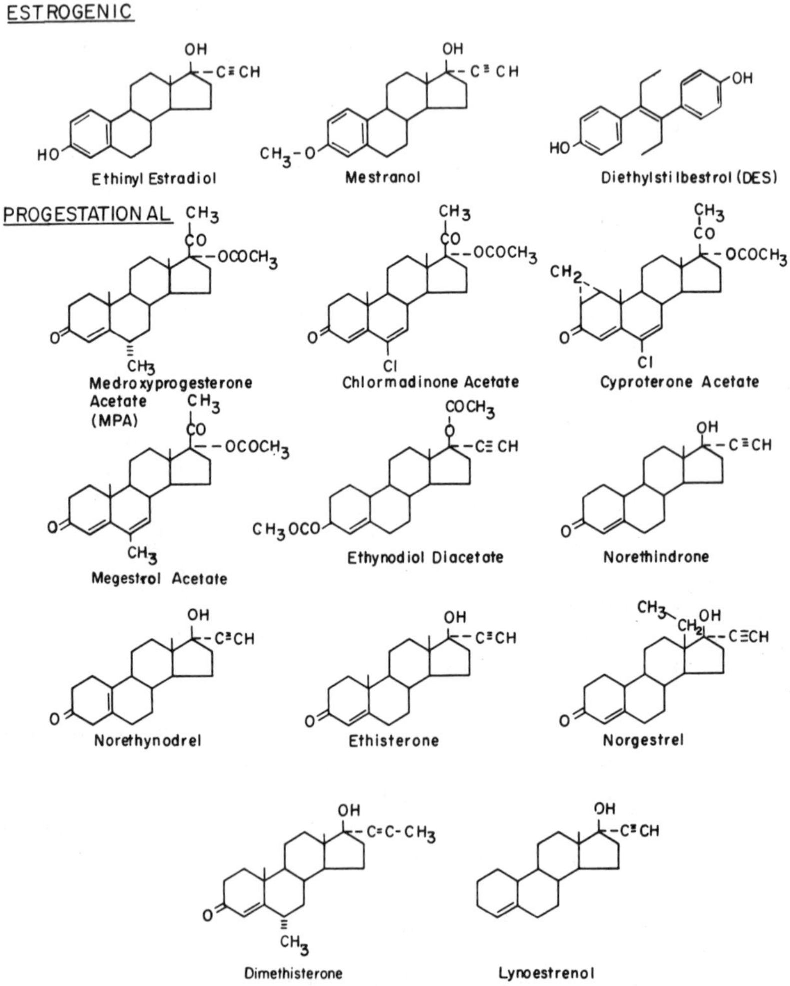

Figure 23   Differences in chemical polarity among common steroids.

point for the concentration of free cortisol in plasma occurs. Since the increase in plasma progesterone concentration in pregnancy is much greater than the rise in cortisol, progesterone gradually displaces cortisol from the CBG, causing the reservoir of bound hormones to shift in favor of progesterone. Since the ratio of bound to free progesterone remains nearly constant or increases slightly (58) during pregnancy, the absolute concentration of this unbound portion similarly increases. Progesterone associates equally with CBG and HSA in late gestation; AAG binding is somewhat decreased.

The rise in TeBG seen in pregnancy protects the fetus and mother against rapid fluctuations in testosterone levels during pregnancy. This ability of the protein to reduce the biological activity of a steroid prevents its entry into the extravascular spaces, where its physiological effect may be deleterious. The most marked increase in TeBG appears in the first trimester of gestation, when it reaches a maximum of three to four times the normal concentration in blood. As with CBG, TeBG rapidly returns to prepregnancy levels immediately following parturition.

Estradiol binding to TeBG increases with advancing gestation, such that unbound levels of estradiol remain uniform throughout pregnancy (56, 57). Increased binding of DHA also occurs.

Although HSA shows moderate increases in serum concentrations throughout pregnancy (55), a decreased affinity of HSA for steroids may be noted. This is probably due to the numerous types of lipids present in pregnancy serum that may inhibit the steroid binding. Decreases in HSA have also been reported (57); this is probably due to increased fetal demand for the protein.

Oral contraceptives, primarily because of their estrogenic components, also have significant effects on serum binding proteins (55, 57, 59) (see Fig. 23). Estrogen action on the liver produces altered binding-globulin production, resulting in increased CBG and TeBG concentrations. The concentrations of these proteins are significantly increased, but lower than those in late pregnancy.

Patients with liver cirrhosis, nephrotic syndromes, or hypoalbuminemias have been shown to have decreased CBG concentrations (55), although low serum HSA is not consistently associated with lowered CBG concentration. A decrease in CBG concentration also accompanies toxemia of pregnancy, but the concentration remains higher than in the nonpregnant state.

REFERENCES

1 IUPAC-IUB Revised Tentative Rules of Nomenclature of Steroids, *Arch. Biochem. Biophys.* **136**, 13 (1970); *Biochem. J.* **113**, 5 (1969); *Biochemistry* **8**, 2227 (1969); *Steroids* **13**, 277 (1969) ; Amendments to the Steroid Rules, *Arch. Biochem. Biophys.* **147**, 4 (1971).

2 Fieser, L. F., and Fieser, M., *The Chemistry of Natural Products Related to Phenanthrene*, 3rd ed., Reinhold, New York (1949).

3 Cahn, R. S., Ingold, C., and Prelog, V., Specificity of molecular chirality. *Angew. Chem. Int. Ed.* **5**, 385 (1966).

4 Fotherby, K., The biochemistry of progesterone. *Vit. Horm.* **22**, 153 (1964).

5 Courrier, R., Interactions between estrogens and progesterone. *Vit. Horm.* **8**, 179 (1950).

6 Rothchild, I., Interrelations between progesterone and the ovary, pituitary and central nervous system in the control of ovulation and the regulation of progesterone secretion. *Vit. Horm.* **23**, 209 (1965).

7 Corner, G. W., and Allen, W. M., Physiology of the corpus luteum. *Am. J. Physiol.* **88**, 326 (1929).

8 Migeon, C. J., Adrenal androgens in man. *Am. J. Med.* **53**, 606 (1972).

9 Dorfman, R. I., and Shipley, R. A., *Androgens. Biochemistry, Physiology and Clinical Significance.* Wiley, New York (1956).

10 Brooks, R. V., Androgens. *Clin. in Endocr. Metab.* **4**, 503 (1975).

11 Besch, P. K., and Buttram, V. C. B. Steroidogenesis in the Human Ovary, *in* Balin, H., and Glasser, S. (eds.), pp. 552–571. *Reproductive Biology.* Excerpta Medica, Amsterdam (1972).

12 Morris, R., Oestrogens *in* Butt, W. R., (ed.), *Hormone Chemistry*, 2nd ed., Halsted, New York (1976).

13 Griffiths, K., and Cameron, E. H. D., Steroid biosynthetic pathways in the human adrenal. *Adv. Ster. Biochem. Pharmacol.* **2**, 223 (1970).

14 Williams, G. H., and Dluhy, R. G., Aldosterone biosynthesis—interrelationship of regulatory factors. *Am. J. Med.* **53**, 595 (1972).

15 Brown, J. J., Fraser, R., Lever, A. F., *et al.* Aldosterone: Physiological and pathological variations in man. *Clin. in Endocr. Metab.* **1**, 397 (1972).

16 Dempsey, M. E., Regulation of steroid biosynthesis. *Ann. Rev. Biochem.* **34**, 967 (1974).

17 Goad, L. J., Sterol biosynthesis. *Biochem. Soc. Symp.* **29**, 45 (1970).

18 Burstein, S., and Gut, M., Biosynthesis of pregnenolone. *Rec. Prog. Horm. Res.* **27**, 303 (1971).

19 Boyd, G. S., Browne, A. C., Jeffcoate, C. R., *et al.* Cholesterol hydroxylation in the adrenal cortex and liver. *Biochem. Soc. Symp.* **34**, 207 (1972).

20 Estabrook, R. W., Baron, J., Peterson, J., *et al.*, Oxygenated cytochrome P-450 as an intermediate in hydroxylation reactions. *Biochem. Soc. Symp.* **34**, 159 (1972).

21 Schleyer, H., Cooper, D. Y., Levin, S. S., *et al.* Haem protein P-450 from the adrenal cortex: interaction with steroids and the hydroxylation reaction. *Biochem. Soc. Symp.* **34**, 187 (1972).

22 Savard, K., The biochemistry of the corpus luteum. *Biol. Reprod.* **8**, 183 (1973).

23 Eik-Nes, K. B., Production and secretion of testicular steroids. *Rec. Prog. Horm. Res.* **27**, 517 (1971).

24 Lipsett, M. B., Steroid Secretion by the Testis in Man, *in* James, V. H. T., Serio, M., and Martini, L. (eds.), *The Endocrine Function of the Testis*, Vol. 2, pp. 1–11, Academic, London (1973).

25 Lipsett, M. B., Production of testosterone by prostate and other peripheral tissues in man. *Vit. Horm.* **33**, 209 (1975).

26 Siiteri, P. K., Gant, N. F., and MacDonald, P. C., Synthesis of steroid hormones by the placenta, *in* Moghissi, K. S., and Hafez, E. S. E. (eds.), *The Placenta, Biological and Clinical Aspects*, pp. 238–257, Thomas, Springfield (1974).

27 Diczfalusy, E., Recent Progress in the Feto-Placental Metabolism of Steroids, *in* Vokaer, R., and DeBock, G. (eds.), *Reproductive Endocrinology*, pp. 3–19, Pergamon, New York (1975).

28 Diczfalusy, E., Endocrine functions of the human fetus and placenta. *Am. J. Obstet. Gynecol.* **119**, 419 (1974).

29 Jeffery, J., and Klopper, A., Steroid metabolism in the feto-placental unit. *Adv. Ster. Biochem. Pharmacol.* **2**, 71 (1970).

30 Oakey, R. E., A progressive increase in estrogen production in human pregnancy: an appraisal of the factors responsible. *Vit. Horm.* **28**, 1 (1970).

31 Levitz, M., and Young, B. K., Estrogens in pregnancy. *Vit. Horm.* **35**, 109 (1977).

32 Ebling, F. J., Steroid hormones and sebaceous secretion. *Adv. Ster. Biochem. Pharmacol.* **2**, 1 (1970).

33 Longcope, C., Kato, T., and Horton, R., Conversion of blood androgens to estrogens in normal adult men and women. *J. Clin. Invest.* **48**, 2191 (1969).

34 Tait, J. F., and Burstein, S., *In Vivo* Studies of Steroid Dynamics in Man, *in* Pincus, G., Thimann, K. V., and Astwood, E. B. (eds.), *The Hormones*, Vol. 5, pp. 441–557, Academic, London-New York (1964).

35 Gurpide, E., Mathematical analysis for the interpretation of *in vivo* tracer infusion experiments. *Acta Endocr.* Suppl. **158**, 26 (1971).

36 Gruber, J. S., and Lucas, C. P., Endocrinology of Puberty, *in* Hafez, E. S. E., and Peluso, J. J. (eds.), *Sexual Maturity: Physiological and Clinical Parameters*, pp. 123–140, Ann Arbor Science, Ann Arbor (1976).

37 Swerdloff, R. S., and Odell, W. D., Hormonal mechanisms in the onset of puberty. *Postgrad. Med. J.* **51**, 200 (1975).

38 Gupta, D., Changes in the gonadal and adrenal steroid patterns during puberty. *Clin. in Endocr. Metab.* **4**, 27 (1975).

39 Winter, J. S. D., Faiman, C., Reyes, F. I., *et al.*, Gonadotropins and steroid hormones in the blood and urine of prepubertal girls and other primates. *Clin. in Endocr. Metab.* **7**, 513 (1978).

40 VandeWiele, R., Bogumil, J., Dyenfurth, I., *et al.*, Mechanisms regulating the menstrual cycle in women. *Rec. Prog. Horm. Res.* **26**, 63 (1970).

41 Moghissi, K. S., Syner, F. N., and Evans, T. N., A composite picture of the menstrual cycle. *Am. J. Obstet. Gynecol.* **114**, 405 (1972).

42 Henzel, M. R., and Segre, E. J., Physiology of the human menstrual cycle and early pregnancy. A review of recent investigations. *Contraception* **1**, 315 (1970).

43 Ross, G. T., Cargille, C. M., Lipsett, M. B., *et al.*, Pituitary and gonadal hormones in women during spontaneous and induced ovulatory cycles. *Rec. Prog. Horm. Res.* **26**, 1 (1970).

44 Van Leusden, H. A., Hormonal changes in pathological pregnancy. *Vit. Horm.* **30**, 282 (1972).

45 Goebelsmann, V., Protein and steroid hormones in pregnancy. *J. Reprod. Med.* **23**, 166 (1979).

46 MacNaughton, M. C., Abnormalities of ovarian function. *Clin. in Obstet. Gynecol.* **3**, 181 (1976).

47 Franchimont, P., Valcke, J. C., and Lambotte, R., Female gonadal dysfunction. *Clin. in Endocr. Metab.* **3**, 533 (1974).

48 Dewhurst, C. J., Primary and secondary amenorrhea. *Clin. in Obstet. Gynecol.* **1**, 619 (1974).

49 Baker, H. W. G., and Hudson, B., Male gonadal dysfunction. *Clin. in Endocr. Metab.* **3**, 507 (1974).

50 London, D. R., Medical aspects of hypogonadism. *Clin. in Endocr. Metab.* **4**, 597 (1975).

51 Horton, R., Biosynthetic Disorders of the Adrenal Cortex and Gonad, *in* Bergsma, D. (ed.), *Genetic Forms of Hypogonadism*. Birth Defects: Original Article Series, Vol. 11, No. 4, pp. 63–68, National Foundation March of Dimes, New York (1975).

52 Steinbeck, A. W., and Theile, H. M., The adrenal cortex. *Clin. in Endocr. Metab.* **3**, 557 (1974).

53 Bongiovanni, A., Eberlein, W., Goldman, A. S., *et al.* Disorders of adrenal steroid biogenesis. *Rec. Prog. Horm. Res.* **23**, 375 (1967).

54 Hamilton, W., Congenital adrenal hyperplasia. *Clin. in Endocr. Metab.* **1**, 503 (1972).

55 Westphal, U., *Steroid-Protein Interactions*, Springer-Verlag, New York (1971).

56 Mercier-Bodard, C., Alfsen, A., and Baulieu, E. E., Sex steroid binding plasma protein (SBP). *Acta. Endocr.* Suppl. **147**, 204 (1970).

57 Goldie, D. J., Hasham, N., Kean, P. M., Pearson, J., and Walker, W. H. C., Cortisol binding characteristics measured by dialysis. *J. Endocrinol.* **43**, 127 (1969).

58 Batra, S., Bengtsson, L. P., Grundsell, H., and Sjöberg, N-O., Levels of free and protein-bound progesterone in plasma during late pregnancy. *J. Clin. Endocrinol. Metab.* **42**, 1041 (1976).

59 Mishell, D. R., Current status of oral contraceptive steroids. *Clin. Obstet. Gynecol.* **19**, 743 (1976).

CHAPTER TWENTY

CYTOPLASMIC RECEPTORS FOR STEROID HORMONES— CHARACTERISTICS AND ASSAY

WILLIAM J. KING

The Ben May Laboratory for Cancer Research
University of Chicago
Chicago, Illinois

1 MECHANISMS OF STEROID ACTION—THE RECEPTOR CONCEPT

The previous chapter discussed mechanisms by which steroid hormones are transported from their sites of synthesis to the target tissues on which they act. This chapter therefore considers the biochemical features of these target tissues that enable them to recognize and respond to such hormones. The model to be presented represents a consensus viewpoint, many details of which are still debated. Although we have attempted to present alternative viewpoints, an exhaustive critique is beyond the scope of this chapter. Instead, we attempt to present an overview of the field as it now stands and refer the reader to the several reviews cited at the end of this chapter for additional discussion of these matters.

The search for the mechanisms by which hormones exert their effects on their target tissues has spanned several decades, but received special impetus during the 1940s and 1950s, following the elucidation of the critical role that steriod hormones play in the growth and induction of several human and experimental cancers. This work, as well as the pioneering studies on contraceptive steroids that were also taking place at the time, focused special attention upon the ovarian steroids. A critical breakthrough occurred in the late 1950s at the University of Chicago, where Jensen and Jacobson became the first investigators to synthesize tritiated estradiol of high specific activity. They subsequently observed that administration of physiological amounts of labeled steroid to experimental animals was followed by accumulation of tritium in estrogen target tissues against a concentration gradient (Fig. 1), showing that estrogen target tissues differed from nontarget tissues with respect to their retention of the hormone. This phenomenon was investigated more fully by Jensen's group and by that of Gorski at the University of Illinois, and eventually led to the generalized model presented in Fig. 2.

Although originally proposed to explain the action of estrogens in the uterus, the essential features of this scheme have been applied to the progestins, androgens, glucocorticoids, mineralocorticoids, vitamin D_3, and insect hormones upon their respective targets. According to this model, free steroid passes (*1*)

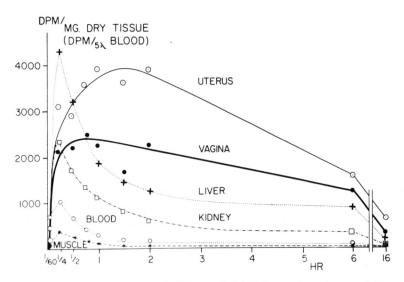

Figure 1 Uptake and retention of [^{3}H]estradiol-17β by various tissues of immature rats. Reproduced with permission from Jensen and Jacobson, *Rec. Prog. Horm. Res.* **18**, 390 (1962).

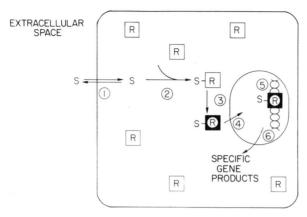

Figure 2   Generalized scheme for steroid hormone–receptor interaction. See text for details.

from the extracellular space to the cytoplasm (2) probably by simple diffusion. Steroid binding (3) permits the receptor to become *activated* and (4) acquire the ability to bind to the nucleus (*translocation*), (5) interacting with nuclear *acceptor* sites, (6) altering gene expression. This model implies that although the presence of receptors does not guarantee steroid responsiveness (since control may be exerted at any of the steps subsequent to steroid binding), the absence of such binders ensures steroid insensitivity.

   Criteria have since been established that must be met in order to demonstrate conclusively the presence of a physiologically relevant steroid receptor in biological materials. The proposed receptor should, first of all, bind the appropriate steroid with a *high affinity* such that at physiological hormone concentrations ($10^{-12} - 10^{-9} M$) a significant proportion of the receptor will be occupied. Furthermore, there should be a finite number of binding sites such that *saturation* of steroid binding can be observed upon addition of increasing amounts of steroid. The receptor should be *specific* with respect to the types of steroid it will bind and the tissues in which the receptor is found. A prospective estrogen receptor, for instance, should be found in higher concentrations in estrogen target tissues than in nontarget tissues and should have a higher affinity for estrogens than for nonestrogenic steroids. Finally, if possible, it should be demonstrated that alterations in the levels of receptor within the target tissues are accompanied by alterations in a steroid-mediated *biological response.*

2 METHODS OF RECEPTOR ASSAY

Many methods have been developed for receptor quantification and characterization that are suited to particular tissues or receptors; most of these may be categorized as adsorbtive techniques, centrifugation techniques, or some combination of these (see Refs. 1–3). These are summarized in Table 1. Typically,

Table 1   Review of Commonly Employed Methods for Assay of Steroid Receptors[a]
(Reproduced with permission from McCarty, K. S., Jr., and McCarty, K. S., Sr. (1977)
American J. Pathol. **86:715, Harper and Row).**

Method	Rationale
Charcoal	Method rapid, reproducible, and inexpensive. Charcoal used to absorb free steroid. Sensitivity .001→.8 nM. Problem: charcoal may differentially strip nonsteroid and steroid-binding components. Difficult to interpret when competing multiple-binding components are present.
Hydroxylapatite	All known steroid receptors bind. Since elevated salts do not interfere, can be used for nuclear receptors. Problem: some nonspecific binding, high background binding.
DEAE filter	Used for glucocorticoid receptor absorption binding. Advantage for dilute solutions. Major drawback: absorption of free steroids, resulting in high background.
Protamine precipitation	Useful for both bound and free receptors, good for first steps in purification. Stabilizes receptor protein against proteolytic activity. Disadvantage: nonspecific binding proteins may precipitate with receptor fraction.
Gel filtration	Gentle technique, gives good fractionation, dependent on salt concentration. Disadvantage: technically difficult for multiple assays. Requires higher protein concentrations and should be performed in parallel with several varying concentrations of competing steroid.
Gel electrophoresis	Elegant test, gives good fractionation. Problem: stripping and coelectrophoresis of labeled steroids; pH. Works well with progesterone receptor; its use for estrogen receptor requires further development.
Affinity labeling	Use of 4-mercuri-17β-estradiol to form a pseudocovalent bond with -SH of receptors. Problem: complexity of assay and fractionation of product.
Immobilized antibody	Antiserum to estrogen receptor is immobilized to vinylidene fluoride film. Competition of tritiated estradiol binding with the binding protein. Stable 2 months. Problem: interpretation of complex binding kinetics with multiple equilibrium constants.
Sucrose gradients	Excellent for identification of receptor and separation of specific from nonspecific binding proteins. Method of choice for clinical identification of receptor. Problem: quantitation complicated by lack of equilibrium conditions during centrifugation; time and equipment required.

[a] Reproduced with permission from McCarty, Jr., and McCarty, Sr., *Am. J. Pathol.* **86**, 715 (1977).

tissue is homogenized in slightly alkaline hypotonic buffer and centrifuged at high speed (e.g., 105,000 × g, 1 hr) to obtain a clear supernatant, termed *cytosol*. Steroid bound to cytosolic components is separated from unbound or free steroid either by sucrose density gradient centrifugation or by employing any number of adsorbant materials capable of selectively binding either free steroid or receptor. The most commonly employed adsorbant is probably dextran-coated charcoal, which efficiently adsorbs large amounts of free steroid and is subsequently easily removed by low-speed centrifugation. Many investigators prefer protamine sulfate or hydroxyapatite precipitation of the receptor for this assay step, since these agents effectively separate the receptor from proteolytic enzymes that reduce the steroid-binding capacity of certain cytosols (4).

In addition to the high-affinity, limited-capacity binding displayed by hormone receptors (often termed *specific binding*), steroids may also bind to lower-affinity $K_d \geq 10^{-6}$ M), high-capacity substances (*nonspecific binding*). To some

extent this low-affinity binding represents contamination of cytosol preparations with serum proteins such as corticosteroid-binding globulin (CBG) and
testosterone-estradiol-binding globulin (TeBG); however, it is also likely that
intracellular elements contribute to the nonspecific binding. Whether such intracellular low-affinity sites play any role in the physiological activity of the
steroids is unknown.

Methods to distinguish specific from nonspecific binding usually exploit four
properties of the former: its limited capacity, its specificity, its sedimentation
characteristics, and its temperature lability. In a typical assay, specific binding
of labeled steroid to aliquots of cytosol is determined by the difference in
binding in the absence and in the presence of an excess of unlabeled steroid.
Assay conditions are adjusted with respect to time and temperature to permit
the binding reactions to reach kinetic equilibrium (based on previous experiments). The unlabeled steroid may be the same as the labeled steroid in which
case it is assumed that radioactivity bound in the presence of excess steroid is
associated primarily with the high-capacity (i.e., nonspecific) binding sites. Specific binding is estimated by subtracting the nonspecifically bound radioactivity
from radioactivity bound in the absence of added, unlabeled steroid. If this
procedure is repeated using various concentrations of labeled steroid, data such
as those seen in Fig. 3 are generated. Nonspecific binding increases linearly as a
function of the labeled steroid concentration—or, more accurately, the *free*
steroid concentration (see below)—whereas specific binding to the steroid receptor approaches a plateau, reflecting its limited capacity.

In some cases, especially clinical samples, the amount of tissue available is
insufficient to perform analyses with more than one concentration of labeled
steroid. Under such circumstances a few of the larger samples are used to
determine the amount of labeled steroid required to saturate the receptor.

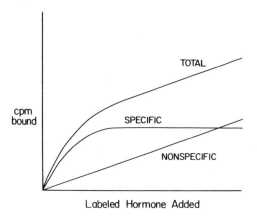

Figure 3 Relationship between total, nonspecific, and specific binding. *Total*, The cpm
bound in absence of excess cold steroid; *nonspecific*, the cpm bound in presence of excess
cold steroid; *specific*, Total − nonspecific.

Subsequent analyses are then performed at one steroid concentration sufficient to ensure such saturation.

The precision of these determinations is sometimes improved by employing synthetic compounds that bind to a receptor as well as or better than the appropriate natural steroid, but have little or no affinity for serum proteins. Such synthetic steroids may be used both as the unlabeled and labeled compound or, as is more common, in combination with a natural compound. Thus, for example, diethylstilbestrol (DES) when present in 100-fold excess will displace tritiated estradiol from the estrogen receptor, but since DES has little affinity for TeBG, labeled steroid associated with this serum protein will be included as part of the nonspecifically bound fraction. This is of particular value if one encounters samples with relatively small amounts of nonspecific binding. In this case, when using the unlabeled counterpart of the radioactive natural steroid as the competitor, displacement of radioactivity from nonspecific sites may occur, resulting in overestimation of receptor sites.

The third distinguishing characteristic of steroid-hormone receptors, sedimentation velocity, may also be used to discriminate between specific and nonspecific binding. As discussed below in more detail, all of the steroid receptors characterized to date form high-molecular-weight aggregates in low-ionic-strength sucrose gradients. These aggregates sediment predominantly in the

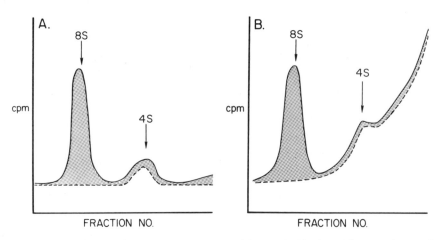

Figure 4 (*A*) Sedimentation analysis of specific steroid binding. Aliquots of cytosol are incubated with radiolabeled steroid in the absence (—) or presence (---) of an excess of unlabeled competitor. After removing unbound steroid with dextran-coated charcoal (DCC), the cytosols are layered on identical sucrose gradients and centrifuged to equilibrium. The stippled area represents specific binding. Since the unlabeled competitor sometimes does not displace all the steroid bound to serum proteins, some binding in the 4 S region may persist in the presence of excess steroid. (*B*) Since DCC may sometimes strip the 4 S form of the receptor of labeled steroid to some degree, gradients are sometimes run without prior DCC treatment. In this case, however, one may obtain a large peak of free steroid near the top of the gradient that may trail into the 4 S peak.

6–8 S region of such gradients, whereas serum proteins sediment at approximately 4 S under the same conditions. In many preparations, however, lower-molecular-weight forms of the receptor exist even under low-salt conditions. Discrimination of these receptors from serum proteins and other nonspecific binding proteins cannot be made without duplicate preparations containing an excess of unlabeled steroid (Fig. 4).

A fourth method, warming the cytosol to 37°C, will often destroy the binding activity of the receptor while not affecting nonspecific binding.

Claims of altered steroid-binding capacity in experimental populations should, of course, contain data confirming that such changes are not due to changes in the equilibrium constant.

3 METHODS OF DATA ANALYSIS

Having established the presence of specific hormone receptors in cytosol preparations from the tissue in question, the investigator is next concerned with characterizing the receptor. The two most commonly studied binding parameters are the equilibrium dissociation constant K_d and the total number of binding sites B_{max}. Again, many methods exist for their determination (1, 5). Re-examining Fig. 3, it is obvious that the data for steroid binding are analogous to the relationship expressed by the Michaelis-Menton equation for enzyme kinetics. Bound steroid (cpm bound) substitutes for initial velocity and total steroid (labeled steroid added) is analogous to the initial substrate concentration, with the relationship expressed as

$$\text{Bound} = \frac{B_{max} \cdot K_a \cdot \text{free}}{1 + K_a \cdot \text{free}}$$

where

$$\text{Free} = \text{Total steroid added} - \text{Total bound}$$

Most of the transformations applied to enzyme data may thus be applied to observations concerning steroid binding. The double-reciprocal plot of Lineweaver-Burke is one such transformation commonly employed (Fig. 5). Expressed as

$$\frac{1}{\text{Bound}} = \frac{1}{B_{max} \cdot K_a \cdot \text{free}} + \frac{1}{B_{max}}$$

the resulting plot is a straight line (assuming that the cytosol preparation in question contains only a single class of binding sites) with slope $= 1/B_{max} \cdot K_a$ and that intercepts the ordinate at $1/B_{max}$. This plot is, of course, especially useful in analyses involving competing steroids.

Perhaps the most commonly employed transformation in receptor studies is that of Scatchard, where

$$\frac{\text{Bound}}{\text{Free}} = K_a \, (B_{max} - \text{Bound})$$

This is also a straight-line plot whose slope $= -K_a$ and whose intercept on the horizontal axis $= B_{max}$. The equilibrium dissociation constant K_d is either estimated from this data as $1/K_a$, as is most commonly done, or may be determined directly in separate experiments from the rate constants for association and dissociation of the steroid. Some discrepancy in the K_d values obtained by these two methods is sometimes noted for reasons that are unclear at this time. The dissociation constant determined from Scatchard analysis tends to be higher than that determined from measured rate constants.

In practice, Scatchard plots of steroid-receptor data rarely, if ever, are linear when plotted correctly (see Ref. 5) because of the presence of lower-affinity, higher-capacity binding sites. As work in this field has progressed, several fine discussions have been presented concerning the interpretation of nonlinear Scatchard plots and resolution of complex (multisite) binding systems. Interested readers should consult references (6, 7, 8) at the end of this chapter.

Using the above procedures, the investigator may demonstrate the other defining characteristics of the receptor under study, that is steroid and tissue specificity and correlation of receptor levels with biological responses.

It should be noted that these methods are usually employed to measure unoccupied receptor sites, that is, receptors not already occupied by endogenous steroid. Although steroid–receptor complexes (occupied receptors) are rapidly translocated to the nucleus *in vivo*, cell homogenization may expose unoccupied receptors to steroid that was previously located intercellularly, in serum, and/or bound to low-affinity, high-capacity sites. *Exchange assays* have therefore been developed for many steroid receptors to ensure the measurement of all receptors regardless of their state of occupancy. The strategy of these assays, shown in Fig. 6, consists of first removing all endogenous *free* steroid by preincubation with dextran-charcoal in order to minimize dilution of the la-

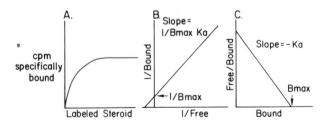

Figure 5   Transformation of data for specific binding (*A*) according to the methods of Lineweaver-Burke (*B*) or Scatchard (*C*).

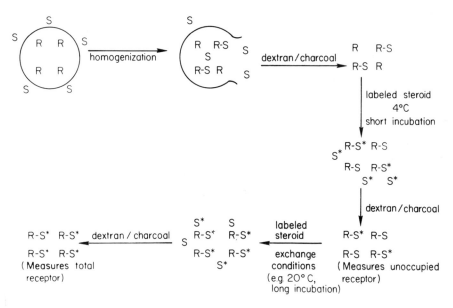

Figure 6 Typical exchange protocol. *R*, Receptor; *S*, steroid; *S**, radiolabeled steroid.

beled steroid used in the next step. Unoccupied sites can be measured after a short incubation, as shown in Fig. 6. To measure the total receptor capacity aliquots of this cytosol are then incubated with saturating concentrations of labeled steroid under conditions that permit exchange of bound steroid with the excess of labeled free steroid. Bound and free hormone are then separated, taking care to minimize steroid dissociation from the receptor, for example by recooling. Nonspecific binding is evaluated by adding a 100-fold greater quantity of unlabeled than labeled steroid in the exchange step. When the extent of receptor occupation by endogenous steroid is large, as may be the case when circulating steroid levels are high, it may be necessary to correct the final binding data for dilution of the isotope by unlabeled steroid.

4 PHYSICOCHEMICAL CHARACTERISTICS OF STEROID RECEPTORS/STEROID BINDING

All of the receptors characterized to date are acidic ($P_I \sim 5.0$), asymmetric (frictional ratio, ~ 1.5–2.0), globular proteins, possibly metalloproteins, that sediment in low-salt (0.01 M) sucrose gradients at 6–8 S. Under conditions of high salt (0.3–0.4 M), a shift in the sedimentation pattern to 3–4 S is usually observed (9, 10, 11). Some investigators report the presence of some 4 S receptor even under low-salt conditions. A number of studies that have been performed under isotonic conditions have yielded sedimentation values in the range of 4–6 S (estrogen receptor), although this value may be affected by many variables, including the protein concentration of the cytosol preparation (11).

Receptors have been purified to various degrees using most of the conventional methods, including ion-exchange chromatography on DEAE-cellulose or phosphocellulose, adsorption to hydroxyapatite, affinity chromatography using DNA-cellulose, steroid-agarose or heparin-agarose, or by sizing techniques such as gel filtration, sucrose density gradient centrifugation, or gel electrophoresis (12, 13). Most of the receptors seem to be precipitable by low (30%) concentrations of ammonium sulfate, a procedure that conveniently separates receptors from most of the steroid-binding globulins of serum.

Extensive analysis of receptors has been hampered in some cases by the lability of these proteins under conventional assay conditions. This is especially true of receptors not occupied by steroids, although progressive changes in binding activity or sedimentation behavior of filled receptors have also been noted during extended incubations. Inclusion of sulfhydryl-protecting reagents such as 2-mercaptoethanol, monothioglycerol, or dithiothreitol in the assay buffer often retards this loss. Protection is also sometimes provided by protein-stabilizing agents like glycerol or by proteinase inhibitors, including leupeptin or iodoacetate. In contrast, receptors may be inactivated by sulfhydryl-blocking agents such as N-ethylmaleimide or p-chloromercuriphenylsulfonic acid (14). Some preparations (e.g., from liver) seem to contain sufficient reducing activity by themselves to mitigate the requirement for sulfhydryl-protecting reagents.

The nature of the steroid–protein interaction has been extensively studied with respect to several receptors; the primary force promoting binding is a hydrophobic interaction between the receptor and both planar faces of the steroid (15). The receptor seems to envelop the steroid within a hydrophobic "pocket." Significant contributions are also made by hydrophobic side groups on the steroid, the size of such side groups, and the conformation (planarity) of steroid ring A. These latter factors are the determinants of binding specificity.

Glucocorticoid binding to the cytosolic receptor of mouse fibroblasts has been proposed to occur in two stages, an initial rapid but weak interaction followed by a slow conversion to a tighter-binding complex (16). Since the specific guanidino reagent 1,2-cyclohexanedione competitively inhibits the binding of this steroid, arginine residues appear to be involved (14).

As recently shown by Sherman (17), the steroid-binding region of this steroid receptor appears to be confined to a relatively small (2–4×10^4 daltons), basic portion of the receptor molecule.

Steroid binding may be subject to control by phosphorylation-dephosphorylation reactions. It has been demonstrated that the ability of glucocorticoid and progesterone (but not estrogen) receptors to bind their respective steroids is destroyed by protein phosphatases such as highly purified alkaline phosphatase (18). Furthermore, phosphatase inhibitors such as fluoride, arsenate, or molybdate have been shown to retard the progressive loss of binding activity usually observed in cytosol preparations (19), although in the case of molybdate a direct interaction with the receptor has also been suggested (20). It is also still unclear whether the putative endogenous phosphatases act on the receptor itself or on some other regulatory molecule.

Many reports have recently appeared in the literature concerning receptor heterogeneity. It seems likely that the sedimentation peaks observed following sucrose density gradient centrifugation represent a number of different molecular species, since subsequent subfractionation of these peaks, for instance, by ion-exchange chromatography, usually yields several peaks of binding activity (21). Whether these represent distinct receptors with differing physiological roles or are simply artifacts induced by the homogenization and fractionation procedures is still unknown in most cases. Evidence has accumulated from other sources, however, reinforcing the notion of structural/functional heterogeneity. Agarwal and a number of other workers have demonstrated that hormone–receptor complexes involving synthetic steroids such as dexamethasone display different chromatographic behavior than those involving natural steroids (22). Although this may represent preferential binding to a subspecies of corticoid receptor, one cannot rule out the possibility that binding of the synthetic steroid induces distinct conformational changes in the receptor that alter its subsequent mobility on ion-exchange resins. Receptor heterogeneity has also been proposed on the basis of physiological studies of glucocorticoid target cells.

Another kind of heterogeneity has been described by Clark and his coworkers with respect to the uterine estrogen receptor (23). Detailed Scatchard analyses over a broad range of estrogen concentrations revealed the presence of a second high-affinity steroid-binding entity in both the cytoplasm and nucleus of uterine cells. The second cytoplasmic binder, termed a type II receptor, has a somewhat lower affinity $K_d = 33$ nM compared with 0.8 nM) and higher capacity than the "classical" type I receptor. Furthermore, type II receptors do not appear to undergo translocation to the nucleus, so that the secondary nuclear receptors (also termed, confusingly, type II receptors) are unlikely to be related to the lower-affinity cytoplasmic receptors. Type II nuclear receptors do, however, appear to increase in number following estrogen administration to the intact animals. The functional significance of these different receptor types is unresolved, but Clark has suggested several possibilities, including one in which the type II receptor is a cytosolic precursor of the type I receptor. Alternatively, the cytosolic type II receptor may facilitate estrogen retention in the uterus or may even be an extracellular contaminant of uterine cytosol preparations.

The most convincing description of receptor heterogeneity at this time is that of the avian progesterone receptor, which has been extensively characterized by Schrader and O'Malley (12). This receptor, which has been purified to homogeneity, consists of two separable subunits, both of which bind steroid, but which appear to serve somewhat different functions in the transduction of the steroid signal.

5   RECEPTOR ACTIVATION*/TRANSLOCATION

Binding of the steroid to its receptor alters the receptor in some way that permits it to interact with nuclear *acceptor* sites (see Chapter 21). *In vitro* this process may be assayed by monitoring the ability of the receptor to bind to isolated nuclei or to anionic matrices such as ATP-Sepharose, DNA-cellulose, or phosphocellulose. The exact nature of this *activation* process remains a mystery, although for all the receptors studied to date it seems to be a temperature-dependent phenomenon, inhibited at 0–4°C and permitted at 25–37°C (9, 10, 11, 12, 16). Activation of the progesterone receptor, however, may be brought about at 4°C by brief exposure to high salt concentrations (e.g., 0.5 M KCl) (24). Salt activation of the receptor can occur even in the absence of steroid. Activation of steroid receptors has been observed following dilution of the cytosol, gel filtration, or dialysis, suggesting the presence in cytosol of a negative modulator of activation (25).

In some cases, for example, uterine estrogen receptors, activation is accompanied by changes in the sedimentation behavior of the receptor (4 S converted to 5 S) and may involve an association of two receptor subunits (Notides in Ref. 12). It is unclear at this time whether the estrogen receptor monomers are identical proteins. Recent evidence suggests that estrogen receptor activation is independent of, and may precede, 4 to 5 S transformation (26).

Although the progesterone receptor of chick oviduct also appears to consist of two subunits, designated A and B, association of these subunits is not involved in progesterone-receptor activation (12). In fact, progesterone binding promotes dissociation of the dimer into its subunits, a step that has been speculated to occur in the nucleus and that may be necessary for the expression of the DNA-binding ability of the A subunit. The B subunit binds only to target-cell chromatin and is thought to direct the dimer to specific chromatin sites in the vicinity of genes regulated by progesterone (Fig. 7, see Chapter 21 for further details). The A subunit may then bind to the DNA of this region, modifying gene transcription. Both the A and B subunits appear to bind a molecule of progesterone.

Activation of the progesterone receptor is accompanied by sedimentation changes suggesting disaggregation of 6–8 S entities to 4 S form(s). Toft and his co-workers have observed that activation of this system may be inhibited by either pyridoxal-5'-phosphate or by molybdate, a phosphatase inhibitor (27). It is possible, then, that activation of this receptor involves dephosphorylation of either the receptor itself or some naturally occurring inhibitor, for example,

---

* The nomenclature adopted here is that which is most popularly accepted with regard to these processes. However, some authors prefer the term *transformation* to denote the acquisition of nuclear binding ability, reserving *activation* to describe those changes (dephosphorylation, reduction) that enable the receptor to bind steroid. In the literature concerning estrogen receptors, on the other hand, *transformation* refers to the sedimentation changes accompanying the heat-induced acquisition of nuclear binding activity.

pyridoxal-5'-phosphate. Direct molybdate–receptor interactions may also contribute to these observations.

Toft has also shown that progesterone-receptor activation may be inhibited by low concentrations of nucleotides, but stimulated by concentrations above 1 mM (28). This suggests the possibility that ATP binding (which is also a quantitative indicator of receptor activation) stabilizes the activated receptor. The significance of these findings *in vivo* has yet to be demonstrated.

DiSorbo and Litwack have proposed a model for activation of glucocorticoid receptors that pictures pyridoxal-5'-phosphate as a modulator that binds to the receptor, probably through a Schiff base linkage through basic amino acid residues (14). Such an association is thought to prevent the binding of the receptor to DNA. Activation is proposed to involve both a conformational change due to the binding of the steroid as well as the dissociation of the modulator molecule. Such activation may expose a region of positive charge on the receptor surface that facilitates receptor interaction with the negatively charged phosphates of DNA. Electrostatic interactions are thus proposed to be the driving forces behind receptor translocation in this model. The guanidino reagent 1,2-cyclohexanedione, besides affecting glucocorticoid binding, may also inhibit binding of activated receptor to DNA-cellulose, further suggesting the participation of (positively charged) arginine residues in receptor binding to DNA (14).

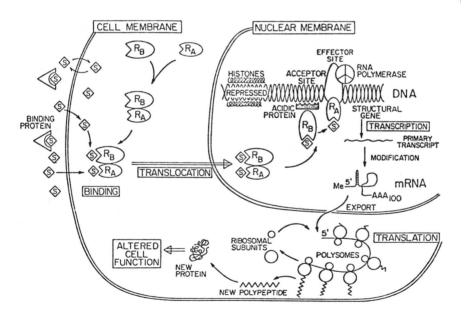

Figure 7 Hypothetical model for mechanism of progesterone action. Reproduced with permission from Schwartz, Chang, Schrader, and O'Malley (1977) *Ann. N.Y. Acad. Sci.* **286**, 158.

Most of the work done to date suggests that translocation of the receptor to the nucleus is a passive process resulting from the interaction of nuclear acceptor sites with sites on the activated receptor (11). Inhibitors of microtubule or microfilament formation that have been shown to affect the intracellular transport of many macromolecules and organelles have no affect on translocation of the estrogen receptor. The same is true for inhibitors of protein or RNA synthesis. Although translocation of the glucocorticoid receptor may indirectly require energy input (because of the need for a phosphorylation step prior to steroid binding), translocation of the estrogen receptor is not affected directly by metabolic inhibitors. It is, however, temperature-dependent. Any discussion of translocation must therefore be directed toward the nature of the activated receptor and the nuclear acceptor sites. As the preceding discussion and Chapter 21 point out, these are still matters of great controversy.

An alternative model for translocation of the estrogen receptor was suggested by Gorski (11). He envisions the receptor as initially restricted to the gel-like cytoplasm, perhaps because of interactions with membranous components of the cytoplasm. Estrogen binding is proposed to release the receptor from these restrictions and to permit equilibration of the receptor with the more fluid nucleoplasm on the basis of phase partitioning. Whether such mechanisms operate independently of, instead of, or in addition to the previously mentioned translocation mechanisms is not evident from the meager experimental data now available.

6   MODULATION OF STEROID-MEDIATED PROCESSES: STEROID STRUCTURE

Binding of steroids to their receptors is influenced by stereochemical modifications of the ring structure as well as by substituent functional groups. Obviously, any alteration in these features such as occurs during metabolism of the steroid will affect the efficiency with which the steroid influences transcriptional events. Metabolism of the most active steroids (e.g., estradiol, progesterone, cortisol) almost invariably decreases the affinity of the steroid for its receptor and diminishes its potency.

Androgens are notable exceptions to this scheme. Although testosterone is the principal androgen produced by the testes and found in plasma, its androgenic potency is usually increased by its conversion to 5α-dihydrotestosterone (5-DHT) by the enzyme 5α-reductase. Androgen receptors have a higher affinity for 5-DHT than for testosterone.

In addition to the so-called androgenic effects of testosterone, this steroid may act on other targets (e.g., kidney, liver, muscle) to stimulate body weight gain and nitrogen balance, the "anabolic" effects of testosterone. Little or no 5α-reductase activity is found in these tissues, and testosterone itself is therefore thought to be the principal anabolic androgen (see Bardin et al., in Ref. 12). Earlier studies seemed to indicate that androgen receptors from nonreproductive tissues had higher affinities for testosterone than receptor from, for instance, prostate tissue. More-recent investigations, however, have suggested that the androgen receptor may be the same in all tissues (see Wilson *et al.*, Bardin *et al.* in Ref. 12; 29, 30). The apparent lower affinity of the prostate receptor for testosterone may have been due to the dissociation of [³H] testosterone from the androgen receptor during the prolonged centrifugation employed for the sucrose density gradient assay. If an assay method is used that allows less time for dissociation, no appreciable differences are found between androgen receptors from reproductive and nonreproductive tissues. Consistent with this finding, mice that are androgen-insensitive because of an inherited absence of androgen receptor—testicular feminization syndrome (tfm)—fail to show both androgenic and anabolic responses to administered steroid (Bardin *et al.*, in Ref. 12), again suggesting the equivalence of receptors in various androgen targets.

It should be noted, however, that dissociation of [³H]testosterone from the renal androgen receptor does not occur during such prolonged centrifugations (Bardin *et al.*, Ref. 12), suggesting that some heterogeneity in androgen receptor interactions may indeed be present.

The vitamin D_3 receptor is another example of a receptor that binds metabolites with greater efficacy than the parent compound. In this case, however, metabolism occurs outside the target tissues themselves. It has been shown that hydroxylation of vitamin D_3 at the C-25 position enhances its effectiveness in target organs such as bone or intestinal mucosa. Such a hydroxylation has been found to occur as the vitamin passes through the liver. In the kidney, 25-$(OH)D_3$ is further metabolized to 1,25-di-$(OH)D_3$, whose even greater potency is related to the higher affinity of the vitamin D_3 receptor for the dihydroxy form (Norman and Wecksler in Ref. 12). Receptors for 25-$(OH)D_3$ have also been found in many tissues, but their function is unknown.

The ability of a steroid receptor to bind a hormonal ligand with preferential affinity does not preclude the possibility that other hormones may utilize the receptor. This may be especially true during *in vitro* studies or following hormonal administration *in vivo* in which the secondary steroid may attain sufficient concentrations to compensate for its lower affinity. In such cases, of course, the response is receptor-specific rather than hormone-specific. Whether such interactions result in mimicry or antagonism of the effects of the primary steroid–receptor complex may depend on the ability of the secondary steroid-receptor complex to successfully complete the entire sequence of activation, translocation, and initiation of transcription.

Thus depending upon dose and structure, a number of synthetic progestins have been demonstrated to have androgenic, synandrogenic, or antiandrogenic effects (31). Little is known concerning the mechanism of synandrogenic effects (e.g., potentiation of the testosterone induction of renal β-glucuronidase by cyproterone acetate or megestrol acetate); however, at high doses these compounds were found to be antiandrogenic. Progestins exerting either androgenic or antiandrogenic effects have been found to bind to the androgen receptors, suggesting a receptor-mediated process. These effects are not observed in tfm mice (31).

Progesterone itself is only marginally effective in maintaining the male reproductive tract (32) and shows variable competition for the androgen receptor (McEwen in Ref. 11), however progesterone may interact strongly with glucocorticoid or mineralocorticoid receptors (Baxter and Ivarie, Anderson and Fanestil in Ref. 12; 1, 32, 33). Such binding has been shown to inhibit the ability of corticoids to bind to their receptors and/or to interact with nuclear binding sites (32, 34). Similar mechanisms have been shown to account for the antagonistic properties of a number of synthetic antiestrogens, antiglucocorticoids and antimineralocorticoids (Anderson and Fanestil, Baxter and Ivarie in Ref. 12; 35).

The most extensively investigated model illustrating the relationship between the affinity of steroids for their receptors and the ability of these steroids to influence gene transcription is the induction of tyrosine aminotransferase (TAT) by glucocorticoids in hepatoma tissue culture cells. Tomkins and his co-workers found that steroids fell into four major functional classes based on their TAT-inducing abilities (36). Optimal inducers were capable of eliciting maximal TAT responses, whereas suboptimal inducers evoked characteristic but submaximal enzyme levels. Anti-inducers did not induce the enzyme themselves, but competitively inhibited the responses to optimal and suboptimal inducers. A number of steroids were inactive with respect to both induction and inhibition. An allosteric model was proposed (32, 36) based on the enzyme model of Monod, Wyman, and Changeux (although an "induced-fit" model based on the allosteric theory of Koshland could have been chosen as well). The steroid receptor was pictured as existing in two conformational states, termed R and T (Fig. 8), which were thought to be in equilibrium. In the absence of steroid, the equilibrium was proposed to favor the T form of the receptor. The R form, on the other hand, was thought to be the only form capable of interacting with nuclear sites to induce TAT synthesis. According to the model, binding of the steroid to either form of the receptor stabilizes that form and thus pulls the conformational equilibrium toward the form being bound. The ability of a steroid to induce TAT synthesis therefore was said to depend upon its relative affinity for the R and T receptor conformations. Optimal inducers bind primarily to the R form, while suboptimal inducers bind to both R and T forms. Since the hypothetical R and T forms are indistinguishable by conventional assay methods, however, no difference in the overall affinity of optimal, suboptimal, and anti-inducers for the receptor is observed. Within a group (optimal or anti-inducer), on the other hand, there is a good correlation between binding affinity and biological effectiveness (32, 33, 37). Suboptimal inducers do not exhibit this relationship (37, 38).

Anti-inducers are thought to stabilize the T form and, in the process, inhibit the binding of inducers by sequestering the receptor in this inactive form. Therefore, whereas interaction of optimal or suboptimal inducers with the glucocorticoid receptor results in a complex capable of binding to nuclear prepara-

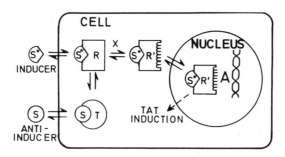

Figure 8 First steps in glucocorticoid hormone action. S^*, Inducer steroid; S, anti-inducer steroid; R and T, active and inactive conformations of the receptor; X, activation process whereby R becomes capable (R') of binding to the nuclear acceptor A. Taken with permission from Rousseau, *J. Steroid. Biochem.* **6**, 84, copyright 1975, Pergamon Press, Ltd.).

tions, anti-inducers (which again may be indistinguishable from inducers with respect to *overall* binding affinity) do not (32). Inactive steroids have little affinity for either conformation.

Support for this model has been provided by the observation that progesterone, an anti-inducer, binds more rapidly to the receptor than dexamethasone (optimal inducer) despite the former's lower affinity (37, 39). This seems consistent with the hypothesis that the T form of the receptor predominates in uninduced cells. Furthermore, the thermostability of progesterone–receptor complexes is similar to that of unbound receptor but lower than those involving corticosterone (optimal inducer), despite the similar overall affinities of the two steroids (37, 39). This is again consistent with the notion of multiple receptor forms.

An alternative model for anti-induction arose from the inability of the above studies to present any evidence of the predicted cooperativity with respect to either inducer or anti-inducer binding, or in the competition between anti-inducers and inducers. According to Pratt and co-workers (16), anti-induction may instead arise from the inability of anti-inducers to effect the transition from weak to tight interaction with the glucocorticoid receptor such as described in Section 4.

6.1 Modulation of Steroid-Mediated Processes: Steroid Transduction

As one might expect, a number of models have also arisen in which hormonal processes are altered, not as a result of alterations in the steroids or interactions between the steroids and their receptors, but because of modification in the receptor itself. Such models not only validate the receptor concept but also sometimes provide insight into the physiological regulation of endocrine systems.

At least two models have been developed in which hormone insensitivity appears to be related to the absence of an effective steroid receptor. One of these models, the tfm mouse, was mentioned in the previous section. The second model, described by Sibley and Tomkins (40, 41) is based on the ability of glucocorticoids to kill cultured mouse lymphoma cells. Variants of a line of such cells were developed that were spared from glucocorticoid killing by virtue of a defect in the binding ability of the glucocorticoid receptor (r−). Intriguingly, other steroid-insensitive strains could be selected in which steroid binding was normal but nuclear transfer was absent (r+, nt−), or enhanced (r+, nt^i), or in which the presumed defect appeared to be located at a point beyond the process of activation/translocation (r+, nt+, d) (40, 41). It was presumed that the steroid insensitivity of the nt^i mutants was a result of increased affinity of the steroid–receptor complex for nonfunctional genome binding sites (41).

Modulation of steroid-binding capacity also appears to be a mechanism by which hormones interact with one another to control the steroid sensitivity of target tissues. Such regulation may be either positive or negative. Following administration of a large dose of estrogen to rats, the uterus undergoes a period of insensitivity to further estrogen treatment that can be correlated with the depletion of uterine cytoplasmic estrogen receptor consequent to translocation (Fig. 9) (42). Within 24–48 hr, however, the uterus becomes hyperreactive to estrogen, and uterine cytosol preparations display estrogen-binding activity that exceeds that present prior to estrogen treatment.

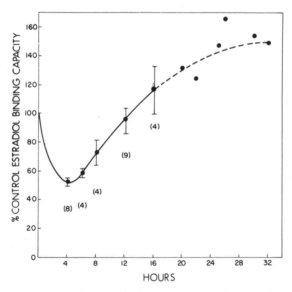

Figure 9 Time course of estrogen-binding capacity in the cytoplasm after estrogen injection. 17β-Estradiol (0.1 μg) was injected intraperitoneally into rats at 0 hr. Results are expressed as the mean (plus and/or minus standard error) of percentage of saline-injected control. Numbers in parentheses are the number of experiments used in calculating the mean plus and/or minus standard error. Points shown beyond 16 hr are single determinations. Reproduced with permission from Sarff and Gorski, *Biochemistry* **10**, 2560, 1971.

Replenishment of uterine cytoplasmic estrogen receptor following estrogen treatment is a rather complex process. Sarff and Gorski have shown that protein and RNA synthesis are necessary only between 1 and 6 hr after hormone exposure in order to observe normal receptor replenishment (42). Administration of actinomycin D or cycloheximide from 6 hr on, that is, from the time that replenishment normally is seen *to begin*, has no effect on the cytoplasmic binding capacity observed at 16 hr (when control binding levels are normally reattained). It was speculated that the receptor is resynthesized during the first 6 hr in a form incapable of binding steroid, which slowly acquires this ability. Alternatively, an enzyme might be synthesized during the early stages that would later be required for reacquisition of binding ability in a recycled receptor molecule. A third possibility, possibly related to the second, is that macromolecular synthesis is required for the nuclear "processing" of the estradiol receptor.

Mester and Baulieu, on the other hand, found that cycloheximide inhibited cytoplasmic-receptor replenishment when the inhibitor was administered at either 0.5 or 6 hr following administration of 0.1 μg of estradiol (43). This is the same dose that was given in the Sarff and Gorski study. If the dose of estradiol was increased to 1.0 μg (which results in more-rapid and extensive translocation and an earlier onset of replenishment), cycloheximide administered at 0.5 hr did not alter the amount of replenishment occurring by 6 hr. By 11 hr, however, the cycloheximide-treated group had less cytoplasmic estrogen receptor in their uteri than did control animals.

Resolution of this question may be forthcoming with the recent development by Jensen and co-workers of antibodies to purified estrogen receptor. These may prove useful in the detection of covert receptor (9, 44). Replenishment of cytoplasmic estrogen receptor does not occur when uteri are exposed to a pulse of estradiol *in vitro* (42). The reason for this is unknown.

A different scheme has arisen concerning replenishment of the cytoplasmic glucocorticoid receptor. Studies performed on thymus cells, mouse fibroblasts, or hepatoma tissue culture cells have shown that replenishment of this receptor occurs *in vitro* and does not require protein synthesis (reviewed in Ref. 33). However, ATP or some other energy source is required. ATP may be required in part for phosphorylation of the receptor since more-recent work has shown that dephosphorylation of the progesterone receptor or glucocorticoid receptor destroys its steroid-binding abilities. The scheme proposed by Munck and his colleagues based on studies with intact thymus cells (as opposed to the broken-cell or cytosol preparations used for more-recent studies) appears to agree with most of the available experimental data (Fig. 10) (45). While this plan describes the short-term regulation of glucocorticoid receptors, long-term regulation such as occurs over a period of days may require macromolecular synthesis. No data is available concerning this latter point. A summary of the current scheme for glucocorticoid transduction is presented in Fig. 11.

The second phase of estrogen-receptor replenishment, the overshoot phase (Fig. 9), illustrates another of the mechanisms by which receptor levels may be

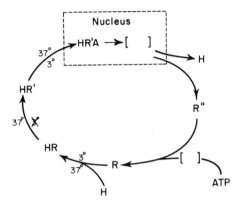

Figure 10 Hypothetical glucocorticoid receptor cycle. *H*, Hormone; *R, R', R"*, different forms of the receptor; and *A*, nuclear acceptor site. Temperatures give the conditions under which the various reactions can proceed. The reactions do not proceed at temperatures that are crossed out. Reproduced with permission from Munck, Wira, Young, Mosher, Hallahan and Bell, *J. Steroid. Biochem.* **3**, 574, copyright 1972, Pergamon Press, Ltd.).

controlled, that is, by hormonal induction. Receptors may be regulated by their own hormones (estrogen induction of uterine estrogen receptor) (42) or by other hormones (estrogen induction of uterine progesterone receptor) (46). This latter type of regulation may be responsible for many of the synergistic phenomena observed with respect to hormone action. Receptor induction, in possible contrast to the earlier phase of replenishment, does require protein synthesis. (43, 46).

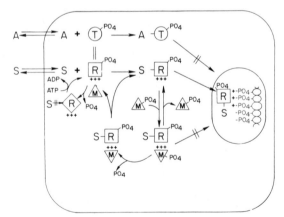

Figure 11 Glucocorticoid-receptor mechanisms. *R*, Active receptor; *T*, inactive receptor; *S*, optimal inducer steroid; *A*, anti-inducer steroid; *M*, modulator molecule. An alternative scheme can be proposed in which the phosphorylated modulator is associated with the receptor prior to steroid binding. The dephosphorylations that affect steroid binding and/or receptor activation may occur on the modulator or on the receptor itself. The (probably) additional requirement of reduced sulfhydryl groups on the receptor and/or modulator is not shown.

Hormones may also antagonize one another's activities through negative regulation of receptor levels. Progesterone, for example, antagonizes most of the actions of estrogen on a variety of target tissues, yet estradiol and progesterone show negligible competition for binding to each other's receptor (Leavitt, *et al.* in Ref. 12; 47). Furthermore, progesterone does not appear to interfere with estrogen uptake, estrogen receptor activation, translocation, or the amount of estrogen receptor that can bind to nuclei (48). Clark and co-workers have demonstrated that progesterone interferes with the replenishment of cytoplasmic estrogen receptor in the rat uterus and that this reduction can be correlated with impaired estrogen sensitivity (Fig. 12). As can be seen, it is primarily the second phase of replenishment, the overshoot, that is affected by progesterone treatment. The resulting loss of cytoplasmic estrogen receptor means that less receptor is available for translocation and nuclear retention, which are both necessary for estrogen-induced responses. The decrease in cytoplasmic estrogen receptor brought about by progesterone may be rather modest in view of the relatively limited percentage of the receptor that appears to be required for the full expression of estrogenic effects (49). The significance of these findings with regard to the normal physiological function of the uterus is therefore still uncertain.

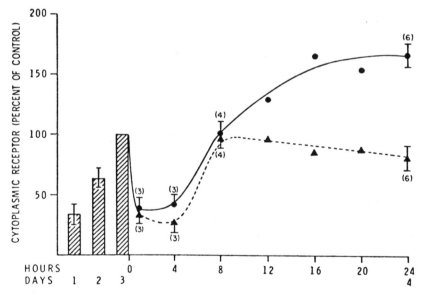

Figure 12 Effect of progesterone on the estrogen-induced depletion and replenishment of the cytoplasmic estrogen receptor. Immature rats received 2.5 μg of estradiol for two days and on the third day were injected with either estradiol (●) or estradiol plus progesterone (Δ). The quantity of cytoplasmic estrogen receptor was determined on days 0–2 and at various times after estrogen or estrogen plus progesterone treatment on day 3. All values are expressed as the percentage of control based on the receptor concentration on day 3. Numbers in parentheses represent the numbers of experiments with four animals per experiment. Reproduced with permission from Clark, Hseuh and Peck, Jr., *Ann. N.Y. Acad. Sci.* **286**, 163 (1977).

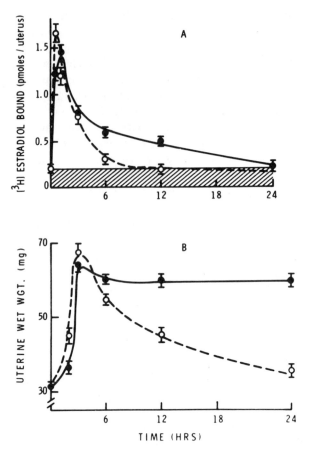

Figure 13 The concentration-time parameters of nuclear receptor–estrogen binding and uterine response following an injection with 1.0 μg of either estradiol (closed circles) or estriol (open circles). At various times after treatment the uteri were weighed to the nearest 0.1 mg (*B*) and the concentration of the nuclear RE complex (*A*) was determined by the [³H]estradiol exchange assay. The cross-hatched portion in (*A*) represents the control level of nuclear RE in saline-injected controls. Reproduced with permission from Anderson, Peck, Jr. and Clark, *Endocrinology* **96**, 163 (1975).

    Regulation of the receptor-activation process is poorly understood although, as mentioned earlier, the potential for such regulation exists through modulation of the availability of cofactors or of kinase-phosphatase systems.

    The nuclear binding of steroid–receptor complexes is fully discussed in the next chapter. For the present it is sufficient to note that the control of nuclear binding of the steroid–receptor complex appears to be a potential means of regulating steroid-mediated processes. Long-term steroid effects (e.g., increases in uterine dry weight 24 hr after estradiol administration, DNA synthesis at 18–24 hr, glucose metabolism at 18–24 hr) appear to be related to the length of time that steroid–receptor complexes remain associated with the nucleus as well as to the amount of receptor translocated to the nucleus (50). Estriol, for instance, can bind the uterine estrogen receptor (albeit with lower affinity than estradiol), activate it, and translocate it to the nucleus. When the dose of estriol is adjusted to permit an equivalent amount of receptor to be translocated, estriol and estradiol show similar potencies with respect to short-term (0–6 hr, postinjection) effects such as stimulation of uterine blood flow, fluid imbibition, glucose transport and utilization, histamine release, and stimulation of RNA and protein synthesis (50). Estriol cannot, however, induce long-term uterine growth. This difference appears to be related to the inability of estriol to maintain elevated titers of nuclear receptor for a prolonged period of time (Fig. 13). However, if estriol is administered in multiple doses such that nuclear receptor levels are maintained, long-term growth effects are also observed (50).

Paradoxically, a number of synthetic compounds with long-term antiestrogenic activity have been shown to *prolong* the association between receptor and nucleus and to similarly sustain the depletion of cytoplasmic estrogen receptor (35). Nuclear occupancy per se is thus insufficient to guarantee long-term estrogenic responses, implying that some "processing" step beyond nuclear binding is required for such responses (51) and that antiestrogen–receptor complexes are incapable of such operations. Whether control of receptor processing is exerted under physiological conditions is unknown at this time.

7 ALTERNATIVE MODES OF STEROID-HORMONE ACTION

In the wake of the studies of Sutherland and others in the 1940s and 1950s demonstrating the crucial role of cyclic nucleotides in the expression of peptide hormone effects, considerable effort was expended attempting to link the action of steroid hormones to alterations in cyclic AMP (cAMP) levels. In many of the systems that were investigated, alterations in cAMP levels were noted following steroid exposure (reviewed in Ref. 37). However, the full spectrum of steroid-mediated events could be elicited in other targets in the absence of any changes in cyclic nucleotide concentrations. Furthermore, cAMP administration rarely mimics the complete response evoked by steroids, and may in some cases have quite opposite effects. There is also evidence that cAMP- and steroid-mediated events are under separate control, as evidenced by their independent modification in a number of differentiating systems. (37).

Estrogen-induced increases in uterine cGMP levels have also been reported, although this phenomenon has not been extensively investigated (52).

There is at least one case in which a steroid appears to act at the surface of the cell rather than by a soluble receptor. Progesterone may hasten the maturation of amphibian oocytes only if present in the incubation medium surrounding the oocytes (53). Injection of progesterone into the eggs themselves has no effect.

A plasma membrane–associated binder of glucocorticoid hormones has been described in a mouse pituitary tumor cell line (54). The specificity of this binder differs from that of the cytosolic glucocorticoid receptor in that the membrane component shows little affinity for synthetic glucocorticoids. Although the function to these receptors has not been established, steroid hormones have a number of rapid effects on membrane permeability to ions and substrates, and several authors have speculated that such effects are likely to be modulated by membrane-associated receptors.

Membrane-linked interactions of another kind have been proposed by Szego to explain the mechanism by which estrogen acts on its target tissues (55). She maintains that the initial interaction between steroid and target cell may occur in the primary lysosomes, resulting in a "destabilization" of the lysosomal membrane. The resulting release of limited amounts of hydrolytic enzymes are proposed to initiate a cascade of events that constitute the target-cell response. Furthermore, following exposure to steroids, lysosomes may assume a perinuclear position (translocation?) that enables them to influence intranuclear events, perhaps through limited hydrolytic cleavage of components of chromatin and of the nuclear membrane.

Over the last several years, evidence has accumulated that steroid hormones may exert some of their effects independent of alterations of gene transcription, for example, by affecting the translation of preexisting messenger RNA. Liang and Liao, for example, have shown that in the rat ventral prostate, androgens promote the binding of initiator transfer RNA ($[^{35}S]$Met-tRNA$_f^{Met}$) to an initiation-factor-like protein(s) (56). Although this effect is blocked by antiandrogens known to compete for the androgen receptor, the rapidity of the response and its independence from protein or RNA synthesis suggest that the critical events may be occurring in the cytoplasm itself. Further suggestions of transcription-independent steroid-induced events have been made with respect to other androgen-induced systems (growth of the chick comb, NADH increase in the rat prostate, increase in amino acid incorporation into proteins, and increased arginase activity in mouse kidney), as well as estrogen-induced (increase in peptide elongation rate in the uterus), and progesterone-induced (protein synthesis in enucleated amphibian oocytes) systems. An alternative model for glucocorticoid induction of TAT synthesis that involves posttranscriptional controls has also been proposed and is discussed more fully in the next chapter.

8   SUMMARY

The current model for the mechanism of action of steroid hormones includes binding of steroids to high-affinity, limited-capacity protein receptors in the cytoplasm of target cells. Subsequent receptor activation permits the hormone–receptor complex to enter the nucleus and to alter the expression of specific genes.

Methods have been developed to quantify and characterize hormone receptors; the methods distinguish hormone–receptor interactions from those involving low-affinity, high-capacity sites. Receptors seem to share a number of general physicochemical characteristics, although the mechanisms by which they transduce hormonal signals may involve important differences. The development of the present model for steroid-hormone action has facilitated our understanding of some of the mechanisms by which hormones interact to promote or inhibit one another's effectiveness.

REFERENCES

1 King, R. J. B., and Mainwaring, W. I. P., *Steroid-Cell Interactions*. University Park Press, Baltimore (1974).

2 McCarty, K. S., Jr., McCarty, K. S., Sr., Steroid hormone receptors in the regulation of differentiation. *Am. J. Pathol.* **86**, 705 (1977).

3 Wagner, R. K., Extracellular and intracellular steroid binding proteins. Properties, discrimination, assay and clinical application. *Acta Endocrinol.* Suppl. **218**, 7 (1978).

4 Garola, R. F., and McGuire, W., An improved assay for nuclear estrogen receptor in experimental and human breast cancer. *Cancer Res.* **37**, 3333 (1977).

5 Braunsberg, H., and Hammond, K. D., Methods of steroid receptor calculations: an interlaboratory study. *J. Steroid Biochem.* **11**, 1561 (1979).

6 Rodbard, D., Mathematics of Hormone-Receptor Interaction. I. Basic Principles, in O'Malley, B. W., and Means, A. R. (eds.), *Receptors for Reproductive* Organs. *Adv. Exp. Med. Biol.* **36**, 289 (1973).

7 Rosenthal, H. E., A graphic method for the determination and presentation of binding parameters in a complex system. Anal. Biochem. **20**, 525 (1967).

8 Chamness, G. C., and McGuire, W. L., Scatchard plots: common errors in correction and interpretation. *Steroids* **26**, 538 (1975).

9 Jensen, E. V., Interaction of steroid hormones with the nucleus. *Pharmacol. Rev.,* **30**, 477 (1979).

10 Jensen, E. V., and DeSombre, E. R., Mechanism of action of the female sex hormones. *Ann. Rev. Biochem.* **41**, 203 (1972).

11 Gorski, J., and Gannon, F., Current models of steroid hormone action: a critique. *Ann. Rev. Physiol.* **38**, 425 (1976).

12 O'Malley, B. W., and Birnbaumer, L. (eds.), *Receptors and Hormone Action*, Vols. 1 and 2, Academic, New York (1977).

13 Sica, V., and Bresciani, F., Estrogen-binding proteins of calf uterus. Purification to homogeneity of receptor from cytosol by affinity chromatography *Biochemistry* **18**, 2369 (1979).

14 DiSorbo, D. M., and Litwack, G., Molecular Aspects of Glucocorticoid Receptors in Sharma, R. K., and Criss, W. E., (eds.), *Endocrine Control in Neoplasia*, p. 249. Raven, New York (1978).

15 Wolf, M. E., Baster, J. D., Kollman, P. A., *et al.* Nature of steroid-glucocorticoid receptor interactions: Thermodynamic analysis of the binding reaction. *Biochemistry* **17**, 3201 (1978).

16 Pratt, W. B., Kaine, J. L., and Pratt, D. V., The kinetics of glucocorticoid binding to the soluble specific binding protein of mouse fibroblasts. *J. Biol. Chem.* **250**, 4584 (1975).

17 Sherman, M. R., Pickering, L. A., Rollwagen, F. M., *et al.,* Mero-receptors: Proteolytic fragments of receptors containing the steroid-binding site. *Fed. Proc.* **37**, 167 (1978).

18 Nielsen, C. J., Sando, J. J., and Pratt, W. B., Evidence that dephosphorylation inactivates glucocorticoid receptors. *Proc. Nat. Acad. Sci.* **74**, 1398 (1977).

19 Sando, J. J., Hammond, N. D., Statford, C. A., *et al.,* Activation of thymocyte glucocorticoid receptors to the steroid binding form. *J. Biol. Chem.* **254,** 4779 (1979).

20 Leach, K. L., Dahmer, M. K., Hammond, N. D., *et al.,* Molybdate inhibition of glucocorticoid receptor inactivation and transformation. *J. Biol. Chem.* **254,** 11884 (1979).

21 Kute, T. E., Heideman, P., Wittliff, J. L., Molecular heterogeneity of cytosolic forms of estrogen receptors from human breast tumors. *Cancer Res.,* **38,** 4307 (1978).

22 Agarwal, M. K., Evidence that natural vs synthetic steroid hormones bind to physicochemically distinct cellular receptors. *Biochem. Biophys. Res. Comm.* **73,** 767 (1976).

23 Clark, J. H., Hardin, J. W., Upchurch, S., *et al.,* Heterogeneity of estrogen binding sites in the cytosol of the rat uterus. *J. Biol. Chem.* **253,** 7630 (1978).

24 Miller, J. B., and Toft, D., Requirement for activation in the binding of progesterone receptor to ATP-Sepharose. *Biochemistry* **17,** 173 (1980).

25 Sato, B., Noma, K., Nishizawa, Y., *et al.,* Mechanism of activation of steroid receptors: involvement of low molecular weight inhibitor in activation of androgen, glucocorticoid and estrogen receptor systems. *Endocrinology* **106,** 1142 (1980).

26 Bailly, A., LeFevre, B., Savouret, J.-F., *et al.,* Activation and changes in sedimentation properties of steroid receptors. *J. Biol. Chem.* **255,** 2729 (1980).

27 Nishigori, H., and Toft, D., Inhibition of progesterone receptor activation by sodium molybdate. *Biochemistry* **19,** 77 (1980).

28 Toft, D. Moudgil, V. Lohmar, P., *et al.,* Binding of ATP to progesterone receptors: properties and functional significance of this interaction. *Ann. N.Y. Acad. Sci.* **286,** 29 (1977).

29 Liao, S., Molecular Actions of Androgens, in Litwack, G., (ed.), *Biochemical Actions of Hormones,* Vol. 4, Academic, New York (1977).

30 Verhoeven, G., Heyns, W., DeMoor, P., Ammonium sulfate precipitation as a tool for the study of androgen receptor proteins in rat prostate and mouse kidney. *Steroids* **26,** 149 (1975).

31 Bullock, L. P., Bardin, C. W., Androgenic, synandrogenic and antiandrogenic actions of progestins. *Ann. N.Y. Acad. Sci.* **286,** 321 (1977).

32 Rousseau, G. G., Interaction of steroids with hepatoma cells: molecular mechanisms of glucocorticoid hormone action. *J. Steroid Biochem.* **6,** 75 (1975).

33 Cake, M. H., and Litwack, G., The Glucocorticoid Receptor, in Litwack, G. (ed.), *Biochemical Actions of Hormones,* Vol. 3, p. 187, Academic, New York (1975).

34 Kaiser, N., Mayer, M., Milholland, R. J., *et al.,* Studies on the antiglucocorticoid action of progesterone in rat thymocytes: early *in vitro* effects. *J. Steroid Biochem.* **10,** 379 (1979).

35 Katzenellenbogen, B. S., Basic Mechanisms of Antiestrogen Action, *in* McGuire, W. L. (ed.), *Hormones, Receptors and Breast Cancer,* p. 135, Raven, New York (1978).

36 Samuels, H. H., and Tomkins, G. M., Relation of steroid structure to enzyme induction in hepatoma tissue culture cells. *J. Mol. Biol.* **52,** 57 (1970).

37 Higgins, S. J., and Gehring, U., Molecular Mechanisms of Steroid Hormone Action, in Klein, G., and Weinhouse, S. (eds.), *Advances in Cancer Research,* Vol. 28, p. 313, Academic, New York (1978).

38 Rousseau, G. G., Schmidt, J. P., Structure-activity relationships for glucocorticoids. I. Determination of receptor binding and biological activity. *J. Steroid Biochem.* **8,** 911 (1977).

39 Rousseau, G. G., Baxter, J. D., and Tomkins, G. M., Glucocorticoid receptors: relations between steroid binding and biological effects. *J. Mol. Biol.* **67,** 99 (1972).

40 Sibley, C. H., and Tomkins, G. M., Mechanisms of steroid resistance. *Cell* **2,** 22 (1974).

41 Yamamoto, K. R., Gehring, U. Stampfer, M. R., *et al.,* Genetic approaches to steroid hormone action. *Rec. Prog. Horm. Res.* **32,** 3 (1976).

42 Sarff, M., and Gorski, J., Control of estrogen binding protein concentration under basal conditions and after estrogen administration. *Biochemistry* **10,** 2557 (1971).

43 Mester, J., Baulieu, E. E., Dynamics of oestrogen-receptor distribution between the cytosol and nuclear fractions of immature rat uterus after oestradiol administration. *Biochem. J.* **146**, 617 (1975).

44 Greene, G. L., Close, L. E., DeSombre, E. R., *et al.*, Estrophilin: Pro and anti. *J. Steroid Biochem.* **12**, 159 (1980).

45 Munck, A., Wira, C., Young, D. A., *et al.*, Glucocorticoid-receptor complexes and the earliest steps in the action of glucocorticoids on thymus cells. *J. Steroid Biochem.* **3**, 567 (1972).

46 Leavitt, W. W., Chen, T. J., and Allen, T. C., Regulation of progesterone receptor formation by estrogen action. *Ann, N.Y. Acad. Sci.* **286**, 210 (1977).

47 Anderson, J., Clark, J. H. and Peck, Jr., E. J., Oestrogen and nuclear binding sites. *Biochem. J.* **126**, 561 (1972).

48 Hsueh, A. J., Peck, Jr., E. J., Clark, J. H., Control of uterine estrogen receptor levels by progesterone. *Endocrinology,* **98**, 438 (1976).

49 DeSombre, E. R., and Lyttle, C. R., Steroid hormone regulation of uterine peroxidase activity. *Adv. Exp. Med. Biol.* **117**, 157 (1979).

50 Anderson, J. N., Peck, Jr., E. J., Clark, J. H., Estrogen-induced uterine responses and growth: relationship to receptor estrogen binding by uterine nuclei. *Endocrinology* **96**, 160 (1975).

51 Horwitz, K. B., and McGuire, W. L., Estrogen control of progesterone receptor in human breast cancer: correlation with nuclear processing of estrogen receptor. *J. Biol. Chem.* **253**, 2223 (1978).

52 Nicol, S. E., Sanford, C. H., Kuehl, F. A., Jr., *et al.*, Estrogen-induced elevation of uterine cyclic GMP concentrations. *Fed. Proc.* **33**, 284 (abstr.) (1974).

53 Smith, L. D., and Ecker, R. E., The interaction of steroids with *Rana pipiens* oocytes in the induction of maturation. *Dev. Biol.* **25**, 232 (1971).

54 Harrison, R. W., Balasubramanian, K., Yeakley, J., *et al.*, Heterogeneity of AtT-20 cell glucocorticoid binding sites: evidence for a membrane receptor. *Adv. Exp. Biol. Med.* **117**, 423 (1979).

55 Szego, C. M., The lysosome as a mediator of hormone action. *Rec. Prog. Horm. Res.* **30**, 171 (1974).

56 Liao, S. Cellular receptors and mechanisms of action of steroid hormones. *Int. Rev. Cytol.* **71**, 87 (1975).

CHAPTER TWENTY-ONE

PROPERTIES AND FUNCTIONS OF STEROID–RECEPTOR COMPLEXES

SHEILA M. JUDGE

The Ben May Laboratory for Cancer Research
University of Chicago
Chicago, Illinois

RANDAL C. JAFFE

Department of Physiology and Biophysics
University of Illinois at the Medical Center
Chicago, Illinois

1 MODELS FOR EUKARYOTIC GENE ACTIVATION
BY STEROID HORMONES

Steroid receptors are a valuable tool for the investigation of genetic control mechanisms. Because of the high-affinity binding of radioactively labeled steroid molecules to receptor proteins, receptor–steroid complexes can be monitored as they interact with regulatory elements in the cell nucleus. They can therefore be utilized as molecular probes to examine the organization of transcriptionally active eukaryotic chromatin. In addition, nuclear binding by receptors can be correlated with gene activation and biological response.

Transcriptional control of eukaryotic gene expression appears to be exerted mainly through positive activators that release transcriptional units of DNA from the conformational strictures of the supercoiled nucleosomes. Although histones play a fundamental role in the structural organization of DNA into nucleosome filaments, it seems unlikely that they alone are responsible for the entire spectrum of gene regulation in eukaryotic cells. Histones exhibit little tissue or species specificity, they are conserved through evolution, and there are only six chemically distinct classes of histones. Repression of gene transcription by histones *in vitro* occurs only in a nonspecific fashion (1). Chemical modification of histones through acetylation or phosphorylation can evoke changes in chromatin structure or function. However, these changes are insufficient to account for specific gene activation. Other regulatory elements such as steroid receptors, nonhistone proteins, and other chromosomal elements are probably required for alterations of specific gene loci.

It has been suggested that class H1 histones are involved in the mechanism of reversible condensation of chromatin through site-specific phosphorylation of the H1 molecules. H1 histones are located in the spacer regions between nucleosomes, and they interact with DNA, other H1 molecules, and nonhistone proteins. Removal of H1 molecules from chromatin alters the conformation and physical properties of chromatin. However, researchers have been unable to demonstrate a role for H1 histones in *in vitro* gene transcription experiments.

More likely candidates for the role of positive activators of gene transcription are chromosomal nonhistone proteins (NHPs) (2, 3). There are many classes of NHPs, which vary markedly in their chemical and functional properties. The relative abundance of each class can vary markedly. NHP binding to DNA appears to exhibit tissue and species specificity. However, although qualitative differences in the NHP population of chromatins from different sources would be expected if NHPs are to serve as positive activators of transcription, such differences have not been detected. This may be because tissue- and species-specific NHPs probably represent a very small percentage of the total NHP in chromatin, making them difficult to detect with the relatively insensitive analytical methods at hand.

The hypothesis that NHPs modulate the transcriptional activity of eukaryotic chromatin has received significant attention in the literature (4). Control of tissue-specific patterns of repetitive DNA transcription by chromatin templates has been attributed to NHPs, and homologous NHPs are required for the transcription of tissue-specific gene products.

Experimental evidence supports the contention that NHPs act as positive modulators of gene expression, whereas histones are responsible for nonspecific gene repression. How the various chromosomal proteins interact and their sequence of interaction for modulation of gene transcription is not known. The models discussed below demonstrate the importance of these regulatory elements in steroid modulation of gene transcription.

1.1 The Acceptor Hypothesis

The acceptor hypothesis is primarily the result of work performed with progesterone–receptor complexes from chick oviducts. This theory was proposed by O'Malley and his co-workers in 1971 and in its present form contains only minor modifications of the original concept (5, 6).

An important feature of this model is the postulate that steroid–receptor complexes interact with specific nuclear components (*acceptors*) in their target cells that are present in limited quantities. According to O'Malley's hypothesis, the nuclear acceptors consist of DNA plus NHPs. The NHPs act as signals to "attract" the steroid–receptor complex (RS) to specific portions of the genome, thereby preventing time-consuming and meaningless interactions of the receptor complex with DNA by facilitating its "search" for specific gene loci (see Fig. 7 of Chapter 20). There may be multiple or single acceptor sites for the R·S on chromatin. Acceptor redundancy may provide another level of regulation by reducing the minimal dose level of R·S–acceptor interaction for a given response.

The acceptor hypothesis evolved from the following lines of evidence: (1) the requirement for high salt concentrations to extract nuclear-bound receptor, with implied high-affinity binding of the receptor to some nuclear component; (2) the tissue-specific interactions of steroid–receptor complexes with chromosomal NHPs; (3) saturable nuclear–R·S interactions observed *in vitro*; and (4) the differences in the nuclear binding characteristics of the two progesterone-receptor subunits.

The chick oviduct progesterone receptor is a 6 S dimer whose subunits (A and B) can be resolved by DEAE-cellulose chromatography (7, 8). Subunit B of the progesterone receptor binds only to chromatin (9, 10) and not to purified DNA, whereas subunit A binds only to DNA (11). According to the acceptor model, the B receptor subunit (the *specifier* protein) interacts with NHPs and positions the R·S on the genome. Once the gene binding site and the A subunit of the R·S are correctly aligned, the A-B receptor dimer dissociates, thereby freeing the A subunit to interact with DNA (12). O'Malley and Schrader (12) hypothesized that the progesterone-receptor subunit A acts at a specific site on DNA to promote gene activation. The mechanism for receptor activation of gene transcription is not known, but they have speculated that the A subunit binds to specific deoxynucleotide sequences on the DNA and alters the local structure or conformation of a portion of DNA, thereby allowing a specific region of the genome to undergo transcription.

1.2   The Two-Site DNA-Binding Model

The keystone of the two-site binding model for receptor modulation of gene expression is the concept that an entire transcriptional unit must be altered for gene activation to occur. The mechanism, proposed by Yamamoto and Alberts in 1974, accounts for both specific and nonspecific binding of the receptor to DNA (13, 14). Support for the two-site concept is derived primarily from studies of the uterine estrogen receptor. The data that provide a basis for this model are the following: (1) separate investigations showed that a limited number of nuclear estrogen receptors are more resistant to salt extraction than the general population (15–17), suggesting the presence of two classes of nuclear receptors, one of which is more tightly bound than the other; (2) these salt-resistant nuclear receptors appear to be involved in long-term responses to estrogens (15, 17); (3) only high levels of low-affinity binding of R·S to DNA can be detected *in vitro* under low-salt conditions (18–20), and (4) biological-response data indicate that the steroid responsiveness of a target cell is a linear function of nuclear steroid–receptor content (21, 22).

According to their model, activator proteins (steroid–receptor complexes) interact with target-tissue DNA in two ways: by low-affinity binding and by specific, high-affinity binding (Fig. 1). Both classes of interaction generate a wave of alterations in chromosomal proteins that results in the transduction of the original signal generated by the binding of the R·Ss. However, high-affinity, specific binding of the receptor to multiple sites on DNA is required to alter the conformation of contiguous chromatin. These conformational shifts are induced through cooperative interactions, and they result in the generation of an *active patch* of DNA that is more accessible for transcription than inactive regions of the genome. Binding of multiple R·Ss to specific gene sequences is responsible for the selectivity of gene response. The sites of the high-affinity interactions between DNA and the R·S are specific; therefore, different steroids could induce different responses from the same target cell.

The two-site model also proposes that chromosomal proteins, along with steroid–receptor complexes, are required for gene activation. Yamamoto and Alberts suggest that a rapid transient increase in the rate of histone acetylation

---

Figure 1   A model for eukaryotic gene activation. (*A*) Chromosomes are nonhomogeneous in structure, with active regions constituted of patches 6000–7000 base pairs in length that are physically distinct from inactive chromatin. This physical difference is detected by transcription with *E. coli* RNA polymerase, which preferentially synthesizes short pieces of RNA from decondensed chromatin without regard to physiological initiation sites or polarity. Steroid receptors interact with the genome at (*B*) nonspecific loci (where the hormone effect is not elicited) as well as (*C*) specific loci (where the hormone effect occurs). At both types of sites, the receptor induces a local structural

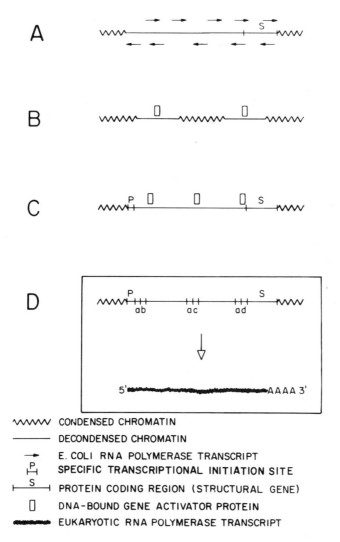

A

B

C

D

∿∿∿∿ CONDENSED CHROMATIN

——— DECONDENSED CHROMATIN

⟶ E. COLI RNA POLYMERASE TRANSCRIPT

P SPECIFIC TRANSCRIPTIONAL INITIATION SITE

⊢—S—⊣ PROTEIN CODING REGION (STRUCTURAL GENE)

▯ DNA-BOUND GENE ACTIVATOR PROTEIN

━━━▶ EUKARYOTIC RNA POLYMERASE TRANSCRIPT

alteration, which is then chemically transduced a short distance along the chromosome. Specific hormone effects require clustered receptor binding (*C*), since only this leads to formation of a large enough patch to include both a specific promotor and its structural gene. To achieve such clustered receptor binding requires specific DNA sequences that bind the receptor with relatively high affinity. Note that the interspersed arrangement of receptor binding is not an absolute requirement of the model, since binding sites for regulatory proteins might be clustered at one end of a patch, with polarity of the spreading effect achieved by binding of the proteins to a specific strand or groove. (*D*) Regulatory flexibility is easily achieved with relatively few activator elements by varying the combination of inputs required for activation of different genes. In the specific case shown, activation would require either the activator-protein-recognizing site *a*, or three different activator-proteins-recognizing sites *b*, *c*, and *d*, respectively. Reprinted with permission, from Yamamoto and Alberts, (13), *Annual Review of Biochemistry*, Vol. 45, © 1976 by Annual Reviews, Inc.

(seen upon estrogen stimulation of rat uterus) acts as the first transducer of the steroid signal upon binding of the receptor to DNA. However, their model does not exclude other chromosomal regulatory elements from playing a similar role.

This model of gene activation is derived from several concepts of prokaryotic gene regulation that were developed from studies on the *lac* operon. These studies showed that gene-regulatory proteins interact with prokaryote DNA in both nonspecific and specific fashions. The regulatory protein is in equilibrium between the two classes of DNA binding sites. Similarly, steroid receptors could establish an equilibrium between the cytoplasmic binding sites, low-affinity DNA binding sites, and high-affinity sites on the DNA molecule.

Both models of steroid-hormone modulation of gene transcription are based on the concept that receptor-complex binding to DNA is responsible for gene activation whereas they differ in their descriptions of the nature of this interaction. They also differ in the roles they ascribe to chromosomal proteins. In both cases, the investigators have attempted to account for the rapidity of steroid-induced gene activation; in the two-site theory, amplification of the incoming signal is created by local alterations in chromosomal proteins that lead to cooperative perturbations in chromatin configuration, whereas in the acceptor model, nonhistone proteins act as "magnets" for the receptors and rapidly "draw" them into the appropriate region of the genome where gene activation is to occur. The acceptor model has been more widely accepted than the two-site theory, and it has benefited from the extensive study of the chick oviduct progesterone receptor. The true mechanism of steriod-hormone-induced gene activation may represent a combination of the two presented here, or a variation thereof. It is likely that a single fundamental mechanism of action underlies all long-term steroid-hormone effects.

2   INTERACTION OF STEROID RECEPTORS WITH TARGET-CELL NUCLEI

Binding of steroid–receptor complexes to target-cell nuclei is integral to both models of gene activation presented in this chapter and to the two-step theory of steroid-hormone action discussed in the previous chapter. Evidence for nuclear binding by R·S first appeared in studies in which radioactivity was recovered from salt extracts of nuclei isolated from uteri that had been exposed to radioactively labeled estrogen *in vivo* (23). Subsequent studies showed that nuclear localization of steroid–receptor complexes was a general phenomenon and could be detected in target tissues for any steroid hormone following the administration of that hormone (24–26). The characteristics of the interaction between R·S and nuclei were examined *in vitro* several steroid-hormone–target-cell systems using unpurified, radioactively labeled R·S. There appeared to be a limited number of binding sites in the nuclei for steroid-receptor complexes (7, 27). Measurement of binding affinities indicated that the R·S complexes were tightly bound to nuclei, although this claim has been disputed (18, 28). The specificity of R·S interaction with nuclei has not yet been established. The uterine estrogen receptor appears to be able to bind only to uterine nuclei (29, 30), whereas others have been unable to observe such tissue-specific binding (31).

Subsequent examination of the nuclear-binding properties of the chick oviduct progesterone receptor demonstrated that the salt concentration utilized in these *in vitro* studies was critical (32). Under conditions of low ionic strength (0.01–0.05 M KCl), nonspecific and nonsaturable binding of the steroid–receptor complex to nuclei occurred, whereas under physiological salt conditions (0.15 or 0.2 M KCl), saturable, tissue-specific binding was observed. When nuclear binding of receptors is examined under low-salt conditions, cytosol contaminants may cause artificial saturation (18) because of inhibition by the contaminants of receptor binding to nuclei. These results may explain the observations of low-affinity, nonsaturable binding of estrogen– and glucocorticoid–receptor complexes to nuclei published by some investigators (27, 33, 34).

However, when binding studies were performed at high salt concentrations with purified and crude progesterone receptors and oviduct chromatin, the binding properties of the receptor preparations were identical, illustrating that competition for binding sites by cytosol contaminants does not occur at high salt concentrations (32, 35).

Clark *et al.* (36, 37) have postulated that there may be more than one class of nuclear estrogen receptor. Their postulate was based on the finding that approximately 10% of the nuclear estrogen-receptor population was resistant to salt extraction. The increased resistance to salt extraction exhibited by these receptors may indicate that this "class" of receptors binds to nuclei with greater affinity than the majority of receptors. These "salt resistant" nuclear binding sites may be associated with the uterine nuclear matrix, which is a residual structure in the nucleus (16).

Induction of uterine growth by estrogens requires prolonged retention of the estrogen receptor in the nucleus (38). There was a correlation between the levels of salt-resistant nuclear estrogen receptor and the number of estrogen receptors in the nucleus for a prolonged period of time (36). The salt-resistant receptor fraction was implicated in long-term growth effects of estrogens upon the rat uterus. It has been postulated that loosely bound salt-extractable receptors undergo a time-dependent transformation that results in nuclear receptors that are more resistant to disruption by salt treatment (37). Other investigators have disputed this theory, claiming that the non-salt-extractable R·S actually represents physical entrapment of the receptor complex in the gelatinous nuclear pellet created by salt extraction (39).

The interaction of steroid receptors with target-cell nuclei is a composite of receptor interactions with various nuclear components (DNA, histones, and other chromosomal proteins).

2.1   Interaction of Steroid Receptors with Target-Cell Chromatin

The binding of steroid–receptor complexes to chromatin is a characteristic shared by all steroid-sensitive tissues. Some have claimed that steroid–receptor complexes bind more extensively to their respective target-tissue chromatins (40, 41). Others have found no such specificity (28, 42).

NHPs may impart species and tissue specificity upon interactions of R·S complexes with nuclei. For example, one of the two components of the progesterone receptor binds to a specific class of acidic NHPs. Binding of the progesterone–receptor complex to chromatin requires these specific NHPs (43). These conclusions were drawn from experiments in which chromatins from various tissues were fractionated into histone, DNA, and four NHP fractions, and then reconstituted into either "native" (components were all from the same same tissue) or "hybrid" (using fractions from two different tissues) chromatins. The ability of various native and hybrid chromatins to bind [^{3}H]progesterone–receptor complexes was then compared. The presence of the NHPs from target tissue greatly enhanced progesterone–receptor binding by the reconstructed chromatin. Unpurified nuclear glucocorticoid–receptor complexes also bind to NHPs. In addition, binding of uterine estrogen–receptor complexes to uterine NHPs (44) and androgen receptors to prostate NHP (41) has been demonstrated by affinity chromatography.

2.2   Interaction of Steroid Receptors With DNA

All classes of steroid receptors bind to purified DNA, as measured by DNA-cellulose chromatography of R·Ss and sucrose density gradient centrifugation of DNA–R·S complexes (45–47). The interaction between glucocorticoid receptors and hepatoma tissue culture (HTC) cell DNA has been extensively examined by the Tompkins group (48, 49). Co-elution of receptor complexes with DNA was used to compare dexamethasone (Dx)–receptor complex binding to native and denatured HTC cell DNA (49). Native DNA bound approximately three times more R·Dx than denatured DNA. The binding of R·Dx to DNA in this system was characterized by large binding capacity, lack of saturation, and the absence of species specificity.

Indirect observations of receptor–DNA interactions were obtained from studies that measured the effect of enzymatic degradation of DNA on receptor binding (48). Treatment of DNA with pancreatic DNAse I and staphylococcal endonuclease resulted in a 90% reduction of the capacity of HTC cell nuclei to bind R·Dx (48).

The importance of charge–charge interactions between estradiol–receptor complexes and rat uterine DNA was demonstrated by experiments in which polyanions, dextran sulfate, and heparin inhibited receptor binding to DNA (50). Other studies have employed intercalating agents to study the nature of receptor binding to DNA. These studies have demonstrated that binding of the estradiol–receptor complex is inhibited by the treatment of DNA with ethidium bromide and 9-hydroxyellipticine (51). Their results suggest that the binding of R·S by DNA represents the hydrophobic interaction of R·S with the double–stranded helix along with ionic interactions. Similar results were obtained from studies that measured competition between soluble double-stranded DNA and DNA-cellulose for cytosol preparations containing [³H]-labeled estradiol–receptor complexes (52).

Some of the current methods used for measuring receptor binding to DNA, such as DNA-cellulose chromatography and sucrose density gradient centrifugation, are subject to experimental artifacts (53). This problem has contributed to the discrepancies found among the results obtained by different laboratories regarding target-tissue specificity and binding-site saturation of R·S binding to DNA. It has been suggested (14) that the techniques used to measure R·S binding to DNA are unable to detect high-affinity binding of R·S to DNA because this binding is "masked" by the high levels of low-affinity binding of R·S to DNA observed in most systems.

To further explore the nature of the specific interactions between R·S and DNA, researchers are cloning structural genes that code for specific hormone-induced mRNAs by insertion of these genes into plasmids. This has already been done for the ovalbumin gene from the chick oviduct (54). Utilizing these "minichromosomes," researchers will be able to examine the interactions between R·S and specific nucleotide sequences in the region of the cistron. They may also be able to measure the number of DNA binding sites for R·S and examine the effects of steroid–receptor binding on gene transcription.

3 STEROID HORMONES AND EUKARYOTIC GENE EXPRESSION

Several lines of evidence suggest that the expression of eukaryotic genes is under transcriptional control. The first of these centers upon studies by Ashburner and his colleagues (55) on polytene chromosome "puffing" in larval Diptera. These "giant" chromosomes were readily visualized, and a correlation between puffing and the degree of condensation of the insect chromatin was soon established. Sites of transcriptional activity on chromatin correspond to the areas of puffing. Further study revealed the presence of tissue-specific patterns in puffing activity, and alterations in these patterns corresponded to developmental changes in the tissues. These studies demonstrated that gene transcription, as represented by chromosome puffing, could be directly related to several features of gene expression, such as tissue specificity and developmental alterations. Further data to support the theory of transcriptional control of gene activity came from a similar set of studies in which chromatin was examined directly by electron microscopy. Again, the degree of chromatin condensation could be inversely related to transcriptional activity. A third line of evidence was derived from the analysis of repetitive and unique-sequence messenger (mRNA) populations from various sources that are the products of gene transcription. Several features of the mRNA populations were established.

First, mRNA populations from within the same animal exhibited tissue specificity. Second, the composition of the mRNA populations found in various tissues was related to the developmental state of the tissue. In addition, exposure to steroid hormones elicited changes in the mRNA populations that reflected differential gene transcription.

Modulation of gene expression by steroid hormones is realized through hormone-stimulated increases in specific gene transcription that result in the stimulation of "induction" of specific protein synthesis. Current investigations on this topic have been extensively reveiwed by Chan *et al.* (54) and by Higgins and Gehring (56). A direct, rate-limiting effect of steroid hormones on gene transcription has been demonstrated in several systems, most notably stimulation of ovalbumin synthesis by estrogens in the chicken. Steroids may also influence gene expression by altering the processing and translation of mRNA molecules and by modulating the rate of protein turnover, although these effects of steroid hormones are not as well documented as those on transcription.

Studies of steroid modulation of gene transcription have been limited primarily to those systems from which specific mRNA molecules for hormone-induced gene products could be isolated and purified. Contemporary theories of steroid-hormone action are therefore derived from evidence accumulated from systems in which the synthetic capacity of the hormone-sensitive cell is devoted primarily to the production of a single protein.

The development of several new experimental techniques has played an important role in the study of steroid-hormone regulation of gene expression. The first of these techniques includes several systems for cell-free protein synthesis. These systems have provided an important tool for the identification of mRNA molecules that are induced by hormones. The development of the methodology for fractionating nucleic acids also aided these studies. Nucleic acid fractionation techniques have permitted researchers to isolate specific mRNA molecules that correspond to hormone-induced proteins. A third development, the discovery of reverse transcriptase (RNA-dependent DNA polymerase), implemented the synthesis of complementary DNA molecules (cDNAs) to purified mRNAs *in vitro*. The highly specific cDNAs were then used to determine gene frequencies and for precise measurements of cellular mRNA levels.

The following review of several steroid-hormone-responsive systems illustrates the similarities of steroid effects on target cells and may suggest to the reader some basic molecular mechanisms of steroid action.

3.1 Estrogen Action in Mammalian Uterus

Administration of estrogen to an immature animal results in a number of marked changes in the uterus (57). These include a pronounced increase in uterine dry weight, increases in metabolic activities, water uptake, vascularization, and accelerated cell division. On a molecular level, estrogen stimulates rapid increases in RNA synthesis (within 10 min), stimulates RNA polymerase activity (within 1 hr) and alters metabolic parameters (within 1 hr). Within 2–4 hr the rates of protein synthesis and peptide-chain elongation have increased. Cellular mRNA levels start to rise within 8–12 hr after estrogen treatment. Long-term effects of estrogen, such as increases in DNA and histone synthesis and in dry weight, do not appear until 18 hr after estrogen administration. Twenty-four hours after estrogen treatment, cell division is stimulated. This pattern of cell response to estrogen treatment correlates with changes in cell morphology.

"Induced protein" (IP), an acidic protein with a molecular weight of approximately 42,000, is associated with uterine maturation and increases during the proestrous phase of the normal estrous cycle, and has been identified as an isozyme of creatine kinase. Since it can be readily detected by polyacrylamide-gel electrophoresis, IP has been a useful product for measuring correlations between estrogen-receptor action and cellular response. Induction is rapid, both *in vivo* and *in vitro*. IP is particularly well suited as a model for biological response to estrogens, since its induction is hormone specific and correlates well with both receptor binding and estrogenic potency (58). Estrogen induction of IP appears to depend on the concentration of nuclear estrogen receptors.

Estrogen treatment also provokes a rapid increase in RNA polymerase II activity. This surge of polymerase II activity is followed 2–4 hr later by an increase in RNA polymerase I activity. This suggests that estrogen treatment first stimulates the synthesis of RNA molecules, followed by increased synthesis of ribosomal RNA.

Delayed effects of estrogen treatment include stimulation of DNA and histone synthesis and histone acetylation. Alterations in the rate of histone synthesis occur concomitant with or slightly preceding DNA synthesis. Estrogen specifically alters the activity of the enzymes responsible for histone acetylation. Selective alterations of the various classes of histones by estrogens has been demonstrated.

A model for estrogen action in the uterus has been developed that is referred to as the *sustained-output model* (37, 57). This model proposes that "late" uterotrophic responses to estrogen depend not only on earlier events such as increased RNA synthesis, but upon the continued binding and action of the hormone itself. This model is based on observations of estrogenic potency that showed that those estrogens capable of evoking long-term growth effects, such as cell division, DNA synthesis, and uterine growth, are accompanied by prolonged retention of the estrogen receptor in the cell nucleus (58). This theory supersedes the notion that these late responses to estrogen were simply an outcome of earlier events; this has been referred to as the *cascade* or *domino* theory (57). More recent work with antiestrogens suggests that sustained occupancy of the nuclear receptor is insufficient by itself for the induction of these growth effects. Replenishment or recycling of the estrogen receptor to the cytoplasmic compartment may also be required for the stimulation of uterine growth by estrogens. (17).

3.2   Progesterone Effects in Chick Oviduct

The chick oviduct has been a popular system for studying the effects of estrogen and progesterone on gene expression (2). Under the influence of estrogen, the primitive homogeneous-appearing epithelial cells lining the oviduct mucosa differentiate into distinct cell types. Estrogen also causes the immature oviduct to grow more than 1000-fold in four weeks, and a net increase in total protein synthesis occurs. Estrogen-induced maturation of the gland is required for the potentiation of progesterone action in the oviduct, that is, progesterone has little effect on the undifferentiated oviduct cells. Progesterone selectively promotes synthesis of a secretory protein, avidin, in the estrogen-primed oviduct. Progesterone also stimulates avidin synthesis *in vitro* (59); the first appearance of avidin molecules occurs 6 hr after progesterone addition. Cellular avidin levels reach a peak level within 48–72 hr after hormone administration *in vitro*. This induction requires some protein factors, but not DNA synthesis or mitosis. An increase in the level of nuclear RNA polymerases I and II occurs prior to the induction of avidin *in vivo*.

Progesterone treatment of estrogen-primed chicks stimulates the accumulation of intracellular avidin by increasing the level of avidin mRNA molecules, which are used for protein synthesis (60). The increase in the avidin-specific mRNA concentration of the cell occurs prior to avidin accumulation and coincident with the increased rate of avidin synthesis. This elevation of avidin mRNA levels appears to be due to hormone-induced acceleration of the rate of initiation of mRNA transcription, since progesterone does not alter RNA chain elongation rates or average RNA chain lengths. Quantification of the number of RNA polymerase initiation sites demonstrated that progesterone increases the number of polymerase initiation sites *in vitro* (61).

3.3   Glucocorticoid Action in HTC Cells

Glucocorticoids induce the synthesis of various gluconeogenic enzymes in the liver, including tyrosine aminotransferase (TAT). The property of glucocorticoid induction of TAT is retained in the hepatoma (HTC) cell line that was derived from a Morris hepatoma (62). Dexamethasone, a potent synthetic glucocorticoid, induces an increase in the rate of synthesis of TAT within 30 min (63) and an increase in cellular TAT levels occurs 6–7 hr after Dx treatment (64). Removal of Dx causes a rapid deinduction of TAT (drop in cellular TAT levels). These effects are mediated through the glucocorticoid receptor; there is a linear relationship between TAT induction and the relative saturation of the receptor (11).

The HTC cell system has been an important tool for studies of the relationship between binding of different glucocorticoids to the glucocorticoid receptor and induction of TAT (65, 66). Several lines of evidence support the concept that nuclear binding of the glucocorticoid–receptor complex is required for stimulation of TAT synthesis: (1) there is a close correlation between the levels of nuclear glucocorticoid receptors and the induction of TAT (65), (2) the ability of a given steroid to translocate the glucocorticoid receptor to the nucleus parallels its ability to induce TAT, and (3) the rapidity of nuclear uptake and deinduction of TAT (65).

Studies that have examined the kinetics of induction and deinduction and the stability of the TAT-specific mRNA in enucleated cells (67) support a direct role for glucocorticoids in stimulating TAT synthesis through induction of TAT mRNA synthesis (11). An older and less-viable hypothesis suggested that cellular TAT levels are regulated through posttranscriptional modulation of TAT mRNA levels (22). It was hypothesized by Tomkins (68) that the TAT gene was transcribed at a constant rate (constitutive gene expression) and that regulation of TAT mRNA levels was then achieved through two effectors: (1) a negative effector (a "repressor"), which was proposed to be an mRNA or protein that suppressed the expression of another gene product, destabilizing the TAT mRNA; and (2) a positive effector, which would counteract the repressor. This "posttranscriptional" model was proposed to explain studies in which superinduction or maintenance of induction of TAT was obtained throughout the cell cycle in the presence of actinomycin D (an inhibitor of transcription). Studies on TAT mRNA translation in heterologous systems (69) have shown that TAT induction is due to the hormone-induced accumulation of its specific mRNA. The phenomenon of superinduction, which formed the basis for the posttranscriptional theory, appears to be the result of the differential stability of the mRNAs and competition of these RNAs for rate-limiting factors in the translation process.

3.4 Androgen Action in Rat Ventral Prostate

Androgens are responsible for the maintenance and function of the prostate. On a molecular level, androgens stimulate a rapid increase in prostatic nuclear and ribosomal RNA levels (70) along with increased synthesis and phosphorylation of nuclear proteins. Later effects stimulated by androgen treatment include increases in nuclear RNA (nRNA) synthesis and alterations that lead to the synthesis of the major portion of prostatic proteins (71). Growth-related events such as stimulation of DNA synthesis and cell division occur after several days. As was discussed in Chapter 20, these effects appear to be mediated through the actions of the intracellular metabolite of testosterone, 5α-dihydrotestosterone (DHT), which binds with high affinity to the androgen receptor.

The mechanism through which androgens (specifically DHT) modify gene expression appears to be similar to those described previously, although some investigators have claimed that DHT stimulation of RNA polymerase activity rather than changes in chromatin-template activity are responsible for DHT-induced increases in protein synthesis (72). Increased activity of chromatin-bound RNA polymerase is detected within 1 hr of androgen treatment (73). In addition, nucleolar RNA polymerase activity is stimulated *in vitro* by DHT–receptor complexes (74). It has also been suggested that androgens may exert some degree of posttranscriptional control of gene expression in the prostate (75). This regulation may occur through association of the (DHT–receptor)–nuclear-acceptor complex with mRNAs to form RNA-protein particles (RNPs). Interaction between receptor molecules, NHPs, and mRNAs have been demonstrated *in vitro* (75), although these results have not yet been confirmed. An effect of androgens at the translational level of protein synthesis has also been claimed (75). These translational effects may be implicated in long-term androgen growth effects.

More recent reports have indicated that androgens act primarily by altering the nature and relative abundance of the mRNAs synthesized by rat prostate (76). Testosterone appears to regulate the abundance of specific mRNA species, in particular three highly abundant mRNA species (77). The discovery of prostatic binding protein (PBP) by Heyns and coworkers (78) has led to the development of a marker for androgen action in the prostate gland. PBP has been characterized as a sex-steroid-binding protein that binds steroids with a relatively low degree of affinity, and has a molecular weight of approximately 51000, and sediments at 3.7 S. Its function is not yet understood. The PBP content of the prostate gland decreases after castration, as does the level of PBP-specific mRNA (79). PBP mRNA activity is restored by androgen treatment of castrated rats.

As this review has indicated, several characteristics are common to most steroid-responsive systems: (1) steroid binding to a cytosol receptor is correlated with steroid action, (2) nuclear uptake of that steroid–receptor complex parallels induction of specific or general protein synthesis, and (3) steroid hormones rapidly alter general or specific mRNA levels and RNA polymerase activity in their target cells.

REFERENCES

1 Paul, J., Gilmour, R. S., Organ-specific restriction of transcription in mammalian chromatin. *J. Mol. Biol.* **34**, 305 (1968).

2 Gilmour, R. S., Paul, J., Role of non-histone components in determining organ specificity of rabbit chromatins. *FEBS Lett.* **9**, 242 (1970).

3 Spelsberg, T. C., Hnilica, L. S., and Ansevin, A. T., Protein of chromatin in template restriction. III. The macromolecules in specific transcription of the chromatin DNA. *Biochim. Biophys. Acta* **228**, 550 (1970).

4 Gilmour, R. S., Gene Expression in Eukaryotic Cell, in O'Malley, B. W., Birnbaumer, L. (eds.), *Receptors and Hormone Action,* Vol. I, p. 331, Academic, New York (1977).

5 O'Malley, B. W., and Means, A. R., Female steroid hormones and target cell nuclei. *Science* **183**, 610 (1974).

6 Schrader, W. T., and O'Malley, B. W., Molecular Structure and Analysis of Progesterone Receptors, *in* O'Malley, B. W., and Birnbaumer, L. (eds), *Receptors and Steroid Hormones*, Vol. 2, p. 189, Academic, New York (1978).

7 Kuhn, R. W., Schrader, W. T., and Smith, R. G., *et al.*, Progesterone-binding components of chick oviduct. X. Purification by affinity chromatography. *J. Biol. Chem.* **250**, 4220 (1975).

8 Coty, W. A., Schrader, W. T., and O'Malley, B. W., Purification and characterization of the chick oviduct progesterone receptor A subunit. *J. Steroid Biochem.* **10**, 1 (1979).

9 Schrader, W. T., Toft, D. O., and O'Malley, B. W., Progesterone-binding protein of chick oviduct. VI. Interaction of purified-receptor components with nuclear constituents. *J. Biol. Chem.* **247**, 2401 (1972).

10 O'Malley, B. W., Spelsberg, T. C., Schrader, W. T., *et al.*, Mechanism of interaction of a hormone-receptor complex with a genome of a eukaryotic target cell. *Nature* **235**, 141 (1972).

11 O'Malley, B. W., Schrader, W. T., and Spelsberg, T. C., Hormone receptor interactions with the genome of eukaryotic target cells. *Adv. Exp. Med. Biol.* **36**, 174 (1973).

12 O'Malley, B. W., and Schrader, W. T., The receptors of steroid hormones. *Sci. Am.* **234**, 32 (1976)

13 Yamamoto, K. R., and Alberts, B. M., Steroid receptors: elements for modulation of eukaryotic transcription. *Ann. Rev., Biochem.* **45**, 721 (1976).

14 Yamamoto, K. R., and Alberts, B., The interaction of estradiol-receptor protein with the genome: an argument for the existence of undetected specific sites. *Cell* **4**, 301 (1975).

15 Clark, J. H., and Peck, E. J., Jr., Nuclear retention of receptor-oestrogen complex and nuclear acceptor sites. *Nature* **260**, 635 (1976).

16 Barrack, E. R., Hawkins, E. F., Allen, S. L., *et al.*, Concepts related to salt resistant estradiol receptors in rat uterine nuclei: nuclear matrix. *Biochem. Biophys. Res. Commun.* **79**, 829 (1977).

17 Ruh, R. S., and Baudendistel, L. J., Different nuclear binding sites for antiestrogen and estrogen receptor complexes. *Endocrinology,* **100**, 420 (1977).

18 Chamness, G. C., Jennings, A. W., and McGuire, W. L., Estrogen receptor binding to isolated nuclei. A nonsaturable process. *Biochemistry* **13**, 327 (1974).

19 Milgrom, E., and Atger, M., Receptor translocation inhibitor and apparent saturability of the nuclear acceptor. *J. Steroid Biochem.* **6**, 487 (1975).

20 Rousseau, G. G., Higgins, S. J., Baxter, J. D., *et al.*, Binding of glucocorticoid receptors to DNA. *J. Biol. Chem.* **250**, 6015 (1975).

21 Katzenellenbogen, B. S., Dynamics of steroid hormone receptor action. *Ann. Rev. Physiol.* **42**, 17 (1980).

22 Gorski, J., and Gannon, F., Current models of steroid hormone action: a critique. *Ann. Rev. Physiol.* **38**, 425 (1976).

23 Jensen, E. V., DeSombre, E. R., Jungblut, P. W., *et al.*, Biochemical and Autoradiographic Studies of ³H-Estradiol Location, in Roth, L. J., and Stumpf, W. E. (eds.), *Autoradiography of Diffusible Substances*, p. 89, Academic, New York (1969).

24 Rousseau, G. G., Baxter, J. D., Higgins, S. J., *et al.*, Steroid-induced nuclear binding of glucocorticoid receptors in intact hepatoma cells. *J. Mol. Biol.* **79**, 539 (1973).

25 Noteboom, W. D., and Gorski, J., An early estrogen effect on protein synthesis. *Proc. Nat. Acad. Sci.* **60**, 280 (1963).

26 Fang, S., Anderson, K. M., and Liao, S., Receptor proteins for androgens. *J. Biol. Chem.* **224**, 6584 (1969).

27 Rousseau, G. G., Interactions of steroids with hepatoma cells: molecular mechanisms of glucocorticoid hormone action. *J. Steroid Biochem.* **6**, 75 (1975).

28 Simons, S. S., Martinez, H. M. Carcea, R. L., *et al.*, Interactions of glucocorticoid receptor steroid complexes with acceptor sites. *J. Biol. Chem.* **251**, 334 (1976).

29 Gschwendt, M., and Hamilton, T. H., The transformation of the cytoplasmic oestradiol receptor complex into the nuclear complex in a uterine cell-free system. *Biochem. J.* **128**, 611 (1972).

30 Musliner, T. A., Chader, G. J., and Villee, C. A., Studies on estradiol receptors of the rat uterus. Nuclear uptake *in vitro*. *Biochemistry* **9**, 4448 (1970).

31 Chamness, G. C., Hennings, A. W., and McGuire, W. L., Oestrogen receptor binding is not restricted to target nuclei. *Nature* **241**, 458 (1973).

32 Jaffe, R. C., Socher, S. H., and O'Malley, B. W., An analysis of the binding of the chick oviduct progesterone-receptor to chromatin. *Biochim. Biophys. Acta* **399**, 403 (1975).

33 Giannopoulos, G., Glucocorticoid receptors in the lung. Mechanisms of specific glucocorticoid uptake by rabbit fetal lung nuclei. *J. Biol. Chem.* **250**, 2896 (1975).

34 Giannopoulos, G., and Gorski, J., Estrogen-binding protein of the rat uterus: different molecular forms associated with nuclear uptake of estradiol. *J. Biol. Chem.* **246**, 2425 (1971).

35 Spelsberg, T. C., Webster, R., Piler, G., *et al.,* Nuclear binding sites ("acceptors") for progesterone in avian oviduct: Characterization of the highest-affinity sites. *Ann. N.Y. Acad. Sci.* **286**, 43 (1977).

36 Clark, J. H., Erikkson, H. A., and Hardin, J. W., Uterine receptor-estradiol complexes and their interaction with nuclear binding sites. *J. Steroid Biochem.* **7**, 1039 (1976).

37 Clark, J. H., Peck, E. J., Jr., Hardin, J. W., *et al.,* The Biology and Pharmacology of Estrogen Receptor Binding: Relationship to Uterine Growth, *in* O'Malley, B. W., and Birnbaumer, L. (eds.), *Receptors and Hormone Action,* Vol. 2, p. 189, Academic, New York (1978).

38 Katzenellenbogen, B. S., Iwamoto, H. S., Heiman, D. F., *et al.* Stilbestrols and stilbestrol derivatives: estrogenic potency and temporal relationships between estrogen receptor binding and uterine growth. *Molec. Cell Endocrinol.* **10**, 103 (1978).

39 Traish, A. M., Muller, R. E., and Wotiz, H. H., Binding of estrogen receptor to uterine nuclei: salt-extractable versus salt-resistant receptor estrogen complexes. *J. Biol. Chem.* **252**, 6823 (1977).

40 Steggles, A. W., Spelsberg, T. C., and O'Malley, B. W., Tissue specific binding *in vitro* of progesterone-receptor to the chromatin of chick tissues. *Biochem. Biophys. Res. Comm.* **43**, 20 (1970).

41 Mainwaring, W. I. P., Symes, E. R., and Higgins, S. J., Nuclear components responsible for the retention of steroid-receptor complexes, especially from the standpoint of the specificity of the hormonal responses. *Biochem. J.* **156**, 129 (1976).

42 McGuire, W. L., Huff, K., and Chamness, G. C., Temperature-dependent binding of estrogen receptor to chromatin. *Biochemistry* **11**, 4562 (1972).

43 Spelsberg, T. C., Steggles, A. W., and O'Malley, B. W., Progesterone-binding components of chick oviduct. III. Chromatin acceptor sites. *J. Biol. Chem.* **246**, 4188 (1971).

44 Puca, G. A., Nola, E., Hibner, U., *et al.* Interaction of the estradiol receptor from calf uterus with its nuclear acceptor sites. *J. Biol. Chem.* **250**, 6452 (1975).

45 King, R. J. B., and Gordon, J., Involvement of DNA in the acceptor mechanism for uterine estradiol receptor. *Nature (London) New Biol.* **240**, 185 (1972).

46 Clemens, L. E., and Kleinsmith, L. J., Specific binding of the oestradiol-receptor complex to DNA. *Nature (London) New Biol.* **237**, 204 (1972).

47 Yamamoto, K. R., and Alberts, B. M., *In vitro* conversion of estradiol-receptor protein to its nuclear form: dependence on hormone and DNA. *Proc. Nat. Acad. Sci.* **69**, 2105 (1972).

48 Baxter, J. D., Rousseau, G. G., Benson, C., *et al.,* Role of DNA and specific cytoplasmic receptors in glucocorticoid action. *Proc. Nat. Acad. Sci.* **69**, 1892 (1972).

49 Rousseau, G. G., Higgins, S. J., Baxter, J. D., *et al.,* Binding of glucocorticoid receptors to DNA. *J. Biol. Chem.* **250**, 6015 (1975).

50 Yamamoto, K. R., Characterization of the 4S and 5S forms of the estradiol receptor protein and their interaction with deoxyribonucleic acid. *J. Biol. Chem.* **249**, 7068 (1974).

51 Andre, J., Vic, P., Humeau, C., *et al.,* Nuclear translocation of the estradiol receptor: partial inhibition by ethidium bromide. *Molec. Cell Endocrinol.* **8**, 225 (1977).

52 Kallos, J., Hollander, V. P., Assessment of specificity of oestrogen-receptor DNA interaction by competitive assay. *Nature* **272**, 177 (1978).

53 Yamamoto, K. R., Alberts, B., On the specificity of the binding of the estradiol receptor protein to deoxyribonucleic acid. *J. Biol. Chem.* **249**, 7076 (1974).

54 Chan, L., Means, A. R., and O'Malley, B. W., Steroid hormone regulation of specific gene expression. *Vit. Horm.* **36**, 259 (1978).

55 Ashburner, M. Chihara, C., Meltzer, P., *et al.,* Temporal control of puffing activity in polytene chromosomes. *Quant. Biol.* **38**, 655 (1974).

56 Higgins, S. J., Gehring, U., Molecular mechanisms of steroid hormone action. *Adv. Cancer Res.* **28**, 313 (1978).

57 Katzellenbogen, B. S., and Gorski, J., Estrogen actions on syntheses of macromolecules in target cells. *Biochem. Actions Horm.* **3**, 187 (1975).

58 Clark, J. H., Paszko, Z., and Peck, Jr., E. J., Nuclear binding and retention of the receptor estrogen complex: relation to the agonistic and antagonistic properties of estriol. *Endocrinology* **100**, 91 (1977).

59 O'Malley, B. W., and Kohler, P. O., Hormonal induction of specific proteins in chick oviduct cell cultures. *Biochem. Biophys. Res. Comm.* **28**, 1 (1976).

60 Chan, L., Means, A. R., and O'Malley, B. W., Rates of induction of specific translatable messenger RNAs for ovalbumin and avidin by steroid hormones. *Proc. Nat. Acad. Sci.* **70**, 1870 (1973).

61 Schwartz, R. J., Tsai, M.-J., Tsai, S., *et al.,* Effect of estrogen on gene expression in the chick oviduct. V. Changes in the number of RNA polymerase binding and initiation sites in chromatin. *J. Biol. Chem.* **250**, 5175 (1975).

62 Thompson, E. B., Tomkins, G. M., Curran, J. F., Induction of tyrosine α-ketoglutarate transaminase by steroid hormones in a newly established tissue culture cell line. *Proc. Nat. Acad. Sci.* **56**, 296 (1966).

63 Granner, D. K., Thompson, E., B., and Tomkins, G. M., Dexamethasone phosphate-induced synthesis of tyrosine aminotransferase in hepatoma tissue culture cells. *J. Biol. Chem.* **245**, 1472 (1970).

64 Tomkins, G. M., Thompson, E. B., Hayashi, S., *et al.,* Tyrosine transaminase induction in mammalian cells in tissue culture. *Quant. Biol.* **31**, 349 (1966).

65 Higgins, S. J., and Gehring, U., Molecular mechanisms of steroid hormone action. *Adv. Cancer Res.* **28**, 313 (1978).

66 Baxter, J. D., and Ivarie, R. D., Regulation of Gene Expression by Glucocorticoid Hormones. Studies of Receptors and Responses in Cultured Cells, in O'Malley, B. W., and Birnbaumer, L. (eds.), *Receptors and Hormone Action,* Vol. 3, p. 252, Academic, New York (1978).

67 Ivarie, R. D., Fan, W. J.-W., and Tomkins, G. M., Analysis of the induction and deinduction of tyrosine aminotransferase in enucleated HTC cells. *J. Cell Physiol.* **85**, 357 (1975).

68 Tomkins, G. M., Gelehrter, T. D., Granner, D., *et al.,* Control of specific gene expression in higher organisms. *Science* **166**, 1474 (1969).

69 Diesterhaft, M., Noguchi, T., Hargrove, J., *et al.,* Translation of tyrosine aminotransferase mRNA in a modified reticulocyte system. *Biochem. Biophys. Res. Comm.* **79**, 1015 (1977).

70 Isotalo, A., and Santi, R. S., Changes in labelling of ribonucleic acid species fractionated from the rat ventral prostate in response to testosterone. *Acta Endocrinol,* **78**, 401 (1975).

71 Mainwaring, W. I. P., Wilce, P. A., and Smith, A. E., Studies on the form and synthesis of messenger ribonucleic acid in the rat ventral prostate gland, including its tissue-specific stimulation by androgens. *Biochem. J.* **137**, 513 (1974).

72 Bardin, C. W., Bullock, L. P., Mills, N.C., *et al.,* The Role of Receptors in the Anabolic Sections of Androgens, in O'Malley, B. W., and Birnbaumer, L. (eds.), *Receptors and Hormone Action,* Vol. 3, p. 83, Academic, New York (1978).

73 Coffey, DS, Shimazaki, J., and Williams-Ashman, H. G., Polymerization of deoxyribonucleotides in relation to androgen-induced prostatic growth. *Arch. Biochem. Biophys.* **124**, 184 (1968).

74 Davies, R., and Griffiths, K., Stimulation in vitro of prostatic ribonucleic acid polymerase by 5α-dihydrotestosterone-receptor complexes. *Biochem. Biophys. Res. Comm.* **53**, 373 (1973).

75 Tymoczko, J. L., Liang, T., and Liao, S., Androgen Receptor Interactions in Target Cells: Biochemical Evaluation, in O'Malley, B W., and Birnbaumer, L., (eds.), *Receptors and Hormone Action,* Vol. 2, p. 122, Academic, New York, (1978).

76 Nyberg, L. M., Hua, A.-L., Loor, R. M., *et al.,* Androgen-induced gene activation in the rat prostate. *Biochem. Biophys. Res. Comm.* **73**, 330 (1976).

77 Parker, M. G., and Scrace, G. T., The androgenic regulation of abundant mRNA in rat vental prostate. *Eur. J. Biochem.* **85**, 399 (1978).

78 Heyns, W., Verhoeven, G., DeMoor, P., A comparative study of androgen binding in rat uterus and prostate. *J. Steroid Biochem.* **7**, 987 (1976).

79 Heyns, W., Peeters, B., Mous, J., Influence of androgens on the concentration of prostatic binding protein (PBP) and its mRNA in rat prostate. *Biochem. Biophys. Res. Comm.* **77**, 1492 (1977).

CHAPTER TWENTY-TWO

INACTIVATION AND EXCRETION OF STEROIDS

ROBERT T. CHATTERTON, JR.

Department of Obstetrics and Gynecology
Northwestern University Medical School
Chicago, Illinois

1 SERUM BINDING AS A PROCESS OF INACTIVATION

Enzymatic modification of active steroids is almost always an essentially irreversible process leading to inactivation of the hormone, but serum binding globulins make the steroid only temporarily unavailable to the target organ. Binding of steroid hormones to serum proteins occurs immediately after their elaboration and diffusion through the interstitial fluid into blood vessels. Two types of binding globulins predominate: corticosteroid-binding globulin (CBG or transcortin) and testosterone-estradiol-binding globulin (TeBG), which is also known as sex-hormone-binding globulin (SHBG) and as 17β-hydroxysteroid-binding globulin. These are both glycoproteins with molecular weights of approximately 50,000 and are produced by the liver. The steroid specificity of the binding globulins is different from that of the intracellular receptors for the same hormones. CBG binds cortisol, corticosterone, and progesterone with similar affinity, but the fluorinated cortisol derivatives with high glucocorticoid activity such as dexamethasone do not bind, nor does the biologically potent progestin R5020 (Roussel) that is used in receptor studies. Similarly, TeBG binds the biologically active estrogens and androgens that have the 17β-hydroxyl, but it does not bind the potent estrogen diethylstilbestrol or the androgen R1881 (Roussel). A distinct difference between binding to the serum binding globulins and structural requirements for biological activity is thus apparent. The lack of serum binding of the synthetic hormone analogues is also a factor in their biological potency, since after absorption they are immediately available to receptors in target cells. Some evidence also has been presented that supports the concept that steroids bound to serum binding globulins are more available to the liver than to other potential target organs (1), and therefore may be metabolized by the liver to inactive products more rapidly than they are made available to other target organs. In support of this concept are observations that the metabolic clearance rate frequently does not decrease in proportion to increases in the concentrations of CBG- or TeBG-bound steroids. Whether the serum binding globulins actually serve to shunt steroids away from target organs toward the liver, where they are metabolized and cleared, or not, is a matter for further study.

Elevated levels of binding globulins in serum would appear to compete with receptors in target tissues. The binding affinity of serum binding globulins and cytosol receptors is similar for many steroids, although binding of estradiol-17β to TeBG is relatively weak. The availability of steroids to tissue receptors must take into account the equilibrium between the serum and tissue binding substances. This is particularly important, because concentrations of steroids in serum very rarely exceed the binding capacity of the circulating binding globulins. The properties of the binding globulins are described in Chapter 19 and more completely by Westphal (2).

2 SITES OF METABOLISM OF STEROIDS

The liver is the major site of metabolism of steroids, although transformation to active as well as inactive forms may occur in many other organs, particularly the kidney, the gut, and the skin. Each organ may have characteristic pathways for metabolism of the secreted steroid. Some differences have been described for metabolism of progesterone in the liver (3), kidney (4), and mammary gland (5) of the rabbit, for example. Active and inactive forms occur in most tissues in redox pairs that are interconvertible, depending upon the availability of the dehydrogenase and the reduced or oxidized cofactor. These pairs include cortisol/cortisone, estradiol-17β/estrone, testosterone/androstenedione, and 20α-dihydroprogesterone/progesterone.

3 REDUCTIVE METABOLISM OF Δ^4-3-KETO STEROIDS

The conjugated-double-bond system of the 3-ketone and double bond at C-4 is required for biological activity of glucocorticoids, mineralocorticoids, and progestins. The androgens are unusual in that although testosterone (which has the Δ^4-3-keto structure) may be considered the complete androgen, some reduced metabolites of testosterone apparently are the final active products that inter-act at the cellular level (6). Thus, the double bond in ring A must be reduced to form 17β-hydroxy-5α-androstan-3-one (5α-dihydrotestosterone) before effects of testosterone can be elicited in accessory sex organs. Even the tetrahydro product androsterone (3α-hydroxy-5α-androstan-17-one) retains a high degree of androgenic activity in some tissues. Reduction of testosterone to form metab-olites with the 5β configuration, however, results in inactivation of the steroid as far as reproductive function is concerned. The liver is capable of forming both the active and inactive metabolites. According to work of Gustafsson and others (7), the rat liver has a male or female pattern of metabolism that can be induced by neonatal exposure of the animal of either genotype to male or female hormones. The male liver forms a larger proportion of metabolites with the 3-hydroxyl in the β or equatorial position, but less of 5α-dyhydrotestoster-one. Nevertheless, the testosterone concentration in serum of male and female rats as in other species is an order of magnitude greater than that of 5α-dihydrotestosterone. Most of the 5α-dihydrotestosterone available to its target organs is formed by reduction of testosterone within the responsive tissues. 5α-Pregnane-3,20-dione also retains some progestogenic activity in supporting decidual growth in rats, but it cannot substitute for progesterone in mainte-nance of pregnancy.

The metabolites with a 5β configuration have some properties that are unlike the A/B *trans* products. Both 5β-androstan-3α-ol-17-one (etiocholano-lane) and 5β-pregnan-3α-ol-20-one (pregnanolone) have thermogenic proper-ties that their 5α isomers do not have. The fact that an increase in serum progesterone is associated with an increase in body temperature in women is not a direct effect of progesterone, but of its metabolite pregnanolone. Thus the first steps in the "inactivation" pathway do not result in a loss of all activity even though specific functions of the respective parent compounds are lost.

Generally, reduction of the double bond occurs first and then reduction of the 3-ketone to a 3-hydroxyl. The 20-ketone of progesterone also undergoes reduction in formation of its major urinary product, 5β-pregnane-3α,20α-diol. Pregnanediol can be formed after initial reduction of either the 3- or 20-ketones of progesterone; both pregnanolone and 20α-dihydroprogesterone (20α-hy-droxy-4-pregnen-3-one) can be intermediates. Catabolism of the secreted an-drogen testosterone to androsterone and etiocholanolone requires, in addition to ring A reduction, oxidation of the 17β-hydroxyl to a ketone.

C-20 reduction of cortisol and cortisone after initial reduction of the Δ^4-3-keto structure also occurs to form cortols and cortolones respectively, but this is more limited than C-20 reduction of progesterone. Consequently, the initial reduction products of cortisol, that with reduction of the double bond (dihy-drocortisol) and that with additional reduction of the 3-ketone (tetrahydrocor-tisol), are those that are usually measured in urinary assays of glucocorticoids.

The other major transformations that occur in catabolism of steroids are hydroxylation at specific sites on the steroid nucleus and conjugation of the steroid with glucuronic or sulfuric acids through various hydroxyl groups. Fig. 1 gives an example of the structure of each of these conjugates. In addition to glucuronic acid, other saccharides have been shown to be conjugated to steroids through a glucosidic bond (8).

Hydroxylation products, like reduction products of secreted steroid hormones, may retain certain activities of the parent hormone. Whether this is because the separate activities of the products are attributed to the parent steroid or because each of the products is capable of eliciting only part of the actions of the parent compound is not known for many compounds.

Steroids that are unbound and unconjugated diffuse rather freely between vascular and extravascular fluids because their lipid solubility permits diffusion across membranes. Binding to serum proteins only shifts equilibrium concentrations, but conjugation to glucuronic or sulfuric acids results in much greater water solubility and complete loss of biological activity. That is, these conjugates, being ionized at physiological pH, do not readily cross membranes and do not bind to soluble receptors that are present in the cytoplasm of target cells. The biochemical process by which conjugation of steroids to glucuronic acid from uridine diphosphoglucuronic acid occurs in the presence of uridine diphosphoglucuronyl transferase has been fairly well characterized for a number of steroids (8). Likewise, sulfurylation of hydroxylated steroids by transfer of active sulfate from 3'-phosphoadenylyl sulfate has been described. The rate of conjugation of steroids as a factor in the biological activity of secreted hormones has not been studied extensively, since conjugation is generally considered a secondary process in inactivation subsequent to a reduction or hydroxylation step. Nevertheless, estradiol-17β and testosterone are found in urine as their respective 17β-glucuronides, and thus the rate of conjugation may be an important determinant in the biological activity of the secreted steroid. De-

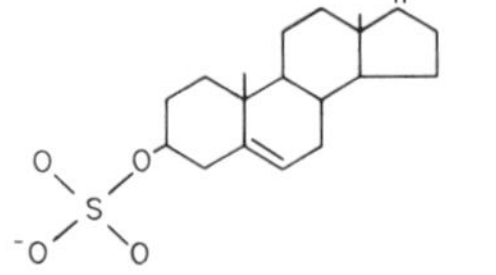

Dehydroepiandrosterone
Sulfate

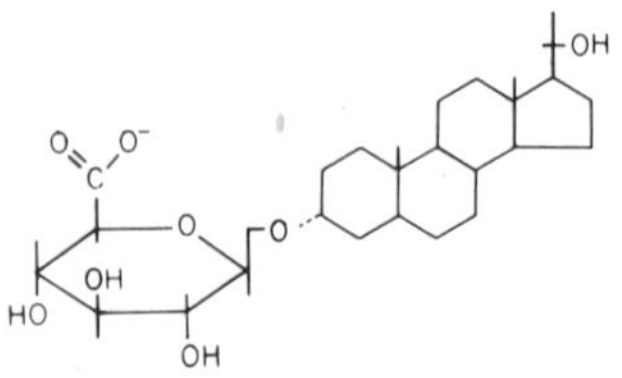

Pregonediol-3α-yl-β-D-glucopyranosiduronic acid

Figure 1 Two common forms of conjugated steroids found in urine.

creases in pituitary secretions, especially those that influence carbohydrate metabolism in the liver, can decrease the rate of conjugation (9).

Other effects of hormones, particularly thyroid hormones and estrogens, have been shown to alter metabolism of steroids in the liver (7, 10). Diseases that influence liver function generally also decrease metabolism of steroids, and may result in excessive hormone concentrations in blood (11).

4 ASSAY OF STEROIDS IN URINE AND SERUM

Most urinary assays require that the glucuronide or sulfate be hydrolyzed before the steroid is subjected to quantitative measurement. The only exceptions are some newer radioimmunoassays for estriol glucuronide and pregnanediol glucuronide, in which these compounds are measured directly by the use of specific antisera. The antisera have been produced in animals immunized with the steroid glucuronide that was covalently bonded through the carboxylic acid moiety to an immunogenic protein (12). Most typical assays of estriol or pregnanediol have been carried out after hot acid or enzymatic hydrolysis of the conjugates and either colorimetric or gas-chromatographic measurement of the steroid after some purification. The C_{19} steroids derived largely from the adrenal, but also including androgens of ovarian or testicular origin, have been measured primarily by use of the Zimmermann reagent as 17-ketosteroids.

Assay of the glucocorticoids as 17-hydroxycorticosteroid (17-OHCS) products in urine has the advantage that only the adrenal produces steroids with the dihydroxyacetone side chain that is required for color development in the Porter-Silber method. The assay will measure cortisol, cortisone, 11-deoxycortisol, or the ring-A-reduced metabolites of these compounds in urine after hydrolysis of conjugates. The primary shortcoming of the method is that it will not measure the metabolites in which the 20-ketone has been reduced, that is, cortol and cortolone. Therefore, it must be assumed that the ratio of C-20 reduced to unreduced metabolites remains constant in order to relate increases in 17-OHCS to increases in cortisol production.

Measurements of steroids in blood provide an assessment of the exposure of target tissues to the hormones, especially if both the total concentration in serum and the concentration of unbound steroid is determined. However, blood is usually drawn for assays over a very short interval of time, and this can lead to substantial errors in estimating the actual exposure of target organs because of episodic pulsatile secretion (13) and diurnal and other variations in blood levels of hormones.

REFERENCES

1 Keller, N., Richardson, U. I., and Yates, F. E., Protein binding and the biological activity of corticosteroids: *in vivo* induction of hepatic and pancreatic alanine aminotransferases by corticosteroids in normal and estrogen-treated rats. *Endocrinology* **84**, 49 (1969).

2   Westphal, U., Steroid-Protein Interactions, *in* Gross, F., Labhart, A., Mann R, *et al.* (eds.), *Monographs on Endocrinology*, Vol. 4, Springer-Verlag, New York (1971).

3   Chatterton, Jr., R. T., Chatterton, A. J., and Hellman, L., Metabolism of progesterone by the rabbit liver. *Endocrinology* **87**, 941 (1970).

4   Chatterton, Jr., R. T., Chatterton, A. J., and Hellman, L., Metabolism of progesterone by the rabbit kidney. *Endocrinology* **84**, 1089 (1969).

5   Chatterton, Jr., R. T., Chatterton, A., J., and Hellman, L., Metabolism of progesterone by the rabbit mammary gland. *Endocrinology* **85**, 16 (1969).

6   Wilson, J. D., Griffin, J. E., and George, F. W., Sexual differentiation: early hormone synthesis and action. *Biol. Reprod.* **22**, 9 (1980).

7   Gustafsson, J. A., and Stenberg, A., Specificity of neonatal, androgen-induced imprinting of hepatic steroid metabolism of rats. *Science* **191**, 203 (1976).

8   Roy, A. B., Enzymological Aspects of Steroid Conjugation, in Bernstein, S., and Solomon, S. (eds.), *Chemical and Biological Aspects of Steroid Conjugation*, pp. 74–130, Springer-Verlag, New York (1970).

9   Anderson, C. H., and Chatterton, Jr., R. T., Effects of hypophysectomy on estrogen conjugation and on plasma and tissue concentrations of tritium after administration of ^{3}H-estradiol-17β. *Steroids* **28**, 785 (1976).

10 Bradlow, H. L., Hellman, L., Zumoff, B., *et al.,* Interaction of hormonal effects: influence of triiodothyronine on androgen metabolism. *Science* **124**, 1206 (1956).

11 Zumoff, B., Fishman, J., Gallagher, T. F., *et al.,* Estradiol metabolism in cirrhosis. *J. Clin. Invest.* **47**, 20 (1968).

12 Parker, C. W., Nature of Immunological Response and Antigen-Antibody Interaction, in Odell, W. D., and Daughaday, W. H., (eds.), *Principles of Competitive Protein-Binding Assays*, pp. 25–48, Lippincott, Philadelphia (1971).

13 Hellman, L., Nakada, F., Curti, J., *et al.,* Cortisol is secreted episodically by normal man. *J. Clin. Endocr.* **30**, 411 (1970).

CHAPTER TWENTY-THREE

ACTIONS OF PROLACTIN AND CHORIONIC SOMATOMAMMOTROPIN (CS, PLACENTAL LACTOGEN) IN MOTHER AND FETUS

JOHN B. JOSIMOVICH

Department of Obstetrics and Gynecology
College of Medicine & Dentistry of New Jersey
Newark, New Jersey

As described by Nicoll (1) in his review of the physiological actions of prolactin (PRL), this hormone has a greater variety of effects than other hormones, and seems to serve a modulatory role in the action of other hormones.

1 ROLES IN LIPID MOBILIZATION AND STEROIDOGENESIS

The monomer of human chorionic somatomammotropin (hCS) is capable of mobilizing peripheral fat with consequent elevation in fasting plasma free fatty acids and generation of glycerol. This effect is accomplished by serum levels of hCS seen in the second half of pregnancy, and has been produced by administering such high doses to hypopituitary (growth hormone deficient) dwarfs (2). Although this effect of hCS appears beneficial to the fetus in that it spares glucose for fetal needs, there is evidence that the consequent ketosis that develops during fasting may have adverse effects on late fetal neurological development (4). Pituitary PRL also appears to have some fat-mobilizing effects *in vitro* and *in vivo* when the hormone is tested in lower mammals (2), but does not uniformly have this effect in humans. Since the human studies were carried out with porcine PRL, studies on the metabolic effects of human PRL are awaited with interest.

It is uncertain whether either hCS or PRL plays a role in the mobilization of cholesterol from intracellular storage pools in the placenta for production of the progesterone and estrogen precursor Δ^5-pregnenolone. Recent studies (4) have suggested that the nocturnal surge in PRL secretion in the rat on the sixth day of induced pseudopregnancy is responsible for cholesterol ester hydrolase activity in the corpora lutea.

It is clear that both hormones (especially hCS) are potent in inhibiting 20α reduction of progesterone in the rodent corpus luteum, thus preventing conversion of progesterone to the essentially biologically inactive compound 20α-dihydroprogesterone (5). These effects are thought to be responsible for the so-called luteotropic effect of PRL in the rodent, for which the hormone was called at one time luteotrophic hormone (LtH). The obvious luteolytic effects of PRL in other species, especially the rabbit, resulted in the dropping of this nomenclature (1). Nevertheless inactivation of progesterone does not occur to any great extent in the human corpus luteum, as evidenced by the failure of hCS alone, or in combination with hCG, to prolong progesterone secretion (3). Instead, it is possible that hCS within the placenta may be responsible for suppressing too great a conversion of progesterone to the reduced compound, since 20α-reductase has been found to be present in the placenta (as well as in various fetal tissues) (2). Thus the net amount of progesterone secreted by the placenta may depend on the balance of inactivation and reactivation of progesterone, a process known to be active in the human fetoplacental unit (2). The extent to which fetal and/or maternal pituitary PRL affects the metabolism of progesterone and other steroids remains unknown. If important, they do so in the fetal compartment, where PRL levels are high and 20α-hydroxysteroid dehydrogenase activity has been found to be most prominent.

2   ROLE IN ELECTROLYTE BALANCE

Prolactin has been shown to have at least two physiological effects on water and salt balance in the developing mammalian fetus.

In the developing pregnancy of the rhesus monkey, it has been shown that high levels of amniotic fluid PRL markedly reduce the amniotic fluid volume (5). Such a reduction in amniotic fluid volume is also seen in humans suffering hypertensive disorders during pregnancy and in whom serum PRL levels are higher than in normal pregnant women. It has also been shown that the normally high amniotic fluid PRL concentration is probably derived from both maternal pituitary and endometrial decidual sources (5, 6). More recently, it was shown that the use of bromoergocryptine to reduce maternal serum PRL, and thereby amniotic PRL, caused an increase in amniotic fluid in a pregnant woman suffering from anuria and severe oligohydramonios (reduction of amniotic fluid) in association with fetal renal agenesis (7). The mechanism for the action of PRL in regulating net amniotic volume appears to be through binding of the hormone to the amnion. This binding is followed by a reduction in flux of chloride, and secondarily water, in both directions across the chorio-amnion, but more so in the fetal-to-maternal direction. Studies by Frantz, Horrobin, Tyson and their co-workers (8) have shown this process to occur *in vitro* with both guinea pig and human fetal membranes. The biochemical mechanisms for these effects are not known.

The second effect on fetal salt and water conservation was predicted by Friesen and colleagues (7, 8), namely, that since other workers had shown an important conserving effect of PRL in certain fish and larval amphibia when these forms passed from saltwater to freshwater environments (1), such a process might occur in the developing mammalian embryo. Indeed, it has been shown in the pregnant rhesus monkey (8) that as the osmolarity of the amniotic fluid is altered from that found in the circulation, the fetus suffers from dehydration or overhydration if amniotic PRL levels are low; but that high levels of PRL reverse or prevent isosmotic shifts of water into or out of the fetal extracellular fluid. The sites of action of PRL in bringing about this effect could involve amnion, umbilical cord, and/or fetal lung, since specific high-affinity PRL receptors have been described in these tissues. Again, the biochemical mechanisms of action of PRL in exerting these protective effects on the fetus are unknown at present.

A review of the involvement of PRL in mammalian water and electrolyte conservation by the mammalian kidney (1) suggests that this hormone plays a role in sodium- and potassium-ion conservation and that serum osmolarity in turn affects serum PRL levels. Chronic reduction of serum PRL in rats results in increased Na^+ and K^+ excretion; administration of PRL to rats reduces urinary Na^+ and water excretion; water and Na^+ are retained to a greater extent in cat renal perfusion experiments when PRL is added; and PRL protects the merino ewe from reversal of the effects of aldosterone on renal Na^+ excretion brought about by high salt loads, and also against the reversal of the effects of antidiuretic hormone (ADH) by cortisol during low salt loading. As to the effects of serum osmolarity on PRL levels, it has been shown that dehydration followed by water loading increases serum PRL levels in the human, and dehydration in rats lowers serum PRL levels. The overall changes seen in studies of the effects of PRL on human salt balance and the feedback of serum tonicity on PRL release, suggest that PRL plays less of a role in osmoregulation in adult mammals than in the fetus.

3   BEHAVIORAL EFFECTS OF PROLACTIN

Although earlier studies on the effects of PRL on behavior in mammals some-times lacked adequate controls, Nicoll (1) has reviewed the evidence suggesting a role for this hormone on the central nervous system. Maltz and colleagues showed that PRL eventually causes maternal brooding behavior in virgin female rats, but estrogen and progesterone accelerate the appearance of the response. Zarrow and co-workers showed similar effects in castrated female rats. Finally, nesting behavior in pregnant rabbits depends in part on the presence of pituitary PRL.

4   GROWTH-HORMONE-LIKE EFFECTS

In addition to the lipid-mobilizing effects of PRL and hCS cited above, there have been recorded cases of positive nitrogen balance achieved in hypopitui-tary humans (8) with certain batches of ovine of porcine PRL (3). The effects are not consistent, however, raising the possibility that contamination by pitui-tary growth hormone and/or chemical alterations of the PRL molecule may be responsible for these occasional findings (1). Grumbach and colleagues (2, 8) have clearly shown positive nitrogen balance in hypopituitary humans given large amounts of hCS, although smaller amounts given by other workers failed to show somatotropic effects.

In regard to anti-insulin effects, Paul Beck and colleagues and others have shown that some increased resistance to the hypoglycemic effect of insulin can be brought about by hCS. Heterologous PRL (e.g., ovine) has been shown to synergize with growth hormone (1), as did hCS when human growth hormone was tested in the tibial epiphysial cartilage growth bioassay in hypophysecto-mized rats (9). Thus PRL appears to have little somatotropic effect itself, except perhaps with growth hormone. The weak somatotropic effects of hCS become physiologically evident by midpregnancy in humans because of the high blood levels attained (one $\mu g/ml$ or greater).

Recent studies have suggested that fetal pituitary PRL may be an important fetal growth factor in sheep (10).

5   FUNCTION IN THE MAMMARY GLAND

As shown by Turkington, Topper, and others (5) in mouse mammary explants *in vitro*, PRL promotes synthesis of lactose, α-lactalbumin, and casein proteins. The initial step is the high-affinity, specific binding of PRL to cell membrane receptors. It is also evident that hCS binds to the same receptors. In fact, although HCS is lactogenic (2) and may shorten the time needed for PRL to act on protein synthesis (5), some authors (11) believe that hCS and the corre-sponding hormone in subprimate placentas may competitively inhibit the binding and full lactogenic expression of rising levels of PRL. Of recent interest is the finding that PRL binding may lead to a decrease in availability of further PRL-receptor activity in mammary tissue (12), while leading to an increase in testicular and ovarian luteinizing hormone receptors (13). Conversely, follicle-stimulating hormone may induce PRL receptors in rat ovarian granulosa cells (14).

Although Turkington (1) has suggested that PRL may stimulate production of cAMP-binding protein, which in turn inhibits protein kinase, there is yet no direct evidence for generation of cyclic nucleotides as an early action of this hormone as there is for other protein hormones (see Chapter 13). Nevertheless, it is clear that there is increased transcription of mRNA responsible for casein synthesis as well as a decreased half-life of these mRNAs (15).

Clearly, galactosyltransferase activity, lactose synthesis, and α-lactalbumin synthesis are increased by prolactin (1, 11). As Nicoll noted for the generally modulatory action of the hormone, PRL alone cannot cause increased synthesis of α-lactalbumin in monolayer cultures of rat mammary cells. It, along with insulin, can only augment the stimulatory effects of corticosterone, as shown by Meites et al. (5) and confirmed by others (16). Thus the early work of Lyons and colleagues (5), employing the whole rat, on the cooperation of PRL with other hormones in initiating galactopoiesis has been confirmed by in vitro studies.

Maintenance of lactation is less well understood. Removal of milk, which contains lactation-inhibitory factors such as citrate and malonate, appears essential for continuing lactation (5). In the cow, maintenance of lactose production does not appear to require PRL (17).

The inhibitory effects of progesterone and estrogen on the galactopoietic expression of PRL has been well known for years in a variety of mammals (1). Thus, although hCS and PRL are considerably elevated in women's serum by the end of pregnancy, there is absence of full lactation until a decline in progesterone and estrogen after delivery permit the full effect of PRL. Vorherr (11) also believes that hCS may prevent full lactation. Although this hormone has been shown to induce early lactation in monkeys and full lactation in rodents and rabbits, it has been shown not to induce lactation in women whose breasts have been primed by an hCG-induced prolonged luteal phase followed by a decline in estrogen and progesterone (5). At any rate, recent studies suggest that the blocking effect of progesterone may be complex. Progesterone added to rat mammary gland explants in organ culture has been shown to block PRL-mediated transcription in the formation of casein mRNA (15).

The cellular sites at which PRL exerts its effects on protein synthesis appear to be centered in the endoplasmic reticulum and the Golgi apparatus as well as the microvilli at the luminal cell surfaces of the mammary epithelium (5). And hCS appears to have similar effects, both by light and electron microscopic studies in rabbit, rodent, and human.

In summary, both prolactin and chorionic somatomammotropin, after binding to mammary epithelial cells, affect numerous intracellular processes in casein, lactose, and α-lactalbumin synthesis. Blockade of their effects is maintained until parturition by progesterone (and perhaps estrogen) inhibition of mRNA transcription and perhaps by other mechanisms.

REFERENCES

1   Nicoll, C. S., Physiological Action of Prolactin, *in* Greep, R. O., and Astwood, E. B., (eds.), *Handbook of Physiology,* Sect. 7, Vol. 4, *Pituitary Gland and Its Neuroendocrine Control* Part 2, p. 253, American Physiological Society, Washington, DC (1974).

2   Josimovich, J. B., Placental protein hormones in pregnancy. *Clin. Obstet. Gynecol.* **16**, 46 (1973).

3   Klemcke, H. G., and Brinkley, H. J., Endogenous rhythms of luteal and adrenal cholesterol ester hydrolase and serum PRL, LH and progesterone in mature pseudopregnant rats. *Biol. Reprod.* **22**, 1022 (1980).

4   Beck, J. C., Gonda, A., Hamid, N. A., *et al.* Some metabolic changes induced by primate growth hormone and purified ovine prolactin. *Metabolism* **13**, 144 (1964).

5   Josimovich, J. B., and Archer, D. F., Primate Lactogenic Hormones and Functions, *in* Larson, B. L., (ed.), *Lactation: A Comprehensive Treatise,* Vol. 4, p. 281, Academic, New York (1978).

6   Josimovich, J. B., Filkins, K., and May, P., Unpublished observations.

7   Healy, D. L., Kimpton, W. G., Muller, H. T. C., *et al.*, Synthesis of immunoreactive prolactin by decidua—chorion. *Brit. J. Obstet. Gynaecol.* **86**, 307 (1979).

8   Josimovich, J. B., Hormonal Physiology of Pregnancy: Steroid Hormones of the Placenta, and Polypeptide Hormones of the Placenta and Pituitary, in Gold, J. J., and Josimovich, J. B., (eds.), *Gynecologic Endocrinology,* Vol. 3, p. 147, Harper and Row, Hagerstown (1981).

9   Josimovich, J. B., Potentiation of somatotrophic and diabetogenic effects of growth hormone by human placental lactogen (HPL). *Endocrinology* **78**, 707 (1966).

10   Takahashi, K., Burd, L. I., Scommegna, A., *et al.,* Lack of relationship between prolactin and adrenal steroidogenesis in the ovine fetus. *Am. J. Obstet. Gynecol.* **139**, 427 (1981).

11   Vorherr, H., Human Lactation and Breast Feeding, in Larson, B. L., (ed.), *Lactation: A Comprehensive Treatise,* Vol. 3, p. 181, Academic, New York (1978).

12   Djiane, J., Delouise, E., and Kelly, P. A., Prolactin receptors in organ culture of rabbit mammary gland: Effect of cycloheximide and prolactin. *Proc. Soc. Exp. Biol. Med.* **162**, 342 (1979).

13   Sequin, C., and Auclair, C., Role of Prolactin in the Control of Testicular and Ovarian LH Receptors in the Rat, abstract 165, in *Program and Abstracts of the 61st Annual Meeting of the Endocrine Society, 1979.*

14   Wang, C., Hsueh, A. J. W., and Erickson, G. F., FSH introduction of functional prolactin receptors by follicle-stimulating hormone in rat granulosa cells *in vivo,* and *in vitro. J. Biol. Chem.* **254**, 11330 (1979).

15   Matusik, R. J., Guyette, W. A., and Rosen, G. M., Prolactin-mediated transcriptional and post-transcriptional control of casein gene expression. *Cell* **17**, 1013 (1979).

16   Ray, D. B., Horst, I. A., Jansen, R. W., *et al.*, Normal mammary cells in long term culture. I. Development of hormone-dependent functional monolayer cultures and assay of α-lactalbumin production. *Endocrinology* **108**, 573 (1981).

17   Jones, E. A., Lactose Biosynthesis, *in* Larson, B. L., (ed.), *Lactation: A Comprehensive Treatise,* Vol. 3, p. 371 Academic, New York (1978).

CHAPTER TWENTY-FOUR

NEUROHYPOPHYSEAL HORMONES

PAULA L. HOFFMAN

Department of Physiology and Biophysics
University of Illinois at the Medical Center
Chicago, Illinois

1 INTRODUCTION*

Extracts of the neural lobe of the pituitary (posterior pituitary, neurohypophysis) were shown in the early 1900s to possess five distinct biological activities: mammalian pressor and antidiuretic activities, avian vasodepressor activity, and uterotonic and milk-ejecting activities. These actions can be attributed to the two peptide hormones secreted by the neurohypophysis, vasopressin (primar-

I am grateful to Dr. L. Wilson for helpful discussion. This work was supported in part by USPHS Grant AA-2696.

* Hormones are designated according to the IUPAC–IUB Tentative Rules on Biochemical Nomenclature, *J. Biol. Chem.* **247**, 577 (1972).

CYS–TYR–*ILE*–GLN–ASN–CYS–PRO–*LEU*–GLY (NH$_2$)

OXYTOCIN

ARG

CYS–TYR–*PHE*–GLN–ASN–CYS–PRO–   –GLY (NH$_2$)

LYS

VASOPRESSIN

Figure 1   Primary structures of oxytocin and vasopressin.

ily pressor and antidiuretic) and oxytocin (primarily avian vasodepressor, uterotonic, milk-ejecting). By the 1950s, both of these peptides had been isolated,
their sequences determined, and their structures confirmed by synthesis (1). The
hormones each consist of nine amino acid residues, with a disulfide bridge
connecting the Cys residues in positions 1 and 6 (Fig. 1). Oxytocin and vasopressin differ in primary structure only in positions 3 and 8 (Fig. 1); however,
their actions are relatively selective. Vasopressin is generally regarded as the
mammalian antidiuretic hormone (ADH) and its pressor activity as a pharmacological effect. Although exogenously administered oxytocin is much more
potent than vasopressin in inducing contraction of the smooth muscle of the
uterus and mammary gland, its physiological role is still under investigation.
However, there are two instances, both having to do with reproduction, in
which oxytocin seems to be important. First, the hormone plays a key role in
the milk-ejection reflex by its action on the myoepithelial cells of the mammary
gland. Second, although the role of oxytocin as an endogenous initiator of
labor is not clear, oxytocin is a very powerful stimulant of myometrial activity

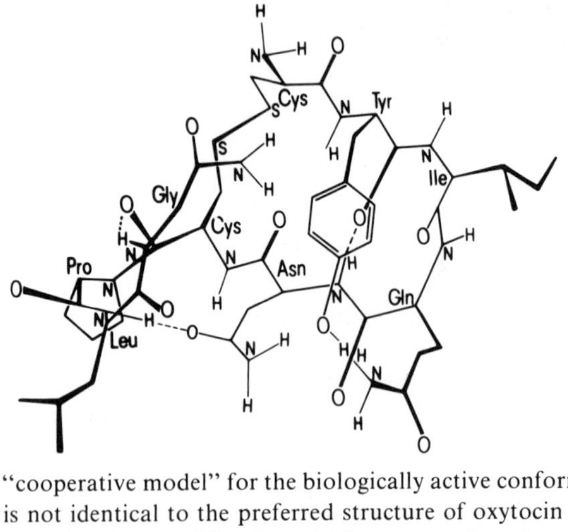

Figure 2   The "cooperative model" for the biologically active conformation of oxytocin. This model is not identical to the preferred structure of oxytocin as determined by
NMR measurements, but is a modification of this structure based on structure–activity
data. It includes a relatively featureless hydrophobic surface and a hydrophilic surface
postulated to contain the "active center" of oxytocin.

when given at the correct time during labor (2). At all other times, there seems to be a large store of oxytocin in the hypothalamus and posterior pituitary that has no apparent function, although oxytocin or its fragments have been proposed to have behavioral activities (3) and to play a role in regulation of anterior pituitary function (4).

An explanation for the specificity of neurophypophyseal hormone action, in spite of the similarities in primary structure, may reside in the recently elucidated three-dimensional conformations of these peptides in solution (5) (Fig. 2). It is these conformations that ultimately determine the nature of the hormone–receptor interactions and the intensity of the stimuli arising from such interactions.

2 BIOSYNTHESIS, SECRETION, AND METABOLISM OF NEUROHYPOPHYSEAL HORMONES

In general, this chapter emphasizes the biochemical and physiological actions of oxytocin. Many studies, particularly with respect to biosynthesis, however, have centered on vasopressin (ADH). The results of these studies will also be discussed, with the understanding that although it is likely that the same concepts apply to oxytocin, this has not always been proved.

It was originally believed that oxytocin and vasopressin were synthesized within the posterior pituitary. The elegant work of Bargmann and the Scharrers demonstrated that the hormones are, in fact, produced in the cell bodies of the supraoptic and paraventricular nuclei of the hypothalamus, and transported by axoplasmic flow to the neurohypophysis, where they are stored and released (6). The hypothalamoneurohypophyseal system is often described as a classic *neurosecretory* system, in which fully functional neurons, in addition to transmitting impulses, also secrete a blood-borne chemical mediator (6).

Throughout the hypothalamoneurohypophyseal system, oxytocin and vasopressin are found in noncovalent association with proteins having a molecular weight of about 10,000; they are known as neurophysins (7). In most species there are two major neurophysins, each of which appears to be associated specifically with oxytocin or vasopressin.

The biosynthesis of vasopressin has been postulated to occur by way of a higher-molecular-weight precursor (8, 9). Synthesis occurs by normal ribosomal mechanisms in the hypothalamic cell body, and the molecules are packaged into granules (*neurosecretory granules*) that are transported to the neurohypophysis by axoplasmic flow. The enzymatic liberation of hormone from precursor can occur both prior to and during axonal transport. It appears that neurophysin also represents a portion of the putative precursor molecule (9).

Vasopressin, oxytocin, and the neurophysins have been demonstrated by immunocytochemical techniques in large granules all along the axonal pathways of the hypothalamus, in the zona interna of the median eminence, and along the pituitary stalk to the storage and secretory granules of the posterior pituitary. There also appear to be hormone-containing neuronal projections to the zona externa of the median eminence, which may end on the capillaries of the pituitary portal plexus. These fibers may arise either in the paraventricular or suprachiasmatic nuclei. Some studies also indicate that vasopressin can be secreted directly into the cerebrospinal fluid (CSF) (10).

Light and electron microscopic immunohistochemical studies in a number of species support the concept that vasopressin and oxytocin are produced in separate neurons (11). Both vasopressin and oxytocin neurons occur in the supraoptic and paraventricular nuclei, but species differences are apparent with regard to the ratio of vasopressin- to oxytocin-containing neurons within each nucleus (11).

Once oxytocin and vasopressin reach the neurohypophysis, there may be heterogeneity of storage, that is, the newly arrived hormone is released first (12). The hormones are probably secreted by the process of exocytosis, in which the membrane of the neurosecretory granule fuses with the plasma membrane of the axon terminal. Rupture occurs at the point of fusion, and the granule contents are extruded directly into the extracellular fluid, whereas the remaining granule membrane becomes part of the neuronal plasma membrane. This is later removed by micropinocytosis (Fig. 3) (12). The evidence supporting this mode of hormone release includes morphological findings, the Ca^{2+} dependence of neurohypophyseal hormone release (exocytosis is a Ca^{2+}-dependent process), and the demonstration that hormone release is invariably accompanied by neurophysin release, suggesting that the entire content of the neurosecretory granule is released in response to various stimuli (12). The immediate stimulus for neurohypophyseal hormone release is the propagation of an action potential along the neurosecretory neuron. This results in depolarization of the axon terminal, influx of Ca^{2+}, and exocytotic secretion of hormone. The factors that initiate this sequence of events have been studied in some detail.

The major stimuli for oxytocin release in the female are suckling and genital stimulation (13). These stimuli generate a neuroendocrine reflex arc, in which afferent impulses ascend in the spinal cord, relay in various brain areas and finally excite the hypothalamic neurons, which release oxytocin (14).

It has been demonstrated that each milk ejection in the rat is preceded by an increase in firing rate of antidromically identified neurosecretory neurons in the supraoptic and paraventricular nuceli (2). Thus there is periodic accelerated neuronal activity that results in bursts of oxytocin release during suckling. The milk-ejection reflex in rats could also be initiated by injections of oxytocin or by periodic electrical stimulation of the neurohypophysis or hypothalamoneurohypophyseal tract, but such treatments did not affect the frequency or periodicity of normal neurosecretory activation (2).

It has also been possible to show that vaginal dilatation increases the electrical activity of supraoptic neurosecretory neurons, although with a different pattern than that seen during suckling, and causes release of oxytocin, as evidenced by contraction of the myoepithelial cells of the mammary gland. There was no evidence to indicate that oxytocin release caused by one stimulus facili-

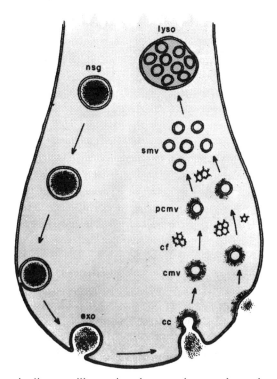

Figure 3   Schematic diagram illustrating the steps in neurohypophyseal hormone release. The entire content of the neurosecretory granule (*nsg*) is released by exocytosis
(*exo*) and the granule membrane is retrieved from the plasma membrane by micropinocytosis (formation of coated caveolae [cc], which pinch off as coated microvesicles
[cmv], shed coat fragments [cf] to become partially coated microvesicles [pcmv], and
finally smooth microvesicles [smv]) to be eventually incorporated into lysosomal bodies
(*lyso*). From Douglas (12) with permission.

tated that by the other, although both stimuli appeared to act on the same
population of neurosecretory neurons (2).

The neurotransmitters responsible for mediating oxytocin release are not
well defined. Cholinergic and noradrenergic systems appear to play, respectively, stimulatory and inhibitory roles, as they apparently do in controlling the
secretion of vasopressin (14). It has been noted that β-adrenergic blocking
agents can facilitate the milk-ejection reflex in animals in which it has been
blocked (2).

The secretion of oxytocin can also be controlled by higher centers in the
CNS. The inhibitory influence of emotional stress on milk ejection in lactating
women has been recognized for many years. Stressful stimuli will also inhibit
milk ejection in other animals, such as the rat (13). On the other hand, milk
ejection can be stimulated in the human by the sight or sound of an infant, and
in the cow by showing the calf to the mother. In animals trained to a strict
milking routine, transient release of oxytocin was detected when the milker

entered the stall. Thus the reflex control of oxytocin secretion can be modulated by various external factors (13).

It is not clear whether the same stimuli are effective for neurohypophyseal hormone release by pathways other than the hypothalamoneurohypophyseal tract, for example, to the external zone of the median eminence, the CSF, or various brain areas. The proposed roles of oxytocin as a mediator of memory-related processes (3) or drug tolerance, and as a precursor for hypothalamic regulatory factors (4), suggest that central excitatory or inhibitory stimuli may be important factors for certain aspects of oxytocin release.

Following its secretion and action, oxytocin is rapidly removed from the organism by metabolism and excretion. The plasma clearance of oxytocin and vasopressin has been studied in detail in a number of species, as has hormone half-life (15). Using a two-compartment model, the apparent early half-life for oxytocin was found to be approximately 1.5–2 min in a number of species. The average half-life for the second component in rats was calculated to be 15 min (15).

There are enzymes capable of inactivating peptides, including oxytocin and vasopressin, in all tissues (16); however, the relative importance of particular organs for the *in vivo* metabolism of the neurohypophyseal hormones has not been established. Although the liver and kidney do inactivate oxytocin *in vivo,* in a study in which the contribution of both of these organs was excluded, the residual clearance in rats was still equal to one third of the total clearance in intact control animals (15).

Oxytocin and vasopressin are not significantly inactivated by plasma, except in the case of women and some primates during pregnancy. At this time, an aminopeptidase capable of degrading oxytocin originates in the placenta of these species and spills over into the general circulation. This enzyme has been called oxytocinase, but is not specific for the metabolism of oxytocin or vaso-pressin. Instead, the enzyme, which cleaves the first five peptide bonds of the neurohypophyseal hormones (without prior reduction of the disulfide bond), can hydrolyze a large number of aminopeptidase substrates (16), and is more properly referred to as cystine aminopeptidase (CAP). Much attention has been focused on the occurrence, structure, and mechanism of action of CAP (15, 16), but it is now believed that the enzyme has little effect on the clearance of oxytocin in pregnant women, and does not represent a major fraction of the oxytocin-inactivating capacity of human placenta extracts or human colostrum (16). Thus the physiological role of CAP is not clear.

There appear to be four major modes of enzymatic attack on the neurohy-pophyseal hormones leading to hormone inactivation (16). One of these is the stepwise degradation of the peptides from the N-terminus by aminopeptidases, either by CAP without disulfide-bond reduction, or by other aminopeptidases following the opening of the hormone ring structure. The hormones can also be hydrolyzed in the acyclic portion to release either the C-terminal glycinamide residue or the C-terminal dipeptide amide. Finally, the disulfide bond can be reduced by thiol:protein-disulfide oxidoreductases. These may not be the only pathways for neurohypophyseal hormone metabolism, since the demonstration of inactivating enzymes appears to depend heavily on the conditions used during extraction and assay procedures (16). However, considering only these four established modes of hormone degradation, it appears that under normal conditions, enzymatic hydrolysis of the acyclic portion of oxytocin (cleavage of peptide bonds between residues 7 and 8, as well as 8 and 9, see Fig. 1) is quantitatively the most important mechanism of oxytocin inactivation (16). Glycinamide is probably released from oxytocin more rapidly than the C-terminal dipeptide, although this may vary in different organs (16). The enzymes responsible for these modes of oxytocin degradation have been identified in uterus, kidney, brain, and liver (16).

3 NEUROHYPOPHYSEAL HORMONE RECEPTORS

In order for oxytocin and vasopressin to exert their peripheral effects, the hormones must interact with specific receptors in their target organs. The characterization of vasopressin receptors in the plasma membranes of kidney medullary cells has been extensive (17), and the role of cyclic adenosine-3′,5′-monophosphate (cAMP) as a mediator of vasopressin action in the kidney is well established (18).

More recently, specific oxytocin-binding sites have been identified in the uterus and mammary gland (19). The binding sites in the uterus have been localized to plasma membranes of myometrial cells in several species (19, 20) and, in sheep, to particulate fractions of endometrial cells as well (21). The dissociation constant for oxytocin in the myometrial preparations correlates well with the concentration necessary for eliciting half-maximal contraction of the uterus *in vitro* (Table 1). In addition, various analogs compete for oxytocin binding to myometrial membranes in proportion to their agonist or antagonist potencies with respect to contraction of the uterus (19). The influence of several ions on oxytocin binding is also compatible with ionic influences on oxytocin-induced uterine contraction, for example, both are enhanced by Mg^{2+} and Mn^{2+}. Furthermore, estrogen treatment was shown to increase the affinity and possibly the number of oxytocin binding sites in the rat uterus, compatible with the changes in sensitivity of the myometrium to oxytocin during the estrous cycle. These results have all been taken to indicate that the myometrial binding sites for oxytocin do represent the "recognition" portion of the uterine oxytocin

Oxytocin binding to mammary gland cells has also been reported. Binding was saturable and of relatively high affinity and was reported to be specific with respect to the ligand (19); however, these sites have not been as well characterized as the uterine sites, nor has the myoepithelial response to oxytocin been as well studied as the uterine response.

Although the myometrial oxytocin receptor has yet to be isolated and biochemically or physicochemically defined, many of its characteristics can be

Table 1   Comparison of Hormone Binding and Pharmacological Parameters in Rat Uterus[a]

Peptide	B_{max} (fmol/mg protein)	K_D (nM)	ED_{50} (nM)	pA_2 (K_D,nM)[f]
Oxytocin (OT)	330[b]; 390[c]; 180[d]	10[b]; 2.4[c]; 1.7[d]; 2.0[e]	1.2	
(NαAc,OMeTyr[2])OT		105		7.03 (93)
[Deaminopenicillamine[1], Thr[4]]OT		27		~7.7 (~20)
		Displacement[g]	Relative Activity[h]	
[Pro[4]]OT		0[a, e]	1.6×10^{-5}	
(Mpr[1])OT		1.5[e]	1.6	

[a] All values are taken from Soloff et al. (19) unless otherwise noted. Further references are given in that article.

[b] Ovariectomized rat.

[c] Ovariectomized rat treated with estrogen 24 hr before testing.

[d] Intact rat treated with estrogen for two days before testing.

[e] Crankshaw et al. (20).

[f] The pA_2 is the negative log of the K_D for the antagonist-receptor interaction.

[g] Displacement represents the % displacement of [³H]oxytocin specifically bound to rat uterine tissue by the indicated analogue, *relative* to the % displacement obtained with an equal amount of unlabeled oxytocin.

[h] Activity, relative to oxytocin, in U/mg, using the *in vitro* rat uterus assay. Activity of [Pro[4]]OT is from Sawyer et al. (31), and of [Mpr[1]]OT from Ferrier et al. (32).

deduced from both structure–activity studies and conformational considerations (5). Oxytocin, the first peptide hormone to be synthesized chemically and the first for which a conformation in solution was proposed, has been the focus of intense study with regard to correlations between structure and biological activity.

Based on conformational considerations, a biologically active model of oxytocin was proposed (Fig. 2) in which one surface of the molecule encompassed a hydrophobic edge, constituted of the side chains of the residues in positions 3, 4, 7, and 8—the corners of the "β-turns" in oxytocin (5). This surface of oxytocin was proposed to include the binding sites for the hormone. Thus in structure–activity studies, the side chains of the residues in positions 3, 4, 7, and 8 have been shown to contribute to the affinity of oxytocin for the uterine receptor, and the aromatic ring of the tyrosine in position 2 also enhances affinity (5).

The other plane of the peptide molecule was thought to be characterized by a hydrophilic area that contained the "active center" of oxytocin, the area that influences the intrinsic activity or efficacy of the hormone. At the uterine receptor, this parameter is determined to a great extent by the hydroxyl group of the tyrosine chain, which was proposed to be folded over the ring of oxytocin and to act cooperatively with the asparagine carboxamide group (5). The intrinsic activity of oxytocin is also influenced by hydrophilic groups of the side chains of residues in positions 4 and 9, which also contribute to the active center of oxytocin in the conformational model (5). The correlation of peptide conformation with activity provides a unifying framework for the multitude of structure–activity studies (22) and also provides, in effect, a mirror-image of the elements constituting the oxytocin receptor *in vitro*.

It has recently been proposed that oxytocin receptors may regulate the response of target organs to circulating hormone. The concentration of oxytocin-binding sites in the ewe myometrium was shown to vary during the estrous cycle, reaching a maximal value at estrus, when the uterus is most sensitive to oxytocin. Similarly, the concentration of oxytocin receptors in rat myometrium increased several hours before labor and reached a maximal value during labor. These results led to the hypothesis that labor could be initiated by a sudden increase in uterine sensitivity to oxytocin, without a change in circulating hormone levels (23). The increased number of oxytocin receptors may result from changes in the estradiol:progesterone ratio that occurs during gestation, since estrogen treatment was shown to increase the number of uterine oxytocin receptors (19, 23).

The binding of oxytocin to mammary gland preparations also increased during pregnancy in the rat and was maximal during lactation, suggesting that the increased sensitivity of the myoepithelial cells to oxytocin during pregnancy and lactation could also be due to changes in affinity or number of mammary gland oxytocin receptors (23).

These intriguing findings require confirmation; however, they suggest a mechanism by which oxytocin may participate in or initiate various reproductive processes without requiring prior changes in hormone release or metabolism.

4 ACTION OF OXYTOCIN ON THE MYOMETRIUM AND MYOEPITHELIAL CELLS OF THE MAMMARY GLAND

4.1 Myometrium

The major function of the myometrium is to accommodate the fetus during gestation and to facilitate its expulsion at the appropriate time. Consequently, uterine smooth muscle activity is sensitive to and regulated by hormonal and physical factors that are associated with the estrous cycle and pregnancy. For example, the resting membrane potential of myometrial cells is within the range reported for other visceral muscles, but varies with the hormonal state of the animal and the amount of stretch imposed on the muscle. The lowest values have been observed in immature animals, and administration of estrogen to such animals elevates the resting potential toward the threshold, with a concomitant increase in uterine excitability (24).

Oxytocin increases the frequency and force of spontaneous contractions of the uterus. In quiescent uteri, the hormone acts by initiating action potentials.

In active uteri, oxytocin increases the frequency of spontaneous bursts of action potentials and increases the rate and duration of action-potential trains in each burst. Oxytocin also induces an increase in the rate of rise and the height of individual action potentials (24). It is apparent that any agent that enhances uterine excitability, such as estrogen, would also tend to enhance the stimulatory effect of oxytocin.

The ability of oxytocin to depolarize the myometrial cell membrane and initiate action potentials is reduced or abolished in solutions deficient in K^+, Na^+, or Ca^{2+} ions (24). However, in Na^+-deficient solution, some of oxytocin's effects can be restored by raising the external Ca^{2+} concentration. It appears that uterine action potentials may utilize Ca^{2+} as a second current carrier during spike generation—although action potentials disappear in Na^+-free solutions (24). A relatively recent investigation using the sucrose-gap technique, indicated that oxytocin increased mainly the Ca^{2+} component of the inward current, whereas the Na^+ component was hardly affected (25). Excitation–contraction coupling in uterine smooth muscle is generally assumed, by analogy to striated muscle, to involve an increase in intracellular Ca^{2+} during depolarization. Therefore, the ability of oxytocin to initiate or enhance uterine contractile activity may well be related to its modulation of Ca^{2+} influx.

Mediation of the antidiuretic effect of vasopressin by cAMP is well established (18). There is also an adenylyl cyclase in the myometrium, and its highest specific activity occurs in the microsomal fraction. However, oxytocin did not stimulate adenylyl cyclase activity in cell-free myometrial extracts and did not alter cAMP levels in intact tissue (18). Instead, myometrial adenylyl cyclase appears to be coupled to a β-adrenergic receptor (26), and β-agonist-induced uterine relaxation may be associated with increased levels of cAMP.

Cyclic AMP has been suggested to activate a myometrial protein kinase that phosphorylates membrane-bound proteins, resulting in an increased sequestration of Ca^{2+}. The resulting decrease in free Ca^{2+} levels is responsible for uterine relaxation (26). According to this hypothesis, oxytocin and other uterine stimulants interfere with the activation of adenylyl cyclase by β-agonists—oxytocin was reported to *reduce* isoproterenol-elevated cAMP levels in rat uterus (26)— and thus, assuming tonic β-adrenergic stimulation, increase the frequency of uterine contraction (Fig. 4). This hypothesis must still be regarded with caution, since prostaglandin E_1 (PGE$_1$), which stimulates uterine contraction, has been reported to increase myometrial cAMP levels (25). It is possible, however, that PGE$_1$ activates an endometrial adenylyl cyclase distinct from the myometrial enzyme (26).

The role of prostaglandins in the myometrial action of oxytocin has also received some attention (27). Prostaglandins (PGE$_1$, PGE$_2$, and PGF$_{2\alpha}$) stimulate uterine contractions in pregnant and nonpregnant humans and animals, and the human myometrium responds to prostaglandins throughout gestation. The demonstration that uterine muscle releases prostaglandin in response to oxytocin, both *in vitro* and *in vivo*, led to the suggestion that prostaglandins were mediators of the contractile action of oxytocin. However, the administra-

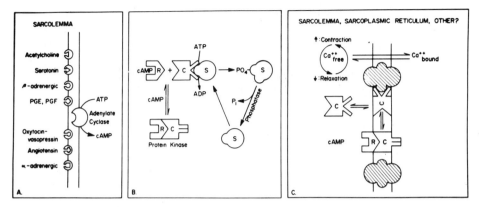

Figure 4 Hypothetical mechanism to which myometrial adenylyl cyclase regulates uterine contractility. (*A*) A variety of agents can modify cAMP levels. (*B*) The cAMP activates soluble protein kinase by binding to the regulatory (*R*) subunit and liberating the catalytic (*C*) subunit, which phosphorylates specific proteins (*S*). (*C*) In myometrial cells, the free catalytic subunits are translocated to the membrane, where they phosphorylate specific membrane-bound proteins that regulate Ca^{2+} transport and sequestration. From Korenman and Krall (26) with permission.

tion of prostaglandin-synthesis inhibitors *in vivo*, which prevented the release of $PGF_{2\alpha}$ in response to oxytocin, did not prevent oxytocin-induced uterine contractions (27). In an *in vitro* study, oxytocin was able to induce contraction of the pregnant rabbit uterus in the presence of high concentrations of indomethacin (a prostaglandin-synthesis inhibitor), although oxytocic action in the nonpregnant uterus was reduced under these conditions (28). The results indicate that prostaglandins are not obligatory intermediates of oxytocin action on the myometrium, but the interaction between the compounds may be influenced by the endocrine status of the experimental subject.

It is generally accepted that in several species circulating oxytocin levels are increased during the expulsive phases of labor as a result of the reflex release of the hormone due to vaginal dilatation. However, in humans and rats, the occurence of such reflex release has been questioned. In addition, the role of oxytocin as an inducer of labor has been controversial (23, 28). At term, the uterus is very sensitive to oxytocin, but no definite changes in maternal oxytocin levels have been demonstrated that could be associated with the onset of labor (29). As discussed above, the recent finding that in the rat, uterine oxytocin binding increases abruptly at the time of parturition, may resolve the problem. The change in myometrial resting potential and the increase in oxytocin receptors, both of which appear to be facilitated by high estrogen levels, may combine to rather suddenly increase myometrial sensitivity to the point that normal circulating levels of oxytocin can induce labor and parturition.

The lack of evidence for a change in maternal oxytocin levels at the onset of labor has also led to the proposal that fetal oxytocin could be involved in the increase in uterine activity. The finding that prolonged pregnancy was associated with anencephaly was thought to provide evidence for a role of the fetal pituitary in labor, but more recent studies indicate that the mean gestation length for anencephalics is normal (29). On the other hand, both oxytocin and vasopressin, as well as neurophysin, are consistently found in human umbilical plasma at the time of delivery, and hormone levels are higher in the umbilical artery than in the umbilical vein (29). Oxytocin can cross the placenta, and fetal oxytocin could also be excreted into the amniotic fluid, so that the hormone could well influence myometrial activity. It remains to be determined whether fetal neurohypophyseal hormones actually play a role in the onset or progress of labor, or are released secondarily, perhaps in response to the stress of labor.

4.2   Myoepithelial Cells

The action of oxytocin on the mammary gland is to elicit contraction of the myoepithelial cells, which results in ejection of milk from the alveoli and small ducts into the larger ducts and sinuses (30). Although milk ejection is commonly accepted as the physiological role for oxytocin, little is known regarding the mechanism of oxytocin action in the mammary gland. Suckling initiates the milk-ejection reflex, and without oxytocin, it is impossible for milk to be removed from the mammary glands, even though milk synthesis and secretion may proceed normally.

Specific binding sites for oxytocin have been reported in particulate fractions of rat mammary gland and the concentration of these sites was found to increase steadily throughout pregnancy, becoming maximal during lactation (23). Such a finding is in line with the increase in sensitivity of mammary tissue to oxytocin during pregnancy.

It is reasonable to suppose that oxytocin's effect on contraction of myoepithelial cells is similar to its action in the myometrium, although very little work has been carried out in this area. It is of interest that catecholamines can directly inhibit the milk-ejecting activity of oxytocin, both *in vivo* and *in vitro*, by actions at β-receptors, at least in some species (30). Such antagonism is similar to that seen in the uterus and may imply a similar mediation by a catecholamine-sensitive adenylylcyclase. Catecholamines also appear to antagonize oxytocin action by a central inhibition of hormone release, which may be related to the well-characterized emotional inhibition of milk ejection (30).

It is apparent that although certain areas of neurohypophyseal hormone research have been extremely fruitful, particularly with regard to the relationship of hormone conformation to activity and the recent characterization of hormone receptors, the biochemical mechanism of oxytocin action has only recently begun to be elucidated. Much work is still necessary in order to understand the biochemical mechanisms underlying the complex regulation of the action of this peptide hormone.

REFERENCES

1 duVigneaud, V., Hormones of the posterior pituitary gland: Oxytocin and vasopressin. *Harvey Lectures (1954-1955), Ser. 50*, 1 (1956).

2 Lincoln, D. W., Clarke, G., Mason, C. A., *et al.*, Physiological Mechanisms Determing the Release of Oxytocin in Milk Ejection and Labour, *in* Moses, A. M., and Share, L. (eds.), *Neurohypophysis*, p. 101, Karger, Basel (1977).

3 de Wied, D., Bohus, J. M., van Ree, I., *et al.*, Neurohypophyseal Hormones and Behavior, *in* Moses, A. M., and Share, L. (eds.), *Neurohypophysis*, p. 201. Karger, Basel (1977).

4 Walter, R., Oxytocin and Other Peptide Hormones as Prohormones, *in* Hatotani, N. (ed.), *Psychoneuroendocrinology*, p. 285, Karger, Basel (1974).

5 Walter, R., Identification of sites in oxytocin involved in uterine receptor recognition and activation. *Fed. Proc.* **36**, 1872 (1977).

6 Lederis, K., Neurosecretion and the Functional Structure of the Neurohypophysis, *in* Knobil, E., and Sawyer, W. H. (eds.), *Handbook of Physiology*, Sect. 7, Vol. 4, Part 1, p. 81, Waverly, Baltimore (1974).

7 Walter, R. (ed.), Neurophysins: carriers of peptide hormones. *Ann. N.Y. Acad. Sci.* **248** (1975).

8 Sachs, H., Fawcett, Y., Takabatake, Y., *et al.*, Biosynthesis and release of vasopressin and neurophysin. *Rec. Prog. Horm. Res.* **25**, 447 (1969).

9 Brownstein, M. J., Russell, J. T., and Gainer, H., Synthesis, transport and release of posterior pituitary hormones. *Science* **207**, 373 (1980).

10 Zimmerman, E. A., and Robinson, A. G., Hypothalamic neurons secreting vasopressin and neurophysin. *Kidney Int.* **10**, 12 (1976).

11 Sofroniew, M. V., Weindl, A., Schinko, I., *et al.*, The distribution of vasopressin, oxytocin- and neurophysin-producing neurons in the guinea pig brain. *Cell Tissue Res.* **196**, 367 (1979).

12 Douglas, W. W., Mechanism of Release of Neurohypophysial Hormones: Stimulus-Secretion Coupling, *in* Knobil, E., and Sawyer, W. H. (eds.), *Handbook of Physiology*, Sect. 7, Vol. 4, Part 1, p. 191, Waverly, Baltimore (1974).

13 Tindal, J. S., Stimuli that Cause the Release of Oxytocin, *in* Knobil, E., and Sawyer, W. H. (eds.), *Handbook of Physiology*, Sect. 7, Vol. 4, Part 1, p. 257, Waverly, Baltimore (1974).

14 Cross, B. A., Dyball, R. E. J., Central Pathways for Neurohypophysial Hormone Release, *in* Knobil, E., and Sawyer, W. H. (eds.), *Handbook of Physiology*, Sect. 7, Vol. 4, Part 1, p. 269, Waverly, Baltimore (1974).

15 Lauson, H., Metabolism of Neurohypophysial Hormones, *in* Knobil, E., and Sawyer, W. H. (eds.), *Handbook of Physiology*, Sect. 7, Vol. 4, Part 1, p. 287, Waverly, Baltimore (1974).

16 Walter, R., Simmons, W. H., Metabolism of Neurohypophyseal Hormones: Considerations From a Molecular Viewpoint, *in* Moses, A. M. and Share, L. (eds.), *Neurohypophysis*, p. 167, Karger, Basel (1977).

17 Roy, C., Barth, T., Jard, S., Vasopressin-sensitive kidney adenyl cyclase. *J. Biol. Chem.* **250**, 3157 (1975).

18 Dousa, T. P., Cyclic nucleotides in the cellular action of neurohypophyseal hormones. *Fed. Proc.* **36**, 1867 (1977).

19 Soloff, M. S., Schroeder, B. T., Chakraborty, J., *et al.*, Characterization of oxytocin receptors in the uterus and mammary gland. *Fed. Proc.* **36**, 1861 (1977).

20 Crankshaw, D. J., Branda, L. A., Matlib, M. A., *et al.*, Localization of the oxytocin receptor in the plasma membrane of rat myometrium. *Eur. J. Biochem.* **86**, 481 (1978).

21 Roberts, J. S., McCracken, J. A., Gavagan, J. E., *et al.*, Oxytocin-stimulated release of prosta-glandin F_2 from ovine endometrium *in vitro*: Correlation with estrous cycle and oxytocin-receptor binding. *Endocrinology* **99**, 1107 (1976).

22 Berde, B., and Boissonas, R. A., Basic Pharmacological Properties of Synthetic Analogues and Homologues of the Neurohypophysial Hormones, *in* Berde, B. (ed.), *Handbook of Experimental Pharmacology*, Vol. 23, p. 802, Springer-Verlag, Berlin (1968).

23 Soloff, M. S., Regulation of oxytocin action at the receptor level. *Life Sci.* **25**, 1453 (1979).

24 Marshall, J. M., Effects of Neurohypophyseal Hormones on the Myometrium, *in* Knobil, E., and Sawyer, W. H. (eds.), *Handbook of Physiology*, Sect. 7, Vol. 4, Part 1, p. 469, Waverly, Baltimore (1974).

25 Mirroneau, J., Effects of oxytocin on ionic currents underlying rhythmic activity and contraction in uterine smooth muscle. *Pflügers Arch.* **363**, 113 (1976).

26 Korenman, S. G., and Krall, J. F., The role of cyclic AMP in the regulation of smooth muscle cell contraction in the uterus. *Biol. Reprod.* **16**, 1 (1977).

27 Fuchs, A.-R., Prostaglandins, *in* Fuchs, F. and Klopper, A. (eds.), *Endocrinology of Pregnancy*, 2nd ed., p. 294, Harper and Row, Hagerstown (1977).

28 Laudanski, T., Akerlund, M., and Batra, S., Differences in the effects of vasopressin and oxytocin on rabbit myometrial activity and a possible mediation of prostaglandins. *J. Reprod. Fert.* **51**, 355 (1977).

29 Chard, T., Neurohypophysial Hormones, *in* Fuchs, F., and Klopper, A. (eds.), *Endocrinology of Pregnancy*, 2nd ed., p. 271, Harper and Row, Hagerstown (1977).

30 Bisset, G. W., Milk Ejection, *in* Knobil, E., and Sawyer, W. H. (eds.), *Handbook of Physiology*, Sect. 7, Vol. 4, Part 1, p. 493, Waverly, Baltimore (1974).

31 Sawyer, W. H., Wuu, T. C., Baxter, J. W. M., *et al.*, 4-Proline analogues of neurohypophysial hormones: Hypothetical intermediates in peptide evolution. *Endocrinology* **85**, 385 (1969).

32 Ferrier, B. M., Jarvis, D., du Vigneaud, V., Deamino-oxytocin: Its isolation by partition chromatography on Sephadex and crystallization from water, and its biological activities. *J. Biol. Chem.* **240**, 4264 (1965).

INDEX

527